D1250862

The Chemistry of the Semiconductor Industry

Edited by
S. J. MOSS
Molecular Sciences
Aston University
Birmingham, UK

and

A. LEDWITH
Group Research and Development Laboratories
Pilkington Brothers plc
Ormskirk, UK

Blackie

Glasgow and London

Published in the USA by
Chapman and Hall
New York

Blackie & Son Limited,
Bishopbriggs, Glasgow G64 2NZ
7 Leicester Place, London WC2H 7BP

Published in the USA by
Chapman and Hall
in association with Methuen, Inc.
29 West 35th Street, New York, NY 10001

First published 1987

British Library Cataloguing in Publication Data

The chemistry of the semiconductor industry.
1. Integrated circuits—Design and construction
I. Moss, S. J. II. Ledwith, A.
621.381'73 TK7874

ISBN 0-216-92005-1

Library of Congress Cataloging-in-Publication Data

Main entry under title:

The Chemistry of the semiconductor industry.

Bibliography: p.
Includes index.
1. Semiconductors—Design and construction.
2. Integrated circuits—Design and construction.
I. Moss, S. J. II. Ledwith, A.
TK7871.85.C477 1986 621.3815'2 86-978

ISBN 0-412-01321-5

Photosetting by Thomson Press (India) Limited, New Delhi
Printed in Great Britain by Bell & Bain (Glasgow) Ltd.

Contributors

P. S. Bagus IBM Almaden Research Center, 650 Harry Road, San Jose, CA 95120–6099, USA.

K. G. Barraclough Royal Signals and Radar Establishment, St Andrews Road, Great Malvern WR14 3PS, UK.

J. Q. Broughton Materials Science Department, SUNY, Stony Brook, NY 11794, USA.

D. L. Flamm A. T. & T. Bell Laboratories, 600 Mountain Avenue, Murray Hill, New Jersey 07974, USA.

M. Fryd E. I. du Pont de Nemours & Co., Marshall Research and Development Laboratory, Philadelphia, Pennsylvania 19146, USA.

I. R. Grant ICI Wafer Technology, Unit 34, Maryland Road, Tongwell, Milton Keynes MK15 8HJ, UK.

P. D. Greene STC Technology Ltd, London Road, Harlow CM17 9NA, UK.

P. John Chemistry Department, Heriot-Watt University, Riccarton, Currie, Edinburgh EJ14 4AS, UK.

B. L. Jones The General Electric Company plc, Hurst Research Centre, East Lane, Wembley, Middlesex HA9 7PP, UK.

W. Kern Device Physics and Reliability, RCA Laboratories, David Sarnoff Research Center, Princeton, NJ 08540, USA.

A. Ledwith Group Research and Development Laboratories, Pilkington Brothers plc, Hall Lane, Lathom, Ormskirk L40 5UF, UK.

Y. K. Lee E. I. du Pont de Nemours & Co., Marshall Research and Development Laboratory, Philadelphia, Pennsylvania, 19146, USA.

N. Mason Clarendon Laboratory, Physics Department, Oxford University, Oxford OX1 3PU, UK.

S. J. Moss Molecular Sciences, Aston University, Aston Triangle, Birmingham B4 7ET, UK.

J. A. Mucha A. T. & T. Bell Laboratories, 600 Mountain Avenue, Murray Hill, New Jersey 07974, USA.

R. W. Munn Department of Chemistry and Centre for Electronic Materials, UMIST, P. O. Box 88, Manchester M60 1QD, UK.

E. D. Roberts 54 Woodcrest Road, Purley, Surrey CR2 4JB, UK (formerly Philips Research Laboratories, Redhill, Surrey RH1 5HA).

G. L. Schnable Device Physics and Reliability, RCA Laboratories, David Sarnoff Research Center, Princeton, NJ 08540, USA.

J. Speight B. T. & D. Technologies Ltd., Whitehouse Road, Ipswich IP1 5PB, UK.

J. Woods Department of Applied Physics and Electronics, University of Durham, South Road, Durham DH1 3LE, UK.

Preface

The transistor was invented in 1947 and incorporated into primitive planar integrated circuits in 1959, since when there have been most spectacular developments in integrated circuit design and in the required basic process technologies. Computers formed an early and natural target for applications since they rely on integrated circuits for memory, processing, and many other functions. Nowadays, however, almost every variety of electronic system, process control unit, and consumer appliance requires such devices in increasing numbers and complexity. The explosive development of the telecommunications industry could surely not have happened without the sophistication of integrated circuits.

The basic technologies utilized in manufacture of the earliest planar circuits (silicon surface oxidation, diffusion doping, optical lithographic exposure, wet chemical etching) all have substantial chemical content, continue to constitute important manufacturing processes, and are likely to remain dominant for at least another decade. A recent survey (1) of microelectronics business opportunities suggests that the world markets for chemicals required in integrated circuit manufacture in the year 2000 will be approximately (US dollars) 1.3 billion for wet chemicals, 2.5 billion for gases, 1.1 billion for deposition materials, and 0.9 billion for photoresists. These figures compare with a projected market of 16.2 billion US dollars for semiconductor wafers and represent a compound annual growth rate approaching 20%. A report by the UK Chemical Industries Association (2) also emphasizes the major contributions from chemicals research and development to the electronics and optoelectronics industries expected by the mid-1990s.

It can be concluded therefore that both present and future processes for the manufacture of integrated circuits have a substantial chemistry input, although this is relatively little known outside the semiconductor industry. Indeed, it is fair to say that the successful manufacture of today's complex microcircuits represents a major achievement in applied chemistry, a fact now recognized by the recent establishment of a new journal *Chemtronics* (Butterworths), devoted to research at the interface of chemistry and electronics.

Various books are available on aspects of manufacturing processes in the semiconductor industry, but none focuses primarily on the role of chemistry over the whole field. The principal objective of the present book is to help create a wider appreciation of the role of chemistry, in a crucially important area, by means of a survey of the major chemical processes used in the manufacture of integrated circuits. In general, the individual contributors describe the major processes, define the current state of knowledge of the underlying chemistry, and identify areas of uncertainty where further research is being carried out. We hope that the book will be a convenient and valuable source of information for all those involved in research and development in the semiconductor industry and related industries, and in universities, polytechnics, and colleges.

Integrated circuits are popularly known as 'silicon chips' since silicon is the starting material for the overwhelming majority of such devices. The dominance of silicon-based circuits is likely to continue to at least the end of the century (1), and much of the book necessarily concerns processes using silicon. However, other materials are already of considerable importance because of specific advantages arising from their use, and are likely to be of increasing importance in the future. The book thus includes substantial accounts of the chemistry of Group III–V and Group II–VI materials as used in the semiconductor industry.

The layout of the chapters in the book broadly follows the order of the manufacturing process with some exceptions. After an overview chapter, the preparations of bulk silicon, III–V materials, and II–VI materials are discussed. Then follow chapters on film-forming technologies, i.e. chemical vapour deposition, metal-organic vapour-phase epitaxy, and liquid-phase epitaxy of III–V compounds. The chemistry of the lithographic process is the subject of two chapters dealing with photoresists, and electron-beam and x-ray resists respectively, followed by a chapter on wet etching. For technical reasons relating to the production of the book, two chapters on plasma etching appear at the end. A separate chapter deals with polyimides which are becoming increasingly important as resist materials and as passive components in finished devices.

Future developments in the electronics field will be manifold, but it is already clear that molecular materials will become increasingly important. This arises because of both intrinsic properties and relative ease of fabrication. A major US survey of opportunities in chemistry, the Pimentel report (3), notes that chemistry will occupy a central position in the design of molecular-scale devices. Characterization of molecular materials for potential applications is currently at the research stage, but a number of likely developments are indicated in the chapter on molecular electronics. Here again, we also note the emergence of the new *Journal of Molecular Electronics* (Wiley) as a measure of current interest.

Computer modelling is widely used in research relating to the semiconductor industry and continually provides new insights into the details of complex processes. A discussion of the results of such modelling exercises is to be found in several chapters. However, we felt it appropriate to include a separate chapter outlining some aspects of computer modelling, at the atomic level, of processes and reactions at semiconductor interfaces.

Some topics of great importance within the industry have had to be omitted to keep the book within reasonable limits. In some cases topics, although discussed *inter alia* within chapters, have not been given special attention because we judge the role of chemistry to be relatively minor (e.g. sputtering, ion implantation, molecular beam epitaxy). In other cases the material was judged less central to the main thrust of the book which focuses on the chemistry of processes leading to the more electronically active components of integrated circuits. Thus packaging, purification of chemicals and safe handling of chemicals have been omitted, although we freely acknowledge their importance and feel strongly that methods of purification and analysis will become increasingly important if the expansion of the semiconductor business is to proceed as projected.

On the basis of the materials and technologies covered by the contents of this book, the semiconductor industry is destined to expand dramatically (1, 4). Current research highlights the future potential of devices using Groups III–V semiconductor layers on silicon, silicon–germanium strained lattice structures, and, most importantly, ultrathin multilayer semiconductor structures with properties quite different from those of related bulk solids (4). Process requirements for the formation of such low-dimensional structures have already stimulated considerable interest in new thermally and photochemically activated vapour deposition techniques using organometallic compounds, and provide a suitable indication of the increasing value of chemistry to the semiconductor industry.

SJM
AL

References

1. *Microelectronics to the Year 2000*, SRI International, Business Intelligence Program, Report no. 739, Fall 1986.
2. *Priority Areas for Chemicals R & D*, Chemical Industries Association (U.K.), 1986.
3. *Opportunities in Chemistry*, US National Research Council (National Academy Press, Washington DC, 1985).
4. *Advanced Processing of Electronic Materials in the United States and Japan, A State-of-the-Art Review* (National Academy Press, Washington DC, 1986).

Contents

1 Overview

J. SPEIGHT

Introduction

Few aspects of our lives remain untouched by the semiconductor industry. The all-pervasive use of digital electronics, and therefore semiconductors, in our social and business lives is now taken for granted. In data processing, communications, consumer goods and transportation, many of the key advances of the last decade have resulted from the application of digital integrated circuitry. Already a significant contributor to the economies of the USA, Western Europe and Japan, semiconductors are destined to overtake many of the traditional industries on which these economies were based. In financial terms the impacts are equally impressive. The current European market alone for integrated circuits is $5bn. By 1996 this will have grown to around $20bn, and will be part of a world market likely to be worth more than $150 bn. All of this has been achieved by the products of an industry less than 20 years old.

An essential part of the manufacturing activity which will service these markets will be based on the preparation of the semiconductor raw materials, the processing technology required to form the integrated circuits and the packaging of the final chip products. For all of these areas, chemistry has played, and will continue to play, a leading role. In the present volume many aspects of this vital contribution will be reviewed. The present chapter takes an overview of the key elements of industry and speculates on the roles likely to be central to the chemist and chemistry over the next decade. This progress will be examined by considering how the more common transistor types have been reduced in size so as to achieve circuitry of extremely high complexity. A brief examination of the basic technologies involved in transistor and circuit fabrication will allow the key current areas of chemistry-related activities to be identified, for example layer deposition, pattern definition and doping techniques. The recent progress in these areas will form the main part of this overview, which will be concluded with some thoughts on where devices, and hence materials chemical research, may move over the next two decades.

Devices

Chip complexity

The foundation of the semiconductor industry may be traced to the earliest of silicon integrated circuits in 1958. From the integration of a few transistors, resistors and other simple components on the surface of a silicon chip, progress in the integration of larger numbers of transistors and more complex circuit elements has been extremely rapid (see Figure 1.1). Throughout the 1960s and early 1970s the number of transistors integrated per chip doubled every year, driven by what seemed at the time to be an almost insatiable appetite for memory chips. Accompanying the increase in circuit complexity and reduction in size of circuit elements, the performance of circuits, particularly the access time and power dissipation for memories and the cost per function, have all fallen rapidly. In the 1980s, progress has continued, but at a less hectic pace, with integration levels doubling approximately every other year. As may be

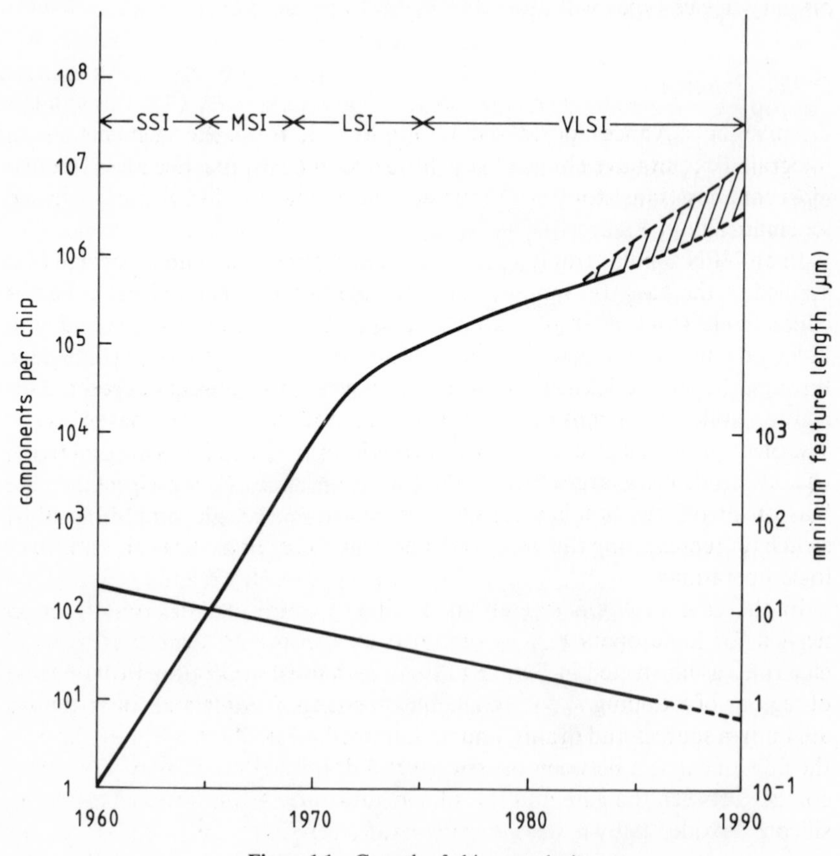

Figure 1.1 Growth of chip complexity

seen from Figure 1.1, this rate, though less spectacular than in the 1960s and 70s, will lead us to silicon integrated circuits chips with 10^7 transistors by the end of this decade. These high levels of integration have been achieved by progressive reduction in the size of individual transistor elements and the interconnection materials used on these circuits. The feature sizes corresponding to particular levels of integration achieved over the period 1960–80 are also shown in Figure 1.1. It is interesting to note that the log-linear plot of feature size does not show any change in rate as did the integration *v.* time plot, and that the silicon devices of the 1990s will be composed of submicron features. At this size the devices are still far from the fundamental limits imposed by physics.

In addition to achieving the impressive increases in integration and reduction in linewidth, the individual elements of the chip, namely the transistors, have also undergone many changes over the same period. Many of these changes have been influenced by, or dependent upon, the chemistry of a transistor fabrication step. A brief consideration of the operation of the primary device types will allow this point to be amplified.

Device evolution

Despite the advances portrayed in Figure 1.1, the basic elements of an integrated circuit have changed very little over the past two decades. The two most common transistor types in current use, bipolar and MOS, are illustrated schematically in Figure 1.2.

In an NPN bipolar transistor as shown in Figure 1.2*a*, with a positive bias applied to the base (0.7V higher than the emitter voltage) and the collector contacts, electrons (as shown) are injected from the emitter into the base and holes flow in the opposite direction. Some of the injected electrons migrate through the total thickness of the base material to reach the collector. This emitter–collecter current can be much larger than the emitter–base current. The principle which gives device gain arises because a small bias applied to the base electrode has a large effect on the collector current. By switching the base bias on or off, the bipolar transistor is able to act as a current-controlled switch for representing the ‘zero’ and ‘one’ states needed as the basis for binary logic operations.

In the case of MOS (metal-oxide–silicon) transistor, the on/off action needed for logic operations is obtained by controlling the lateral flow of electrons, as illustrated in Figure 1.2*b*. An *n*-channel MOS transistor consists of regions of *n*-doping in *p*-type silicon substrates. Contacts to the *n*-regions are known sources and drains, and an intermediate contact, the gate, controls the flow of carriers between the source and drain regions. (There is no direct contact between the gate and the silicon substrate. A thin layer of insulator, silicon dioxide, known as the gate oxide, separates the metal from the substrate.) With no bias applied to the gate, there is no conduction between

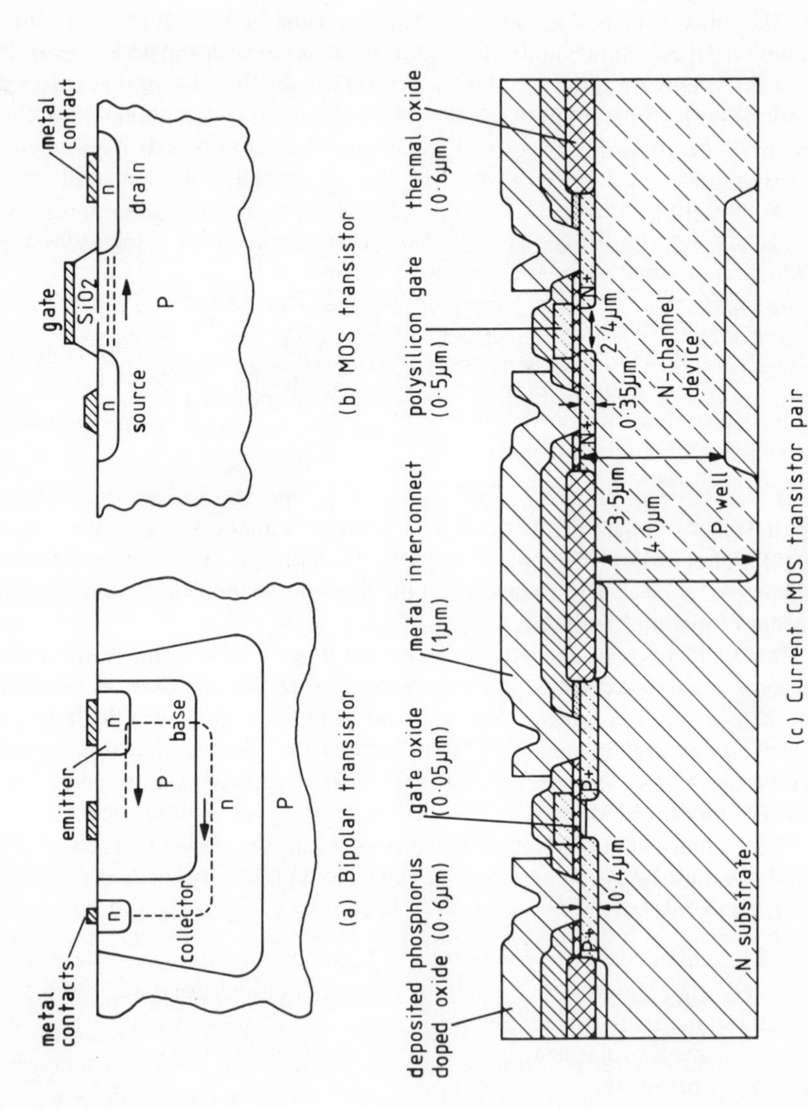

Figure 1.2 The challenge of device evolution

source and drain. A small bias on the gate leads to a field at the silicon surface sufficient to invert *p*-type material to *n*-type, thus creating a conduction on-state channel between source and drain. Variation of the gate bias in the MOS transistor thus permits the device to be switched on and off.

The transistors used in the integrated circuits of the early 1970s were closely similar to those shown in Figure 1.2*a,b*. In present circuits, the basic elements remain predominantly bipolar, and MOS transistors and the switching actions are no different from the ones described above; the major changes have been made in order to gain greater switching speed and lower power dissipation.

A present-day CMOS device is illustrated in Figure 1.2*c*. Some of the main reasons for the additional layer structures which have been incorporated are:

(i) Better control of device parameters (for example via polysilicon layer inclusion in gate dielectric)
(ii) Additional doping introduced under gate electrodes and outside devices to provide more precise turn-on voltage control
(iii) Higher density of integration achieved by use of multi- rather than single layers of conductor materials.

A CMOS pair consists of linked *n*- and *p*-type MOS transistors. CMOS pairs offer significant advantages in power dissipation over conventional MOS integration. Typical dimensions for current devices are shown in Figure 1.2. As noted in Figure 1.1, these dimensions are likely to shrink by an order of magnitude over the next decade.

The reductions in linear dimensions and hence carrier transit distances in silicon integrated circuits have resulted in impressive increases in switching speed. An alternative method for achieving higher speeds is to realize the same device structures in materials offering higher carrier mobility than silicon ($1300\,\mathrm{cm}^2\,\mathrm{V}^{-1}\,\mathrm{s}^{-1}$ for electrons), such as gallium arsenide ($7800\,\mathrm{cm}^2\,\mathrm{V}^{-1}\,\mathrm{s}^{-1}$) or indium phosphide ($4600\,\mathrm{cm}^2\,\mathrm{V}^{-1}\,\mathrm{s}^{-1}$).

The achievement of MOS and bipolar transistors in materials such as GaAs and InP does, however, present a greater technical challenge than in silicon for two major reasons:

(i) The creation of *p*– *n* junctions by diffusion on implantation in GaAs, InP and other compound semiconductors is not straightforward—on raising these materials to temperatures above 400° C loss of some constituent elements by volatilization occurs unless the surface is encapsulated by an oxide or nitride
(ii) Surface encapsulation by nature oxides, a well-established and practical technique for silicon, is not possible in compound semiconductors, where nature oxides are mixed oxides of variable composition and dielectric properties.

Simpler device structures such as the FET (Field Effect Transistor) are therefore usually preferred for GaAs and InP. The FET electrode configuration is closely similar to the source, drain and gate outlined for the MOS device in Figure 1.2*b*, except that the FET gate has the metal in direct contact with the semiconductor channel material. Such devices do have a number of disadvantages compared to MOS structures but have been used successfully to exploit the high mobility of GaAs, etc., for very high-speed devices (> 1GHz switching). The general trend for devices based on compound semiconductors—to decrease features size to achieve higher switching speeds and integration density—is closely similar to the trends outlined for silicon devices.

Device technology

The basic steps in the fabrication of transistors and circuits are the growth of thin layers of material on the chosen semiconductor substrate, formation of patterns to define the shape of the transistors and the electrical interconnections, and the introduction of dopants to create the near-surface regions in the semiconductor responsible for the switching actions underlying logical operations. Figure 1.3 is a schematic representation of how these three major processes are used in a simple, but very representative, fabrication step, that of defining a contact hole in a layer of SiO_2.

In the example shown, the layer is SiO_2, but identical operations are carried out on all layers involved in device and circuit fabrication (see Figure 1.2), such as polysilicon, metal layers such as aluminium, platinum, gold or titanium, or, more recently, conductive metallic silicides. Pattern definition of the required shape in the layer is carried out by the microprinting process known as photolithography. An outline is again provided by Figure 1.3. The feature or shape to be transferred to the layer is created on a glass photomask plate by photoreduction techniques. Transfer of the mask shape feature to the layer is by selective exposure of the photosensitive coating applied to the substrate. Subsequent development of the resist creates the required shape in the resist layer. Final definition of the shape in the layer is achieved by selective etching, either by wet chemical or active-gas plasma etching of the layer exposed in the resist holes. During the process, the resist provides protection of the surrounding area of the material.

For some applications, the patterns and holes created by photolithography will be overlaid by other layers such as conducting materials, with the defined hole serving as a local contact to the underlying silicon via the dielectric layer. In other cases, it may be necessary to introduce dopants such as boron, arsenic or phosphorus into the hole to form shallow junctions of *n* (as illustrated) or *p*-regions in the near surface of the semiconductor. Doping is currently carried out by one of two common methods.

In *high-temperature diffusion*, the wafers are raised to a high temperature in a gas stream containing the appropriate dopant, which enters the holes and

GROWTH/DEPOSITION OF LAYERS

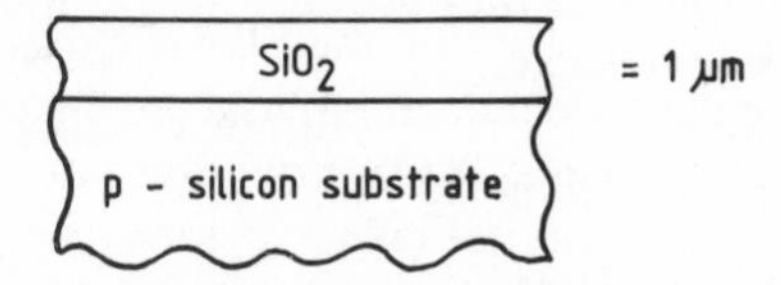

PATTERN DEFINITION BY PHOTOLITHOGRAPHY

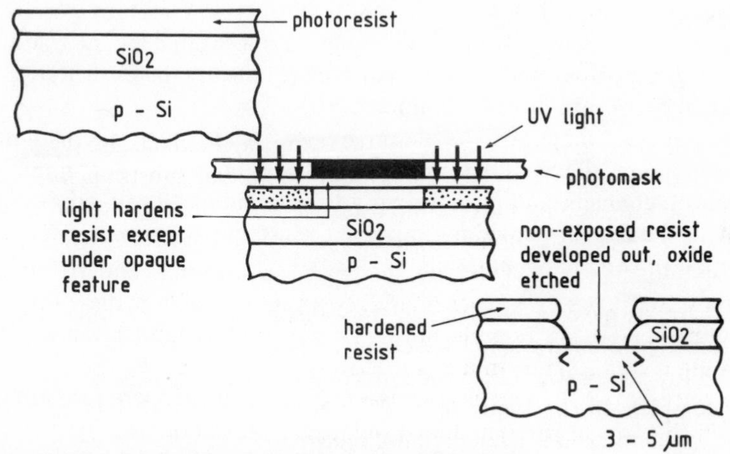

CONTROLLED INTRODUCTION OF DOPANTS

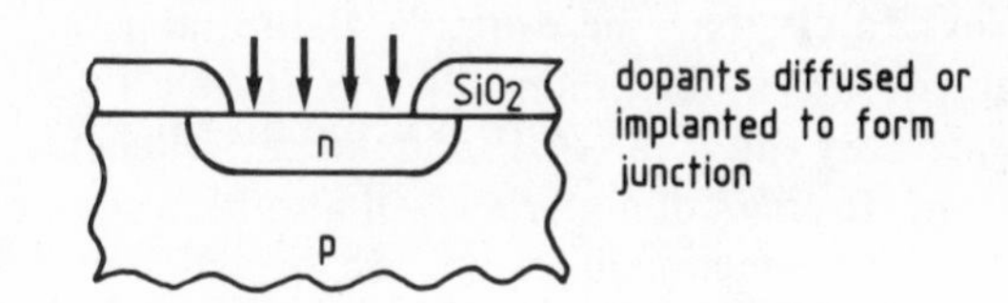

Figure 1.3 Elements of device fabrication

diffuses into the near surface of the silicon. Depth and concentration are determined by time, temperature and pressures.

In *ion implantation* of the dopant species by exposure to keV-beams, typical conditions are 100 keV and 10 μA beams for boron and 400 keV and up to 10 mA for arsenic. Penetration at these beam energies is limited, and depth of junction is controlled and surface damage is removed in a post-implantation heat treatment known as a drive-in (typically $\sim 1000^\circ$ C for 30 minutes).

For a modern VLSI circuit, up to 14 photolithography steps, six layer depositions and four ion implantations may be used in the fabrication sequence.

Device material

From Figure 1.3 it is clear that as integrated circuit chip size increases and transistor dimensions decrease, the transistor also becomes more complex as a material system. Some of the more prominent changes seen over the past two decades have been the following.

(i) The use of polysilicon/oxide composite gate dielectrics to enhance reliability and give superior turn-on voltage control in MOS transistors
(ii) Doped oxides using either phosphorus or boron to give surface smoothing by reflow at high temperatures
(iii) Polymers such as polyimides (Chapter 12) substituted for the traditional inorganic oxides and nitrides, particularly for interlayer dielectric isolation between multilayer conductors
(iv) Conductor materials have progressed from pure aluminium to aluminium–silicon alloys, inhibiting silicon out-diffusion at aluminium–silicon contacts, and more recently to aluminium–silicon–copper alloy, with the copper providing improved resistance to current-aided migration of conductor materials.

Similar trends have taken place in non-silicon ICs. For example, the gate structure in gallium arsenide FET circuits is likely to be an Au–Pt–Ti or Au–W–Ti alloy composite. The basic semiconductor materials have also grown more complex. In the 1960s and 1970s the industry was firmly centred on silicon, gallium arsenide for digital circuits, and GaAlAs for light-emitting devices. During the coming decade, device materials will become increasingly complex. To cite but one example, the standard material system for light-emitting devices in the practically important 1.3–1.5 μm range is likely to be the quaternary alloy GaAsInP, and optical switches will be based on $LiNbO_3$. Increasing diversity of materials will also be accompanied by other developments such as reduction in feature size, reduction in layer thickness, and larger wafer sizes. All these factors (Table 1.1), which are described in detail elsewhere, will conspire to present an unprecedented challenge to chemists and materials scientists, and also to the analysts responsible for providing the vital feedback to the device development teams. Significant development in the

Table 1.1 The analytical challenge

	1980s	1990s
Device dimensions (μm)		
Lateral	3–1	1–0.1
Vertical	1.0–0.05	0.05–0.005
Materials	Si, GaAs, InP	Si, InP-Ga AlAs $LiNbO_3$, polymers
Wafer size (mm)	500–750	750–1000
Complexity	10^4–10^5 transistors	$> 10^6$ transistors

currently available analytical techniques such as SIMS, Auger and SEM will also be required.

There will need to be progress in our understanding of the surface chemistry which will play an increasingly important role in the 1990s as a consequence of the dramatic increase in the surface: volume ratio produced by the scaling of transistor dimensions. Not shown in Table 1.1, but of equal importance, is the form of material likely to be encountered in the next decade. Traditionally, our understanding in semiconductor science has been restricted to the chemistry of single and polycrystalline forms of material. Increasingly it will be necessary to extend our understanding to amorphous, polymorphic or polymeric material and to quaternary or quinternary alloys.

The trends shown in Table 1.1, whilst challenging, do not represent the fundamental limits of device dimensions or performance. The ultimate device or integrated circuit is not, in fact, governed by a single limit, but by a hierarchy of limits, starting with effects dictated by the laws of nature, at the most fundamental, through to circuit limits which prescribe how the devices will operate in practical configurations. This hierarchy is shown in more detail in Table 1.2. Within each category there may be several possible physical mechanisms, as follows.

Fundamentals. Primarily thermodynamic limits define the minimum energy needed to form a bi-state device. This is of the order of a few tens of RT, that is, sufficiently different to avoid switching by thermal agitation, but low enough to ensure switching by an input stimulus.

Materials and technology. Materials parameters which are likely to affect device performance include bandgap, carrier mobility, or breakdown fields. To be of practical use, potential device materials need to be capable of undergoing all the operations experienced in device fabrication technology, as in the sequences outlined in Figure 1.3 and in Table 1.2.

Table 1.2 Hierarchy of limiting effects in integrated circuits

Fundamentals	
Materials technology	Energy gap
	Mobility
	Breakdown field
	Thermal conductivity
	Diffusion coefficients
	Doping characteristics
	Layer deposition
	Pattern definition
	Doping techniques
	Design and test
	Economics
Devices	
Circuits	
Systems architecture	

Devices. Real transistors have greater power dissipation, by two to three orders of magnitude, than the fundamental level referred to above. The limiting feature size of the bipolar transistor is the base width, for which the minimum theoretical dimensions are around 0.05 μm. Such a device would offer potential switching speeds of a few picoseconds.

MOS transistor action could be sustained with gate widths of around 0.1 μm, leading to switching speeds similar to the bipolar device, although it should be noted that both MOS and bipolar devices offer specific advantages for circuit applications beyond mere speed of switching.

Circuits. Using the minimum size of transistor, it should in theory be possible to achieve integrated circuit chips with 10^7–10^{10} transistors with picosecond switching. Practical problems of power dissipation, the interconnection losses between transistors and chip or package losses all conspire to give a real circuit performance much inferior to theoretical limits. A practical limit may be represented by a 10-million-transistor chip operating at megabit per second rate. Such an answer may seem disappointing compared to the fundamental values, but represents a capability far greater than can currently be achieved.

Systems architecture. A system is a configuration of circuits. Many of the limiting problems encountered at device or circuit level re-appear in systems design, for example, power dissipation and interconnect losses. Collectively, the system is slower than the device and circuits. Novel systems processing will be used to overcome these limitations.

The role of chemistry in the industry's progress towards these objectives will be largely devoted to removing materials and technology limitations. Some aspects of how these two areas of activity are being addressed will now be examined in more detail. The subject is vast, and the presentation will therefore be selective.

Device materials research

Some of the central materials parameters which control device performance are listed on the right-hand side of Table 1.2. All of these parameters, in a host of semiconductor materials, are currently being explored in research and development establishments throughout the industry with a view to enhancing them for specific device applications.

An excellent example here is the current spate of work on silicon–germanium alloys. For the past two decades it was generally assumed that the mobility and energy gap of silicon was unlikely to be influenced significantly, and that progress in ICs would mainly depend on attention paid to other limiting areas such as device scaling, circuit design and systems architecture. Recent work by Kasper and colleagues has shown that silicon–germanium alloys with bandgaps intermediate between silicon and germanium can exhibit mobilities up to seven times as high as that observed in bulk silicon. To achieve these exceptional results, thin multilayer sandwiches of silicon, silicon–germanium and silicon, known as super-lattices, need to be grown. This work is in its infancy, but is an excellent example of the way in which research on materials and chemical fundamentals may still remove limitations in even the most well-established semiconductor materials. Alternative dielectrics, silicides and polymer materials, and also compound semiconductors such as gallium arsenide and indium phosphide, are known to offer superior basic properties in terms of bandgap and mobility. Research in the compound semiconductor area has two main thrusts:

(i) Optimization of the band properties to give specific wavelength emissions, for instance in the use of GaAs–InP for lasers and receivers
(ii) Examination of the fundamental parameters, such as diffusion coefficients, thermal conductivity and the chemistry of practical dopants.

Broadly, research in the first area seeks to extend the performance of devices by offering superior materials characteristics, while in the second, the material facets which will form the foundations of a practical device technology are investigated. Research on the traditional and newer semiconductors will no doubt continue unabated. A good example of this is recent work demonstrating the fabrication of GaAs-based devices on silicon substrates, whose performance rivals state-of-the-art GaAs substrates. Equally importantly, it has been shown that silicon devices such as MOS transistors formed on the surrounding material are not degraded by the GaAs growth processes.

While ion implantation (see Figure 1.3) will probably continue to be the predominant method of introducing dopants into semiconductors, our knowledge of the interaction of ion-implanted species with the lattice defect structure in materials other than silicon and gallium arsenide is relatively limited. Even more rudimentary is our understanding of dopant behaviour in the ternary and quaternary alloys finding increasing use in opto-electronics.

Research in device technology

In parallel with the work designed to establish new materials with enhanced properties, the other main theme of materials work in electronics is the search for improvements in device technology. The driving forces for most of this work are those depicted in the outline of device technology in Figure 1.3, namely layer deposition, pattern definition and doping techniques (the other areas noted in Table 1.2, such as device testing, design and the economics of device fabrication, do have key roles in device technology but do not have a strong chemical or material involvement and will not therefore be considered here). In these three areas the main work is focused on problems standing in the way of the next generation of devices. This can be illustrated by a brief survey of current trends in these areas.

Layer deposition

A current CMOS integrated circuit fabrication process may contain 6–8 process steps involving the growth or deposition of thin layers of oxides, polysilicon, conductor metallization and other dielectrics such as phosphorus-doped oxides or silicon nitrides. These layers are used for device-to-device or interlayer isolation. Advances in device integration have imposed a number of new demands on the methods used for layer deposition.

(i) Device scaling to increase switching performance has resulted in reduced lateral dimension, and to maximize performance, some layers such as the gate oxide have to be reduced in thickness
(ii) Smaller feature sizes of devices can only be achieved using thinner constituent layers because of the difficulty of photolith operations on thick layers
(iii) Wafer sizes have increased, as noted earlier, demanding uniform layer formation over larger areas
(iv) Control of diffusion between layers becomes crucial as layer thicknesses decrease, and lower-temperature methods of deposition are actively being sought.

Perhaps the most active field of work in silicon IC technology has been the extension of chemical vapour deposition low-pressure techniques as a means of producing a variety of commonly-used layers, for instance polysilicon, silicon nitride and silicon oxide, with improved batch-to-batch and cross-wafer uniformity (Chapter 6). Progress has been made as the result of a greater understanding of the chemistry of deposition, particularly the roles played by homogeneous gas-phase reactions *v.* heterogeneous surface-limited reactions. More recently, the addition of RF plasmas to LPCVD systems, and photo-assisted LPCVD systems with the use of small additional stimuli supplied by light and plasma-generated species, has led to dramatic reduction in deposition temperatures whilst maintaining the excellent uniformity characteristics of low-pressure systems. The deposition of oxide interlayer dielectrics for use in multilayer metallic conductor systems is an area where progress has been greatly aided by these new techniques. Temperatures for SiO_2 deposition can be lowered from the traditional 450° C regime by more than 200° C. Low-pressure, plasma-assisted and photon-assisted deposition has been successfully demonstrated for a wide variety of technologically important layers, such as epitaxial silicon, silicon nitride, metallic silicides and doped oxides. For layer deposition in the sub-0.1 μm regime, the technique of molecular beam epitaxy (MBE) has become increasingly important. In this growth process, the constituent atoms of the exitaxial layer are supplied to the heated substrate as beams generated by evaporating the element or alloy in a Knudsen-type effusion cell. The process is carried out in ultra-high vacuum, so that the effused beam suffers few collisions between cell and substrate. Surface analysis techniques, such as Auger spectroscopy and low-energy electron diffraction, can be used to characterize the as-prepared substrate surface and the deposited layers. The technique has already found application for depositing silicon, GaAs, GaInAsP and related compound semiconductors in ultra-thin-layer configurations. At present, MBE is largely confined to development studies on the production of high-value, specialist microwave device material. As dimensions decrease, this technique seems destined to play an increasingly important role.

Progress in the field of layer deposition will continue to be rapid, although our knowledge of the underlying chemical mechanisms remains scanty. This deficiency is apparent at both macro- and micro-levels. At macro-level, our understanding is insufficient to model many of the new processes. At micro-level, the lack of detailed chemical understanding of the underlying gas and surface properties means that many inherent reliability problems are identified only by the time-honoured methods of trial and failure.

Pattern definition

The ability to print patterns of increasingly fine feature size is central to the advance of electronics. Most of the problems associated with the formation of fineline images to create photomasks or print directly on to wafers are in the domain of physics rather than chemistry. Chemistry is, however, central to vital areas of pattern definition (see Figure 1.3): in the creation of photoresists capable of printing fineline features, and in the use of gaseous plasmas for etching the underlying materials exposed in the photoresist after development.

The current limitations of commonly used photoresists are:

(i) Lack of tolerance to plasma exposure—photoresists must etch significantly more slowly than the underlying layers
(ii) Poor response at deep uv wavelengths—most current resists are designed to operate above 400 nm, but finer lines will demand exposure at 200–350 nm
(iii) The ability to retain printed shapes during bake and hardening processes.

During the past few years most resists have been based on diazoquinone compounds, and a great deal of work has been devoted to the optimization of critical parameters such as development, exposure and contrast of this particular resist system. For electron-beam, rather than optical, exposure, cross-linked polymethyl methacrylate (PMMA) resists have been widely used. As with photoresists, the chemical composition of PMMA has been varied widely to give ranges of sensitivity, enhanced plasma tolerance, and so on.

For lines below 0.5 μm, wavelengths other than the visible must be used, because of the limitation of refracting optics. The common options of electron-beam writing or x-ray printing both permit the printing of linewidths of 0.1 μm. In the case of e-beams, it is possible to write the required pattern directly on to the resist by rapid beam deflection and beam-shape variations, thus removing the need for the intermediate step of mask making. E-beam direct writing has been used for fast turn-around of prototype designs where mask-making would introduce unacceptable delays, but the technique is generally impractical for production quantities because of the lengthy writing time per layer. (It should be noted that e-beam writing of fine-line masks is established as the standard way of creating low-defect mask plates).

For both x-rays and e-beams, resists are still actively being developed. Similar requirements to those noted for photoexposure, with the additional need to optimize exposure times which are currently impractically long, will be needed for these radiations.

Exposure of the developed photoresist patterns to low-pressure plasma etching environments has produced many effects requiring a better understanding of the chemistry involved:

(i) Interaction between plasma species with photo- and other resists
(ii) The role of plasma chemistry in determining whether a given material layer is etched isotropically or anisotropically
(iii) Selectivity of etch rate between the layer being defined and the substrate and the resist.

These points are of great importance in the etching of contact holes, often a limiting feature in high-speed, high-packaging density circuits. Oxide etching is usually carried out using reactive-ion etching or reactive-sputter etch techniques (see Chapter 6) leading to enhanced anisotropy over conventional plasma etching. This continues to be an active field of research, the trade-offs between etching characteristics, wafer heating and device radiation damage from plasma species being areas of particular practical concern. Up to the present time, much of the progress in the application of plasma etching has been achieved by improvements in vacuum chamber furniture and a rudimentary understanding of plasma chemistry. As transistor and interconnection dimensions continue to decrease, there is evidence that our understanding of fundamental mechanisms will need to be increased to obtain the necessary precision and reproducibility for practical circuits. The demands of patterning submicron circuitry will ensure that a more rigorous approach is made to plasma processing (see Chapter 7).

Layer deposition and pattern definition can be combined in the recently-introduced technique of laser direct-writing. The patterns are generally written in conductive metals such as Al, Zn or Cd, which are formed by photolytic, pyrolytic or photothermal processes. The photolytic process is a direct decomposition of an organometallic compound through photon absorption, whereas for pyrolysis a local rise in substrate temperature induced by the laser leads to a CVD process. Photothermal techniques involve both the above mechanisms to some degree. At present, the technique is still in the developmental stage, but seems likely to be extended to other materials.

Doping technology

Over the past decade, ion implantation of phosphorus, boron and arsenic species at energies in the range 50–250 keV has replaced the high-temperature gaseous diffusion doping techniques formerly used in silicon technology. The main advantages offered by ion implantation are precise control of dopant dosage (by measurement of ion current), depth distribution, and uniformity across the wafer. However, the final impurity profile obtained from ion implantation depends strongly on both the implant conditions and the subsequent thermal processing (typically at $>900°$ C for 30 minutes) necessary to activate the dopant and remove lattice damage introduced by the implant process. In order to minimize diffusion of the dopants during the activation-damage relief anneal, much work has been devoted to short-time high-temperature processing techniques. To achieve these short-time high-temperature anneals, lasers, electron beams and, most commonly, high-powered incoherent light sources have been used in place of conventional tube furnaces. Using these techniques, wafers can be raised to temperatures in excess of 1000° C for precise times in the range 0.5–100 s, which is sufficient for activation and damage removal. These techniques have regenerated interest in the subject of dopant activation under what are effectively pulse-heating conditions, and the possible influence of the energy sources in the activation process. To date the techniques have been largely applied to silicon, but the potential advantages in annealing of implantation in compound semiconductors are even greater, as compound semiconductors are more volatile than silicon. As mentioned earlier, our understanding of dopants in compound semiconductors, with the possible exception of gallium arsenide, is still rudimentary. In view of the increasing commercial importance of compound semiconductors, especially in light-emitting applications, this is certain to be an area of active research in the next decade.

The future

For the next decade, the main progress in silicon integrated circuits will continue to be made by adhering to the trends outlined in the present chapter, that is, a continuing scaling-down of the dimensions of bipolar and MOS transistors until the limiting physical dimensions of transistor action are reached. As noted earlier in the chapter, this limit will occur at dimensions in the range 0.1–0.01 μm, depending on the device structure. Thus a regime can be foreseen beyond which it would not be possible to design and build integrated circuits around devices exhibiting the transistor characteristics exploited over the past two decades. This is, however, not the end of the story. A new field of study, based on what are known as Low-Dimensional Structures, is rapidly emerging which will probably lead to devices based on new physical phenomena only observable below the limiting dimensions of transistor action. It seems certain that chemistry and chemists will be involved in at least two key aspects of this progress:

(i) As dimensions decrease, surface and near-surface phenomena become increasingly important to reproducible device performance
(ii) Progress in more detailed chemical understanding of all the processes used in device fabrication has been rapid in the last decade, but further efforts are needed to enable the industry to move competently into the sub-micron regime.

Long before the limiting size of the transistor acts as a brake to the progress of the industry, severe limitations will be imposed by the difficulties of efficiently interconnecting hundreds of thousands of transistors on the surface of the circuit. One method of overcoming both the limiting-size and the inter-connectivity problems is the use of molecular electronics, in which organic polymers or inorganic materials are chemically designed for use as electronic logic gates and switches interconnected by conducting tracks in the molecule (see Chapter 13). Work on molecular electronics is still at an early stage, but seems destined to be increasingly important during the next decade. Current work is largly devoted to conductive polymer systems based on pyrrole, phthalocyanines and polyacetylenes and their derivatives.

A further area of electronics where chemistry seems destined to make a uniquely formative contribution is that of electrochemical sensors for biomedical applications. Such sensors, based on derivatives of existing silicon technology and new and novel materials, could revolutionize *in-vivo* monitoring of many biochemical functions in the next decade.

In summary, the future of the industry will present a wealth of opportunity for chemistry and chemists to influence progress. Some of the more important areas where this progress will occur have been highlighted in this chapter. In subsequent chapters, many of the subjects introduced here will be dealt with in greater detail and a more vivid picture of the rate of change and the exciting work being done in this field will be presented. There can be little doubt that the next forty years of the semiconductor industry will be just as exciting and prolific as the forty years we have experienced to date. Almost unbounded opportunities exist not only for chemists but for workers in almost all scientific disciplines. It is impossible not to be exhilarated by the prospects.

Further reading

1. S. M. Sze, *Physics of Semiconductor Devices*, 2nd edn., Wiley-Interscience, New York (1981).
2. R. W. Keyes, *Proc. IEEE* **63** (1975) 740–767.
3. P. M. Soloman, *Proc. IEEE* **70** (1982) 489–509.
4. A. N. Broers, *Solid State Technol.* **28** (1985) 119.
5. M. J. Mayo, *Solid State Technol.* **29** (1986) 141.
6. E. Kasper, in *Proc. ESSDERC 1986*, Inst. Phys. Conf. Ser., Institute of Physics, London.

2 Semiconductor silicon

K. G. BARRACLOUGH

Introduction

The Group IV semiconductor element silicon occupies a unique position in the development of the solid-state electronics industry. So spectacular have been the applications of large-scale integrated circuits in information systems, that the term microelectronics is virtually synonymous with the silicon chip. Silicon semiconductor devices can also be found in a vast array of other electronic applications such as solar cells, power switching devices and sensors. Undoubtedly, new applications will be developed in the future to confirm silicon's dominant position in the solid-state industry. The basic semiconducting properties of silicon, however, do not reveal any obvious clue to explain this dominance (Table 2.1). Many properties appear inferior to those of other materials. The energy bandgap is intermediate between those of Ge and GaAs; the bandgap is indirect, preventing first-order optical transitions for light-emitting devices; the electron mobility is less than that of GaAs for very high-speed devices. The key to silicon's success is to be found in its chemical nature, notably the stability of its oxide.

Unlike many of its competitors, silicon is readily available as a naturally occurring oxide, SiO_2. The chemical stability of SiO_2 is also turned to advantage during device manufacture when thin oxide layers are used as an integral part of the device structure. Silicon–oxygen interactions are important in many aspects of silicon materials technology and we shall be constantly referring to these in the following sections. Supplies of SiO_2 in the Earth's crust are virtually inexhaustible, but nature demands considerable amounts of thermal energy and human ingenuity for the extraction of elemental silicon in suitable form for semiconductor applications.

Table 2.1 The basic semiconducting properties of silicon compared with GaAs and Ge.

Property	GaAs	Si	Ge
Energy bandgap at 300 K (eV)	1.42	1.12	0.66
Energy bandgap type	Direct	Indirect	Indirect
Electron mobility (drift) at 300 K (cm^2/V-s)	8500	1500	3900
Hole mobility (drift) at 300 K (cm^2/V-s)	400	450	1900
Intrinsic carrier concentration at 300 K (cm^{-3})	1.79×10^6	1.45×10^{10}	2.4×10^{13}
Intrinsic resistivity at 300 K (ohm cm)	10^8	2.3×10^5	47

Silicon is a covalently bonded element which in its purest form exhibits only weak electrical conductivity at room temperature. Substitution of the silicon atoms through doping with either Group III or Group V elements at levels as low as 1 ppb produces free carriers in the silicon lattice which then allows passage of electric current by either negatively charged electrons (*n*-type through Group V donors) or positively charged holes (*p*-type through Group III acceptors)—Figure 2.1. The technology of device manufacture is based on fabrication of precisely defined junctions between these different types of silicon, a subject of later chapters. It can be appreciated, however, that a prerequisite for silicon device manufacture is a starting material in sufficiently pure form for the controlled addition of dopants in the ppb–ppm range. Other impurities also cause undesirable effects in silicon devices, and as a general rule chemical purity is a major criterion. However, if the problem were merely one of chemical purification, the silicon device industry would not have progressed

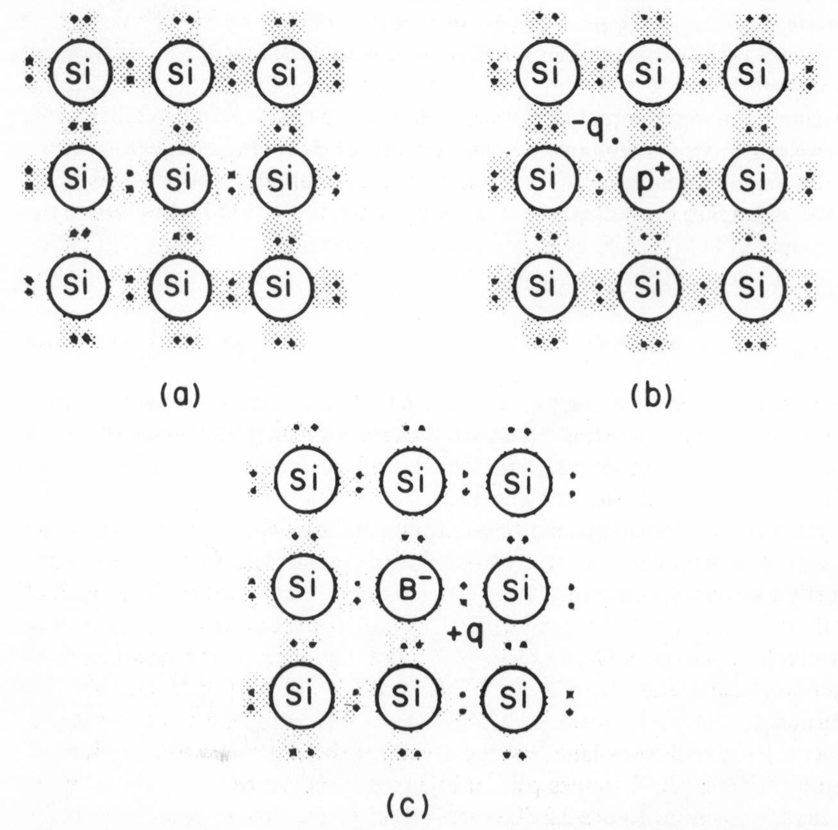

Figure 2.1 Schematic of bonding in (*a*) intrinsic Si with negligible impurities; (*b*) *n*-type Si with donor (phosphorus); (*c*) *p*-type Si with acceptor (boron). From S. M. Sze (1981) *Physics of Semiconductor Devices*, 2nd edn., copyright © Bell Laboratories, reprinted by permission of John Wiley and Sons, Inc., New York

to its current position, which has been achieved mainly without the application of the purest silicon available! To explain this paradox we have to consider other criteria, notably the physical structure of the base material. Most silicon devices have to be made in highly perfect *single crystals*, since the degradation of electrical performance by lattice imperfections cannot be tolerated except in devices such as terrestrial solar cells where performance may be deliberately offset against the lower cost of polycrystalline material. To avoid lattice defects, care must be taken with the processes used in the preparation of the single crystal material, in the wafering of the crystal and in the tailoring of the wafer to the device process. In the following sections examples will be given to show that absolute chemical purity is sometimes compromised to satisfy structural requirements in device applications.

This chapter concentrates on the bulk preparation of semiconductor silicon from basic raw materials to finished single-crystal wafers. The formation of thin films of silicon, epitaxially or otherwise, will be dealt with in Chapters 5 and 6.

A wide range of processes is outlined to illustrate the role of chemistry in the development of semiconductor silicon. Space does not permit extensive discussion of the role that analytical techniques have played in keeping these processes under control, but their importance should not be underestimated. In the final section the areas of research and development which are likely to have an impact on production processes in the future are highlighted.

Extraction and refining of silicon

General

In this section we are concerned with the chemical processes which define the basic purity of starting materials for crystal growth processes which, in turn, define the crystallographic structure and electrical activity of the silicon for use in semiconductor devices.

High-purity polycrystalline silicon, often referred to as polysilicon, is the essential feedstock for the growth of large silicon single crystals, whilst high-purity gaseous chlorosilanes SiH_xCl_{4-x} $(0 \leqslant x \leqslant 4)$ are used in the growth of thin epitaxial layers (Figure 2.2). Polysilicon is produced from the readily available materials SiO_2, H_2 and HCl. The first stage is the production of metallurgical-grade silicon (MG-Si). This is followed by the production and purification of trichlorosilane, $SiHCl_3$. The final stage is the deposition of silicon from trichlorosilane. Practically all of the current world production (approximately 5000 tonnes p.a.) of high-purity polysilicon is derived by the process shown in Figure 2.2. The latter half of the process is known as the Siemens Process (1). Alternative technologies are, however, of great interest, particularly for applications where cost reductions are needed, and these are outlined at the end of the chapter.

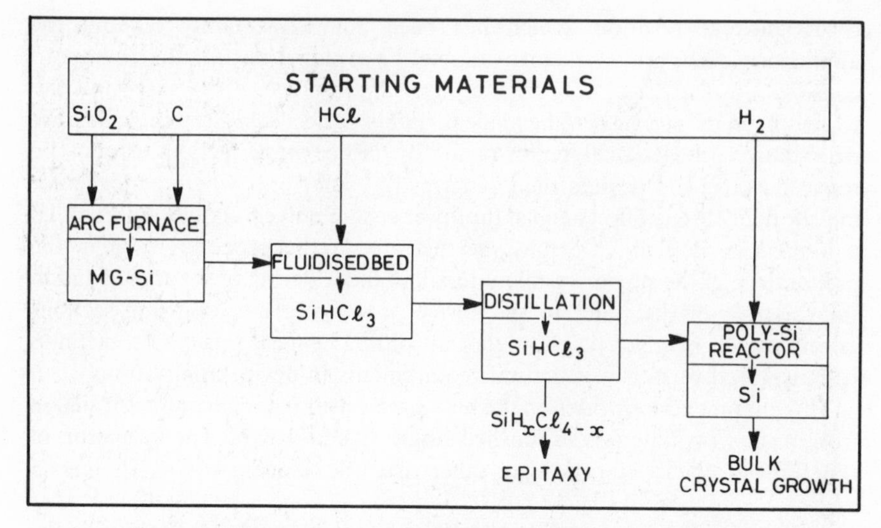

Figure 2.2 Process flow for the production of high-purity polysilicon

Metallurgical-grade silicon

Metallurgical-grade silicon (MG-Si) is produced by the carbothermic reaction between quartz and carbon in an electric arc furnace (Figure 2.3) at temperatures up to 2000° C according to the basic reaction

$$SiO_2 + 2C \rightarrow Si + 2CO \tag{2.1}$$

Impurity levels in the quartz may vary from approximately 1% in quartzite rock to $<0.03\%$ in high-purity quartz veins. Major impurities in the raw starting materials are Fe, Al, B and P. The process is energy-intensive (approximately 14kW h kg^{-1}), and MG-Si is normally produced in countries where there is an abundance of cheap, hydroelectric power, e.g. Norway. Most of the world's annual production of approximately 500 000 tonnes is used in the aluminium and steel industries, and only a small fraction is refined further.

Production and distillation of $SiHCl_3$

Gaseous trichlorosilane, $SiHCl_3$, is produced by the reaction of anhydrous HCl with finely crushed MG-Si in a fluidized bed at temperatures around 300° C:

$$3HCl + Si \rightarrow SiHCl_3 + H_2 \tag{2.2}$$

Most of the $SiHCl_3$ is used in the silicone industry. The purest $SiHCl_3$ is separated from other chlorosilanes and further purified by distillation (b.p., 32° C). This is a critical purification stage, as the major electrically-active impurities boron and phosphorus are removed as BCl_3 (b.p., 13° C) and PCl_3

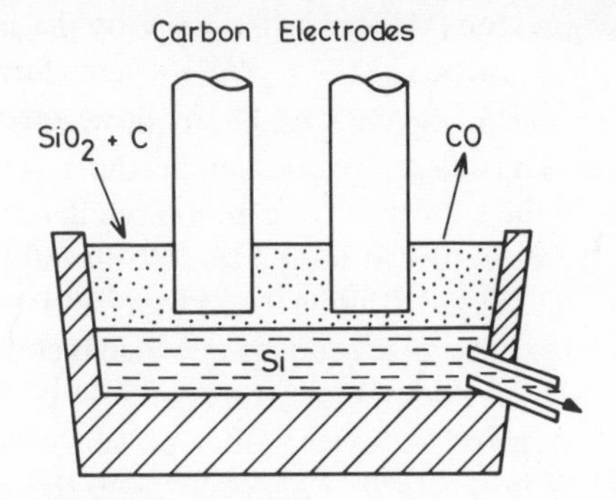

Figure 2.3 Production of Mg-Si in an arc furnace (photograph courtesy of Elkem Ltd)

(b.p., 74° C) respectively. The $SiHCl_3$ now has a purity of < 1 ppba with respect to electrically-active impurities, and is in a suitably pure form for either deposition of bulk polysilicon or thin-layer epitaxial deposition.

Polysilicon deposition

Polysilicon deposition occurs by the hydrogen reduction of trichlorosilane at temperatures around 1100° C according to the reaction

$$SiHCl_3 + H_2 \rightarrow Si + 3HCl \tag{2.3}$$

Deposition takes place on thin, high-purity polycrystalline rods (slim rods)

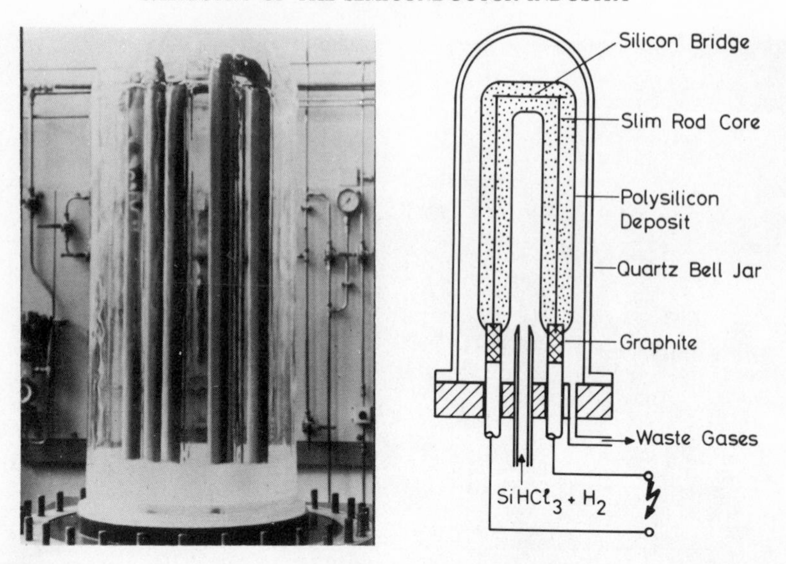

Figure 2.4 Production of high-purity polysilicon by the Siemens process (photograph courtesy of W. Keller, Siemens AG)

produced from the melt by a similar process to that described later. Two slim rods are bridged by a third within a quartz enclosure (Figure 2.4). The rods form a self-heating resistor when pre-heated to allow passage of current from a low-voltage source. The reactant gases enter the base of the reactor and deposition occurs on the slim rods to form a large inverted U-shaped silicon product which may be up to 2 m long and 200 mm in diameter. Deposition occurs over several days at growth rates < 25 mm per day. The reaction occurs at high temperatures to maximize deposition which is controlled by gas flow to the substrate. However, the temperature and deposition rate must be compromised to optimize the yield and power consumption. At high temperatures other side reactions may occur, e.g. the production of $SiCl_2$. $SiCl_4$ is always generated as a by-product of the reaction, and is used in the preparation of high-purity quartz and epitaxial deposition.

It is not possible to use the Siemens deposition process as a vapour growth technique for large-diameter single crystals since it is extremely difficult to exercise control over the crystallographic structure of the rod as its diameter steadily increases. Control of grain size and strain within the polysilicon is, however, critical when growing large-diameter starting material for the floating zone process which requires crack-free polysilicon rods. The Czochralski crystal growth process does not require crack-free starting material, and the polysilicon rods are usually crushed to irregular lumps (Figure 2.5) for loading into crucibles.

Although the polysilicon rods may be deliberately doped with electrically active elements during the Siemens process, most polysilicon is supplied

Figure 2.5 Irregular pieces of polysilicon used as starting material for the Czochralski process

Table 2.2 Typical impurity levels in high-purity polysilicon

Element	Concentration (ppba)
Carbon	100–1000
Oxygen	100–400
Phosphorus	$\leqslant 0.3$
Arsenic	< 0.001
Antimony	< 0.001
Boron	$\leqslant 0.1$
Iron	0.1 – 1.0
Nickel	0.1 – 0.5
Chromium	< 0.01
Cobalt	0.001
Copper	0.1
Zinc	< 0.1
Silver	0.001
Gold	< 0.00001

undoped. At this stage the polysilicon has an electrically active impurity content at the sub ppba level (Table 2.2). Boron levels below 0.1 ppba and phosphorus levels below 0.3 ppba are achieved routinely. The major impurities are carbon and oxygen at the lower end of the 0.1 to 1 ppma range. These impurity levels are adequate for most applications.

The availability of such a high-quality product in tonnage quantities at a cost of only a few pence per gram is a major achievement of the chemical industry.

Bulk silicon crystal growth

General

Silicon chip devices are fabricated in batch processes on wafer substrates prepared from single crystal ingots, and there are considerable economic benefits to be gained by processing large wafers. Wafer diameters have increased steadily from around 25 mm in the 1960s to 150 mm in many modern production lines. Large single-crystal silicon ingots of the desired perfection can only be grown from the melt (Si, m.p. = 1420° C). Two processes have evolved to meet device requirements, namely the crucible-free floating zone technique (FZ-Si) and the crucible-pulling or Czochralski technique (Cz-Si) (Figure 2.6). FZ-Si is used mainly in high-power devices and in applications where high purities are essential. Cz-Si dominates the low-power integrated circuit business and occupies around 90% of the total silicon crystal market.

The crucible-free vertical floating zone technique (2) combines the advantages of purification by zone refining (3) with single-crystal growth

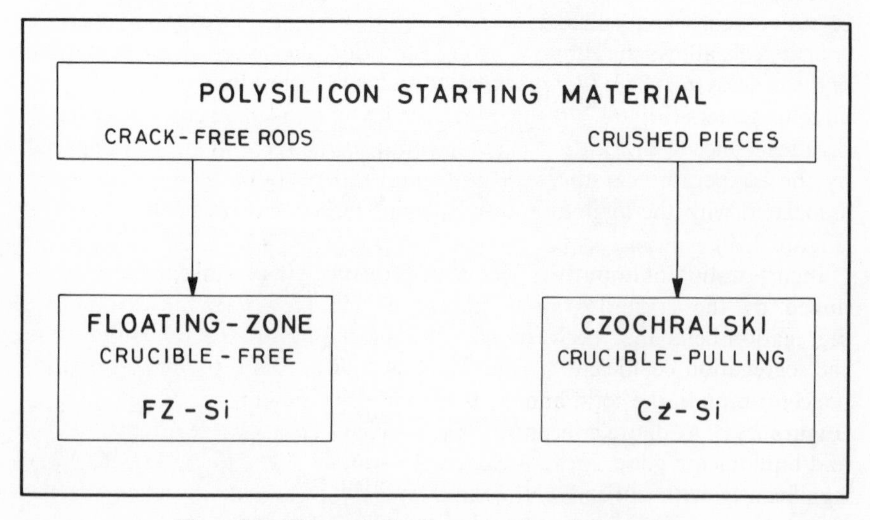

Figure 2.6 The two bulk silicon crystal-growth processes

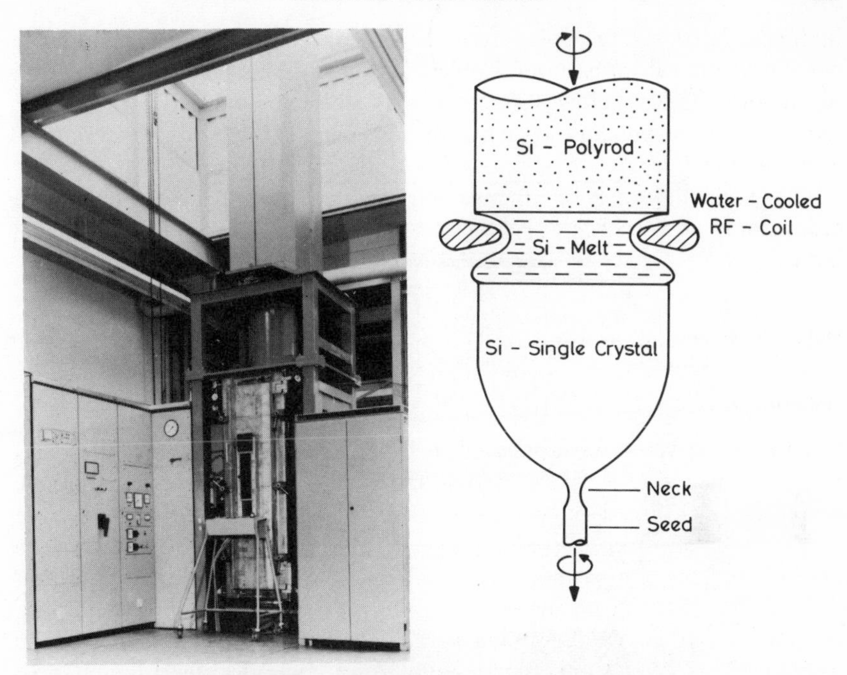

Figure 2.7 Production of FZ-Si single crystals in modern equipment by the needle-eye technique (photograph courtesy of W. Keller, Siemens AG)

(Figure 2.7). Czochralski pulling from a melt contained within a crucible (Figure 2.8) was first applied to metals (4) and much later to silicon (5). In both techniques crystallization takes place on a $\langle 100 \rangle$ or $\langle 111 \rangle$ oriented seed crystal which, when pulled at growth rates of several mm min^{-1} to a thin crystal neck, allows the elimination of dislocation line defects at the beginning of the process (6,7). It is then possible to pull dislocation-free single crystals at large diameters (Figure 2.9) since the covalently bonded silicon lattice has a high resistance to dislocation generation under the thermal stresses imposed by the temperature gradients. Remaining imperfections in the crystal are associated with the agglomeration of point defects and residual impurities (Figure 2.10).

Incorporation of impurities into melt-grown single-crystal silicon is determined by the crystallographic nature of the growth interface and the segregation behaviour between the solid and liquid phases. At equilibrium, the segregation coefficient k_0 is defined as C_S/C_L, where C_S is the impurity concentration in the solid and C_L is the impurity concentration in the liquid (Figure 2.11). At dilute concentrations, as in most practical cases, the solidus and liquidus are good approximations to straight lines, so k_0 does not vary significantly with concentration and temperature. With the exception of boron, k_0 is $\ll 1$ for the major impurities in silicon (Table 2.3), and melt growth

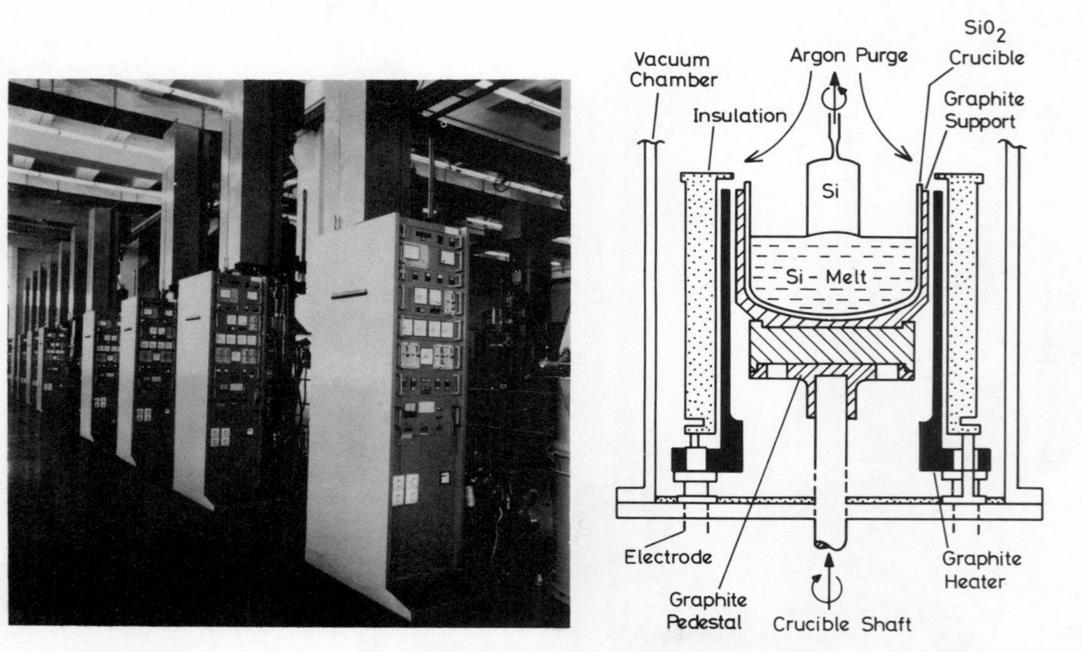

Figure 2.8 Production of Cz-Si single crystals in modern equipment by pulling from a crucible (photograph courtesy of Wacker Chemitronic GmbH)

Figure 2.9 Pulling of large diameter, dislocation-free ⟨100⟩ Cz-Si single crystal using a thin 2–3 mm diameter neck

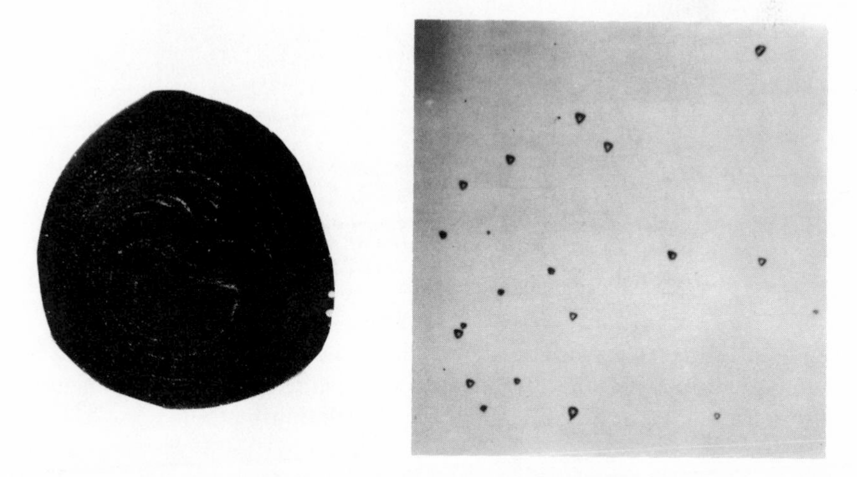

Figure 2.10 Dislocation-free FZ-Si slice showing agglomeration of point defects and impurities in swirl bands (left) which are composed of tiny etch pits (right), revealed by preferential chemical etching. Left, × 1.5; right, × 250

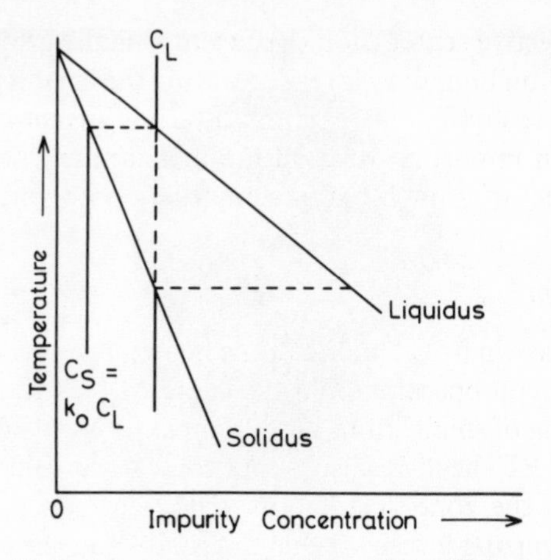

Figure 2.11 Phase equilibria for an impurity in silicon with equilibrium segregation coefficient, $k_0 = C_S/C_L < 1$

Table 2.3 Equilibrium solid/liquid segregation coefficients for impurities in silicon.

Element	Equilibrium solid/liquid Segregation coefficient, k_0
Boron	8×10^{-1}
Phosphorus	3.5×10^{-1}
Arsenic	3.0×10^{-1}
Oxygen	$2.5 \times 10^{-1} - 1$ (controversial)
Carbon	7×10^{-2}
Antimony	2.3×10^{-2}
Lithium	1.0×10^{-2}
Gallium	8×10^{-3}
Aluminium	2×10^{-3}
Nitrogen	7×10^{-4}
Indium	4×10^{-4}
Copper	4×10^{-4}
Nickel	3×10^{-5}
Chromium	3×10^{-5}
Zinc	1×10^{-5}
Titanium	9×10^{-6}
Iron	8×10^{-6}
Cobalt	8×10^{-6}
Silver	1×10^{-6}

processes can, in principle at least, be purification processes. Crystal growth is not an equilibrium process, however, and k_0 depends on growth parameters according to the Burton–Prim–Schlichter (8) relationship:

$$k_{\text{eff}} = \frac{k_0}{k_0 + (1 - k_0)\exp\left(\frac{-v\delta}{D}\right)} \tag{2.4}$$

where k_{eff} is the effective segregation coefficient, v is the growth rate, δ is the thickness of a diffusion boundary layer separating the crystal growth interface from the bulk of the liquid, and D is the liquid diffusion coefficient of the impurity. This is an important relationship and fundamental to the understanding of the effect of growth parameters (v, δ) on dopant uniformity.

Floating-zone silicon

FZ-Si crystals are grown from a molten zone heated by induced currents from a water-cooled RF coil operating in a typical frequency range of 2–3 MHz. This permits melting of small grains of the supply rod and melting of the thin seed to the zone. RF heating also supplies a significant electrodynamic stabilizing force to the zone in addition to that provided by melt surface tension. The destabilizing forces are due to gravity and the centrifugal forces from crystal and feed rod rotations which are normally applied to avoid problems associated with thermal asymmetry. Stable growth of large-diameter crystals, now available up to 125 mm in diameter, requires the coil diameter to be smaller than the feed rod and growing crystal, the aptly named needle-eye technique. Factors affecting zone stability also require large-diameter crystals to be pulled downwards. Growth rates up to several mm min^{-1} are used with feed rod and crystal counter-rotated at rates dependent on diameter and other parameters. Stretching and squeezing the zone by variation of the polyrod feed rate and crystal pull rate provides a way of controlling crystal diameter.

Growth under high-vacuum conditions assists the evaporation of volatile impurities. When combined with the possibility of multipass zone refining for the sweeping of impurities with $k_{eff} < 1$ to the end of the rod, vacuum FZ-Si growth can yield ultrapure crystals with near-intrinsic resistivities of several thousand ohm cm. To achieve these levels the electrically active boron content of the polysilicon has to be carefully controlled as it cannot be readily removed by either zone refining or evaporation. The most abundant impurity in vacuum FZ-Si is electrically inactive carbon, at levels usually below the detection limit of IR absorption analysis (< 0.1 ppma). Vacuum growth is not without its problems, however. Evaporation of silicon from the molten zone results in the colder parts of the equipment collecting silicon deposits which may drop back into the zone and cause breakdown of single-crystal growth. Defects may also nucleate at the rough surface produced by evaporation of silicon from that part of the crystal near the growing interface. FZ-Si growth under vacuum is a particular problem at large diameters. Apart from a few applications, such as detectors, most FZ-Si crystals are required with specific (> 1 ppb) concentrations of dopant which should be as uniform as possible in the axial and radial directions to satisfy device tolerances from wafer to water and within wafers. To meet these requirements in dislocation-free material the FZ-Si crystals are usually grown under an atmosphere of high-purity argon and doped either before, during or after the growth process.

Doping may take place during the preparation of the feed rods either from the slim rod core or from additions of the dopant (e.g. BCl_3, PH_3) to the $SiHCl_3/H_2$ gas feed. Doping may take place during crystal growth, either by gas doping or by addition of the dopant element with low k_{eff} and low volatility to the feed rod (Ga, In). In the above techniques the uniformity of the dopant in the crystal is largely determined by the events in the melt and at the growth interface. Normally, macroscopic axial uniformity is not a major problem, as steady state segregation behaviour can be controlled over a significant fraction of FZ–Si rods. Radial uniformity, however, is a major problem due to the effects of the convective flow in the melt and the crystallographic orientation of the growth interface. Non-uniformities of dopants with $k_{eff} < 1$ are almost inevitable, since the rotational constraints of the FZ-system do not allow ready control of the melt convection and diffusion boundary layer thickness as in the Cz-system. The effect of growth parameters on macroscopic radial uniformity is a complex issue, and has been extensively studied by Keller (9).

A further disadvantage of melt doping is the occurrence of microscopic non-uniformities of dopants. These are usually caused by temperature variations at the solidification interface which cause microscopic growth rate variations and hence variations in the effective distribution coefficient (equation (2.4)). Etched cross-sections of crystals may show spiral patterns delineating variations in dopant distribution. Measurement of the microscopic resistivity by Spreading Resistance can be used to indicate variations in electrically active dopant concentration (Figure 2.12) which are usually caused by the rotation of the crystal in a non-uniform temperature field (rotational striations) and sometimes by melt convection and external vibrations (non-rotational striations).

Macroscopic and microscopic dopant inhomogeneities in *n*-type, phosphorus-doped FZ-Si crystals were of concern in the 1960s and 1970s as diameters increased to meet requirements for power devices. The solution to the problem was provided by neutron transmutation doping (10) which is carried out after the crystal growth process. The stable isotope ^{30}Si is converted to stable ^{31}P by thermal neutrons in a nuclear reactor:

$$^{30}Si(n,\gamma)^{31}Si \xrightarrow{t_{1/2}=2.6h} {}^{31}P + \beta^- \tag{2.5}$$

^{30}Si is distributed uniformly throughout the crystal at concentrations of approximately 3%. The resulting phosphorus distribution is then primarily dependent on a uniform flux density and irradiation time. Typical uniformities are $\pm 3\%$ (Figure 2.12). The main requirement of the crystal growth process is the control of background impurity levels to prevent generation of harmful isotopes. Lattice damage caused by the high-energy irradiation is readily annealed out at temperatures above 750° C. Neutron-transmutation-doped silicon has had a major impact on the manufacture of power devices which are

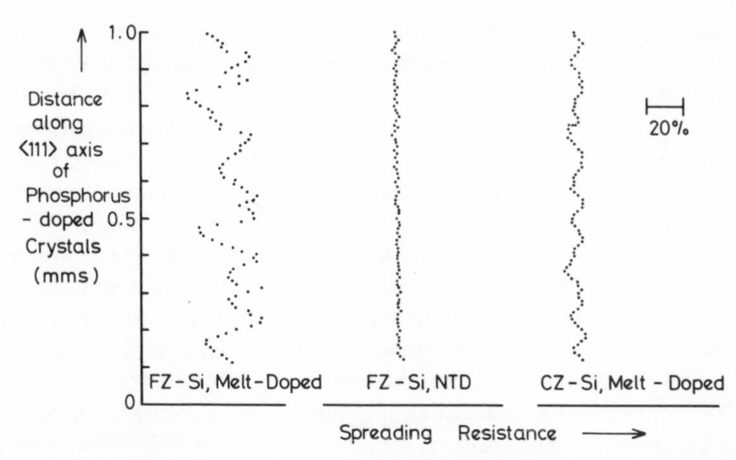

Figure 2.12 Spreading resistance measurements at 10 μm intervals on the same scale along the ⟨111⟩ axis of three different, phosphorus-doped crystals

predominantly made in *n*-type substrates. Practical limits imposed by the irradiation flux and irradiation time usually restrict the doping level to resistivities above 10 ohm cm.

Czochralski silicon

Molten silicon reacts with all crucible materials, and it is not possible to maintain the purity of the polysilicon during growth of Cz-Si crystals. Use of crucible materials containing elements which permanently change the electrical properties of the pulled crystal (e.g. Al from Al_2O_3, B from BN) must be avoided. This narrows the possibilities to refractory silicon compounds. Vitreous silica, SiO_2, has emerged as the most important crucible material. Although the silica dissolves rapidly to produce oxygen in the melt, most of this is continuously evaporated as silicon monoxide from the free melt surface. The solubility limit for oxygen in the melt (approximately 50 ppma) is easily avoided, and dislocation-free crystals can be readily obtained from more than 90% of the initial charge. This is far more difficult with silicon carbide and silicon nitride crucibles since their reaction products, carbon and nitrogen, are concentrated in the melt as a result of their low segregation coefficients (Table 2.3) and small evaporation rates.

Silica softens at temperatures below the melting point of silicon and the crucible has to be supported by a graphite cup (Figure 2.8). Graphite materials are also used in the electrical resistance heater and heat shields.

It is not possible to pull Cz-Si crystals under high vacuum as this causes boiling of the melt from the large amount of SiO. This is avoided by growing under an atmosphere of high-purity argon at pressures in the range 5–1000 mbar. The furnace is continuously purged from the top with argon which is removed near the base. Purging with argon has two major effects. Firstly, silicon monoxide is swept away from the cooler parts of the furnace above the melt to avoid condensation and subsequent flaking into the melt where it causes loss of single-crystal growth. Secondly, carbon monoxide gas generated by the graphite components is also swept away from the melt to avoid carbon contamination. Carbon levels in Cz-Si crystals are typically < 0.1 to 2 ppma. Oxygen levels are typically 10–20 ppma.

Cz-Si crystals are usually pulled with automatic diameter control in which the position of the melt meniscus at the solid–liquid interface is sensed optically and fed into instruments controlling the pull rate and heater temperature in a closed loop system. Pull rates are typically in the range 1–2 mm min^{-1}, depending on crystal diameter, and are generally less than those in FZ-Si growth due to the shallower temperature gradients resulting from the larger mass of molten silicon.

Crystal and crucible are usually counter-rotated at rates between 0 and approximately 30 rpm to reduce the effect of any temperature inhomogeneities in the furnace and melt. A much larger range of rotations is possible in Cz-Si growth than in FZ-Si growth. Forced convection from crystal rotation can be used to counteract the natural convection from buoyancy driven flows by drawing up liquid below the growth interface and imposing a centrifugal flow outward. This can result in good control of the diffusion boundary layer, radial dopant uniformity and microscopic impurity striations compared with melt-doped FZ-Si crystals (Figure 2.12).

The axial distribution of impurities in Cz-Si crystals differs significantly from that in FZ-Si crystals due to the gradual enrichment of the melt ($k_{eff} < 1$) as the crystal is pulled. The concentration of dopant in the crystal, C_S, is related to the initial concentration in the liquid, C_L, the effective segregation coefficient, k_{eff}, and the melt fraction solidified, g, by the normal freeze relationship

$$C_S = k_{eff} C_L (1 - g)^{k_{eff} - 1} \tag{2.6}$$

Note that this relationship is not valid if the concentration of the impurity in the liquid is changed by other means, e.g. continuous contamination from the vapour (carbon) or evaporation of dopant (antimony). The distribution of oxygen in Cz-Si is a particular case where the normal freeze distribution cannot be applied (see below).

Contamination of Cz-Si crystals by electrically active dopants from impurities in the crucible is typically 2 ppba. This limits the controlled growth of conventional Cz-Si crystals to resistivities less than 100 ohm cm, some two orders of magnitude below the level possible in undoped FZ-crystals.

The presence of interstitial oxygen was originally considered to be a major disadvantage of Cz-Si, since the formation of electrically active defects (11) and precipitates (12) during heat treatments is related to the oxygen content (Figure 2.13). However, since most low-power integrated circuits use resistiv-

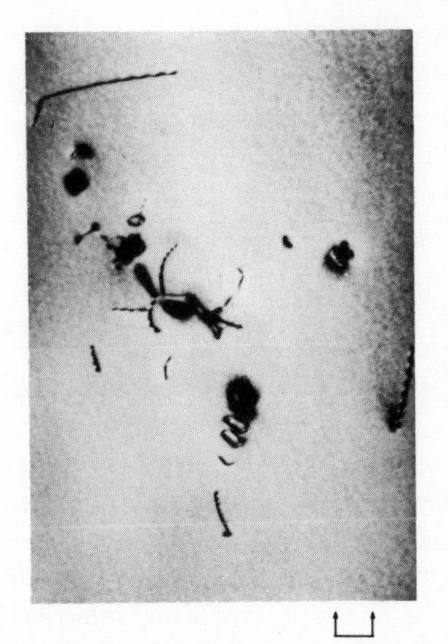

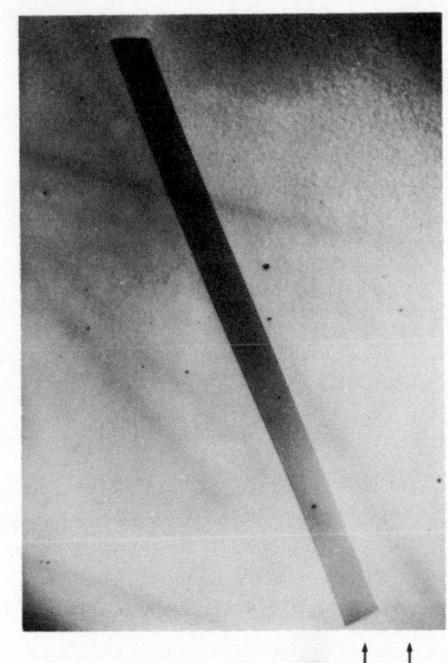

Figure 2.13 TEM micrographs of heat-treated Czochralski silicon crystals with dislocation defects (left) and stacking fault (right) associated with the precipitation of oxygen. Left, 1050° C, 16h; right, 700° C, 16h and 1050° C, 16h. Scale bar 1 μm. (Courtesy of A. G. Cullis, RSRE Malvern.)

ities less than 100 ohm cm the problem of electrically-active oxygen defects can easily be overcome by the use of stabilizing heat treatments. The *controlled* precipitation of oxygen can also be turned to advantage in the gettering of fast-diffusing metallic impurities (see later). A further advantage of interstitial oxygen in Cz-Si is its ability to harden the silicon lattice, which can offer increased resistance to slip deformation and warpage during high-temperature device processing. The lack of intrinsic gettering and the poor resistance to plastic deformation during microcircuit fabrication are major disadvantages of the much purer FZ-Si starting materials. Thus, instead of completely removing oxygen from Cz-Si crystals, tight control at specific levels to optimize the above effects in device processes is now becoming part of silicon product specifications in advanced VLSI (Very Large Scale Integration) applications.

Control of oxygen in Cz-Si melts and crystals is dependent on many factors (13), notably the silica crucible temperature (radial temperature gradient), nature of oxygen transport (convective flow) and evaporation rate (convective flow, surface area). Oxygen levels in commercial crystals are now controlled to approximately $\pm 10\%$ axially and $\pm 5\%$ radially over the range 10–20 ppma.

Wafering and surface modification

General

The crystal growth processes described above establish the bulk properties of the silicon. Of major importance, too, are the processes which are needed to

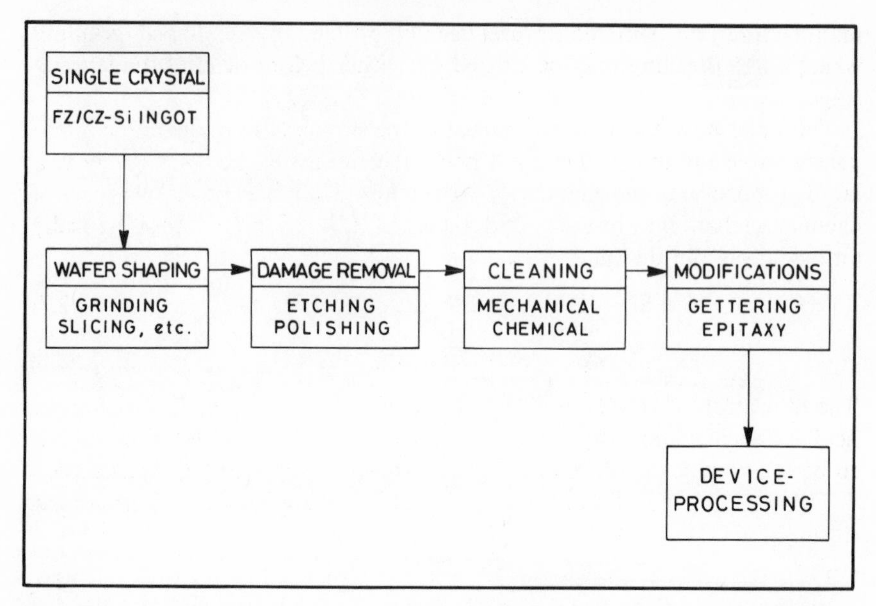

Figure 2.14 Processes used to transform silicon crystal ingots to wafers for device fabrication

present a flat, clean wafer (approximately 0.3–1 mm thick) to the device manufacturer for subsequent processing. The wafer surface may undergo various treatments to change its chemical and physical nature, depending on device requirements. In this section we trace the numerous processes which transform a silicon crystal ingot into a usable starting material for device manufacture (Figure 2.14). Many of the initial stages are mechanical in nature but we shall see how chemical processes play a vital role in wafer etching, polishing, cleaning and in surface modifications such as gettering and epitaxy.

Wafer shaping

The first stages of wafering involve the machining of the silicon to produce a slice of the required shape. The ingot is initially cropped to remove seed and tail sections. This is followed by centre grinding of the ingot to produce the required diameter. Flats are then ground on the crystal at specific crystallographic orientations to denote the surface orientation and dopant type, according to internationally accepted standards (14). The ground ingot is then sliced with a diamond-impregnated annular saw blade to produce a slice of the desired orientation. This is a critical stage in the control of slice bow and taper.

Accurate alignment and tensioning of the blade and correctly adjusted coolant flow are essential. Sawing introduces damage and thickness variations which can be reduced and made more uniform by lapping. In order to prevent edge chipping and photoresist pile-up during subsequent handling in device manufacture, the slice edges are usually profiled by a shaped grinding wheel. Edge profiling may be carried out either before or after the lapping stage.

The wafer now has its basic shape, but the surfaces are rough and contain considerable amounts of damage. Chemical etching is carried out to remove a large proportion of the damage (approximately 30 μm per side). Two types of chemical etchant may be used. Acid etching in HNO_3/HF mixtures is basically an oxidation and dissolution reaction:

$$Si + 4HNO_3 \rightarrow SiO_2 + 4NO_2 + 2H_2O \quad (2.7)$$

$$SiO_2 + 6HF \rightarrow H_2SiF_6 + 2H_2O \quad (2.8)$$

The ionization of HNO_3 and HF can be modified by the addition of acetic acid. A typical volume ratio is 4:1:2 $HNO_3/HF/CH_3COOH$. Extreme care has to be taken with the scrubbing and controlled removal of NO_2 waste gas.

The second type of damage removal etch is based on caustic solutions, normally KOH:

$$Si + 4OH^- \rightarrow SiO_4^{4-} + 2H_2 \quad (2.9)$$

Agitation of both etchants to remove reaction products and gas bubbles from the surface is essential to ensure good uniformity.

The next stage is wafer polishing, where one surface is made extremely smooth, flat and damage-free by a chemical-mechanical process. Device dimensions in integrated circuits are continuously shrinking, and photolithographic printing of features around 1 μm is now required. This limits the maximum deviation from the highest to the lowest part of the polished surface to a few microns over the complete wafer diameter when it is held on a vacuum chuck.

Current polishing technology is based on the Syton Process (15) which uses a colloidal silica slurry in an alkaline medium (pH 9–12). The wafer is held on a metal carrier by either wax or waxless mounting. Careful selection of slice thickness has to be made, as most machines polish several wafers at a time. The carrier plates are accurately aligned to a large rotating polishing plate which supports a polyurethane pad. Polishing takes place by chemical and mechanical action. The chemical reaction is basically an oxidation of silicon to silicate:

$$Si + 2OH^- + H_2O \rightarrow SiO_3^{2-} + 2H_2 \quad (2.10)$$

Generally, polishing is carried out in two stages. The stock removal stage removes about 20–40 μm silicon, including any damage remaining from the etching stage. A second stage uses a dilute slurry, low pH and low pressure on a softer pad to produce the final specular surface. The polishing process has to be carried out in a clean environment in which control of particulates is essential to avoid scratches and surface damage.

Wafer cleaning

The removal of particulates from silicon wafer surfaces is carried out at various stages of wafering. Coarse particulates generated in the mechanical shaping processes are usually removed by ultrasonic agitation in the presence of detergents. Brush scrubbing with or without detergents may be used later in the wafering stage, although the greatest of care must be exercised to avoid scratching and damage of polished surfaces.

Contaminating films on polished wafers are less readily detected than particulates, but their removal is essential before the wafer is suitable for device processing. It is not possible to maintain a perfectly clean, bare silicon surface under normal processing environments, and removal of contaminating films usually involves replacement of an undesirable film by a more desirable one under controlled conditions. There are basically two types of contaminating film which require different chemical treatments. Organic films are removed by an oxidation process, whilst metal films are complexed by chlorination. There are various proprietary techniques used in the industry, but the most commonly used cleaning process is based on the RCA process (16), which involves three stages.

The first stage uses an $H_2O/NH_4OH/H_2O_2$ solution in the volume ratio 5:1:1 to oxidize organics at 75–80° C for approximately 10 minutes. The second stage strips the thin hydrous oxide formed in the first stage with an HF/H_2O solution in the volume ratio 1:50. Plating of metals on to the exposed silicon surface is minimized by limiting the second stage to 10–15 seconds. The third stage uses an $H_2O/HCl/H_2O_2$ solution in the volume ratio 6:1:1 at 80° C to complex metal films. After each stage, quenching of the chemical reaction is achieved by rapidly transferring the wafer to filtered, de-ionized water with a resistivity of 10–20 megohms. Reagents in the first and third stages are kept in high-purity silica containers, and the HF solution of the second stage is kept in high-purity polypropylene beakers. The chemicals of the first and third stages must also be kept in separate fume hoods to avoid the formation of ammonium chloride. It is important during all stages to prevent the wafer from drying, which can result in tenacious deposits. Final drying is carried out by controlled spinning to avoid aerosol droplet formation.

Gettering

Silicon wafers are subjected to considerable thermal, physical and chemical treatments during device processing, and the introduction of unwanted impurities and point defects in the active device region must be prevented if the devices are to function with the desired properties. Since it is virtually impossible to avoid the introduction of some harmful species, the use of gettering techniques to remove them from the active device region to some distant site in the wafer is essential. Gettering relies basically on two mechanisms. The introduction of lattice strain and associated crystallographic defects such as dislocations can provide preferential sites for trapping of mobile impurities which reduce the net strain. It is also possible to introduce regions of different chemical composition such as phosphorus-diffused layers to enhance the solubility of impurities. Gettering sites are introduced either at the wafering stage, during device processing or both, depending on the gettering process.

Extrinsic gettering introduces damage to the wafer back surface. Abrasive processes such as light lapping and modified sand blasting must be carried out at the wafering stage. The introduction of damage by a laser beam or by ion implantation of inert gases are relatively clean, 'cold' processes and can be carried out either before or during processing. Misfit dislocations and grain boundaries provided by deposition of a polysilicon film on the back surface can be introduced at the wafering stage.

The effectiveness of extrinsic gettering techniques is highly dependent on the device process. It is essential that the deliberately created damage does not anneal out early on in the device process. Sufficient control must be exercised to ensure effective gettering of impurities throughout the wafer thickness without allowing the damage to penetrate the active regions. These requirements are becoming difficult to meet, since the flatness specifications of large-diameter wafers necessitate wafers thicker than 0.5 mm. The technique of intrinsic gettering in Cz-Si wafers is an attractive alternative to extrinsic gettering.

Intrinsic gettering in Cz-Si wafers relies on the presence of supersaturated

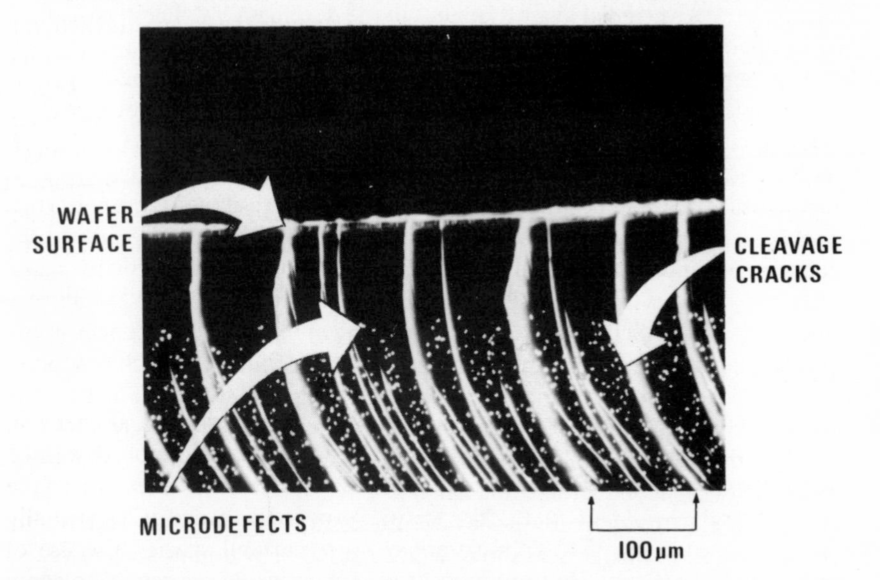

Figure 2.15 Cross-section of a cleaved and preferentially etched Cz-Si wafer after a two-stage anneal to generate a denuded zone near-surface defect-free and bulk microdefects

interstitial oxygen which, when heat treated at typical device processing temperatures below 1150° C, precipitates as silicon oxide in the bulk of the wafer. The volume change produced by a growing oxide precipitate generates micro-defects which act as preferential sites for aggregation of metallic impurities (Figure 2.13). Growth of the oxide precipitate is dependent on the oxygen diffusion rate, hence intrinsic gettering can be effective throughout the device process, depending on the device process steps. Intrinsic gettering was probably operative in all device processing of Cz-Si wafers long before its importance was recognized in the late 1970s (17, 18).

Clearly, the occurrence of oxide precipitates and associated defects must be avoided in the near-surface active regions. The key to effective intrinsic gettering is the generation of a near-surface denuded zone free of oxide precipitates whilst controlling oxide precipitation in the bulk of the wafer by a two-stage high temperature/low temperature heat treatment (Figure 2.15) (19). Unlike micro-defects in oxygen-free FZ-Si (Figure 2.10) oxide micro-defects in Cz-Si are amenable to solid-state defect engineering, thus turning a potential problem into a positive benefit.

Future developments

The major driving forces for the development of processes in silicon technology are cost reduction and quality improvement. Both objectives are often achieved simultaneously when processes are automated, and this is a general characteristic of current developments in many processes. In certain cases cost will be traded against quality in the development of new processes, as in the development of large-area terrestrial silicon solar cells.

The Siemens process is unlikely to be displaced as the mainstream process for production of high-quality polysilicon despite its relatively poor conversion efficiency and high energy cost. Alternative processes of potential importance are based on silane (20) and dichlorosilane (21). By using a low temperature fluidized bed to deposit silicon from chlorosilanes, there exists the potential not only to reduce power consumption but also to introduce continuous processing (22). For the production of solar-grade silicon there are many possibilities for reducing costs. An obvious starting point is to improve the purity of MG-Si in order to eliminate many of the subsequent purification stages. This has been achieved in an experimental arc furnace which produced 99.99% silicon from high-purity silica and graphite feedstock (23). Two recrystallizations realized sufficient refinement by solid/liquid segregation to produce material with solar cell efficiencies within 1% of those prepared from high-purity semiconductor-grade silicon.

Despite the trend to lower device processing temperatures and recent improvements to the thermal slip resistance of FZ-Si by doping with nitrogen (24), advances in Cz-Si will probably strengthen its position in high-performance microelectronic applications. The growth and wafering of

200 mm-diameter Cz-Si crystals, computer control and continuous Cz-pulling processes are already at an advanced stage of development. Using continuous melt replenishment techniques (25), for example, the control of melt convection at constant melt aspect ratios will be beneficial in the control of oxygen and dopant uniformity. This system becomes similar to the steady-state FZ-system with improved axial uniformity of dopant compared with a normal freeze distribution. The range of oxygen content and resistivities of Cz-Si will improve as new techniques are implemented in production. Of these, the use of dc magnetic fields to control melt temperature gradients and fluid flow by eddy current damping may be significant in the production of low-oxygen (< 2 ppm), high-resistivity (> 500 ohm cm) crystals for applications where previously only FZ-Si was suitable (26). At the opposite end of the resistivity scale, the increased use of CMOS (Complementary Metal Oxide Semiconductor) logic in VLSI circuitry will require large-diameter, epitaxial wafers with a high-resistivity epitaxial layer deposited on a low-resistivity wafer substrate to suppress latch-up effects during device operation. This material's structure will require a better understanding of intrinsic gettering properties of heavily doped substrates in which the role of oxygen content and the nucleation and growth of oxide precipitates are poorly understood.

The increase in wafer diameters and stringent flatness and cleanliness requirements will provide the stimulus for new wafering and cleaning techniques. Current wafering techniques involve a considerable number of process steps and significant wastage of valuable material. Growth of single-crystal silicon sheet has been demonstrated (27, 28), but the crystal quality is currently not suitable for high-performance applications. However, sheet growth and melt spinning/casting techniques in a wafer mould (29) are attractive options for low-cost solar grade material, as they avoid the wafering cost incurred by other consolidation processes such as ingot casting (30) and directional solidification (31).

Vapour-phase, liquid-phase and solid-state chemical processes have played major roles in establishing silicon as the material of the microelectronic age. New processes for cost reduction, quality improvement and novel device manufacture are continuously evolving, and they will consolidate silicon's leading position in the VLSI era of the 1980s and ULSI (Ultra Large Scale Integration) era of the 1990s.

References

1. US Patent 3,042,494 (1962); USP 3,146,123 (1964); USP 3,200,009 (1965).
2. US Patent 3,060,123 (1952).
3. W. G. Pfann, *Trans. AIME* **194** (1952) 747.
4. J. Czochralski, *Z. Physik. Chemie* **92** (1917) 219.
5. G. K. Teal and E. Buehler, *Phys. Rev.* **87** (1952) 190.
6. W. C. Dash, *J. Appl. Phys.* **29** (1958) 736.

7. G. Ziegler, *Z. Naturforsch.* **16**a (1961) 219
8. J. A. Burton, R. C. Prim and W. P. Schlichter, *J. Chem. Phys.* **21** (1953) 1987.
9. W. Keller, *J. Cryst. Growth* **36** (1976) 215.
10. K. Lark-Horovitz, *Proc. Conf.*, University of Reading, Butterworth, London (1951) 47.
11. C. S. Fuller and R. A. Logan, *J. Appl. Phys.* **28** (1957) 1427.
12. C. W. Pearce and G. A. Rozgonyi, *Proc. Third Int. Conf. on Silicon Materials Science and Technology*, *'Semiconductor Silicon, 1977'*, eds. H. Huff, E. Sirtl, 606.
13. K. G. Barraclough; *Proc. ECS Satellite Symp. to ESSDERC '82*, Munich, 'Aggregation phenomena of point defects in silicon', PV83-4, eds. E. Sirtl, J. Goorissen and P. Wagner, ECS Inc., 176.
14. *BOSS Book of SEMI Standards*, Semiconductor Equipment and Materials Institute Inc.
15. US Patent 3,170,273 (1965).
16. W. Kern and D. A. Puotinen, *RCA Rev.* **31** (1970) 187.
17. T. Y. Tan, E. E. Gardner and W. K. Tice, *Appl. Phys. Lett.* **30** (1977) 175.
18. G. A. Rozgonyi, R. P. Deysher and C. W. Pearce; *J. Electrochem. Soc.* **123** (1977) 1910.
19. UK Patent GB 2,080,780B (1981).
20. US Patent 4,150,168 (1979).
21. J. R. McCormick, F. Plahutnik, D. Sawyer, A. Avvidson and S. Goldfarb, *Proc. 16th IEEE Photovoltaic Specialist Conf.*, San Diego (1982) 54.
22. US Patent 4,092,466 (1978).
23. J. A. Amick, J. P. Dismukes, R. W. Francis, L. P. Hunt, P. S. Ravishankar, M. Schneider, K. Matthei, R. Sylvain, K. Larsen and A. Schei, *J. Electrochem. Soc.* **132** No. 2 (1985) 339.
24. T. Abe, K. Kikuchi, S. Shirai and S. Muraoka, *Proc. 4th Int. Symp. on Silicon Materials Science and Technology, 'Semiconductor Silicon 1981'*, eds. H. R. Huff, R. J. Kriegler, Y. Takeishi, PV81-5, Electrochem. Soc. Inc., 54.
25. G. Fiegl, *Solid State Technol.* **26** (1983) 121.
26. T. Suzuki, N. Isawa, Y. Okubo and K. Hoshi, in ref. 24, 90.
27. T. Surek, in *Proc. Symp. Electronic and Optical Properties of Polycrystalline or Impure Semiconductors and Novel Silicon Growth Methods*, eds. K. V. Ravi and B. O'Mara, The Electrochemical Society Inc., Pennington (1980) 173.
28. J. J. Brissot, in *Current Topics in Materials Science*, Vol. **2**, eds. E. Kaldis and H. J. Scheel, North Holland (1977) 796.
29. Y. Maeda, T. Yokoyama and I. Hide, *J. Cryst. Growth* **65** (1983) 331.
30. D. Helmreich, in ref. 27, p. 184.
31. T. F. Ciszek, G. H. Schwuttke, K. H. Yang, *J. Cryst. Growth* **46** (1979) 527.

3 Group III–V compounds

I. R. GRANT

Introduction

Compounds of Groups IIIB and VB of the Periodic Table are of increasing technological importance as electronic and optoelectronic materials. These can offer many advantages over traditional Group IV semiconductors such as silicon and germanium in permitting the fabrication of relatively efficient light-emitting or -detecting devices and active components working at higher speed and lower power dissipation and capable of operation at higher temperature.

A suitable compound for a given application is selected by consideration of its electronic properties determined by the electronic energy band structure, in particular the bandgap energy and the electron effective mass. The former governs the spectral characteristics of absorbed and emitted radiation while the latter predicts the carrier drift velocity in low electric fields. An additional flexibility is obtained by alloying with one or more substitutional impurities to modify the band structure. Thus ternary alloys such as GaAlAs, GaAsP and quaternaries such as GaInAsP have important applications in optoelectronics. The versatility offered by these compounds is counterbalanced by their increased cost and more complex material processing technology.

Table 3.1 summarizes the relevant physical properties of the more important III–V compounds. The following discussion will be confined, however, to only two of these materials, gallium arsenide and indium phosphide. These are the most prominent commercially, and each supplies a market in the form of wafer substrates, with gallium arsenide being the dominant material in volume and breadth of applications. The major use of GaAs has lain in the field of microwave-frequency signal processing as well as light-emitting diodes and solid state heterojunction lasers. However, the greatest promise is in the fabrication of very high-speed digital integrated circuits. Indium phosphide wafers are used almost exclusively as substrates for epitaxial growth of alloys where the InP lattice spacing most closely matches that of the required alloy composition.

Many of the properties and techniques considered here for these two compounds are of relevance to the other materials, particularly the methods of preparation and single-crystal growth.

Table 3.1 Properties of major III–V compounds (1)

Property		GaAs	GaP	InP	InAs	InSb
Structure		Zinc blende	Zinc blende	Zinc blende	Zinc blende	Zinc blende
Lattice constant	Å	5.654	5.447	5.869	6.058	6.479
Melting point	°C	1238	1467	1062	943	525
Density	$g\,cm^{-3}$	5.316	4.130	4.787	5.667	5.775
Atomic density	$atom\,cm^{-3}$	4.43×10^{22}	4.96×10^{22}	3.96×10^{22}	3.6×10^{22}	2.94×10^{22}
Energy gap (25° C)	eV	1.40	2.24	1.35	0.356	0.18
Molecular wt		144.64	100.70	145.79	189.74	236.58
Intrinsic resistivity (25° C)	ohm cm	3.8×10^{8}	1.9×10^{16}	8.2×10^{7}	3.2×10^{-1}	5.0×10^{-3}

Crystal structure

III–V compounds crystallize in the zinc blende (ZnS) structure, consisting of two interpenetrating equivalent face-centred cubic lattices, one composed entirely of Group III atoms, the other of Group V atoms. This is illustrated in Figure 3.1 for the case of the GaAs structure (2). Using the conventional system of Miller indices, the Ga and As sub-lattices are translated by a distance $(-1/4, -1/4, 1/4)\,a_0$ where $a_0 = 5.654$ Å is the lattice parameter (3). Because of the different valencies of the constituents the Ga–As bond is partly covalent, partly ionic in nature with an ionicity of 0.3e (4).

Referring to Figure 3.1 it is seen that the (111) plane of the arsenic sublattice consists entirely of As atoms and is known as the (111) As plane. The equivalent (111) plane of the gallium sublattice is similarly composed of Ga atoms only and hence is the (111) Ga plane. Considering the [111] direction in the structure, it is seen that a sequence of parallel planes occurs, consisting of closely spaced pairs of (111) Ga and (111) As planes with a separation of 0.82Å. The tetrahedral bonding arrangement with bond length 2.45 Å leads to strong bonding within the pairs of (111) planes, each Ga atom being bound to three As atoms in the nearest plane and one As atom in the next plane. Any disruption of the structure perpendicular to the [111] would occur by breaking of the single bonds to give a surface consisting of (111) As or (111) Ga planes, resulting in chemically distinct surfaces and hence different reaction behaviour. This is of particular importance in the use of chemical etchants (5).

This type of behaviour has general validity for the III–V compounds and a

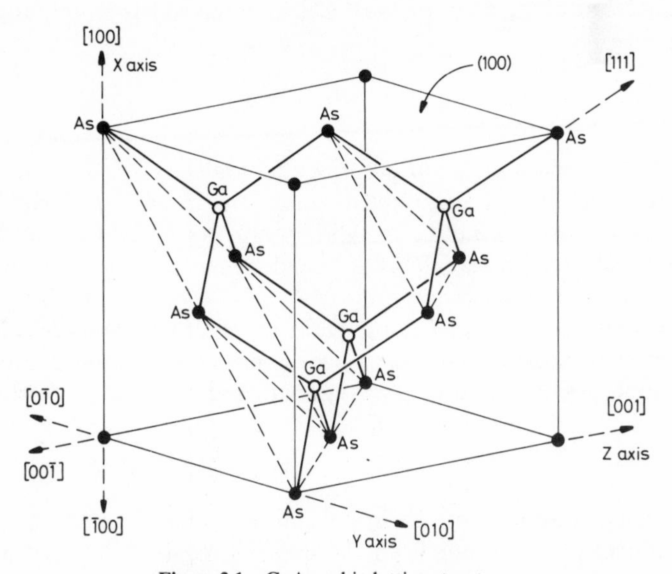

Figure 3.1 GaAs cubic lattice structure

designation of (111) A has been made for the Group III elements and (111) B for Group V.

Raw materials

For most applications the synthesis of the compound material requires the availability of constituent elements of very high purity. The minimum purity used for single-crystal growth is 6N, i.e. 99.9999 atomic % of the element, although 7N quality is now claimed to be available for some elements. Reliable analysis at this sensitivity level is difficult for most of these materials, although spark source mass spectroscopy (SSMS) and secondary ion mass spectroscopy (SIMS) have a general applicability, while atomic absorption analysis may be used for metallic impurities. The increasingly rigorous demands on impurity control are driving the need for improved analytical techniques.

Gallium

Gallium, although not a rare element in terms of total abundance in the Earth's crust, is produced in relatively small quantities as a pure metal and currently only for use in the electronics industry. It is found naturally in association with aluminium-bearing ores such as bauxite, and is produced commercially as a by-product of aluminium and zinc refining. A secondary but increasingly important source of the metal is from the recycling of gallium compound-semiconductor waste.

The metal has a melting point of only 30° C and is normally kept refrigerated. Its boiling point of 2403° C gives it the widest liquid range of all the elements.

Arsenic

Arsenic is a common element, occurring naturally as As_2S_3 contained in copper and lead ores. It is extracted as flue dust during smelting of these metals. High-purity crystalline metallic α-form As is used for compound semiconductor manufacture. This is stored under an inert atmosphere of argon to prevent oxidation.

Arsenic sublimes above 600° C and purification is by successive distillation. The volatile nature of the element at high temperatures is responsible for much of the complexity of synthesis, crystal growth and material processing.

Indium

Indium occurs as an impurity in ores of zinc, lead and copper, principally in sphalerite (ZnS). In common with the other elements considered here, its extraction is a secondary process in the refining of the main ores.

Phosphorus

This is found in the form of orthophosphates such as fluorapatite $3Ca_3(PO_4)_2.Ca(F,Cl)_2$. Extraction is carried out by reduction of the minerals by heating with carbon.

The three principal allotropes of P are white, red and black. The red form is obtained by heating white phosphorus above 400° C, and is much less reactive than white phosphorus. This is the material used in the synthesis of compound semiconductors.

GaAs phase equilibria

GaAs is subject to decomposition by arsenic evaporation at high temperatures. The problem of arsenic containment is most severe during synthesis of the compound due to the very high As vapour pressure at the reaction temperature. This is overcome by the use of sealed pressure balanced ampoules or by the special technique of liquid encapsulation.

Figure 3.2 shows the vapour pressure data for Ga and As over solid GaAs (6) as a function of temperature, indicating a total decomposition pressure of 1

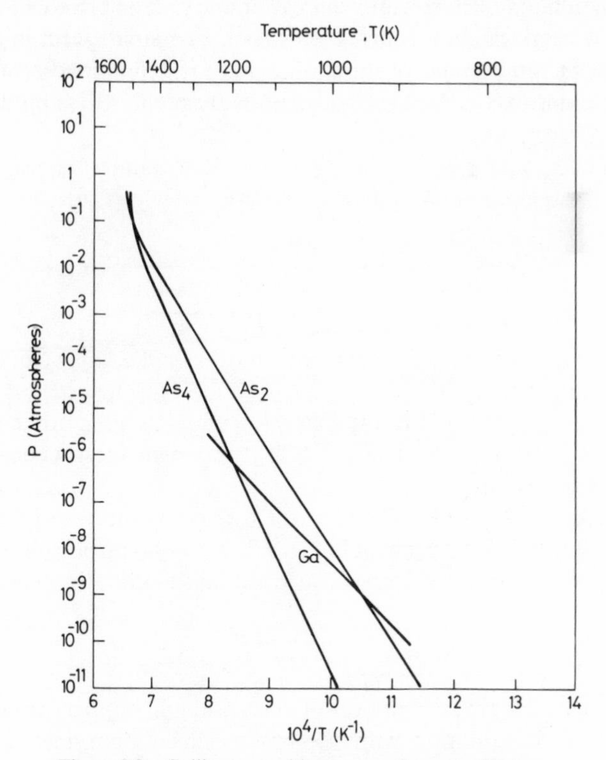

Figure 3.2 Gallium arsenide vapour phase equilibrium

atmosphere at the melting point (1240° C). Control of the gas phase composition is therefore an important consideration during crystal growth from the melt, due to its influence on stoichiometry and hence on defect concentration in the solid crystal.

Horizontal Bridgman growth

Bridgman crystal growth is one of a set of similar techniques involving progressive crystallization by the relative motion of a temperature gradient and a molten bar of the material. The elements of the conventional Bridgman method are illustrated in Figure 3.3.

Synthesis is performed by reaction of gallium contained in a high-purity silica glass 'boat' with arsenic vapour sublimed from a solid arsenic source. The whole system is contained in a sealed ampoule to maintain an arsenic vapour overpressure. By adding an excess weight of arsenic above that required for a stoichiometric composition, a controlled arsenic vapour pressure can be set up. Temperature T_3 (Figure 3.3) is set precisely to control melt stoichiometry, while temperature T_1 is imposed over the melt. T_2 is the temperature of the crystallized solid and determines the temperature gradient for solidification. As the melt isotherm within the gradient moves with respect to the melt, crystallization proceeds from the solid seed crystal. The crystallographic orientation of the grown ingot is determined by that of the seed, and the long axis of the bar normally lies along a ⟨111⟩ crystal direction.

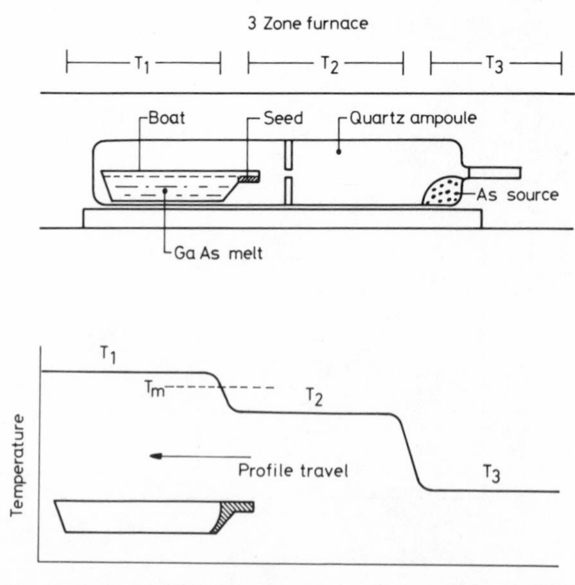

Figure 3.3 Horizontal Bridgman system (top) and temperature profile (bottom)

The use of silica crucibles to contain the GaAs melt leads to the incorporation of silicon into the crystal by the following reaction (7):

$$4Ga + SiO_2(s) \rightleftharpoons 2Ga_2O(g) + Si(l) \quad (3.1)$$

Silicon is a shallow donor dopant in GaAs and the material produced in this way exhibits *n*-type semiconducting electrical behaviour with a carrier concentration around $10^{16}\,cm^{-3}$.

The reaction of the quartz vessel with the melt leads to sticking of the bar to the crucible after solidification, causing stress-induced dislocations in the crystal due to differential contraction on cooling. This effect may be controlled by governing the afterheat temperature of the bar to avoid transport of SiO in the gas phase. In this way increased amounts of Si dopant may be introduced to make the material more conductive and to reduce the dislocation density by 'impurity hardening'. Figure 3.4 shows the effect of Si content, expressed as free carrier concentration, on the dislocation density of horizontal Bridgman material.

By careful control of the temperatures and temperature gradients of the growth apparatus and impurity hardening by doping, 'zero dislocation density' ($< 10^3\,cm^{-2}$) may be obtained by the Bridgman method. This meets the substrate requirements of many optoelectronic applications such as light-

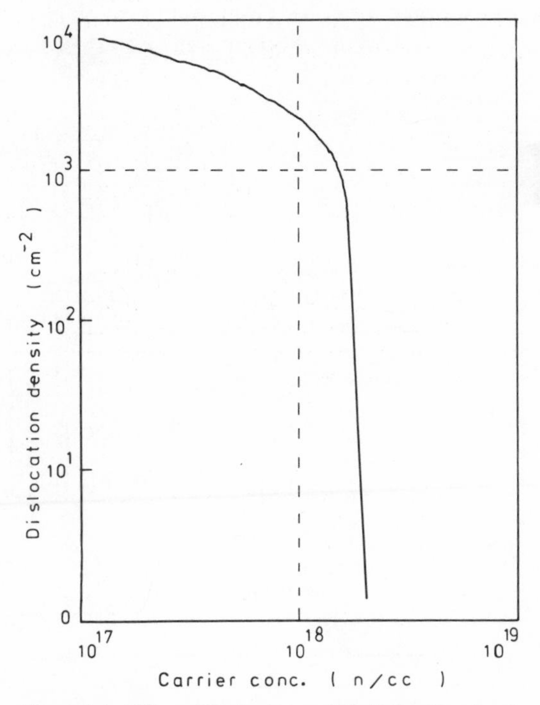

Figure 3.4 Effect of Si doping on HB dislocation density

emitting diodes and lasers. However, semi-insulating ($>10^7$ ohm cm) material can only be obtained by doping, usually with chromium which forms a deep acceptor level in the band gap. Cr is a mobile atom at high temperatures and will readily out-diffuse to the substrate surfaces during processing involving heat treatments (8). This, together with non-availability of circular wafers having the preferred (100) surface orientation, excludes Bridgman-grown material from most device applications aimed at GaAs integrated circuit development.

Liquid encapsulated Czochralski growth

Czochralski crystal growth involving the pulling out of a single crystal from a melt is the main production technique for silicon single-crystal manufacture (see Chapter 2). The method has been so refined for this application that very large ingots, several feet in length and up to 8 inches in diameter can be grown free of dislocations and with good control over impurity incorporation.

The difficulty with melt growth of the dissociable compounds is containment of the Group V component. This may be overcome in a variety of ways by the use of sealed systems maintaining a controlled As vapour pressure over the melt (9).

The liquid-encapsulated Czochralski (LEC) principle was first proposed by Metz *et al.* (10) and adopted for III–V compound growth by Mullin (11). The elements of the growth system are shown schematically in Figure 3.5. GaAs,

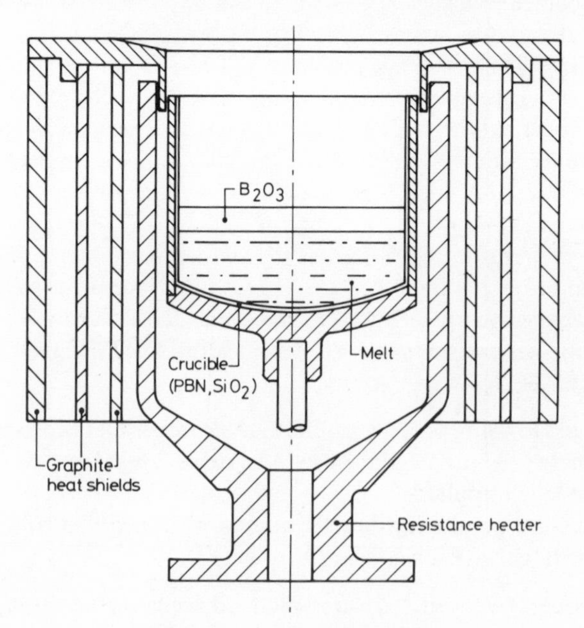

Figure 3.5 LEC growth system

GaP and InP are grown in this way. In the case of GaAs a one-step process involving *in-situ* compounding in the grower is employed. GaP and InP are synthesized in a separate apparatus prior to growth.

The main application for LEC GaAs is as high-purity undoped substrates for epitaxial growth and direct ion implantation. This requires high electrical resistivity for device isolation through the substrate, stable electrical characteristics with heating and very low impurity concentrations to minimize effects of redistribution at the surface during processing. Uniformity across the substrate is a major consideration for integrated circuit fabrication and has provided the driving force to achieve improved crystalline perfection by reducing dislocation densities and dislocation-related effects.

The use of the direct synthesis technique (12, 13) is an important milestone in the development of undoped semi-insulating gallium arsenide. A crucible composed of pyrolytic boron nitride (PBN) is charged with elemental Ga and As in weight proportions to achieve a stoichiometric melt. An excess weight of several grams of As is added to compensate for As lost prior to reaction. The charge weight used for typical growth configurations yielding 80 mm-diameter crystals is 4–4.5 kg.

Overlaying the charge is a pellet of solid boron trioxide (B_2O_3) glass. It is this material which forms the liquid encapsulation over the melt during growth.

The crucible containing the charge is placed inside a graphite resistance heater. The whole assembly is contained inside a stainless steel pressure vessel with water-cooled walls. Special seals allow movable rotating shafts to be introduced at top and bottom of the cylindrical chamber. As the charge is heated the Ga melts first, enclosing the As. The oxide softens and flows above 450° C, to wet the crucible walls and encapsulate the charge. Above 800° C the Ga and As react exothermically and an external overpressure of 60 atmospheres prevents As evaporation from the charge. The temperature is then raised after reaction until the system is stabilized above the melting temperature of GaAs.

In order to perform the direct synthesis compounding a high-pressure growth chamber is required, but the growth itself needs an overpressure of only a few atmospheres. In practice, however, the best growth yields (9) are obtained using gas pressures of $\sim$ 20 atmospheres. The LEC growth process of GaAs currently follows two routes:

(i) Low pressure using polycrystalline GaAs charges and pressures around 5 atmospheres. Many of the growth systems used for this method are modified silicon pullers.
(ii) High pressure, employing direct synthesis, with specialist equipment made by Cambridge Instruments Ltd.

Growth occurs by the introduction of a seed single crystal through the oxide encapsulant layer to contact the melt; the melt temperature is adjusted to avoid both melting of the seed and freezing of the surface on contact. The meniscus between the solid seed and the melt may be monitored by observation through the transparent oxide—a quartz light pipe and sapphire high-pressure window allow direct viewing of the melt surface via a closed-circuit TV system. The seed is pulled out of the melt at a fixed rate of several mm per hour while both it and the crucible are rotated. The melt surface is maintained at a constant position within the heater by raising the crucible during growth. Diameter control is performed by varying the heater temperature. The monitored variable, however, is the weight of crystal measured on a load cell and the instantaneous diameter is calculated from the rate of change of the weight with respect to time. Allowance is made for the anomalous density behaviour of the material which undergoes expansion on freezing (14).

The grown crystal is an extension of the seed and has the same crystallographic orientation, usually a ⟨100⟩ growth axis. Two standard wafer diameters are currently used for processing, two inches and three inches. Crystals are therefore grown slightly larger than these sizes to be ground to exact cylinders and to allow for losses during subsequent wafer production.

The pyrolytic boron nitride crucible is reusable, with an expected life of between 10 and 15 runs. This is a fragile and expensive component and contributes a very significant portion of the cost of the growth, especially for smaller melts. It is made by a low-pressure chemical vapour deposition process involving the pyrolysis of ammonia and BCl_3 at high temperature and the growth of the component over a graphite former. The maximum available crucible size is currently about 8 inches in diameter, but should be scaleable to around 10 inches. The melt capacity is dependent on the availability of large enough crucibles with the largest systems presently working with 8 kg melts and crucibles up to 8 inches diameter. The usual failure mode is by progressive delamination during cleaning after each growth.

Boric oxide

The boron trioxide encapsulant plays a crucial role in the growth process beyond simply providing a liquid seal around the melt. It exhibits a number of properties which are ideal for the application:

(i) It flows at intermediate temperatures before significant As sublimation occurs
(ii) It is non-reactive
(iii) It is available in high-purity grades (99.999%)
(iv) It is non-volatile
(v) It is transparent
(vi) It readily dissolves metallic oxides
(vii) It has a controllable moisture content.

The oxide is manufactured by the dehydration of purified crystalline boric acid on heating in a platinum crucible.

$$2H_3BO_3 \rightarrow B_2O_3 + 3H_2O \qquad (3.2)$$

This reaction results in the formation of the glass with a residual moisture content depending on the temperature, atmospheric pressure and duration of heating.

The vitreous oxide has different structural arrangements (I-III) depending on temperature (15).

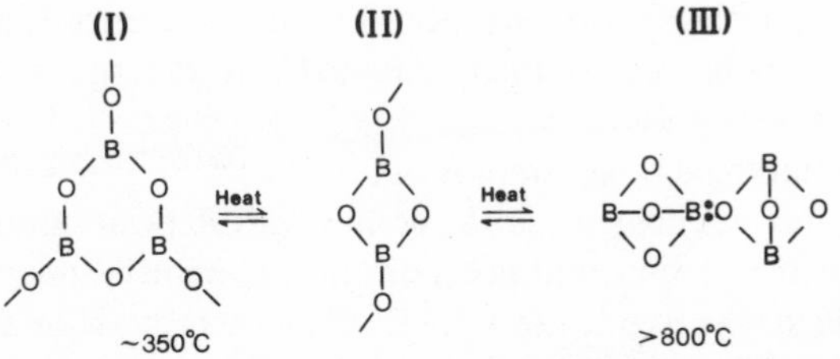

(I) (II) (III)

Heat Heat

~350°C >800°C

The residual water content of the glass has an important influence on the growth behaviour and on the concentration of impurities in the grown crystal. Contained moisture can be present in concentrations from 100 ppm to several thousand ppm. High moisture levels are effective in 'gettering' impurities from the melt into the oxide in the case of GaAs growth, but result in an increased probability of twin formation. It is evident that the measurement and control of the moisture level is an important consideration in the use of this material as an encapsulant. There are two methods of determining the moisture level.

(i) Infra-red absorption analysis. An absorption peak occurs at $3580\,cm^{-1}$ due to oscillation of the OH bond (16). This can be calibrated to yield an equivalent moisture content, but some disagreement persists regarding the value of the molar extinction coefficient used by different manufacturers (17)

(ii) Total evolved moisture. The water vapour evolved on heating the oxide at 1000° C is collected and measured. The measurement is performed by absorbing the water in a P_2O_5 coating in an electrolytic cell and measuring the current flow during electrolysis.

$$H_2O(g) + P_2O_5 \rightarrow P_2O_5.H_2O \rightarrow P_2O_5 + H_2(g) + \tfrac{1}{2}O_2(g) \qquad (3.3)$$

Although there is general disagreement between manufacturers and method on moisture levels, it is possible to correlate them over the usable range of moisture levels.

The oxide is very hygroscopic and readily absorbs atmospheric water vapour. This leads to problems in handling and storage, and also in the accurate determination of bulk moisture where surface absorption may occur during sampling. This is avoided in the infrared analysis, where surface moisture gives rise to a separate absorption peak.

Stoichiometry

Loss of As may occur prior to synthesis of the compound. This is minimized by careful loading of the charge, but the stoichiometry cannot be controlled as precisely as in the Bridgman system where a set As pressure is maintained. There is no established technique for direct measurement of stoichiometry in the solid material, although recent methods have been reported using x-ray diffraction (18) and coulometric titration (19). The starting melt stoichiometry is derived indirectly by performing a mass balance, taking into account all of the components before and after growth. A loss in weight is found, and this is generally assumed to be made up of As which has sublimed and subsequently condensed on the chamber walls. Although wall deposits generally show a 90% As composition, the assumption does not take into account exchange of material between the oxide and melt. It has been shown that Ga is preferentially absorbed into the oxide and this effect depends on the moisture level of the B_2O_3.

Table 3.2 shows the relative Ga:As concentrations found in the residual oxide pellet after single crystal growths, using different moisture levels (20).

The dependence of melt stoichiometry on the oxide moisture level can be confirmed from results of electrical measurements on crystals grown from melts of different Ga:As charge ratios (21).

The proposed mechanism for the reaction is

$$B_2O_3.H_2O + 2Ga \rightarrow Ga_2O + H_2 + B_2O_3 \tag{3.4}$$

The gallium oxide then dissolves in the encapsulant layer.

It is well established that the initial melt stoichiometry has a major influence on the electrical behaviour of the grown crystal (22,23). Figure 3.6 shows the trend in electrical behaviour in crystals grown from melt compositions with [Ga]:[As] atomic ratios between 1.3 and 0.90 (24). The starting ratios were again derived from weight balance calculations. Growth may be performed from melts quite rich in Ga, whereas As excess is difficult to achieve. On the

Table 3.2

B_2O_3 water content	Gallium	Arsenic	Ga/As
2000–5000 ppm-wt	1.2	0.07	18
	2.2	0.05	41
	3.1	0.03	103
200–400 ppm-wt	27.1	24.8	1.1
	9.2	7.9	1.2
	0.08	0.06	1.4

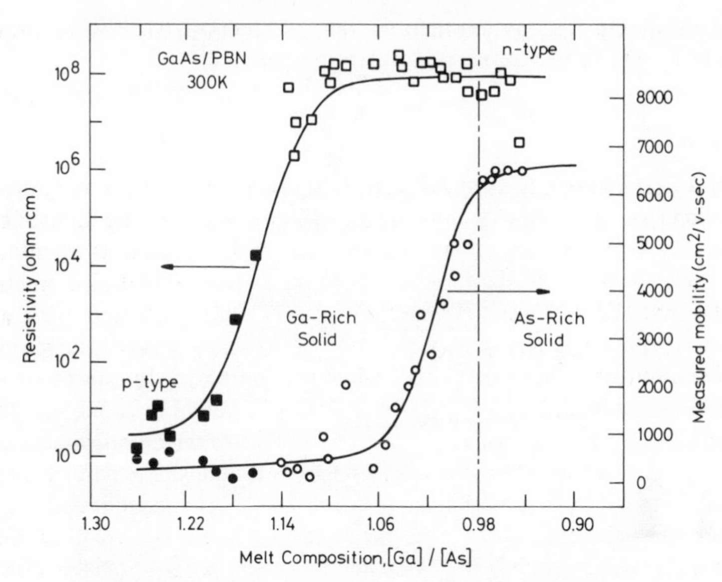

Figure 3.6 Effect of stoichiometry on electrical behaviour

Ga-rich side of the stoichiometric point, the electrical resistivity exhibits a decrease and rapid change to *p*-type conduction. The Hall mobility, however, is most sensitive to composition and achieves its highest values only for stoichiometric or As-rich melts.

This behaviour may be explained by consideration of the electrical compensation mechanism of the undoped material. The conduction is controlled by a deep donor EL2 having an energy close to the middle of the bandgap. This compensates residual acceptors of which the substitutional carbon impurity C_{As} is the dominant one (22). In turn, this compensates residual donor contaminants such as Si and S. The necessary impurity balance for undoped semi-insulating material is therefore $[EL2] > [C_{As}] > N_D$, where N_D is the residual donor concentration.

EL2 is a crystalline defect found in electronic-trap spectroscopic techniques (25) and which gives rise to optical absorption structure close to the band edge of GaAs (26). This latter method may be used to give a quantitative estimate of EL2 in semi-insulating material (27). By this method it can be shown that [EL2] depends on melt stoichiometry (28) reaching a maximum in stoichiometric and slightly As-rich melts. Similarly, [EL2] has been shown to increase with increasing As partial pressure in Bridgman growth (29). For reasons such as these (30), EL2 has been attributed to an intrinsic defect rather than a chemical impurity. The arsenic anti-site defect As_{Ga} is proposed to be at least associated with EL2.

Stoichiometry may also exert an important influence on extended structural defects such as dislocations in the crystal. Precise As pressure control during

Bridgman growth has been shown to influence dislocation density strongly (31).

Impurities

Impurity analysis may be performed by mass spectroscopic methods such as SIMS, SSMS and atomic absorption spectroscopy (AAS) (32). Table 3.3 is a

Table 3.3 Spark source mass spectroscopic analysis of undoped GaAs.

Element	Concentration (ppmA)	Detn. level (ppmA)	Element	Concentration (ppmA)	Detn. level (ppmA)
Li		N.D. < 0.006	Pd		N.D. < 0.02
Be		N.D. < 0.006	Ag		N.D. < 0.015
B	0.7		Cd		N.D. < 0.02
C	⩽ 2.5		In	0.01[2]	
N	⩽ 3.5		Sn		N.D. < 0.02
O	⩽ 30		Sb		N.D. < 0.01
F		N.D. < 0.006	Te		N.D. < 0.02
Na		N.D. < 0.07	I		N.D. < 0.006
Mg		N.D. < 0.007	Cs		N.D. < 0.006
Al	0.006		Ba		N.D. < 0.008
Si		N.D. < 0.006	La		N.D. < 0.006
P	0.07		Ce		N.D. < 0.007
S	0.04–0.2*		Pr		N.D. < 0.006
Cl	0.03[1]		Nd		N.D. < 0.025
K	< 0.006		Sm		N.D. < 0.025
Ca	0.04		Eu		N.D. < 0.01
Sc		N.D. < 0.006	Gd		N.D. < 0.025
Ti		N.D. < 0.008	Tb		N.D. < 0.006
V		N.D. < 0.006	Dy		N.D. < 0.02
Cr	0.008		Ho		N.D. < 0.006
Mn		N.D. < 0.006	Er		N.D. < 0.02
Fe	0.02		Tm		N.D. < 0.006
Ni		N.D. < 0.01	Yb		N.D. < 0.02
Co		N.D. < 0.006	Lu		N.D. < 0.006
Cu		N.D. < 0.01	Hf		N.D. < 0.02
Zn		N.D. < 0.015	Ta	0.25[3]	
Ga	MATRIX		W		N.D. < 0.02
Ge		N.D. < 0.015	Re		N.D. < 0.01
As	MATRIX		Os		N.D. < 0.015
Se		N.D. < 0.015	Ir		N.D. < 0.01
Br		N.D. < 0.015	Pt		N.D. < 0.02
Rb		N.D. < 0.008	Au	0.06[3](TaO?)	
Sr		N.D. < 0.007	Hg		N.D. < 0.02
Y		N.D. < 0.006	Tl		N.D. < 0.008
Zr		N.D. < 0.015	Pb		N.D. < 0.01
Nb		N.D. < 0.006	Bi		N.D. < 0.006
Mo		N.D. < 0.025	Th		N.D. < 0.006
Ru		N.D. < 0.03	U		N.D. < 0.006
Rh		N.D. < 0.006			

Possible sources of contamination: (1) Etchant; (2) previous samples; (3) MS slits. N. D.–Not Detected *–Inhomogeneous complete spark source mass spectroscopic analysis of an undoped GaAs single crystal.

The concentration of carbon, oxygen and nitrogen are not accurately determined, due to residual amounts in the chamber which can only be effectively removed by cryopumping (33). Boron is a significant contaminant in growth from PBN crucibles and may be introduced at concentrations exceeding $10^{18}\,cm^{-3}$. Boron normally substitutes for Ga and is therefore electrically inactive. Under Ga-rich conditions, however, a shift in the defect equilibrium may result in significant concentrations of boron acceptors B_{As} (34). These centres can be observed using local vibrational mode infrared spectroscopy (35).

The same infrared technique is used to measure the concentration of carbon acceptors C_{As} (36), with a sensitivity down to about $1 \times 10^{15}\,cm^{-3}$ for carbon. The spectral line associated with carbon lies close to the two-phonon lattice absorption band, and hence the feature is poorly resolved unless cooling to liquid nitrogen or liquid helium temperatures is used. The calibration of the absorption against carbon concentration is the subject of some disagreement (37).

The moisture content of the boric oxide encapsulant has an important influence on the impurity content of the grown crystal. Figure 3.7 shows the effect of increasing the moisture in the encapsulant from a few hundred ppm (dry) to several hundred ppm (wet). The results shown are the measured

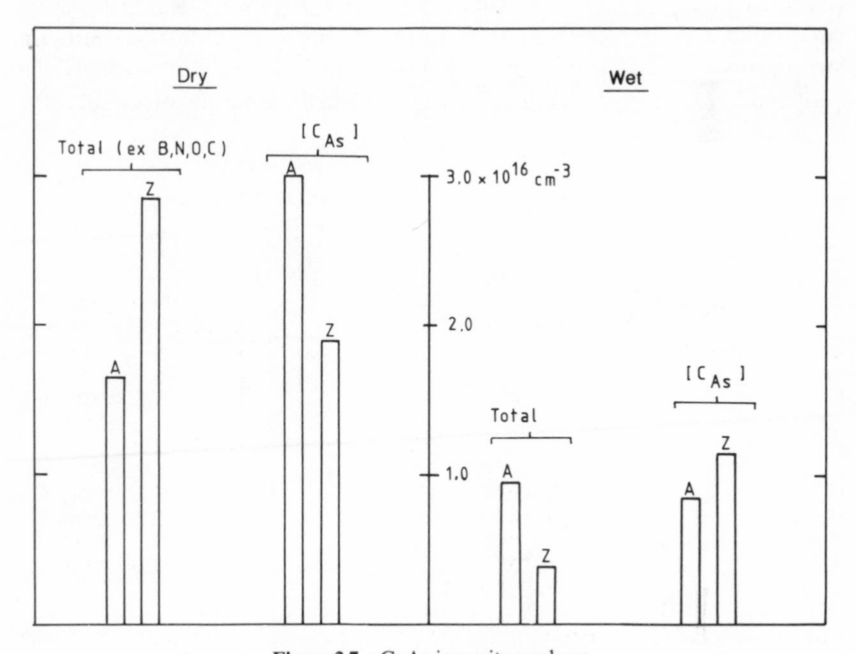

Figure 3.7 GaAs impurity analyses

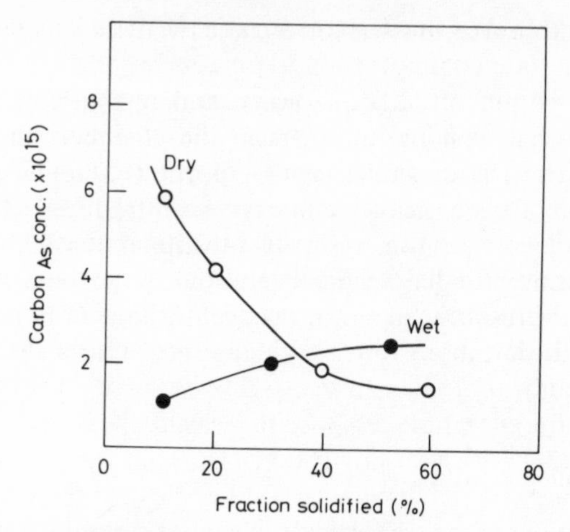

Figure 3.8 Carbon segregation behaviour: dependence on encapsulant moisture

concentrations of impurities at seed and tail ends, designated *A* and *Z* respectively. These exclude C, N, O and boron and are measured by SSMS. They demonstrate the general reduction in total impurity content. Also shown are results for carbon measured by LVM using the calibration factor of ref. 36. In addition to the lower concentration obtained using the wetter oxide, there is a change in the apparent segregation behaviour. Figure 3.8 demonstrates this more fully, showing the switch in segregation from a concentration decreasing with growth to one increasing (38). This effect has been noted widely. The origin of the carbon contamination is not established, but possible sources are

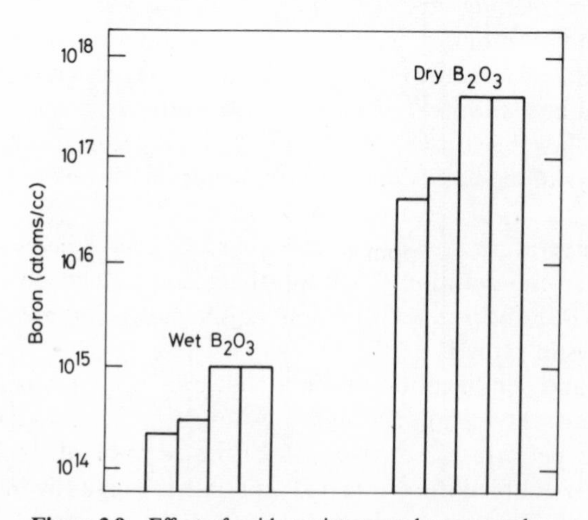

Figure 3.9 Effect of oxide moisture on boron uptake

the graphite within the growth chamber, and the As raw material, where black skeletal residues are commonly found on distillation.

Boron may be present as an impurity in very high concentrations when growth is performed in pyrolytic boron nitride crucibles using dry boric oxide encapsulant. The use of wetter grades of encapsulant is again extremely effective in reducing boron contamination (38), as shown in Figure 3.9.

The device of increasing the encapsulant moisture content to maximize the impurity gettering effect has a limited value due to the onset of twin formation during growth with wet oxides. The reason for this is not known, but may be associated with deviation from stoichiometry towards the As-rich side by preferential removal of Ga into the oxide, as described previously.

Indium phosphide crystal growth

Synthesis

Indium phosphide polycrystalline compound may be synthesized from the elements in a number of different ways. In each case the problem of synthesis is the high vapour pressure of phosphorus over InP at the melting point (27.5 atm). This requirement, coupled with the pyrophoric nature of phosphorus, places more severe demands on the compounding equipment.

The gradient freeze method is commonly employed, in which molten indium is reacted with phosphorus vapour under an equilibrium vapour pressure, and the melt passed through a temperature gradient to induce controlled crystallization (39). The furnace arrangement is qualitatively similar to that described for GaAs horizontal Bridgman growth. The melt is unseeded and maintained at a temperature of around 1100° C. The phosphorus temperature is controlled by a separate furnace and set to maintain the required vapour pressure at the melting point of InP.

After an initial stabilization period of several hours, growth is initiated by moving the sealed ampoule through a fixed temperature gradient between the main furnace and afterheat furnace. Growth rates between 1 mm/hour and 40 mm/hour have been used. The lower rates are necessary to achieve large-grain polycrystalline ingots and to avoid trapping of free indium inclusions (40).

Lower-temperature, lower-pressure techniques have been used involving growth from In-rich solution (41). While this generally leads to undoped InP of higher purity (42), the problem of excess indium makes this material unsuitable for single-crystal growth.

Synthesis and solidification inside a high-pressure chamber, either in the form of an autoclave (43) or a high-pressure LEC crystal grower (44), have been used to perform direct synthesis of the compound. In the latter case phosphorus is bubbled from a heated ampoule through the indium melt at a controlled rate.

Single crystal growth

InP is grown by high-pressure LEC with typical melt sizes of about 1 kg (45). Growth may be carried out using the directionally crystallized bar material. However, if this contains excess concentrations of unreacted In, the probability of single-crystal growth will be low. In this case an additional step is introduced into the process in which the excess In is segregated during a preliminary pulling operation similar to that employed for single crystal growth.

The minimum chamber pressure that can be used to avoid decomposition of the melt is around 30 atm. This necessitates the use of high-pressure growth systems. The growth directions most commonly employed are $\langle 100 \rangle$ or $\langle 111 \rangle$.

The oxide encapsulant again plays an important role in influencing the crystal bulk properties. Although silica is used for the crucible material, Si contamination is avoided by the presence of a high moisture content in the oxide. However, the maximum limit of the moisture level is determined by the occurrence of dislocation clusters in the crystal (46). These dislocation clusters or 'grappes' have been the subject of extensive study (47), but no firm identification has been made for their origin apart from the observed dependence on oxide composition. Stoichiometry-related defects have been suggested as a possible source (48).

Impurities

A major application for InP wafers is as substrates for long-wavelength (1.5 and 1.35 micron) quaternary laser structures. A key requirement is then an absence of crystal dislocations which may propagate from the substrate into the active region and prevent or progressively degrade the lasing action. Dislocation-free InP is fairly readily obtainable by doping with high concentrations of certain impurities, making the material very conductive and modifying its thermo-elastic properties.

Sulphur doping at concentrations greater than $5 \times 10^{18}\,\text{cm}^{-3}$ will yield *n*-type zero-dislocation crystals, as will doping with germanium (49). Zinc may be used to obtain *p*-type zero-D crystals.

Unlike GaAs, high-purity undoped InP contains no analogous dominant mid-gap energy level, and the purest ingots grown exhibit *n*-type conduction with carrier concentrations less than $10^{15}\,\text{cm}^{-3}$ (50).

The purest material is normally the polycrystalline bar with some deterioration in carrier concentration shown on pulling. The main contaminants in undoped material with carrier concentrations greater than $10^{16}\,\text{cm}^{-3}$ are silicon when synthesis is performed in quartz boats, and sulphur when high-purity graphite is used. An excess of In, both during crystallization after synthesis (51) and in single-crystal growth (52), acts as an effective 'sink' for impurities and results in material of lower background carrier concentration and higher electron mobility. The nature of the residual donors in the material of highest purity is not established. Photoluminescence spectroscopy, which may be used for identification of acceptor impurities at very low concentrations, cannot resolve residual hydrogenic donors unless high magnetic fields are employed to provide hyperfine splitting (53). Resolved spectra obtained in this way cannot be fully identified with chemical impurities, and native defects have been suggested as possible candidates.

Semi-insulating material can be obtained by doping with iron (39) and cobalt (54). Iron is commercially the more important for high-resistivity substrates. The maximum Fe content of the crystal is limited by precipitation of FeP_2 inclusions in the solid. For this reason, and in order to minimize the diffusion of Fe from the substrate during epitaxy, charge material of the highest purity is required for semi-insulating crystal growth.

Surface preparation

The chemical preparation of crystal surfaces is a topic of great importance for wafer processing. Wet chemical etching is widely used to define device structures of very fine geometry (55). In the preparation of wafer substrates the final stages involve the use of chemical polishing agents to remove mechanical damage introduced by sawing. In general, the removal of material by etching may be carried out during wafer production for the following functions.

(i) Damage removal—free etching by dissolution in an immersion bath to strip away the sawn surfaces of cut wafers
(ii) Polishing—chemical–mechanical action to controllably remove surface material resulting in an optically flat, reflective and clean surface
(iii) Characterization—to reveal surface inhomogeneities in the form of dislocations which intersect the surface, or residual polishing damage such as scratching.

In most of the etchant systems the chemical attack is by oxidation of the surface using a strong oxidizer such as H_2O_2 or Br_2 and a complexing agent such as H_2SO_4 to aid dissolution of the surface oxide, normally in water. Table 3.4 lists some of the etch combinations used for GaAs and InP.

Table 3.4

Oxidizer	Complexer	Diluent	Ref.
Br_2	CH_3OH	CH_3OH	56
H_2O_2	H_2SO_4	H_2O	57
$AgNO_3$–CrO_3	HF	H_2O	58
HNO_3	HCl	H_2O	59
H_2O_2	NH_4OH	H_2O	60

Dislocation etches

Crystal dislocations have associated with them an elastic strain field, and in some cases an impurity 'atmosphere' as a result of impurity diffusion at growth temperatures. Both of these phenomena may modify the chemical action of an etchant at the point where a dislocation intersects the surface. Under normal circumstances a small pit will be formed. Comparisons of etch pits with X-ray topographic studies show a good correlation for some systems and allow an estimate of surface dislocation density by optical microscopy and etch pit counting.

For GaAs the main dislocation-revealing etch used for (100) surfaces is fused KOH (61) at temperatures between 350° C and 450° C. This produces regular elongated hexagonal etch pits aligned in ⟨110⟩ directions. Figure 3.10 is a photograph of a (100) GaAs wafer after etching in KOH at 350° C for 5 minutes. The complex dislocation structures in the material are clearly visible (62). A KOH/NaOH mixture has been proposed as an alternative etchant for lower temperatures (170° C) (63).

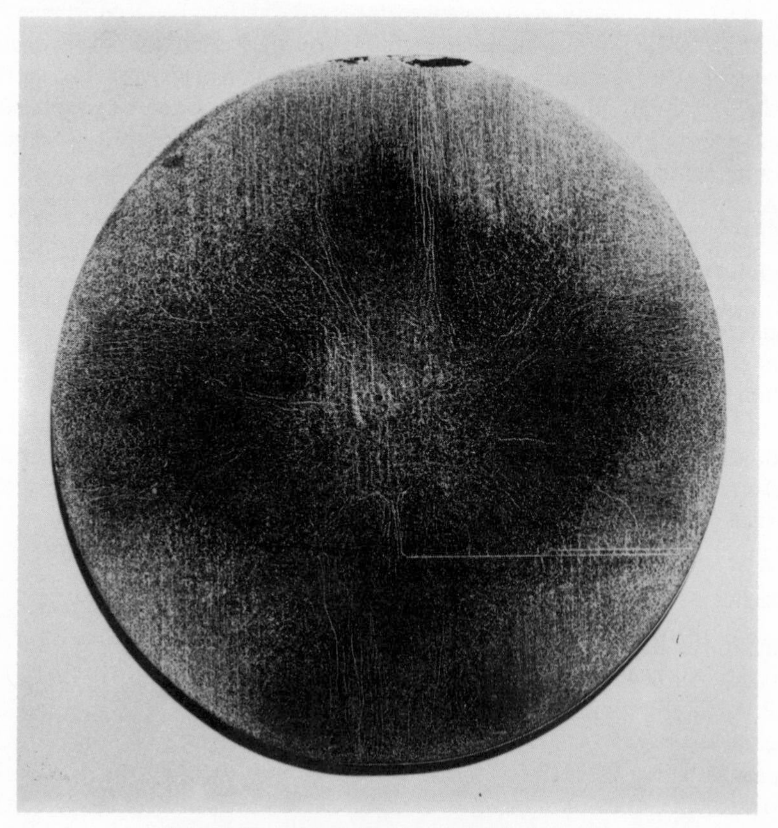

Figure 3.10 KOH dislocation etched GaAs wafer

Conventional dislocation etches exhibit enhanced etch rate at the dislocation, resulting in a pit. The Abrahams and Buiocchi (AB) etch consisting of $CrO_3/AgNO_3/HF/H_2O$ (58) displays an unusual behaviour by preferentially etching the matrix material, leaving ridge features associated with the dislocation. The details of the information thus revealed make this etchant less suitable for dislocation counting, but of great value for microscopy (64).

(100) InP surfaces can be etched for dislocation pits using the H or Huber etch (65) consisting of $HBr/H_3PO_4/H_2O$.

Chemo-mechanical polishing

Free chemical etching cannot produce adequately smooth or flat surfaces for wafer finishing, and some mechanical action is normally applied to control the rate and uniformity of removal. Polishing then proceeds by a combination of abrasion and dissolution. The abrading action is obtained by the use of a polishing pad of a synthetic suede material such as poromeric polyurethane.

Solutions of Br_2 in methanol in strengths between 0.5 and 5% are generally applicable to polishing of the III–V compounds and remain one of the main methods for InP polishing. Equipment used in conjunction with Br_2-methanol must be specially constructed in plastics to avoid the volatile corrosive action of the polishing solution. The rate of attack may be modified by varying the Br_2 content or by changing the viscosity of the solution. Ethylene glycol may be added in ratios up to 40:60 glycol:methanol.

Although Br_2–methanol is suitable for damage-free polishing of more delicate materials, it presents problems in control of solution strength and equipment design. Polishing solutions for wafer finishing of GaAs are normally composed of sodium hypochlorite solution together with an abrasive slurry of colloidal silica. The hypochlorite solution oxidizes the crystal surface. In a free etch a stable surface layer is formed and the etching action is inhibited. The polishing action of the abrasive slurry/pad combination provides continuous exposure of fresh surface for dissolution. The colloidal silica consists of particles with a size distribution between 50 and 100 nm, and control of the slurry in use concerns the maintenance of the particles in suspension without agglomeration or gelling.

Conclusion

This chapter has surveyed the basic chemical considerations in play during the preparation of the important III–V compound semiconductors GaAs and InP. As these materials become more widely used, and as GaAs in particular becomes the basis for high-volume manufacturing processes, control and reproducibility over chemical impurities will become crucial. This has already provided the spur for improved refining and analytical techniques for the various raw materials required in the crystal-growing process.

References

1. Sumitomo Electronic Materials, technical data.
2. S. D. Mukherjee and D. W. Woodard, in *GaAs, Materials, Devices and Circuits*, eds M. J. Howes and D. V. Morgan, John Wiley (1985).
3. J. S. Blakemore, *J. Appl. Phys.* **53** (1982) 10.
4. J. C. Phillips, *Bonds and Bands in Semiconductors*, Academic Press (1973), Chapter 2.
5. B. Tuck, *J. Mater. Sci.* **10** (1975) 321.
6. J. R. Arthur, *J. Phys. Chem. Solids* **28** (1967) 2257.
7. T. Suzuki, S. Akai, K. Kohe, Y. Nishida, K. Fujita and N. Kito, *Sumitomo Electric Tech. Rev.* **18** (1978) 105.
8. F. Simondet, C. Venger, G. M. Martin and J. Chaumont, *Semi-Insulating III–V Materials*, Nottingham; ed. G. J. Rees, Shiva (1980) 100.
9. R. K. Willardson, in *Semi-Insulating III–V Materials*, Kah-Nee-Ta, ed. D. C. Look and J. S. Blakemore, Shiva (1984) 96.
10. E. P. A. Metz, R. C. Miller and R. Mazelsky, *J. Appl. Phys.* **33** (1962) 2016.
11. J. B. Mullin, R. J. Heritage, C. H. Holliday and B. W. Straughan, *J. Cryst. Growth* **34** (1968) 281.
12. E. M. Swiggard, S. H. Lee and F. W. Von Batchelder, *GaAs and Related Compounds, 1976* (St Louis), *Inst. Phys. Conf. Ser.* **33**b (1977) 23.
13. T. R. AuCoin, R. L. Ross, M. J. Wade and R. O. Savage, *Solid State Technol.* **22** (1979) 59.
14. W. Bardsley, G. W. Green, C. H. Holliday, D. T. J. Hurle, G. C. Joyce, W. R. MacEwan and P. J. Tufton, *GaAs and Related Compounds, Inst. Phys. Conf. Ser.* **24** (1975) 355.
15. Tomiyama Chemical Co., Technical Data Sheet.
16. M. R. Shropshall and P. E. Skinner, *Semi-Insulating III–V Materials*, Kah-Nee-Ta, eds. D. C. Look and J. S. Blakemore, Shiva (1984) 178.
17. A. T. Hunter, H. Kimura, J. P. Baukus, H. V. Winston and O. J. Marsh, *Appl. Phys. Lett.* **44** (1984) 74.
18. I. Fujimoto, *Jap. J. Appl. Phys.* **23** (1984) L287.
19. K. Terashima *et al.*, *12th Int. Symp. GaAs and Related Compounds*, Karuizawa, to be published.
20. D. H. Rumsby and R. M. Ware, *GaAs and Related Compounds 1981* (OISO), *Inst. Phys. Conf. Ser.* No. **63** (1982) 573.
21. H. Emori, T. Kikuta, T. Inada, T. Obokata and T. Fukuda, *Jap. J. Appl. Phys.* **24**, No. 5 (1985) L291.
22. D. E. Holmes, R. T. Chen, K. R. Elliot and C. G. Kirkpatrick, *Appl. Phys. Lett.* **40** (1982) 46.
23. L. B. Ta, H. M. Hobgood, A. Rohatgi and R. N. Thomas, *J. Appl. Phys.* **53**, (1982) 5771.
24. H. M. Hobgood, L. B. Ta, A. Rohatgi, G. W. Eldrige and R. N. Thomas, *Semi-Insulating III–V Materials*, eds. S. Makram-Ebeid and B. Tuck, Evian, Shiva (1982) 28.
25. G. M. Martin, A. Mitonneau and A. Mircea, *Elec. Lett.* **13** (1977) 191.
26. G. M. Martin, G. Jacob, A. Goltzene, C. Schwab and G. Poiblaud, *Inst. Phys. Conf. Ser.* **59** (1981) 281.
27. G. M. Martin, *Appl. Phys. Lett.* **39** (1981) 747.
28. D. E. Holmes, R. T. Chen, K. R. Elliott, C. G. Kirkpatrick and P. W. Yu, *IEEE Trans. El. Dev.* **29** (1982) 1045.
29. J. Lagowski, H. C. Gatos, J. M. Parsey, K. Wada, M. Kaminska and W. Walukiewicz, *Appl. Phys. Lett.* **40** (1982) 342.
30. S. Makram-Ebeid, P. Langlade and D. G. Lin, *Semi-Insulating III–V Materials*, Kah-nee-ta, eds. D. C. Look and J. S. Blakemore, Shiva (1984) 184.
31. J. Lagowski, H. C. Gatos, T. Aoyama and D. G. Lin, *Semi-Insulating III–V Materials*, Kah-nee-ta, eds. D. C. Look and J. S. Blackemore, Shiva (1980) 60.
32. J. B. Clegg, *Semi-Insulating III–V Materials*, Evian, eds. S. Markram-Ebeid and B. Tuck, Shiva (1982) 80.
33. J. B. Clegg, I. G. Gale and E. J. Millett, *Analyst* **98** (1973) 69.
34. C. G. Kirkpatrick, R. T. Chen, D. E. Holmes and K. R. Elliott, *Gallium Arsenide, Materials, Devices and Circuits*, eds. M. J. Howes and D. V. Morgan, John Wiley, (1985) 39.
35. J. Woodhead, R. C. Newman, I. Grant, D. Rumsby ane R. M. Ware, *J. Phys. C: Solid State Phys.* **16** (1983) 5523.

36. M. R. Brozel, J. B. Clegg and R. C. Newman, *J. Phys. D.* **11** (1978) 1331.
37. Y. Homma, Y. Ishii, T. Kobayashi and J. Osaka, *J. Appl. Phys.* **57** (1985) 2931.
38. H. Emori, T. Kikuta, T. Inada, T. Obokata and T. Fukuda, *Jap. J. Appl. Phys.* **24** (1985) L291.
39. D. Rumsby, R. M. Ware and M. Whittaker, *Semi-Insulating III–V Materials*, Nottingham, ed. G. J. Reeves, Shiva (1980) 59.
40. W. A. Bonner, *Proc. NATO* Workshop on InP, 1980, p. 43.
41. G. A. Antypas, *Inst. Phys. Conf. Ser.* **33**b (1977) 55.
42. J. E. Wardill, D. J. Dowling, R. A. Brunton, D. A. E. Crouch, J. R. Stockbridge, and A. J. Thompson, *J. Cryst. Growth* **64** (1983) 15.
43. R. O. Savage, J. E. Anthony, T. R. AuCoin, R. L. Ross, W. Harsch and H. E. Cantwell, *Semi-Insulating III–V Materials*, Kah-nee-ta, eds. D. C. Look and J. S. Blakemore, Shiva (1984).
44. J. P. Farges, *J. Cryst. Growth* **59** (1982) 665.
45. R. Coquille, Y. Toudic, M. Gauneau, G. Grandpierre and J. C. Paris, *J. Cryst. Growth* **64**, (1983) 23.
46. B. Cockayne, G. T. Brown and W. R. MacEwan, *J. Cryst. Growth* **51** (1981) 461.
47. P. D. Augustus and D. J. Stirland, *J. Electrochem. Soc.* **129** (1982) 614.
48. B. Cockayne, G. T. Brown and W. R. MacEwan, *J. Cryst. Growth* **64** (1983) 48.
49. G. T. Brown, B. Cockayne and W. R. MacEwan, *Inst. Phys. Conf. Ser.* **60** (1981) 351.
50. J. A. Adamski, *J. Cryst. Growth* **60** (1982) 141.
51. B. Cockayne, G. T. Brown and W. R. MacEwan, *J. Cryst. Growth* **54** (1981) 9.
52. I. Grant, L. Li, D. Rumsby and R. M. Ware, *J. Cryst. Growth* **64** (1983) 32.
53. P. J. Dean, M. S. Skolnick, B. Cockayne, W. R. MacEwan and G. W. Iseler, *J. Cryst. Growth* **67** (1984) 486.
54. B. Cockayne, W. R. MacEwan and G. T. Brown, *J. Cryst. Growth* **55** (1981) 263.
55. S. D. Mukherjee and D. W. Woodard, in *GaAs Materials, Devices and Circuits*, eds. M. J. Howes and D. V. Morgan, John Wiley (1985) 119.
56. Y. Tarui, Y. Komiya and T. Yamaguchi, *J. Jap. Soc. Appl. Phys.* (*Suppl.*) **42** (1973) 78.
57. S. Iida and K. Ito, *J. Electrochem. Soc.* **118** (1971) 768.
58. M. S. Abrahams and C. J. Buiocchi, *J. Appl. Phys.* **36** (1965) 2855.
59. B. Tuck, *J. Mater. Sci.* **10** (1975) 321.
60. E. Kohn, *J. Electrochem. Soc.* **127** (1980) 505.
61. J. G. Grabmaier and C. B. Watson, *Phys. Stat. Solidi.* **32** (1969) K13.
62. M. R. Brozel, I. Grant, R. M. Ware and D. J. Stirland, *Appl. Phys. Lett.* **42** (1983) 610.
63. H. Lessoff and R. Gorman, *Semi-insulating III–V Materials*, Kah-nee-ta, eds. D. C. Look and J. S. Blakemore, Shiva (1984) 83.
64. D. J. Stirland, *GaAs and Related Compounds* (Edinburgh), *Inst. Phys. Conf. Ser.* **33**a, (1977) 150.
65. A. Huber and N. T. Linh, *J. Cryst. Growth* **29** (1975) 80.

4 Group II–VI semiconductors

J. WOODS

Introduction

The II–VI compounds of interest here are those formed between the elements of Groups IIB and VIB of the Periodic Table. They are more ionic than the Group IV or III–V semiconductors, and their bandgaps tend to be direct, so that the transition probabilities for optical absorption and emission processes are large. As a result the wide-bandgap II–VIs, which are composed of the lighter elements, e.g. ZnS, ZnSe, ZnO and CdS, have found extensive application as cathodoluminescent phosphors since about 1930, and today the quantity production of such phosphors is comparable to that of semiconductor grade silicon.

The II–VIs are also highly efficient as detectors of electromagnetic radiation, and CdS and CdSe have been widely used as photoconductive detectors of visible light. The narrower-gap II–VI compounds, which are formed from the heavier elements, have found major applications as infrared detectors. In fact the conduction and valence bands of HgTe and HgSe overlap so that they are classed as semi-metals, but by alloying HgTe with CdTe, a range of narrow-gap semiconductors can be produced. Cadmium mercury telluride photoconductive detectors were first used in prototype thermal imaging systems in 1965. Since then the technology has advanced rapidly, and more resources have been invested in the development of CMT than in any other semiconductor systems, except silicon and gallium arsenide.

Almost all the early work on the II–VI compounds was done on powder samples. Studies of single crystals date essentially from 1947, when Frerichs (1), adapting a much older technique of Lorenz (2), grew platelets and rod-like crystals of CdS from the vapour phase. Since that time, progress has been steady, but modest in comparison with the more exciting developments, first with the elemental semiconductors Ge and Si, and then with the III–V compounds. Although it was soon appreciated that successful characterization of a material requires the preparation of single crystals with a high degree of crystallographic perfection and chemical purity, progress with the II–VIs has nevertheless been slow. The reasons for this are that the II–VI compounds tend to have high melting points and high chemical reactivities. At the same time, comparatively high vapour pressures of the constituents are established above the melts, so that suitable containers are difficult to find, and accidental contamination difficult to avoid. Melting points and dissociation vapour pressures are highest for the sulphides and lowest for the tellurides. As a result there has been a tendency to grow CdS and ZnS from the vapour phase, or from the melt under high pressure. More normal melt-growth methods have been used for the production of CdTe and ZnTe, but crystals have also been grown from the vapour phase.

Although the production of bulk single crystals is very important for materials characterization, many applications of II–VI compounds, in cathodo- and electroluminescence and in solar cells for example, call for large-area devices, so that interest in powdered and thin and thick film layers has been maintained. Also, since the advances in silicon and III–V semiconductor technology have focused attention on epitaxial layers and devices, there has been a corresponding increase in the number and variety of attempts to grow homo- and hetero-epitaxial layers of the II–VI compounds. Most of these avoid the use of high temperatures and pressures, and thus offer possible solutions to most of the materials problems of contamination which currently inhibit the wider application of this group of compounds.

The elements

The elements forming the constituents of the compounds under discussion, namely Zn and Cd from Group II, and S, Se, and Te from Group VI of the Periodic Table all have relatively low melting and boiling points, which are listed in Table 4.1.

In the vapour phase the metals Zn and Cd occur in monatomic form, whereas sulphur vapour contains S, S_2 and S_8 molecules, selenium vapour Se, Se_2 and Se_6, and tellurium vapour exists as Te and Te_2. One of these species usually predominates in a given temperature interval. Relative abundances can be calculated from the thermodynamic data provided by Stuhl and Sinke (3).

Vacuum distillation provides a very convenient method for purifying these elements, although the highest purity is achieved by zone refining. High-purity elements can be obtained from chemical suppliers with impurity contents less than 1 part in 10^5 or 10^6. Oxygen is readily adsorbed by most of the elements;

Table 4.1 Melting and boiling points of the elements measured under 1 atm pressure

Element	m.p. (°C)	b.p. (°C)
Zn	420	908
Cd	321	765
S	119	445
Se	217	685
Te	450	987

high-purity tellurium, for example, is usually stored under nitrogen, and zinc and cadmium are often etched before use to remove oxide layers.

Phase equilibria

In the six II–VI systems derived from the five elements mentioned above, only one compound (MX) is formed from the metal, M, and non-metal, X, components. For growth from the vapour phase the equilibrium between the solid compound (S) and the vapour (V) is of primary interest. Goldfinger and Jeunehomme (4) carried out extensive effusion studies of the S–V equilibrium for all the II–VI compounds, and provided a comprehensive review and set of thermodynamic data. Using mass spectrometric methods they also showed that the compounds dissociated on heating according to the reaction

$$2MX(S) \rightarrow 2M(V) + X_2(V)$$

Any concentration of the molecular species MX in the vapour was below the detection limit of between 10^{-3} and 10^{-5} of the molecules present. Since the vapour does not contain MX, evaporation of the compound can be suppressed by applying an overpressure of one of the elements. This is an important feature of crystal growth from the vapour phase, and of subsequent heat treatment of a grown crystal.

The dissociation reaction above shows that the partial pressures of M and X_2 over the solid are interrelated according to

$$K_p = P_M^2 \cdot P_{X_2} \tag{4.1}$$

where K_p is the equilibrium reaction constant; the total pressure, P_T, over the solid will be $P_M + P_{X_2}$.

At any given temperature, P_T will change if the partial pressure of one of the components is changed, when the partial pressure of the other will be determined by (4.1). In particular, there will be a minimum value of P_T, usually referred to as P_{min}, which occurs when

$$\partial P_T/\partial P_M = \partial P_T/\partial P_{X_2} = 0 \tag{4.2}$$

When $P_T = P_{min}$, this leads to

$$P_M = 2P_{X_2} = 2^{1/3}K_p^{1/3} \tag{4.3}$$

and

$$P_{min} = \tfrac{3}{2} \cdot 2^{1/3}K_p^{1/3} \tag{4.4}$$

Values of K_p can be calculated from the vapour pressure data provided by Goldfinger and Jeunehomme using (4.1). A plot of $\log K_p$ as a function of $1/T$ is shown in Figure 4.1.

Since $P_M = 2P_{X_2}$ at P_{min}, the vapour and the solid have the same composition under these conditions. This is important in crystal growth from the vapour phase, and in sublimation. For example if the solid, containing a

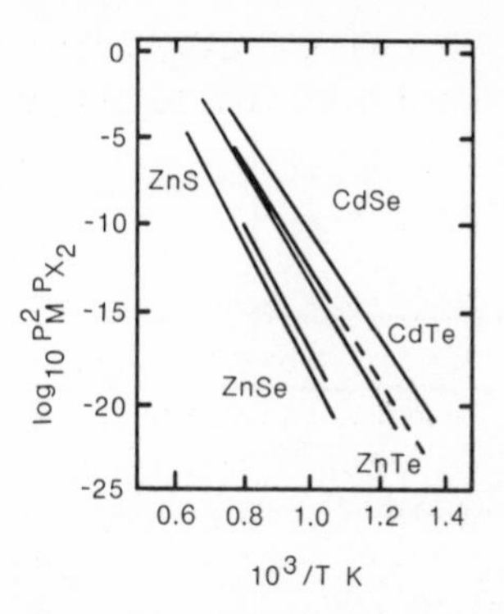

Figure 4.1 The equilibrium constants $K_P = P^2_M \cdot P_{x_2}$ for various II–VI compounds (pressures in atmospheres)

non-stoichiometric excess of one or other of the components, is heated in a system where the vapour can be removed gradually, then the compound will lose that component which is present in excess until the P_{min} condition is reached, and congruent evaporation occurs.

The S–V equilibrium provides essential information for crystal growth from the vapour phase. For growth from the melt, the three-phase equilibrium S–L–V, between solid, liquid and vapour, must be known. Such an equilibrium is usually represented in terms of T–x, P–T and P–x sections through the three-dimensional phase diagram. T is the temperature, x the composition and P the pressure which can be the pressure of one of the components or the sum of them both. The pressure of any inert gas which may be present is usually ignored.

The liquidus curves have been measured for all six of the II–VI compounds under discussion, but the most extensive investigations of the three-phase equilibrium have been carried out on CdTe, which appears to be typical. Useful reviews are provided by Lorenz (5) for all the compounds, and by Strauss (6) and Zanio (7) for CdTe. The liquidus temperature–composition, T–x, curves for the Cd–Te system have been studied by de Nobel (8), Mason and O-Kane (9), Lorenz (10), Kulwicki (11), Steiniger *et al.* (12) and Brebrick (13). Techniques of measurement varied from visual observation of freezing to thermal analysis, while partial pressures of the constituents were established by using a separate reservoir. The results are in good agreement, and are illustrated graphically for Cd-Te in the T–x curve in Figure 4.2. The solidus has a very narrow range of composition about $x = 0.5$ and appears as a vertical line in Figure 4.2. The liquidus reveals that there is a maximum melting point of 1092° C at approximately $x = 0.5$, and eutectic temperatures of $324 \pm 2°$ C on the Cd-rich and $429 \pm 2°$ C on the Te-rich side. The pointed shape of the liquidus at $x = 0.5$ is typical of all the II–VI compounds where measurements are available, and quite unlike the shape of the liquiduses of the III–V compounds, which are broadly parabolic at the stoichiometric composition.

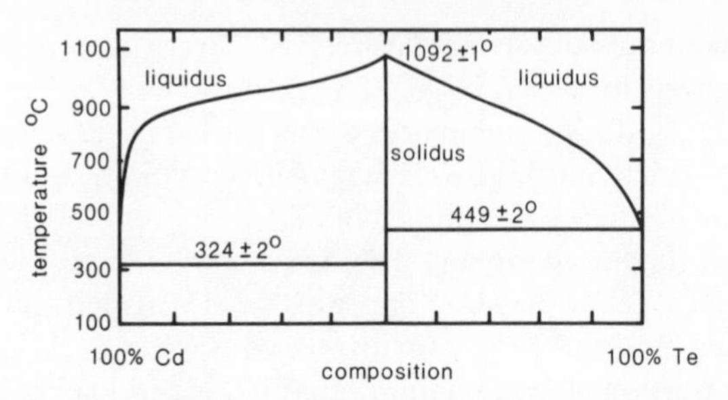

Figure 4.2 Temperature *v.* composition (T–x) diagram for the Cd–Te system

The pointing in Figure 4.2 is attributable to the more pronounced interaction between the atoms in the melt for the II–VIs, which is a function of the larger ionic contribution to the binding energy in the II–VIs.

Portions only of the T–x curves have been measured for the other II–VI compounds. After Cd–Te, the most extensively investigated systems have been Zn–Te (11, 14, 15) and Cd–Se (16, 17, 18). The Zn–Se system has been studied by Fischer (19), Cd–S by Woodbury (20) and Zn–S hardly at all, although Addamiano and Dell (21) have reported a maximum melting point of 1830° C.

The maximum melting points, T_m, of the six compounds are recorded in Table 4.2. Values of P_{min} at T_m as estimated by Lorenz (5), by extrapolating the data of Figure 4.1 to T_m and using equation (4.4) are also included in the table. (Note, however, that P_{min} conditions do not usually obtain at the maximum melting point.) Knowledge of the value of P_{min} is very important for crystal growth. The table shows that only CdS and ZnS develop substantial dissociation pressures, and special techniques are therefore necessary in growing these compounds from the melt.

The extent of the existence region of solid CdTe is less than 1 at % on both the Cd- and Te-rich sides of the stoichiometric composition (22, 23). Smith (24), who made measurements of the Hall coefficient of CdTe at elevated temperatures as a function of partial pressure of the constituents, found that the existence region extended to about 10^{-3} at % of Cd or Te. The maximum melting point, T_m, occurred near the Te-rich limit.

Table 4.2 The maximum melting point and the minimum pressure, P_{min}, at the melting point of the II–VI components (taken from M. R. Lorenz, in *Physics and Chemistry of II–VI Compounds*, eds. M. Aven and J. S. Prener, North-Holland, Amsterdam, 1967)

Compound	m.p. (°C)	P_{min} (atm.)
ZnS	1830	3.7
ZnSe	1520	0.53
ZnTe	1295	0.64
CdS	1475	3.8
CdSe	1239	0.41
CdTe	1092	0.23

The component pressure–temperature, P–T, curves have also been measurred most extensively for CdTe. In the P–T section, the solidus and liquidus are coincident. Strauss (6) has summarized the earlier results of de Nobel (8), Lorenz (10), Brebrick and Strauss (25) and Brebrick (13), and his curves are shown in Figure 4.3. The three-phase boundary is a loop which illustrates the pressures of Cd (upper curve) and Te (lower curve) which co-exist with the solid and liquid CdTe. At the lower pressures on the Cd-rich and Te-rich sides, the loops are asymptotic to the vapour-pressure curves of the pure elements.

In the upper portion of the cadmium vapour loop, the maximum Cd vapour pressure, corresponding to $(dP_{Cd}/d(1/T)) = 0$, is 7 atm at about 1030° C, decreasing to 0.65 atm at the maximum melting point of 1092° C where $(dP_{Cd}/d(1/T)) = \infty$. The maximum vapour pressure of Te is approximately 0.185 atm at 930° C and 5.5×10^{-3} atm at the maximum melting point. The diagram also contains the P_{Cd} and P_{Te} lines corresponding to the P_{min} condition as measured by Brebrick and Strauss (25) and Brebrick (13). The value of P_{min} at the congruent melting point is about 0.18 atm, rather less than the estimate given in Table 4.2, which involved extrapolation to T_m, which is not identical with the congruent melting point. Of the various methods of

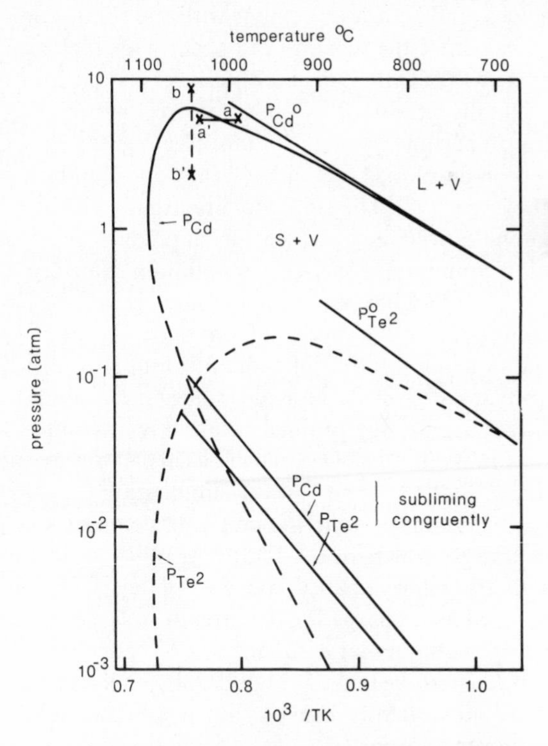

Figure 4.3 Three-phase loop for vapour pressures of Cd and Te_2 in equilibrium with solid CdTe and liquid

measuring partial pressures, which have included Knudsen effusion, Bourdon gauge and transpiration techniques, that employed by Brebrick and Strauss is much the most reliable. This is because their optical absorption technique ensures that the partial pressures are measured under equilibrium conditions.

P–T loops similar in shape to those of the Cd–Te system have been measured for Cd-Se in a number of laboratories. The results are summarized by Burmeister and Stevenson (26). The Zn-Te system has also been investigated (15, 13). Knowledge of the *P–T* loops is vital for crystal growth, since the melting point varies enormously with the applied partial pressure. A vertical line through a loop also defines the limits of stability of a solid in equilibrium with the vapour. This is important for post-growth heat treatment.

Preparation of the compounds

Crystals of the highest purity are most likely to be obtained from a starting charge of the compound synthesized from the component elements of the highest available purity, particularly if this can be done in a single growth system. However, reacting the elements together in a sealed system, especially with the sulphides, and to a lesser extent with the selenides, is a potentially hazardous process, since the reaction can occur with explosive violence. For this reason the growth of CdS, ZnS and ZnSe is often done from a charge of the compound, originally produced by chemical precipitation.

Processes for preparing commercial luminescent grade ZnS, CdS and ZnSe powders have been developed over many years, and standard procedures are described by Leverenz (27). The sulphides are usually produced either by the chloride or sulphate process, in which the sulphide is precipitated from an aqueous solution containing either zinc ammonium chloride or zinc sulphate, using H_2S. Powders of higher purity are prepared by ion exchange processes. ZnSe can be prepared in a number of ways. In one of these, selenium is dissolved in an aqueous solution of sodium sulphite to form $(Na_2SO_3)_2.Se$. ZnSe is then precipitated from an acidic solution by adding zinc sulphate.

Powder charges are usually purified before any attempt is made to grow single crystals from them. Purification can be carried out by sublimation in a stream of inert gas such as argon. During sublimation the volatile impurities are driven off preferentially, while the non-volatile impurities remain in the residue of the charge, which is not totally distilled. With this sublimation technique, small irregularly-shaped rod and platelike crystals are formed. These are then used as a charge for the growth of larger crystals. Chemical suppliers tend to sublime large quantities of powdered material in a continuously evacuated system. A high-density, polycrystalline deposit is formed, which is subsequently broken into pieces and sold as high-purity polycrystalline lump material.

Powder charges are not particularly suitable for growth from the melt because powder when melted occupies about one-fifth of its original volume. For growth from the vapour phase, especially with ZnS and CdS, pressed and sintered pellets have often been used as the charge. Sintering is then usually done in a stream of H_2S.

CdTe and ZnTe can be synthesized from the elements without too much difficulty, although zinc can develop an impervious layer of ZnTe on its surface. Silica ampoules are often coated with a layer of pyrolytic carbon beforehand to reduce any reaction with, or pick-up of impurity from, the silica. ZnSe and CdSe have often been synthesized in sealed, evacuated capsules, heated with an oxy-gas flame played along the silica (28).

Growth from the vapour phase

Techniques of growing single crystals of II–VI compounds from the vapour phase can be divided into three broad categories involving (i) dynamic or open-flow systems, (ii) sealed or nearly-sealed systems or (iii) chemical transport agents. The open-flow technique was revived by Frerichs (1), and used to grow small platelets and rods, principally of CdS, but also of CdSe and CdTe. The method has proved readily adaptable to the formation of all the II–VI compounds of interest, and in providing a wide range of single crystals for appraisal for the first time, stimulated immediate interest in modifications to the technique, and to the growth of larger crystals in sealed systems. Large crystals usually have more uniform electrical and optical properties, and can be cut and oriented into samples as required. In addition, better control over the stoichiometry can be obtained. Nevertheless the open-tube system is valuable because crystals can be grown quickly, in an hour or two, at somewhat lower temperatures. This leads to less accidental contamination from the reactor tube and furnace system. With care, small strain-free platelets can also be grown, and these are of vital importance for photoluminescence studies of shallow donors and acceptors.

Open-flow systems

The simple experimental arrangement used by Frerichs is illustrated in Figure 4.4. A fused silica boat containing molten cadmium was placed in a long silica tube inside an electric furnace. A stream of hydrogen passed over the cadmium, which was held at 800° C, and carried its vapour downstream to the reaction zone where it met and mixed with a stream of H_2S at about 1000° C. Crystals grew by self-nucleation on the walls of the long silica tube. Frerichs also grew crystals of CdSe and CdTe by passing hydrogen over molten Se to produce H_2Se, and over molten Te to produce H_2Te.

Crystals have also been grown in open-flow systems using the two elements involved as the source material. Streams of inert gas then pass over boats containing the molten elements, to be mixed in the reaction zone so that

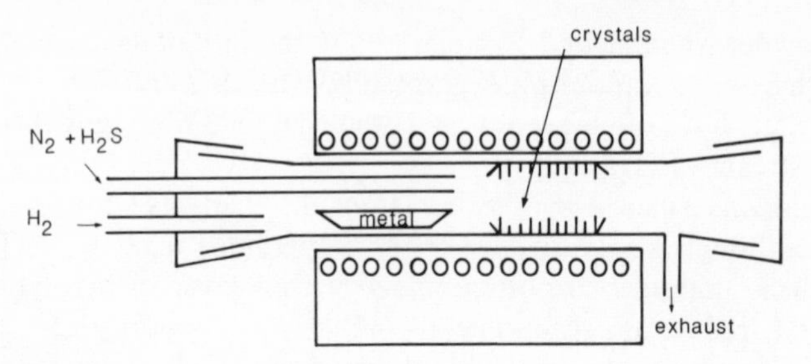

Figure 4.4 Frerichs method of growing crystals of II–VI compounds

growth can occur at slightly lower temperatures. A more reliable, and even simpler, variation of Frerichs' method is to pass a stream of inert gas over the heated compound itself.

Doping in open-flow systems can be achieved in a variety of ways. For example the required dopant might be contained in the starting charge of the compound, or if it has a reasonable vapour pressure it can be contained in a separate boat so that its vapour can be carried to the reaction zone by an inert gas. To dope with chlorine, HCl gas can be added to the carrier gas.

Sealed systems

In the sealed, or nearly sealed, system for crystal growth from the vapour phase, matter transport takes place by sublimation from a charge of the compound down a temperature gradient. The growth capsule is either evacuated, or filled with a suitable gas, such as argon or H_2S, for example. The process is essentially one of dissociative sublimation, in which the charge dissociates as it evaporates, and the elements are transported by diffusion in the gaseous phase to the cooler region of the ampoule where condensation and growth occur. Charge temperatures range from about 1500–1600° C for ZnS to about 950° C for CdTe. Ampoules are usually between 10 and 15 cm long, and the temperature difference between the charge and the growing crystal varies from 20–50° C.

The first reported use of the method was by Reynolds and Czyzak (29), who grew crystals of ZnS and CdS by sublimation from charges of pre-sintered powder in sealed quartz tubes containing 150 Torr of H_2S. The method was substantially improved by Greene *et al.* (30), and used to produce large crystalline boules up to 100 g in mass, which with CdS were sometimes single-crystal, and often contained three or four large polycrystals. For successful growth in a sealed tube, involving reasonable mass transport in a few days, it is essential to produce conditions close to P_{min} in the vapour. This requires zero or little deviation from stoichiometry in the charge. If precautions are not

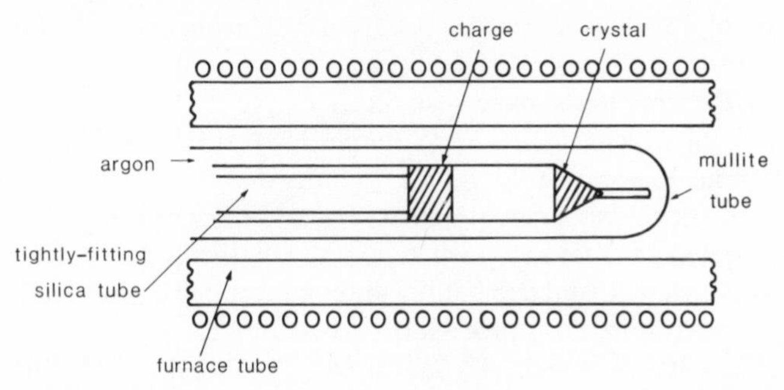

Figure 4.5 Section of furnace arrangement illustrating the Piper–Polich self-sealing method of crystal growth

taken to ensure such conditions, high total pressures can develop in a sealed tube held at temperatures appropriate to crystal growth. In that event matter transport will be minimal, and at worst the tube will explode. Pre-distillation and sintering of the charge are techniques frequently employed to obtain near-stoichiometric charges. Sintering also increases the charge density to about 70% of the true bulk density. Even when these precautions are observed, P_{min} conditions may not obtain in a sealed capsule. To overcome this difficulty, Piper and Polich (31) introduced a self-sealing technique which does ensure that conditions close to P_{min} are established.

The experimental arrangement used by Piper and Polich is shown in Figure 4.5. A quartz ampoule with a conical growth tip at one end was attached to a solid quartz glass rod which acted as a heat sink. At the other end of the capsule a pre-sintered charge of the compound was closely backed by a tightly-fitting quartz tube. The whole ensemble was placed in a mullite tube containing argon. The ampoule was first positioned with the growth tip at the centre of the furnace while the mullite envelope was evacuated and outgassed at 500 °C. The furnace was then brought to the growth temperature while argon was admitted to a pressure of one atmosphere. The growth tip was then moved towards the cooler region at a rate of between 0.3 and 1.5 mm per hour when condensation began and a crystalline boule developed. In many runs with CdS, the boule consisted of one single crystal, and crystals with centimetre dimensions were rapidly produced.

During the heating schedule the growth tip was initially at the hottest zone to ensure that any crystalline debris was evaporated away. As the pulling proceeded and sublimation of the charge began, the capsule was eventually totally sealed by the material deposited between the two tightly-fitting quartz tubes. The essence of the Piper–Polich method is that the system is self-sealing, but sealing does not occur until transport becomes substantial, and this happens when P_{min} conditions prevail. Aven *et al.* (32) used the method to grow ZnSe, and Aven and Segall (33) grew ZnTe in a similar manner. Clark and Woods (34, 35) also grew large crystals of CdS using a modified Piper and Polich system.

Various other methods have also been employed to achieve the self-adjustment of the stoichiometry. For example de Meis and Fischer (36) used an evacuated, sealed ampoule to which a long quartz capillary (~ 1.5 mm diameter) had previously been attached. The capillary extended beyond the furnace and acted as a cold trap where excess volatile material accumulated until P_{min} was reached, and substantial sublimation began and sealed off the capillary. They exploited this procedure successfully to grow crystals of ZnSe and ZnTe. Russell and Woods (37) have used an essentially similar method to grow crystals of ZnS and Zn(S, Se), and Buik and Nitsche (38) have used it successfully to grow crystals of CdTe.

A rather different procedure for preventing the build-up of excess partial pressures, and establishing conditions close to P_{min}, is described in the book by Faktor and Garrett (39). They used a capsule containing a small (0.2 mm dia) effusion hole in its wall, while a vacuum of better than 10^{-5} Torr was maintained outside the hole. The technique employed by Fochs *et al.* (40) is essentially similar. Since the components will leak out in stoichiometric proportions when the material sublimes congruently, their partial pressures on the capsule side of the effusion hole will be determined by their molecular weights. In fact P_{Cd}/P_{S_2} for CdS is about 2.7. The ratio of the partial pressures is then fixed during the whole of the growth, and crystals with uniform compositions are obtained. Dierssen and Gabor (41) have scaled this method up, and have been able to achieve seeded growth of large single crystals of CdS from 400 g charges contained in silica tubes 300 mm long and 30–60 mm in diameter with two 1 mm vent holes. In general less than 1% of the material is lost by effusion through the small effusion hole.

In another modification, somewhat similar to that used by de Meis and Fischer, the growth capsule is connected via a constriction to a reservoir containing one or other of the component elements of the compound. If the reservoir, instead of being at room temperature as in de Meis and Fischer's arrangement, is held at a temperature corresponding to the vapour pressure of the pure element under P_{min} conditions over the evaporating compound, considerable matter transport is obtained, and a substantial boule can be grown in a few days. Prior (42) and Clark and Woods (35) have used the technique to grow CdS, and Burr and Woods (43) adapted it to ZnSe. Kiyosawa *et al.* (44) used the method to measure the transport rate of ZnSe under different partial pressures of zinc and selenium. Similar measurements have been made for the sublimation of CdTe (45, 46, 47).

The growth of crystals of II–VI compounds from the vapour phase has been treated quantitatively by many authors (44, 48–53). Perhaps the most comprehensive discussion has been given by Faktor and his co-workers, which is well summarized in the book by Faktor and Garrett (39). The recent review by Ballentyne (54) is also instructive.

Chemical transport methods

Iodine transport

The technique of chemical transport was first applied to the growth of single crystals of II–VI compounds by Nitsche (55). The material to be grown is placed at one end of a sealed evacuated tube, and the transport agent is added. This is usually elemental iodine, but other halogen-bearing substances have been used, notably HCl gas. The sealed tube is then placed in a temperature gradient, in such a way that the charge end of the tube is some 10–20° C hotter than the end where the crystal is to grow, normally at about 850° C for ZnS, CdS and ZnSe. Taking ZnS as an example, the following reaction takes place at the higher-temperature end of the capsule:

$$ZnS(solid) + I_2(gas) \rightarrow ZnI_2(gas) + \tfrac{1}{2}S_2(gas)$$

The ZnI_2 and S_2 vapours diffuse down their concentration gradients (assisted by Stefan flow) to the cooler end, where the reaction is reversed and solid ZnS is deposited. The iodine released diffuses back to the source, and the whole process repeats itself. The result is that ZnS is deposited at the cooler end of the capsule, while the iodine acts as the transporting agent.

The method is particularly useful because the optimum temperature for the growth of most of the II–VI compounds is 850° C or less. This is substantially lower than the temperatures required in normal vapour-phase growth, or in growth from the melt. With ZnS, which has a phase transition temperature at 1020° C, growth by iodine transport at 850° C leads to a crystal with a homogeneous cubic crystal structure (56, 57, 58). Crystals grown by normal vapour-phase techniques or from the melt have a mixed hexagonal–cubic structure with extensive one-dimensional stacking disorder.

One other important advantage of the method is that it can be used to grow solid solutions. The technique has been used with success in the author's laboratory to grow crystals of ZnS_xSe_{1-x} and $Zn_xCd_{1-x}S$ throughout the entire range of composition, and also by Fujita *et al.* (58) and Palosz (59). A disadvantage, of course, which may be serious for some applications, is that all crystals grown by the method are fairly heavily contaminated with iodine, which is electrically and optically active in these materials.

Chemical vapour deposition

The development of high-power CO_2 lasers has stimulated interest in CdTe, ZnSe and ZnS as laser window materials at 10.6 μm in the infrared. These compounds can have absorption coefficients as low as $10^{-4}\,cm^{-1}$ at this wavelength (60), and indeed optical absorption is very low between the limits set by the bandgap at short wavelengths and by lattice absorption at long wavelengths. CdTe transmits well between 1 and 25 μm, and ZnSe between 0.5 and 22 μm. ZnS can be used in the range 1–12 μm. All three materials can be produced as flat polycrystalline plates, which are mechanically strong, non-hygroscopic, and can take a high optical polish. Plates of ZnSe several mm thick and exceeding 300 mm in diameter can be produced quite readily. The preferred method is by chemical vapour deposition in what is essentially a chemically engineered version of Frerichs' original method (1) of the growth of small platelet crystals (see chapter introduction). Some grades of II–VI materials for infrared optics are produced by hot pressing of purified powders.

Details of CVD processes are sparse. Lewis and Hill (61) and Lewis *et al.* (62) have given brief descriptions of their method of preparing polycrystalline plates of ZnSe, ZnS and ZnS.Se. Zinc vapour is generated by evaporation from a molten charge of 5N purity zinc contained in a carbon crucible. The vapour is entrained in a stream of argon and ducted to a reaction zone where it is mixed with hydrogen selenide or hydrogen sulphide, depending on the compound required. The reaction zone is formed by the four walls of a rectangular graphite substrate, on which deposition occurs. Samples are readily cut from the deposits using a diamond wheel, and can be lapped and polished with diamond abrasive.

Tuller *et al.* (63) have described a system for the growth of polycrystalline plates of CdTe suitable for use as laser window material. In their system the cadmium vapour was transported in a stream of $He + 8\% H_2$ (to prevent oxidation), and tellurium vapour in He. The CdTe was deposited on graphite substrates at temperatures in the range 700–970° C. The plates tended to be rich in tellurium and to have relatively high *p*-type conductivity and accompanying free-carrier absorption. Conductivities as low as $1.5 \times 10^{-6}\,ohm^{-1}\,cm^{-1}$, however, were obtained by annealing in cadmium vapour. Samples deposited at temperatures less than 850° C contained large numbers of voids located preferentially at grain boundaries, which seriously degraded the optical quality. A significant improvement in optical quality was obtained by raising the deposition temperature of the CdTe to 950° C. The material then produced was practically pore-free.

Growth from the melt

The difficulty in growing II–VI compounds from the melt is that they dissociate and establish an appreciable P_{min} pressure at the melting point (Table 4.2). High temperatures are needed to melt ZnS, ZnSe and CdS, and high pressures are developed, so that special precautions are necessary. The remaining three II–VI compounds can be melted at lower temperatures without too much concern for safety and the possibility of explosion, but measures must be taken to eliminate decomposition and the transport of matter to lower-temperature regions of the enclosure. Melt growth therefore poses an interesting set of problems because the compounds should be heated to temperatures some 50° C above their melting points to ensure complete melting before growth is begun.

One possible solution is to contain the compound in a sealed capsule to prevent evaporation. This can be done quite successfully with CdTe and ZnTe. It is important to realize, however, that the charges of II–VI compounds are rarely stoichiometric, and when melted can lead to vapour pressures many times larger than P_{min}. Attempts to melt ZnS, ZnSe, CdS in unsupported capsules would be extremely hazardous. Silica glass, which is a favoured material for capsule fabrication, can withstand pressures of 30 atm at temperatures up to 1000° C. If it is to be stressed beyond this, support or pressure balancing is essential.

An alternative to sealing the material in a capsule is to provide an overpressure of one of the constituents, thus reducing the evaporation rate of the compound by reducing the vapour pressure of the minority component in the vapour. Another approach is to provide a large overpressure of inert gas. In what follows, techniques of growth from the melt are divided into those which employ high-pressure methods in one form or another, and those in which no such extremes are needed.

High-pressure methods

One of the pioneers of melt growth of the II–VIs was A. G. Fischer, who investigated a large number of ways in which ZnS, CdS, ZnSe and ZnTe could be grown from the melt. Most of his work is described in US government contract reports, but he wrote a more readily available review in 1970 (64).

Crystals of the wide gap II–VI compounds were successfully grown by Fischer using a vertical Bridgman technique, with the compound in an unsealed carbon crucible with lid, under a pressure of 100 atm of nitrogen. To grow a crystal, the crucible containing the molten charge was lowered slowly into a cooler region of the furnace. With the high temperatures required, contamination is difficult to avoid, particularly from the carbon crucible. Good single crystals of CdS are often obtained, although they have a tendency to crack, especially if the boule contains more than one polycrystal. Boules of ZnSe and ZnTe are usually more polycrystalline.

A simpler high-pressure technique known as the 'soft ampoule' method utilizes the fact that quartz containers soften at temperatures above 1200° C, and at 1500° C are quite thermoplastic. The procedure, which has been used with some success for ZnSe and CdS, involves sealing the charge in an evacuated quartz ampoule. The ampoule is then placed inside a thick-walled, tight-fitting, graphite 'bomb', which is itself sealed by a threaded graphite plug. At growth temperatures, the pressure in the ampoule drives the quartz walls against the graphite bomb, which resists and counterbalances the outward force of the dissociation pressure in the ampoule. Growth of CdS is carried out in a vertical molybdenum furnace by lowering the bomb from a hot region of 1520° C to a cooler region at about 1400° C. Eventual devitrification of the silica limits the duration of a growth run to a few hours. The method has been used by a variety of workers to grow crystals of all the II–VI compounds. It is, however, always accompanied by the risk of explosion. To eliminate the explosion hazard, the decomposition pressure inside an unsupported ampoule can be balanced by an external gas pressure in a steel autoclave.

Attempts to pull crystals of II–VI compounds using the Czochralski technique (see Chapter 2) have not been very successful. This is because it is essential to use liquid encapsulation, but no entirely suitable encapsulants have been found. Boric oxide is best, and some success has been achieved with CdTe (65).

Melt growth at ordinary pressures

Although ZnTe, CdSe and CdTe can be grown from the melt using high pressures, they can also be grown at more modest pressures. Most work has been done on CdTe, the material with the lowest melting point, and comprehensive reviews have been provided by Strauss (6), Mullin and Straughan (65) and Zanio (7). The three-phase loops of Figure 4.3 show that the partial pressures of Cd and Te vapour over CdTe at the maximum melting point (1092° C) are about 0.7 and 5×10^{-3} atm respectively. To grow crystals from the melt, evaporation and matter transport to cooler parts of the ampoule must be prevented. This can be achieved without recourse to high-pressure techniques, by using a completely sealed ampoule in which the temperature anywhere is not allowed to fall below 728° C, the temperature at which the vapour pressure over pure cadmium reaches 0.7 atm. Alternatively evaporation can be prevented by maintaining a pressure of 0.7 atm of Cd vapour over the melt by using an auxiliary Cd reservoir at a temperature of 728° C or greater. The reservoir must of course be the coolest part of the system.

The vertical Bridgman method affords the simplest method of growing crystals of CdTe from the melt. Here, CdTe, nearly filling a sealed evacuated silica capsule, is placed in a vertical furnace where it is melted at about 1120° C for about 2 h to achieve uniformity. The furnace has a temperature profile such that the top is hotter than the bottom, and a steep temperature gradient midway down ensures a well-defined freezing interface when the capsule is lowered through the furnace, so that the melt solidifies from bottom to top. Crystals 1 cm in diameter and up to 5 cm long can be grown at rates of about 2 cm h^{-1}. One of the chief difficulties in growing CdTe is that it has the lowest thermal conductivity of all the II–VIs, so that it is difficult to conduct away the latent heat of fusion. Condensation of cadmium does not occur with the vertical Bridgman method, because the furnace profile ensures that the walls of the capsule exposed to the vapour are everywhere hotter than the melt.

Attempts to control the conductivity (*n*- or *p*-type) of the resultant crystal have been made by growing from melts containing excess Cd or Te. It is important to realize, however, that a slight excess of Cd leads to a substantial increase in total vapour pressure (Figure 4.3) accompanied by a lowering of the melting point. Constitutional supercooling often results when growing off stoichiometry with excess of either Cd or Te. Dendritic growth and liquid entrapment can then occur, and porous crystals are often obtained.

The simple vertical Bridgman technique offers no sure control over the partial pressures in the ampoule. Triboulet and Rodot (66) and Triboulet and Marfaing (67) improved this situation by maintaining the top of the ampoule at a constant temperature throughout growth. Kyle (68) added a further refinement by attaching a Cd reservoir to the top of the growth ampoule, and maintaining it at a constant temperature. With these variations growth took place under a constant pressure of Cd vapour.

Vertical zone refining has also been used extensively to grow single crystals of CdTe from near stoichiometric melts in completely sealed tubes. Lorenz and Halsted (69) pioneered the method, which was improved by Cornet *et al.* (70) and by Woodbury and Lewandowski (71), who imposed vertical oscillations at 100 cpm on the ingot during the passage of the last zone. They achieved crystals up to 10 cm^3 in volume.

Historically the earliest attempts to grow CdTe by zone melting (8) and Bridgman (10) were carried out in a horizontal configuration. Lorenz's procedure, which is typical of the Bridgman methods, is illustrated in Figure 4.6. A carbon-coated boat containing CdTe is placed in a sealed silica tube, inside a three-zone horizontal furnace. The charge is melted at 1120° C and then directionally frozen at 1 cm h^{-1} by being displaced relative to the furnace. When the displacement is complete the ingot is wholly solid at 1010° C. The silica tube is sufficiently long for the elemental Cd in the reservoir to remain in the third and lowest temperature zone of the furnace throughout the whole process. As long as the temperature of the reservoir exceeds 728° C, evaporation from the CdTe is not excessive. Lorenz used somewhat higher reservoir temperatures and grew ingots containing crystals between 1 and 2 cm long. Most crystals were fairly heavily twinned.

The modification to achieve zone melting in the horizontal configuration is

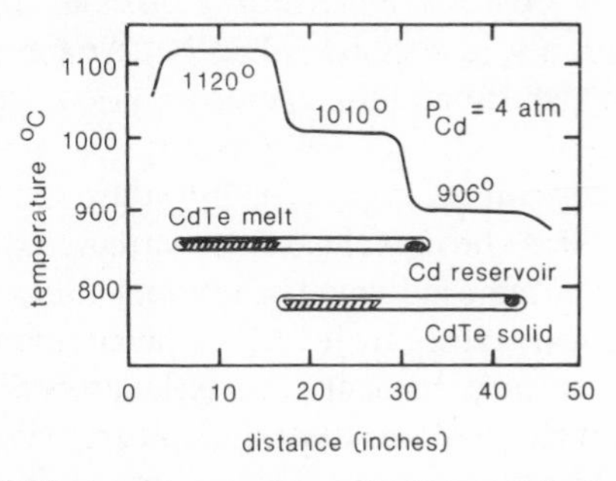

Figure 4.6 Horizontal Bridgman arrangement used by Lorenz (10) to grow CdTe in a controlled vapour pressure of Cd

quite obvious. The method was first employed by de Nobel (8) with a reservoir temperature giving a vapour pressure of cadmium of 1 atm. The technique has also been used by Matveev *et al.* (72) and Prokof'ev and Rud (73). These latter authors grew single crystals up to 15 cm long at a rate of 0.5 cm h^{-1}. The growth rate must be kept slow to allow the bubbles formed by the excess Cd in the melt to rise to the surface.

Growth from solution

Since the growth of II–VI compounds from the melt involves the use of high temperatures of necessity, accidental contamination from the container and other furnace materials is almost unavoidable. One of the advantages of vapour-phase methods is that considerably lower temperatures can be used. In general, even lower temperatures are possible with growth from solution. For example in the Cd-Te system, solutions containing 60% Te by weight freeze at 900° C, and those containing 95% at 550° C. Considerable success has been achieved in growing CdTe from Te-rich solutions in what is known as the travelling heater method (THM). The technique was introduced by Bell, Hemmatt and Wald (74), developed by them (75, 76), and taken up by Triboulet *et al.* (77, 78) and Taguchi *et al.* (79) among others. Large, semi-insulating crystals, suitable for use as nuclear detectors, are produced by this method, which involves passing a molten zone of tellurium-rich liquid from bottom to top through a polycrystalline ingot of CdTe, preferably first produced, according to Triboulet *et al.*, by vertical zone refining. The experimental arrangement is very simple, and best illustrated with reference to the procedure adopted by Bell *et al.* (74).

A cast, near-stoichiometric, 250 gm ingot of CdTe is placed in a silica ampoule, some 1.5 cm in diameter, on top of approximately 10 g of high-purity Te. After evacuation and sealing, the ampoule is lowered through a resistively heated zone heater maintained in the range 650 to 750° C. Rates of lowering range from 0.3 to 2 cm day^{-1}. As the ampoule is lowered through the zone heater the tellurium melts, and CdTe dissolves in it at the upper (hotter) end of the zone. Cd diffuses through the zone and solidifies out at the lower (cooler) end. In most variations a seed crystal is placed at the bottom of the ampoule initially, below the tellurium, to promote good crystal growth on solidification.

Quite large crystals can be produced, but they tend to contain large numbers of tellurium micro-precipitates which detract from their usefulness as nuclear detectors or infrared windows. The crystals also contain rather a high concentration of electron traps, attributed to the compensation of the large number of cadmium vacancies present in material grown from tellurium-rich solution. In an attempt to overcome these difficulties, Triboulet (80) proposed a vapour-phase modification to the THM method in which the molten tellurium zone was simply replaced by an empty space. The space was then passed through the ingot by lowering the ingot through a zone heater. Matter transport through the space took place in the vapour phase. Triboulet grew crystals at 800, 900 and 1000° C at 5–7 mm day^{-1} using a space 12 mm high. The resultant crystals, which were low-resistivity *p*-type, contained no detectable tellurium precipitates. Taguchi *et al.* (81) used sublimation THM to grow crystals of ZnTe.

CdTe crystals have also been grown from solution in excess cadmium, but here higher temperatures are involved. Lorenz (82) grew crystals from a melt containing 61 at % Cd and 39 at % Te in a horizontal Bridgman apparatus, with a reservoir to maintain a constant vapour pressure of cadmium of 5.9 atm throughout the system. In this unique method, crystals were grown at a rate of 1 mm h^{-1} by raising the temperature from 975 to 1025° C. The process can be understood by reference to the three-phase loop in Figure 4.3. Heating starts at point *a*, and continues until *a′* is reached. When the horizontal line crosses the three-phase line, solidification begins, since solid and vapour exist within the loop, and liquid and vapour outside it. The excess cadmium initially present in the melt is removed by evaporation to the reservoir. The method is commonly known, therefore, as the solvent-evaporation technique.

Although Lorenz demonstrated the feasibility of the process, his crystals were of inferior quality and of little practical use. An important alternative procedure was introduced by Lunn and Bettridge (83) who grew crystals from a cadmium-rich melt by holding the temperature constant and reducing the vapour pressure of the cadmium. In this way the three-phase line is crossed by the vertical line *b*–*b′* (Figure 4.3) and once again CdTe solidifies as the excess cadmium evaporates. Solvent evaporation has the major advantage that no moving parts are required, and in Lunn and Bettridge's modification growth is isothermal. Crystals were grown in a vertical configuration, with the charge at the lower end of a carbon-coated silica tube and the cadmium reservoir at the top. Communication between the reservoir and growth zone was via a slot cut in the side of the reservoir. To grow a crystal the charge containing not more than 61 at % of Cd was held at 1080° C while the temperature of the reservoir was reduced from 980 to 750° C by 3° per hour. Although the yield was not high, good crystals 2 cm in diameter up to 6 cm in length were occasionally produced.

The technique has been further refined by Mullin *et al.* (84) who inverted the growth and reservoir chambers to avoid droplets of cadmium falling from the reservoir into the molten charge and setting up thermal fluctuations. They also pointed out the desirability of using a growth temperature in excess of that corresponding to the maximum vapour pressure above the melt (1032° C, according to Lorenz). With their modification, Mullin *et al.* were able to produce a much higher proportion (10%) of completely sound ingots with no porosity.

Thin films

For many practical applications, thin films rather than bulk crystals are more convenient, if not essential. There is, for example, a voluminous literature on the preparation and properties of films of CdS, which in addition to their exploitation in the CdS–Cu_2S photovoltaic cell (18), can be used as ultrasonic transducers (85), photoconductors, phosphors and electroluminescent layers (86), space-charge-limited diodes and triodes (87) and thin-film transistors (88). Films of CdSe have been studied to a lesser extent for applications as photoconductors (89), in photovoltaic cells (90) and in thin-film transistors (91). ZnS and ZnSe have also been investigated extensively for use in cathodo- and electroluminescence (92) and in electro-optical devices (93, 94). In the early work, most films were put down by vacuum deposition on to polycrystalline substrates, although many other techniques have also been used. A review of the earlier techniques of film deposition used specifically with II–VI compounds has been provided by Shallcross (95), and Cusano (92) has described the work on chemical vapour deposition which produced good cathodo- and electroluminescent films of ZnS.

In the 1970s, interest turned increasingly to the growth of epitaxial layers, following the success achieved in this field with the elemental and III–V semiconductors. Initially, most effort was concentrated on vapour-phase techniques (96), but more recently, distinct progress has been made with the growth of ZnSe by liquid-phase epitaxy, and even more recently, work on molecular beam epitaxy, MBE, and organometallic vapour-phase epitaxy, MOVPE (97), has spurred hopes that high-purity, device-quality, epitaxial layers with a high degree of crystallographic perfection will lead to the large-scale commercial exploitation of the diverse interesting properties of II–VI compounds.

Vacuum deposition

Numerous investigators have demonstrated that films of CdS and CdSe can be deposited successfully in a conventional vacuum-coating unit. Thicker layers, with better crystallinity and stoichiometry, are produced if the substrate is heated, and if the source and substrate are enclosed in a 'hot wall chimney' (98), in which communication with the bell jar is restricted. This allows the conditions of thermodynamic equilibrium to be approached a little more closely.

The properties of evaporated films of CdS depend on the rate of evaporation, the substrate temperature and the film thickness (99). The optimum substrate temperature is close to 200° C. It is well known that the resistivity of evaporated films of CdS decreases with increasing thickness, which is attributed to the increasing non-stoichiometry (build-up of excess Cd) of the source as the evaporation proceeds. The CdS films are hexagonal and show some degree of preferred orientation, with the *c*-axis perpendicular to the substrate, for films between 600 and 1000 Å thick (100). The alignment becomes more pronounced in thicker films, and it is for this reason that such layers can be used as high-frequency, longitudinal ultrasonic transducers (85). Incidentally, Foster showed that the *c*-axis tended to align itself in the direction of incidence of the vapour beam. By slowly tilting the substrate relative to that direction he was able to grow layers with the *c*-axis in the plane of the substrate for use as shear wave transducers.

Most CdS films evaporated from a charge of the compound have a resistivity less than 10^4 ohm cm, which decreases to values between 10 and 100 ohm cm for thicknesses exceeding 1 μm. This, coupled with the *c*-axis alignment is exploited in the CdS–Cu_2S cell (18). However, further measures are necessary if useful photoconductors or ultrasonic transducers are to be produced. With photoconductors, a dark resistivity of at least 10^6 ohm cm is required, and ultrasonic transducers need a resistivity of about 10^4 ohm cm. The resistivity is traditionally increased, either by evaporating additional sulphur, or by incorporating or diffusing in silver or copper to form deep acceptors. Separate sources of cadmium and sulphur have often been used to obtain better control of the stoichiometry, and hence of the resistivity (85). Sputtered films have essentially similar properties to vacuum-evaporated layers.

In contrast with CdS, vacuum evaporation of ZnS, traditionally of more interest for its luminescence properties, has not been particularly successful until fairly recently. Polycrystalline, mixed cubic and hexagonal films with extensive one-dimensional stacking disorder are usually produced, and few are luminescent to any degree. It was for this reason that Cusano (92) developed a method of chemical vapour decomposition in which small quantities of Zn, $ZnCl_2$ and a luminescence activator (e.g. $MnCl_2$) were steadily injected into a reactor, where they met a stream of H_2S, and films with good luminescence properties were deposited on to heated glass substrates. More recently, with revived interest in thin-film electroluminescence, and with improved starting materials and vacuum techniques, evaporated films of ZnS doped with Mn have been prepared with good electroluminescence properties.

Vapour-phase expitaxy

Since the late 1960s, attention has been focused increasingly on the possibility of growing epitaxial layers of II–VI compounds on suitable substrates. For best results, good matches between the crystal structures, lattice parameters and coefficients of thermal expansion are required. Table 4.3 lists appropriate values for the II–VI compounds and some commonly used substrates.

With the exception of evaporation in uhv, which is now being replaced by molecular-beam techniques (see p. 87), vapour-phase epitaxy (VPE) is carried out in a gaseous atmosphere or in a gas flow. The function of the gas is to

Table 4.3 Lattice parameters and expansion coefficients of II–VI compounds and some substrate materials

Compound	Structure	Lattice constant at room temperature (A°)	Average linear coefficient of expansion ($10^{-6}\,°C^{-1}$)
ZnSe	ZB	5.6687	9.44
	W	a = 4.003 c = 6.540	
ZnS	ZB	5.4093	4.5–8.5
	W	a = 3.820 c = 6.260	⊥C 5.9–6.5 ‖C 4.4–4.6
CdTe	ZB	6.481	4.9
CdS	ZB	5.820	
	W	a = 4.1368 c = 6.7163	⊥C 5.0 ‖C 2.5
CdSe	ZB	6.05	
	W	a = 4.298 c = 7.0150	
ZnTe	ZB	6.1037	
GaAs	ZB	5.6533	6.72
GaP	ZB	5.4506	5.78
Si	ZB	5.4309	~4.0
Ge	ZB	5.6576	6.2
α-Al_2O_3	Trigonal	a = 4.758 c = 12.991	⊥C 9.2 ‖C 9.8
	Cubic	3.839	
InSb	ZB	6.4798	

promote chemical transport from the compound source, and H_2, HCl, HBr and I_2 have often been used for this purpose. These methods allow films to be produced under conditions approaching thermodynamic equilibrium, and better quality is then obtained with a lower incidence of crystallographic defects. Two fundamental experimental arrangements have been employed: (1) the close-spaced technique, and (2) the open-tube method.

A close-spaced system was used by Hovel and Milnes (101) to study the epitaxial growth of ZnSe on substrates of Ge, GaAs and ZnSe using HCl in H_2 as the carrier gas. The source and substrate were held between oxidized blocks of silicon and were separated by a distance of 0.25 mm only. The silicon blocks were heated radiantly by infrared lamps, so that the source was between 620–700° C while the substrate could be held up to 120° C cooler in the range 550–680° C. The gas-handling arrangement allowed a stream of hydrogen containing 0.01% HCl gas to flow slowly through the system. The very low concentration of HCl was necessary to avoid undesirable etching of the substrate. Growth was 10–40 times faster on (111) Ge than on (111) GaAs and ZnSe, but was accompanied by numerous stacking faults. Yoshikawa and Sakai (102) used a close-spaced technique to grow epilayers of CdS, and Vohl

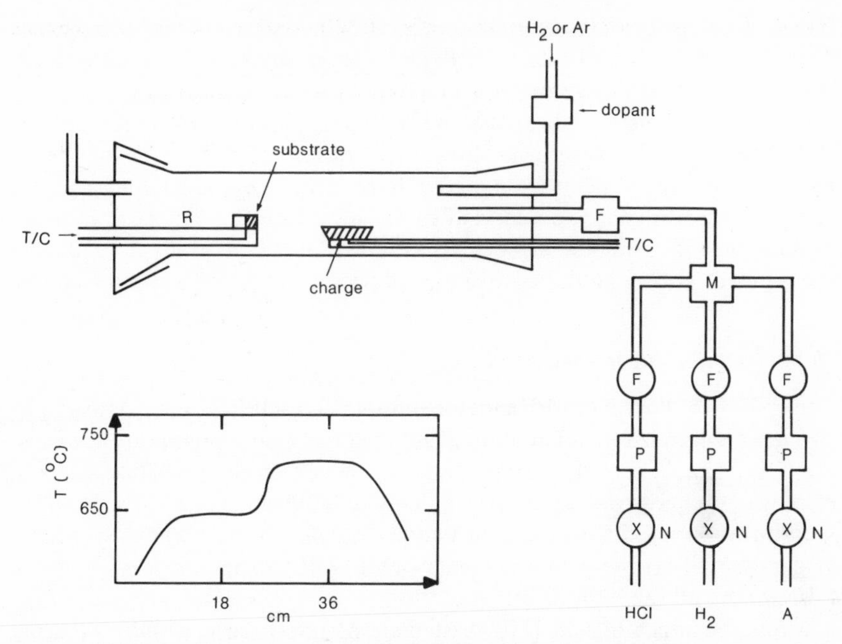

Figure 4.7 Schematic diagram of vapour-phase epitaxy system using chemical vapour deposition from a charge of the compound. F, flow meter: M, mixer value; P, purifier; N, needle valve; R, push rods

et al. (103) used a similar arrangement to grow thick (200 μm) epilayers of ZnSe and ZnS on (100) GaAs. The best epitaxy was in the initially deposited layers. Polycrystalline pockets formed as the layers grew thicker, and fibre growth developed. Kitegawa *et al.* (104) used a close-spaced configuration to confirm that ZnS could be grown on GaP.

A typical open-flow system for VPE is illustrated diagrammatically in Figure 4.7. The gases are mixed before entering the silica reaction tube. If required, dopants can be added via a separate flow line. The reaction tube is contained in a two-zone furnace to control the temperature of the source and substrate. Yim and Stofko (105) used an open-flow system to grow layers of ZnSe, ZnS or CdS on substrates of GaAs, GaP and sapphire. They reacted the appropriate metal vapour with the non-metal hydride, and substrate temperatures ranged from 700° C for CdS and ZnSe to 890° C for ZnS. The quality of epitaxy of ZnSe on GaAs was excellent and attributed to the good lattice and thermal expansion matches, and to the complete solid solution alloying of the two compounds at the interface. The strong cathodoluminescence of the best films of all three compounds was taken as proof of the high chemical purity of the layers. Open-tube techniques have been used by numerous investigators, including Kay *et al.* (94) who tried to grow thick layers of ZnS for electro-optical applications; Besomi and Wessels (106) who prepared layers of ZnSe on GaAs using H_2 and a separate source of Zn metal to obtain low resistivity films; and Matsuda and Akasahi (107) who grew ZnS on GaP. Finally, Scott *et al.* (108) grew epilayers of ZnSe on (100) GaAs in an open-tube system by transporting the elemental vapours in H_2. With ZnSe and GaAs there is a 0.3% lattice mismatch. However with $ZnSe_{1-x}S_x$ with $x = 0.06$ there is exact lattice match with GaAs. Scott *et al.* (109) have grown epitaxial films of this composition using Zn, Se and H_2S in H_2 to reduce the incidence of misfit dislocations. By using the $ZnSe_{1-x}S_x$ as buffer layer, and gradually reducing the input of H_2S, a gradation to more perfect ZnSe could be achieved.

Organometallic vapour-phase epitaxy

The open-tube system of VPE suffers a number of disadvantages. For example, two temperature zones are required, and in consequence premature condensation and reaction of the components can occur. With the halide systems, etching of the substrate can cause problems, and difficulties may arise from the incorporation of the halogen, which is electrically and optically active. Organometallic vapour-phase epitaxy (MOVPE) offers the possibility of overcoming all these disadvantages.

Mansevit and Simpson (110) first demonstrated the feasibility of using organometallics to grow epilayers of II–VI compounds, but it was not until 1978 that reports of further work on the deposition of ZnSe on GaAs appeared (111, 112). In MOVPE, ZnSe is grown from the zinc alkyls, dimethyl zinc, DMZ, or diethyl zinc, DEZ, which are liquids at room temperature, and H_2Se. The alkyl is carried in a stream of H_2 to within a very short distance of the substrate, before it meets and mixes with the H_2Se. Stutius used DMZ where ZnSe is produced by the reaction

$$(CH_3)_2Zn + H_2Se \rightarrow ZnSe + 2CH_4$$

A single-zone furnace only was required, and the substrate temperature was very low, 340–350° C. This is an important advantage because low substrate temperatures should lead to better crystallographic quality and less accidental contamination from the local environment. Glass apparatus can also be used. Stutius worked under reduced pressure to increase the velocity of gas flow, which helps to reduce premature reaction before the components reach the vicinity of the substrate. The use of DMZ rather than DEZ also helped to avoid this problem because the Zn-C bonds are stronger in DMZ so that the reaction rate in the gas phase was reduced. Stutius produced some films of specular quality, but many had a hazy appearance, which was attributed to the presence of small hillocks. The crystallographic properties of these films remain to be evaluated in detail. Bandgap luminescence was observed at 77 K on irradiation with the 3250 A-line of a He-Cd laser. At 4.2 K the I_2-bound exciton line was the dominant feature. No other I_2 lines, or indeed donor–acceptor pair bands were detected in MOVPE ZnSe, which strongly suggests that contamination by unwanted impurities was a minimum. Stutius (113, 114) also recognized that an exact lattice match should be achieved between $ZnSe_{1-x}S_x$ and GaAs, and he therefore grew layers with different values of x. For films with $x = 0.05$ (a value close to exact match if Vegard's law obtains), the width of the I_2 bound exciton line was a minimum (indicating minimum internal strain), and the intensity of the self-activated emission was also a minimum (indicating a small concentration of zinc vacancies). Recently Stutius has prepared low-resistivity ZnSe doped with Al (115, 116) using triethylaluminium. Wright and Cockayne (117) have developed a horizontal reactor in which films of ZnSe and ZnS have been grown by MOVPE at atmospheric pressure. This eliminates the need for a pumping train to handle the toxic and pyrophoric waste products from the system. Most recently there has been a substantial improvement in the quality of MOVPE films of ZnSe prepared at low pressure by Fujita *et al.* (118, 119) and by Fan and Williams (120) using an atmospheric pressure technique. Wright *et al.* (121) have also successfully prepared epilayers of ZnS and ZnSe using heterocyclic compounds of S and Se in reaction with DMZ. The structural quality of such layers is described by Cockayne *et al.* (122).

A brief review of the preparation and properties of layers of II–VI compounds by MOVPE has been given by Mullin *et al.* (123). Kuznetzov *et al.* (124) put down films of CdS by MOVPE, and single-crystal growth occurred at 400 °C. Mullin and his colleagues have studied the homoepitaxy of CdTe on CdTe by allowing dimethyl cadmium to react with diethyl telluride. They have also produced layers of $Cd_xHg_{1-x}Te$ on CdTe (125) using dimethyl cadmium, diethyl telluride and mercury vapour, with hydrogen at atmospheric pressure as the carrier gas.

MOVPE is obviously a powerful technique which offers exciting possibilities of preparing epitaxial films of II–VI compounds with a chemical purity and crystallographic perfection which have not hitherto been reached. Further progress is to be expected, particularly in respect of the development of multilayer quantum-well structures. Improved preparative techniques should allow the numerous optoelectronic properties of the II–VIs to find wider application. However, the hazards to be encountered when using organometallics must be recognized. Although the zinc and cadmium alkyls are toxic, they are very sensitive to oxygen, and spontaneously flammable, so that fire, rather than toxicity, is the major risk. The mercury alkyls are very toxic and more stable, while the tellurium alkyls have a particularly obnoxious odour even in minute concentrations.

Molecular-beam epitaxy

Molecular beam epitaxy (126) is a promising, if expensive, technique, which is beginning to be applied to the preparation of device-quality films. The distinctive appeal of MBE as a technique is that it allows growth to occur at low temperatures, so that interdiffusion from multilayers, or the substrate, can be reduced. In addition, the films grow under conditions that are controlled by surface kinetics.

In the last few years there has been a flurry of interest in the MBE growth of ZnSe, chiefly on GaAs, and attempts have begun to grow device-quality CdTe and $Cd_xHg_{1-x}Te$, mainly on CdTe. The growth of films of ZnTe, ZnSe, $ZnSe_xTe_{1-x}$ and ZnS by MBE has been studied by Yao *et al.* (127–133). Separate zinc and chalcogen effusion sources were used in a uhv system evacuated to 10^{-10} Torr with ion and titanium sublimation pumps. ZnTe was put down on GaAs and Ge; ZnSe on GaAs; and ZnS on GaP and CaF_2. Typical growth temperatures were in the region 350–400° C, and growth rates were dominated by the minority beam flux. The crystal structure of the films was examined using RHEED, and all films showed bandgap luminescence emission under electron bombardment at room temperature, but very little self-activated emission. The luminescence was taken as indicating high chemical purity and the absence of cation vacancies.

Yao *et al.* (134) later fabricated dc electroluminescent devices by MBE deposition on (100) slices of GaAs carrying sputtered layers of SiO_2 in which windows some $1.5 \times 1.5\,mm^2$ in area had been produced by photolithographic etching. The ZnSe layers deposited over the GaAs window areas were monocrystalline, smooth, and had a low resistivity of 10^2–10^3 ohm cm, while the layers deposited on the surrounding SiO_2 were polycrystalline, rough, and had a high resistivity greater than 10^7 ohm cm. Yao *et al.* fabricated Schottky barrier devices by depositing gold over the ZnSe layers. The polycrystalline regions were effective in isolating the windows electrically. A white electroluminescence was readily observed in forward bias. Yao *et al.* (132) have followed this up by demonstrating that the substrate temperature has a strong influence on the resistivity and photoluminescence of the ZnSe. For low resistivity it is necessary to use substrate temperatures between 280 and 370° C, when moderate in-diffusion of Ga from the substrate occurs. Ga then acts as a shallow donor. With substrate temperatures in excess of 400° C, increasing self-activated emission reveals that acceptor-like complexes of Ga and Zn vacancies are introduced. These compensate the shallow donors, and the resistivity rises. Studies of the free and bound exciton emission in photoluminescence at 4.2 K confirm this description. The electrical and optical properties of undoped layers of ZnSe grown by MBE have recently been discussed in some detail by Yao (133). Room-temperature electron mobilities as high as $490\,cm^2\,V^{-1}\,s^{-1}$ are reported and the photoluminescence is dominated by the so-called I_x bound exciton line. Kitagawa *et al.* (135), in a study of films of ZnSe and ZnTe grown with MBE techniques, found that the incorporation of Ga into ZnSe led to low resistivity, but indium led to high resistivity with $\rho > 10^8$ ohm cm. ZnTe films on ZnTe had low resistivity, but on GaAs, ρ increased to $\sim 10^4$ ohm cm. This was attributed to in-diffusion of Ga compensating the *p*-type semiconductivity. Thin film electroluminescent devices of ZnSe:Mn have been fabricated by Mishima *et al.* (136) using MBE.

Faurie and Million (137) have prepared films of CdTe on (111) and (100) surfaces of CdTe by MBE. They have also deposited films of $Cd_xHg_{1-x}Te$ on buffer layers of CdTe (138–140). A single effusion source only was needed when layers of CdTe were required, because CdTe evaporates congruently. Epitaxial growth was obtained on (111) CdTe for substrate temperatures not exceeding 200° C, with RHEED patterns showing uniform streaking for films thicker than 3000 Å. Any defects present in the substrate are propagated in the epilayer. It is difficult to grow single-crystal CdTe free of twins and low-angle grain boundaries, and as a result epilayers grown on CdTe tend to contain these defects. This problem has been addressed by Farrow *et al.* (141) who grew CdTe by MBE on the more crystallographically perfect InSb. InSb is isomorphic with CdTe and has a lattice parameter which is only 0.05% smaller (Table 4.3). Lattice-matched epilayers of high structural quality were produced during growth at the upper end of the range 50–220° C. The CdTe distorted spontaneously during growth, by desorption of the more volatile Cd component. A major interest in epitaxial CdTe is as a buffer layer for the deposition of $Cd_xHg_{1-x}Te$, which is of great importance in infrared technology. In principle $Cd_xHg_{1-x}Te$ can be used as an intrinsic detector over most of the infrared spectrum because its bandgap varies from 1.6 eV to zero as x decreases. Faurie *et al.* (139, 140) have grown layers of $Cd_xHg_{1-x}Te$ on buffer layers of CdTe on (111) CdTe, using separate effusion cells containing Cd, Hg and Te, because evaporation of the ternary compound is not congruent. The buffer layer was deposited from a single source of CdTe. Promising results were obtained with growth rates of $\sim 4 \mu m\, h^{-1}$ at substrate temperatures between 160–180° C, which is sufficiently low to reduce the interdiffusion of Cd and Hg. However, the crystallographic perfection of the $Cd_xHg_{1-x}Te$ films was still limited by the defects present in the CdTe substrate. An interesting aspect of the work was that both *n*- and *p*-type layers were produced for the first time.

Liquid-phase epitaxy

The techniques of liquid-phase epitaxy, LPE, have also been applied to II–VI compounds. The first requirement is to find a suitable solvent for the compound required. All the six II–VI compounds discussed here have an appreciable solubility in Bi and Sn (142, 143), which increases with increasing molecular weight from ZnS to CdTe. Most II–VIs also have reasonable solubility in the metals In, Ga and Pb, and in the Group II halide salts, such as $CdCl_2$ and $ZnCl_2$. The solubility of the chalcogens in the molten Group II metals is low, while the vapour pressure of the metals is high. This explains why LPE has not been attempted from melts containing an excess of the Group II metal. In contrast, the solubility of Cd in Te is quite appreciable and LPE layers of CdTe have been grown from solutions containing excess Te.

One of the earliest reports of the use of LPE with a II–VI compound was that of Widmer *et al.* (144) who grew layers of ZnTe on ZnTe substrates from solutions in In, Ga, Bi, Sn or Pb using a sealed capsule. The slider technique was employed by Kanamori *et al.* (146) who grew mirror-smooth layers of ZnTe from solutions in In, Ga or Sn. They also grew layers of $Zn_{1-x}Cd_xTe$ throughout the entire range of composition to explore the possibility of preparing a II–VI semiconductor with amphoteric properties. Simashovich and Tsuilyana (147) studied the LPE growth of CdSe, ZnSe and $Sn_{1-x}Cd_xTe$ from solution in $CdCl_2$, CdI_2 and $ZnCl_2$ in an attempt to prepare useful heterojunctions, but their ZnTe-CdSe and ZnTe-CdTe devices all had very high interfacial resistances.

The chemical purity of the LPE layers of ZnSe which Fujita *et al.* (148) grew by tilting from solution in Ga and excess Zn was obviously high, as evidenced by the photoluminescence. The best layers were grown on (111) faces of the ZnS_xSe_{1-x} substrate. The LPE technique appears to be particularly suitable for the growth of layers of ZnSe with much improved purity compared with that of bulk crystals. The North American Philips group have made this point quite clearly in a series of investigations in which the sophisticated III–V slider technology used with GaAlAs lasers has been adapted to ZnSe. Having first grown very pure platelets of ZnSe from solution in Ga and excess Zn (149), the group tried to prepare *p*-type LPE layers by growth from solution in Bi containing acceptor-like impurities such as Li, Na, N or P (150). Doping with N or P did yield *p*-type layers, but the resistivity was high, $> 10^5$ ohm cm, and the hole mobility was low, $\sim 5\,cm^2V^{-1}s^{-1}$. It was concluded that the presence of a background impurity content of compensating donors prevented low resistivity *p*-type epilayers from being produced. Lithium was thought to be the most probable donor impurity.

The level of background impurity was examined next (151), using LPE from solution in Sn. Layers between 5 and 10 μm thick were grown at temperatures from 830 to 900° C using a cooling rate of $0.5°\,C\,min^{-1}$. The observation of donor–acceptor emission and self-activated emission in the photoluminescence from the layers revealed that impurities present in the substrate were incorporated in the epilayer. As a result a buffer layer was introduced, with beneficial results, as evidenced by the relatively small number of sharper lines found in the bound exciton region of the photoluminescence spectrum, and the almost total absence of deep-centre emission. The bound exciton lines showed that Al and In donors and Li acceptors were still present. Occasionally Cl and Ga donors and Na acceptors were detected (152). It was estimated that the background level of shallow dopants was better than $5 \times 10^{16}\,cm^{-3}$. No evidence for the existence of native point defects was obtained, which points to the conclusion that all shallow donor and acceptor levels in ZnSe appear to be impurity-related.

In response to the demand for improved infrared detection systems, increasing effort has been directed recently to the problem of growing structurally perfect epilayers of $Cd_xHg_{1-x}Te$ on CdTe. If LPE were to be attempted from solution in excess Hg, vapour pressures of about 8 atm would be encountered at 500° C. This means that an open-tube (slider) system is impracticable. However, by growing from solution in excess Te it is possible to keep the vapour pressure of Hg below 0.1 atm, provided growth is carried out at temperatures below 500° C. An open-tube system is then feasible.

In fact all three LPE techniques of tipping, dipping and sliding have been used to grow layers of $Cd_xHg_{1-x}Te$. Schmit and Bowers (153, 154) and Schmit *et al.* used the slider system to grow layers with $0.2 < x < 0.4$. Growth from a solution of $(Hg_{0.90}Cd_{0.10})_{0.175}Te_{0.825}$ led to an LPE film of final composition $Cd_{0.4}Hg_{0.6}Te$. Both (111) Cd and Te surfaces of CdTe were used as substrates, and layers about 16 μm thick were grown in 30 min. The as-grown layers showed only a slight variation in x by 0.02 from run to run; and in layers with $x = 0.21$ the variation in composition across the surface of the layers was as little as ± 0.001. All as-grown layers were *p*-type with a carrier density of 10^{17} cm^{-3} and a hole mobility of 220 cm^2 V^{-1} s^{-1} at 77 K. The layers were converted to *n*-type by annealing in Hg at 300° C. The carrier density was typically 4×10^{15} cm^3 at 77K with an electron mobility of 4660 cm^2 V^{-1} s^{-1}.

Mroczkowski and Vydyanath (155) first obtained liquidus data for Te-rich solutions by thermal analysis for values of x from 0.2 to 0.4. LPE layers were grown on (111) Cd faces of CdTe by tipping in a sealed capsule. The solution was homogenized at liquidus + 15° C, then tipped over the substrate for 15 minutes, and cooled by 5–7° C min^{-1} for 30–120 minutes. Films up to 40 μm thick were produced with the wider cooling ranges. The layers were smooth, and after annealing in Hg had electron mobilities at 77K of 1×10^5 cm^{2}V^{-1} s^{-1} for compositions with $x = 0.2$. Values of $2 + 10^4$ cm^{2}V^{-1}s^{-1} were obtained with $x = 0.3$. In contrast Wang *et al.* 156, following earlier work by Lanir *et al.* (157), who produced the first back wall CdHgTe-CdTe heterojunction photodiodes, used the dipping method in a sealed system pressurized with 13–20 atm of argon. Mirror-smooth surfaces were produced on (111) Cd faces of CdTe, provided the thickness did not exceed 10 μm. Facets and terraces developed on thicker films. On $(\bar{1}\bar{1}\bar{1})$ Te faces, voids occurred with a density which increased with thickness. The best layers on (111) Cd faces were mirror-smooth, the metallurgical interface with the CdTe was planar, and the major interdiffusion region was only 0.5 μm wide. The crystallographic properties of the epilayers were as good (or bad) as those of the substrate. This is common experience with all investigators, namely that the crystallographic defects in the substrate propagate into the epilayer. Undoped layers produced by this technique were also *p*-type, with a hole mobility of 400 cm^{2}V^{-1}s^{-1} at 77K. Layers annealed in Hg were *n*-type with mobilities of 1×10^5 cm^{2}V^{-1}s^{-1} at 77 K. The quality of the films is undoubtedly limited by the quality of the substrate material.

Before concluding this section it is worth mentioning that Saraie *et al.* (158)

successfully grew layers of CdTe by LPE from a solution of CdTe in $CdCl_2$. The CdTe-$CdCl_2$ system forms a eutectic with 74% $CdCl_2$ at 500 °C. There are no other compounds and the solubility of CdTe in $CdCl_2$ is large. The vapour pressure of $CdCl_2$ is about 10 Torr at 656° C, so that loss of the chloride was prevented by growing in a sealed system in the range 550–600° C. Cooling rates of 0.1–0.5° C min^{-1} and a total temperature change of 20° C led to mirror-smooth films. No other properties were reported.

Chemical purity of II–VI compounds

The chemical purity of bulk crystals and epitaxial layers of the II–VI compounds has increased appreciably in recent years. Chemical assessment of the impurity content, when attempted, has been carried out using emission or mass spectrographic techniques, although atomic absorption spectroscopy, as a means of analysing specific impurities, is becoming more popular. Limits of detection for most elements of interest lie between 0.1 and 1.0 ppma, so that the chemical techniques, although important for revealing large impurity concentrations (in semiconducting terms), are of little help in characterizing impurities which can be extremely effective in the range 10^{15}–10^{16} cm^{-3}, at or below the sensitivity limits of the chemical methods. Use has therefore increasingly been made of the electrical and luminescence properties of a material as a means of assessing and identifying the electrically- and optically-active impurities which are present in semiconducting concentrations.

The majority carrier mobility is often used as a measure of the purity of a sample, and the maximum mobility obtainable at low temperature is frequently taken as a criterion of crystal quality. Quoted maximum mobilities of all the II–VIs have increased steadily through the years.

Photoluminescence techniques, which are now used extensively to identify various shallow donor and acceptor impurities, have come into their own as the quality and purity of the crystals and epi-layers have improved. High resolution photoluminescence emission spectroscopy and magneto-optical studies allowed Merz *et al.* (159, 160) to identify several shallow donors and acceptors in ZnSe, and Henry *et al.* (161) carried out a similar investigation in CdS. Developments in photoluminescence excitation spectroscopy and selective photoluminescence excitation using dye tuned lasers (see for example 162–164) have led to improved resolution, and the measurement of higher excited states, so that more positive identification of donor and acceptor impurities has been obtained. The application of these techniques to ZnSe and other II–VI compounds, notably ZnTe, has recently been reviewed by Bhargava (165). See also the paper by Yao (133) for a discussion of photoluminescence in high-quality films of ZnSe grown by MBE.

In contrast with the other II–VI compounds discussed here, CdTe is a more completely conventional semiconductor, in that it displays amphoteric behaviour, can be zone refined, and grown from the melt without undue difficulty. The impurity content of crystals grown and treated in various ways is well documented (7). Segregation coefficients of a large number of elements between CdTe and its melt have been measured by several authors. Values have been tabulated by Zanio (7).

The resistivity of undoped, as-grown ZnSe, which is very high (in the range 10^{10}–10^{12} ohm cm), can be reduced dramatically to 10–100 ohm cm by prolonged heat treatment in molten zinc at 1000° C. This treatment, known as solvent extraction, was first described by Aven and Woodbury (166), and is thought to work by removing deep acceptors such as those formed by Cu or Ag impurities and zinc vacancies. Aven and Woodbury diffused radioactive ^{64}Cu and ^{110}Ag into ZnTe, ZnSe and CdS, and after heating in molten Zn or Cd measured the remaining activity. They found that the segregation coefficients of Ag and Cu between the molten elements and the compounds were between 2 and 3×10^{-3}. Aven (167) later achieved the highest electron mobility (530 $cm^2V^{-1}s^{-1}$) for ZnSe yet reported at room temperature, by repeatedly heating his samples in fresh, molten zinc.

Thomas *et al.* (168), investigating the possibility of producing low-resistivity ZnS by heating dice with millimetre dimensions in molten mixtures of zinc and aluminium as recommended by Oda and Kukimoto (169) and Kukimoto *et al.* (170), concluded that special precautions were necessary to prevent copper contaminating the ZnS at all times while it was held in fused quartz containers at temperatures of 1000° C and above. When this was done, and small quantities (1% or less) of Al and Ga were added to the melt, resistivities of 100 ohm cm were obtained reproducibly. Thomas *et al.* found no evidence for a reduction in copper content after heating in zinc. In fact the level of contamination increased slowly with prolonged heating. The source of the copper was attributed to the fused quartz tubes and ampoules used to grow and process the ZnS crystals. Only the highest grade silica-ware, such as Spectrosil, contained a low concentration of copper (0.002 ppm). Even so, copper contamination developed at a level of 1 ppm. It was thought that the copper originated from the Kanthal wire with which the furnace was wound (Kanthal contains 2000–3000 ppm Cu), and diffused through the furnace tube and the walls of the silica ampoule to contaminate the ZnS. A similar effect with CdS has been well documented by Russell and Woods (37).

References

Interested readers are referred to the following works on II–VI compounds:

M. Aven and J. S. Prener (eds.) *Physics and Chemistry of II–VI Compounds*. North-Holland, Amsterdam (1967).

B. Ray *II–VI Compounds*, Pergamon, Oxford (1969).

H. Hartmann, R. Mach and B. Selle, 'Wide gap II–VI compounds as electronic materials', in *Current Topics in Materials Science*, ed. E. Kaldis, North-Holland, Amsterdam (1982) 1–414.

1. R. Frerichs, *Phys. Rev.* **72** (1947) 594.
2. R. Lorenz, *Chem. Ber.* **24** (1891) 1509.
3. D. R. Stuhl and G. C. Sinke, Thermodynamic Properties of the Elements, *Advances in Chemistry Ser.* No. **18**, Am. Chem. Soc., Washington DC (1956).
4. P. Goldfinger and M. Jeunehomme, *Trans. Faraday Soc.* **59** (1963) 2851.
5. M. R. Lorenz, in *Physics and Chemistry of II–VI Compounds*, eds. M. Aven and J. S. Prener, North-Holland, Amsterdam (1967).
6. A. J. Strauss, *Proc. Int. Symp. CdTe for γ-ray Detectors*, eds. P. Siffert and A. Cornet, Strasbourg (1971).
7. K. Zanio, *Semiconductors and Semimetals*, Vol. 13, *CdTe*, Academic Press, New York (1978).
8. D. de Nobel, *Philips Res. Rpts.* **14** (1959) 361.
9. D. R. Mason and D. F. O'Kane, *Proc. Int. Conf. on Semiconductor Physics*, Prague (1962).
10. M. R. Lorenz, *J. Phys. Chem. Solids* **23** (1962) 939.
11. B. M. Kulwicki, PhD Dissertation, Univ. of Michigan, Ann Arbor, USA (1963).
12. J. Steiniger, A. J. Strauss and R. F. Brebrick, *J. Electrochem. Soc.* **117** (1970) 1305.
13. R. F. Brebrick, *J. Electrochem. Soc.* **118** (1971) 2014.
14. J. Carides and A. G. Fischer, *Sol. St. Comm.* **2** (1964) 217.
15. R. A. Reynolds, D. G. Strand and D. A. Stevenson, *J. Electrochem. Soc.* **114** (1967) 1281.
16. A. Reisman, M. Berkenblit and M. Witzen, *J. Phys. Chem.* **66** (1962) 2210.
17. L. R. Shiozawa and J. M. Jost, Rept No. ARL 62-365 (Mar. 1962); Aerospace Res. Labs. USAB Rept No. ARL 65–98 (May 1968).
18. L. R. Shiozawa, F. Augustine, G. A. Sullivan, J. M. Smith and W. R. Cole, Aerospace Res. Lab. Contract Report ARL 69-0155, USAF, Wright-Patterson Air Force Base, Ohio (1969).
19. A. G. Fischer: *See* M. Aven and J. S. Prener, *Physics and Chemistry of II–VI Compounds*, North-Holland, Amsterdam (1967), 85.
20. H. H. Woodbury, *J. Phys. Chem. Solids* **24** (1963) 881.
21. A. Addamiano and P. A. Dell, *J. Phys. Chem.* **61** (1957) 1020.
22. H. H. Woodbury and R. B. Hall, *Phys. Rev.* **157** (1867) 641.
23. S. A. Medvedev, Yu. V. Klevkov, K. V. Kiseleva and N. N. Sentyurina, *Inorganic Mats.* **8** (1972) 1064.
24. F. T. J. Smith, *Metallurgy. Trans.* **1** (1970) 617.
25. R. A. Brebrick and A. J. Strauss, *J. Phys. Chem. Sols.* **25** (1964) 1441.
26. R. F. Burmeister and D. A. Stevenson, *J. Electrochem. Soc.* **114** (1967) 394.
27. H. W. Leverenz, *An Introduction to Luminescence of Solids*, John Wiley, New York (1950).
28. A. Libicky, II–VI *Semiconducting Compounds, Int. Conf.*, Brown Univ., R. I. Providence, ed. D. G. Thomas, W. A. Benjamin, New York (1967) 389.
29. D. C. Reynolds and S. J. Czyzack, *Phys. Rev.* **79** (1956) 543.
30. L. C. Greene, D. C. Reynolds, S. J. Czyzack and W. M. Baker, *J. Chem. Phys.* **29** (1958) 1375.
31. W. W. Piper and S. J. Polich, *J. Appl. Phys.* **32** (1961) 1278.
32. M. Aven, D. T. F. Marple and B. Segall, *J. Appl. Phys.* **32** (1961) 2261.
33. M. Aven and B. Segall, *Phys. Rev.* **130** (1963) 87.
34. L. Clark and J. Woods, *Brit. J. Appl. Phys.* **17** (1966) 319.
35. L. Clark and J. Woods, *J. Cryst. Growth* **3/4** (1968) 126.
36. W. M. de Meis and A. G. Fischer, *Mats. Res. Bull.* **2** (1967) 465.
37. G. J. Russell and J. Woods, *J. Cryst. Growth* **47** (1979) 647.
38. P. Buik and R. Nitsche, *J. Cryst. Growth* **48** (1980) 29.
39. M. M. Faktor and J. Garrett, *Growth of Crystals from the Vapour*, Chapman & Hall, London (1974).
40. P. D. Fochs, W. George and P. O. Augustus, *J. Cryst. Growth* **3/4** (1968) 122.
41. G. H. Dierssen and T. Gabor, *J. Cryst. Growth* **43** (1978) 572.
42. A. C. Prior, *J. Electrochem. Soc.* **108** (1961) 82.
43. K. F. Burr and J. Woods, *J. Cryst. Growth* **9** (1971) 183.
44. T. Kiyosawa, K. Igaki and N. Ohashi, *Trans. Jap. Inst. Mat.* **13** (1972) 248.
45. K. Igaki and K. Mochizuki, *J. Cryst. Growth* **24/5** (1974) 1962.
46. K. Igaki, N. Ohashi and K. Mochizuki, *Jap. J. Appl. Phys.* **15** (1976) 1429.
47. K. Mochizuki, *J. Cryst. Growth* **53** (1981) 355.
48. M. M. Faktor, J. Heckingbottom and J. Garrett, *J. Chem. Soc. A* (1970) 2657.
49. M. M. Faktor, J. Heckingbottom and J. Garrett, *J. Chem. Soc. A* (1971) 1.

50. D. W. G. Ballentyne, S. Wetwatana and E. A. D. White, *J. Cryst. Growth* **7** (1970) 79.
51. D. W. G. Ballentyne, M. Rouse and E. A. D. White, *J. Cryst. Growth* **34** (1976) 49.
52. M. Toyama, *Jap. J. Appl. Phys.* **5** (1966) 1204.
53. M. Toyama and T. Sekiva, *Jap. J. Appl. Phys.* **7** (1969) 855.
54. D. W. G. Ballentyne, *Prog. Cryst. Growth & Char.* **6** (1983) 163.
55. R. Nitsche, *J. Phys. Chem. Solids* **17** (1960) 163.
56. H. Hartmann, *J. Cryst. Growth* **42** (1977) 144.
57. E. Lendvay, *J. Cryst. Growth* **10** (1971) 77.
58. S. Fujita, H. Mumuto, H. Takebe and T. Noguchi, *J. Cryst. Growth* **47** (1979) 326.
59. W. Palosz, *J. Cryst. Growth* **58** (1982) 185.
60. B. Bendow and P. D. Gianini, *Optical Comm.* **9** (1973) 306.
61. K. L. Lewis and J. Hill, *Proc. 7th Int. Conf. on Chem. Vapour Deposition* (1980) 629.
62. K. L. Lewis, B. J. Cook and P. B. Roscoe, *J. Cryst. Growth* **56** (1982) 614.
63. H. L. Tuller, K. Uematsu and H. K. Bower, *J. Cryst. Growth* **42** (1977) 150.
64. A. G. Fischer, *J. Electrochem. Soc.* **117** (1970) 410.
65. J. B. Mullin and B. W. Straughan, *Rev. Phys. Appl.* **12** (1977) 105.
66. R. Triboulet and H. Rodot, *C.R. Acad. Sci. Paris, Ser. B* **266** (1968) 498.
67. R. Triboulet and Y. Marfaing, *J. Electrochem. Soc.* **120** (1973) 1260.
68. N. R. Kyle, *Proc. Int. Symp. on CdTe*, eds. P. Siffert and A. Cornet, Strasbourg (1971).
69. M. R. Lorenz and R. E. Halsted, *J. Electrochem. Soc.* **110** (1963) 343.
70. A. Cornet, P. Siffert, A. Coche and R. Triboulet, *Appl. Phys. Letts.* **17** (1970) 432.
71. H. H. Woodbury and X. Lewandowski, *J. Cryst. Growth* **10** (1971) 6.
72. Q. A. Matveev, S. V. Prokof'ev and Yu. V. Rud, *Inorg. Mats.* **5** (1969) 1000.
73. S. V. Prokof'ev and Yu. V. Rud, *J. Cryst. Growth* **6** (1970) 187.
74. R. O. Bell, M. Hammatt and F. Wald, *Phys. Stat. Sol.* **1** (1970) 375.
75. R. O. Bell and F. Wald, *IEEE Trans. Nucl. Sci.* **19** (1972) 334.
76. F. V. Wald and R. O. Bell, *J. Cryst. Growth* **30** (1978) 29.
77. R. Triboulet, Y. Marfaing, A. Cornet and P. Siffert, *Nature (London) Phys. Sci.* **245** (1973) 12.
78. R. Triboulet, Y. Marfaing, A. Cornot and P. Siffert, *J. Appl. Phys.* **45** (1974) 2759.
79. T. Taguchi, J. Shirafugi and Y. Inuishi, *Rev. Phys. Appl.* **12** (1977) 117.
80. R. Triboulet, *Rev. Phys. Appl.* **12** (1977) 123.
81. T. Taguchi, S. Fujita and Y. Inuishi, *J. Cryst. Growth* **30** (1978)
82. M. R. Lorenz, *J. Appl. Phys.* **33** (1962) 3304.
83. B. Lunn and V. Bettridge, *Rev. Phys. Appl.* **22** (1977) 151.
84. J. B. Mullin, C. A. Jones, B. W. Straughan and A. Royle, *Cryst. Growth* **59** (1982) 135.
85. J. de Klerk and R. F. Kelly, *Rev. Sci. Inst.* **36** (1965) 507.
86. A. M. Andrews and C. R. Haden, *Proc. IEEE* **57** (1969) 99.
87. J. Dresner and F. V. Shallcross, *Solid State Electron* **5** (1962) 205.
88. P. K. Weimer, *Phys. of Thin Films* **2** (1964) 147.
89. E. Schwarz, *Nature (London)* **162** (1948) 614.
90. D. Bonnet, *Proc. Photovoltaic Solar Energy Conf.*, Luxembourg, D. Reidel, Dordrecht, Holland (1977) 635.
91. P. K. Weimer, *Proc. IRE* **50** (1962) 1462.
92. D. A. Cusano, *Physics and Chemistry of II–VI Compounds*, eds. M. Aven and J. S. Prener, North-Holland, Amsterdam (1967) 706.
93. P. L. Jones, C. N. W. Litting, D. E. Mason and V. A. Williams, *J. Phys. D. Appl. Phys.* **1** (1968) 283.
94. P. M. R. Kay, P. Lilley and C. N. W. Litting, *J. Phys. D.: Appl. Phys.* **7** (1979) 1206.
95. F. V. Shallcross, *RCA Rev.* **27** (1967) 572.
96. H. Hartmann, *J. Cryst. Growth* **31** (1975) 323.
97. J. B. Mullin, S. J. C. Irvine and D. J. Ashen, *J. Cryst. Growth* **55** (1981) 92.
98. L. R. Koller and H. D. Coghill, *J. Electrochem. Soc.* **107** (1960) 973.
99. J. I. B. Wilson and J. Woods, *J. Phys. Chem. Solids* **34** (1973) 171.
100. N. F. Foster, *J. Appl. Phys.* **38** (1967) 149.
101. H. J. Hovel and A. G. Milnes, *J. Electrochem. Soc.* **116** (1969) 843.
102. A. Yoshikawa and Y. Sakai, *J. Appl. Phys.* **45** (1974) 3521.
103. P. Vohl, W. R. Buchan and J. E. Genthe, *J. Electrochem. Soc.* **118** (1971) 1842.
104. M. Kitegawa, J. Saraie and T. Taneka, *J. Cryst. Growth* **45** (1978) 198.

105. W. M. Yim and E. J. Stofko, *J. Electrochem. Soc.* **119** (1972) 381.
106. P. O. Besomi and B. W. Wessels, *J. Cryst. Growth* **55** (1981) 477.
107. N. Matsuda and K. Akasaki *J. Cryst. Growth* **45** (1978) 192.
108. M. D. Scott, J. O. Williams and R. C. Goodfellow, *J. Cryst. Growth* **51** (1981) 1981.
109. M. D. Scott, J. O. Williams and R. C. Goodfellow, *Thin Solid Films* **72**, (1980) L1.
110. H. M. Mansevit and W. I. Simpson, *J. Electrochem. Soc.* **118** (1971) 644.
111. W. Stutius, *Appl. Phys. Lett.* **33** (1978) 656.
112. P. Blanconnier, M. Cerclet, P. Henoc and A. M. Jean-Louis, *Thin Solid Films* **55** (1978) 375.
113. W. Stutius, *J. Electron. Mat.* **10** (1981) 95.
114. W. Stutius, *Appl. Phys. Lett.* **40** (1982) 246.
115. W. Stutius, *Appl. Phys. Lett.* **38** (1981) 352.
116. W. Stutius, *J. Appl. Phys.* **53** (1982) 284.
117. P. J. Wright and B. Cockayne, *J. Cryst. Growth* **59** (1982) 148.
118. S. Fujita, Y. Matsuda and A. Sadaki, *J. Cryst. Growth* **68** (1984) 231.
119. S. Fujita, T. Yodo and A. Sasaki, *J. Cryst. Growth* **72** (1985) 27.
120. G. Fan and J. O. Williams, *Mats. Lett.* **3** (1985) 453.
121. P. J. Wright, R. J. M. Griffiths and B. Cockayne, *J. Cryst. Growth* **66** (1984) 26.
122. B. Cockayne, P. J. Wright, M. S. Skolnick, A. D. Pitt, J. O. Williams and T. L. Ng, *J. Cryst. Growth* **72** (1985) 17.
123. J. B. Mullin, S. J. C. Irvine, J. Giess and A. Royle, *J. Cryst. Growth* **72** (1985) 1.
124. P. I. Kuznetzov. V. V. Shemet, I. N. Odin and A. V. Novoselova, *Dokl. Akad. Nauk. SSSR* **248** (1979) 879.
125. S. J. C. Irvine, and J. B. Mullin, *J. Cryst. Growth* **55** (1981) 107.
126. A. Y. Cho, *J. Vac. Tech.* **8** (1971) 531.
127. T. Yao, S. Amano, Y. Makita, and S. Maekawa, *Jap. J. Appl. Phys.* **15** (1976) 1001.
128. T. Yao, Y. Makita and S. Maekawa, *Jap. J. Appl. Phys.* **16** (1977) (Suppl.) 451.
129. T. Yao, Y. Makita and S. Maekawa, *J. Cryst. Growth* **45** (1978) 309.
130. T. Yao, Y. Makita and S. Maekawa, *Appl. Phys. Lett.* **35** (1979) 98.
131. T. Yao and S. Maekawa, *J. Cryst. Growth* **53** (1981) 423.
132. T. Yao, M. Ogura, S. Matsuoka and T. Morishita, *Jap. J. Appl. Phys.* **22** (1983) L114.
133. T. Yao, *J. Cryst. Growth* **72** (1985) 31.
134. T. Yao, T. Minato and A. Maekawa, *J. Appl. Phys.* **53** (1982) 4236.
135. F. Kitagawa, T. Mishina and K. Takahashi, *J. Electrochem. Soc.* **127** (1980) 937.
136. T. Mishima, W. Quan-kun and K. Takahaski *J. Appl. Phys.* **52** (1981) 5797.
137. J. P. Faurie and A. Million, *J. Cryst. Growth* **54** (1981) 577.
138. J. P. Faurie and A. Million, *J. Cryst. Growth* **54** (1981) 582.
139. J. P. Faurie, A. Million and G. Jacquier, *Thin Solid Films* **90** (1982) 107.
140. J. P. Faurie, A. Million and J. Piaguet, *J. Cryst. Growth* **59** (1982) 10.
141. R. F. C. Farrow, G. R. Jones, G. M. Williams and I. M. Young, *Appl. Phys. Lett.* **39**
142. M. Rubenstein, *J. Electrochem. Soc.* **113** (1966) 623.
143. M. Rubenstein, *J. Cryst. Growth* **34** (1968) 309.
144. R. Widmer, D. P. Bortfield and H. P. Kleinkrecht, *J. Cryst. Growth* **6** (1970) 237.
146. A. Kanamori, T. Ora and K. Takahashi, *J. Electrochem. Soc.* **122** (1975) 1177.
147. A. V. Simashovich and R. L. Tsuilyana, *J. Cryst. Growth* **35** (1976) 269.
148. S. Fujita, H. Mimoto and T. Noguchi, *J. Cryst. Growth* **45** (1978) 281.
149. B. J. Fitzpatrick, R. N. Bhargawa, S. P. Herko and P. M. Harnach, *J. Electrochem. Soc.* **128** (1979) 341.
150. K. Kosai, B. J. Fitzpatrick, H. G. Grimmeiss, R. N. Bhargava and G. F. Neumark, *Appl. Phys. Lett.* **35** (1979) 1979.
151. C. Werkhoven, B. J. Fitzpatrick, S. P. Herko, R. N. Bhargava and P. J. Dean, *Appl. Phys. Lett.* **38** (1981) 540.
152. B. J. Fitzpatrick, C. J. Werkhoven, T. F. McGee, P. M. Hanack, S. P. Herko, R. N. Bhargava and P. J. Dean, *IEEE Trans.* **ED–28** (1981) 440.
153. J. L. Schmit and J. E. Bowers, *Appl. Phys. Lett.* **35** (1979) 457.
154. J. L. Schmit, R. J. Hagen and R. A. Wood, *J. Cryst. Growth* **56** (1982) 485.
155. J. A. Mroczkowski and R. Vydyanath, *J. Electrochem. Soc.* **128** (1982) 655.
156. C. C. Wang, S. H. Shin, M. Chu, M. Lanir and A. H. B. Vanderwyck, *J. Electrochem. Soc.* **12** (1980) 175.

157. M. Lanir, C. C. Wang and A. H. B. Vanderwyck, *Appl. Phys. Lett.* **34** (1979) 50.
158. J. Saraie, M. Kitagawa, M. Ishiba and T. Taneka, *J. Cryst. Growth* **43** (1978) 13.
159. J. L. Merz, H. Kukimoto, K. Nassau, and J. W. Shiever, *Phys. Rev.* **B6** (1971) 545.
160. J. L. Merz, K. Nassau and J. W. Shiever, *Phys. Rev.* **B8** (1972) 1444.
161. C. H. Henry, K. Nassau and J. W. Shiever, *Phys. Rev.* **B4** (1971) 2453.
162. H. Tews and H. Venghaus, *Solid State* **45** (1979) 204.
163. H. Tews, H. Venghaus and P. J. Dean, *Phys. Rev.* **B19** (1979) 5178.
164. P. J. Dean, D. C. Herbert, C. J. Werkhoven, B. J. Fitzpatrick and R. N. Bhargava, *Phys. Rev.* **B23** (1981) 4888.
165. R. M. Bhargava, *J. Cryst. Growth* **59** (1982) 15.
166. M. Aven and H. H. Woodbury, *Ann. Phys. Lett.* **1** (1962) 53.
167. M. Aven, *J. Appl. Phys.* **42** (1971) 1204.
167. M. Aven, *J. Appl. Phys.* **42** (1971) 1204.
168. A. E. Thomas, G. J. Russell and J. Woods, (1983) *J. Cryst. Growth* **63** (1983) 265.
169. S. Oda and H. Kukimoto, *IEEE Trans.* **ED-24** (1977) 956.
170. H. Kukimoto, S. Oda and T. Nakayama, *J. Luminescence* **18/19** (1979) 365.

5 Chemical vapour deposition

P. JOHN

Introduction

Chemical vapour deposition, CVD, is an essential process in the fabrication of VLSI circuits (1). The deposited layers fulfil many functions, and range, in chemical composition, from metals and alloys through to semiconductors and insulators (2). These materials provide conductive links within the device architecture, maintain electrical insulation between layers, or act as protective coatings during etching or subsequent processing. There is a close relationship between device physics, circuit design, the desired electrical properties of the layers and the chemistry underlying the deposition process. Despite the importance of the latter, the basic chemistry of CVD process is not fully understood. Empiricism has largely prevailed. One of the purposes of this chapter is to amalgamate knowledge of relevant gas-phase reactions, the mechanism of film growth and the structure and morphology of the resultant film. In this way it is hoped to present a more coherent picture of CVD processes with particular emphasis on silicon deposition.

In essence, CVD is a remarkably simple yet versatile technique in which the decomposition of a gaseous precursor, either homogeneously or heterogeneously, produces a solid thin film. Any controllable chemical reaction which deposits a thin film on a substrate would suffice. Indeed, several reaction types have been utilized in semiconductor device fabrication (2) and other coating applications (3). These include thermal decomposition, oxidation and reduction, hydrolysis and ammoniolysis, polymerization and processes which depend on physical mass transport. The microelectronics industry relies predominantly on thermal decomposition at high temperatures for the deposition of epitaxial silicon, polycrystalline and amorphous silicon, silicon dioxide and stoichiometric silicon nitride. The requirement for lower-temperature processing conditions has resulted in the development of plasma-enhanced CVD (see Chapter 6) and photolytic CVD as a means of producing hydrogenated amorphous silicon and related alloys. Recent developments in CVD metallization are encouraging, although they have not yet replaced sputtering or other physical transport processes in high-volume production.

A large body of literature and reviews (2) exists on the topic of this chapter and is regularly supplemented by the proceedings of conferences devoted to CVD technology (4). A comprehensive bibliography (5) of more than 5000 references on CVD and vapour transport processes covering the period 1960–

1980 has been published. In addition, a number of textbooks on microelectronics (6) include chapters on CVD and related techniques.

Scope of chemical vapour deposition

Historically, high-purity metals were prepared by CVD involving the corresponding metal halides, the technique stemming from the work of van Arkel and de Boer (7). The technology was subsequently applied to silicon by Holbing (8). Refinements of this process resulted in the so-called Siemens process (9) in which ultrapure polycrystalline silicon is deposited on to thin silicon rods at 1000–1200° C from the reduction of $SiHCl_3$ in an H_2 diluent. The manufacture of polycrystalline silicon ingots represents the largest single application of CVD; subsequent crystal growth by the Czochralski or the Float Zone method produces the ubiquitous single-crystal silicon wafer (10) (see Chapter 2).

Current CVD techniques for film growth have developed from processes carrried out at atmospheric pressure using thermal heating for reactant dissociation. A number of variations have emerged with the obligatory, although non-standardized, acronyms. These are listed in Table 5.1. The technology of these CVD processes encompasses a wide range of conditions of temperature, pressure and flow rates, each parameter being carefully controlled to produce uniform, coherent and defect-free films exhibiting reproducible characteristics. Substrate temperatures can be as low as 100° C (PECVD) and may extend to above 1000° C (APCVD). Process pressures range from atmospheric to *c*. 0.05 Torr in some LPCVD applications. Traditional thermal dissociation techniques have been supplemented by the use of RF power or photolytic sources. Development work in the use of lasers in materials research is active in many laboratories, although integration into high-volume production has not been accomplished.

This chapter is restricted to thin film deposition processes which are of importance to VLSI fabrication. Emphasis will be placed on the CVD of polycrystalline silicon, amorphous silicon, silicon dioxide and silicon nitride. Reference will be made to vapour phase epitaxy, VPE, in the context of the CVD of the various forms of silicon.

Table 5.1 Chemical vapour deposition (CVD) techniques

Process	Acronym	Energy source
Atmospheric pressure CVD	APCVD	Thermal
Low-pressure CVD	LPCVD	Thermal
Homogeneous CVD	HOMOCVD	Thermal
Metallo-organic CVD	MOCVD	Thermal
Plasma-enhanced CVD	PECVD	RF
Photolytic CVD	UVCVD	Photons
Laser CVD	LCVD	Photons

Whilst silicon processing is relatively well understood and the inherent advantages of silicon chemistry ensure its dominant position for some time, compound semiconductors are emerging as potential rivals for certain applications. The characteristics of III–V semiconductors, notably GaAs, have been known for some time. With the advent of metallo-organic chemical vapour deposition (11), MOCVD, the importance of this technique to the growth of compound semiconductors has grown considerably. Whilst the technique has its roots in conventional CVD, it is worthy of a separate discussion (see Chapter 7).

Improvement in the efficiency, reliability and reproducibility of CVD has resulted from a largely empirical approach. Only recently has a molecular description of the growth processes in polycrystalline and amorphous silicon been tentatively proposed (12). This is indeed fertile ground for further work. Mechanistic evidence of gas-phase reactions, nucleation processes and film growth may now be subsumed within a description of the overall CVD process which, in the past, has concentrated on gas dynamic calculations of temperature, pressure and flow profiles in a horizontal (13) or vertical (14) CVD reactor. The aim of the later sections is to attempt to identify the problems that still remain in CVD processes and, in particular, address the mechanism of *in-situ* doping of polycrystalline silicon (15, 16).

Finally, the applications of laser CVD will be described briefly in the context of the deposition of materials of importance to the microelectronics industry.

Interrelationship of CVD techniques

Epitaxial silicon

Epitaxy is a process in which a crystalline layer is formed on an underlying crystalline substrate. The term is a transliteration of the Greek words *epi*, meaning upon, and *taxis*, meaning ordered. When the deposited layer is the same material as the substrate the process is called 'homotaxy', and when they differ the process is called 'heterotaxy'. During homotaxial growth the layer adopts the crystalline structure of the substrate, and so the crystal lattices of the deposited layer and substrate need to be similar for successful heterotaxy.

Epitaxial silicon layers are utilized in bipolar integrated circuits, junction field-effect transistors and in VMOS technology (1). They are used in preference to bulk single-crystal silicon since the epitaxial layers are less contaminated by oxygen and carbon which are incorporated during crystal growth. Further, *in-situ* doping is more flexible than either ion-implantation or diffusion-source doping techniques which are standard for Si wafers.

In general, the growth of epitaxial layers of silicon by CVD is performed at atmospheric pressure and at temperatures in the range 950–1250° C. Four silicon precursors are commonly employed to grow expitaxial silicon. Silicon

Table 5.2 Epitaxial growth of silicon in hydrogen atmosphere

Chemical deposition	Nominal growth rate (μm/min)	Temperature range (°C)
$SiCl_4$	0.4–1.5	1150–1250
$SiHCl_3$	0.4–2.0	1100–1200
SiH_2Cl_2	0.4–3.0	1050–1150
SiH_4	0.2–0.3	950–1050

tetrachloride, $SiCl_4$, is the preferred industrial source. Table 5.2 lists each, together with the temperature ranges over which the homotaxy of silicon is performed. Details of the type of reactor are deferred until later (p. 106).

Figure 5.1 illustrates the dependence of the growth rates on temperature (17) for the thermal decomposition of each source. Two regimes can be discerned in this figure. At the highest temperatures, the growth rate saturates, and the rate-determining step is the diffusion of the active species towards and away from the wafer surface. In this regime the growth rate is diffusion- or mass-transport-limited, and the slight increase in rate is due to the temperature dependence of the diffusion coefficients. At lower temperatures, gas-phase and surface kinetics determine the deposition rate. For epitaxial growth at atmospheric pressure, the process is thus normally operated under diffusion-limited conditions to minimize the influence of temperature. Whilst atmospheric-pressure CVD has been preferred, operation at reduced pressures (50–100 Torr) reduces autodoping (18) and pattern shift (19). Hydrides of Group III and V elements are employed as dopant sources, e.g. PH_3, B_2H_6 or AsH_3. These gases are diluted in H_2 (20–200 ppm) and metered into the CVD reactor. Further technical details of vapour-phase epitaxy of silicon are comprehensively covered elsewhere (1).

Epitaxial growth of silicon has tended to be described separately from the other CVD techniques. A broad perspective of hydrogenated silicon deposition can be gained from Figure 5.2 where the differing structural forms are related to the deposition conditions (20). This figure is not a thermodynamic phase diagram but portrays the temperature regions over which the growth of epitaxial, polysilicon and amorphous silicon can be experimentally observed. At the highest temperatures, the boundary between epitaxial and polycrystalline silicon growth is a function of both the temperature and growth rates (21), as shown in Figure 5.3. The line represents the maximum growth rate, at a given temperature, that pertains to the deposition of polycrystalline silicon. The activation energy is around 5 eV and is related to the diffusion of Si atoms at the surface. Since VPE is not carried out under uhv conditions, in contrast to molecular beam epitaxy, the value of the activation energy may reflect the surface properties and degree of contamination.

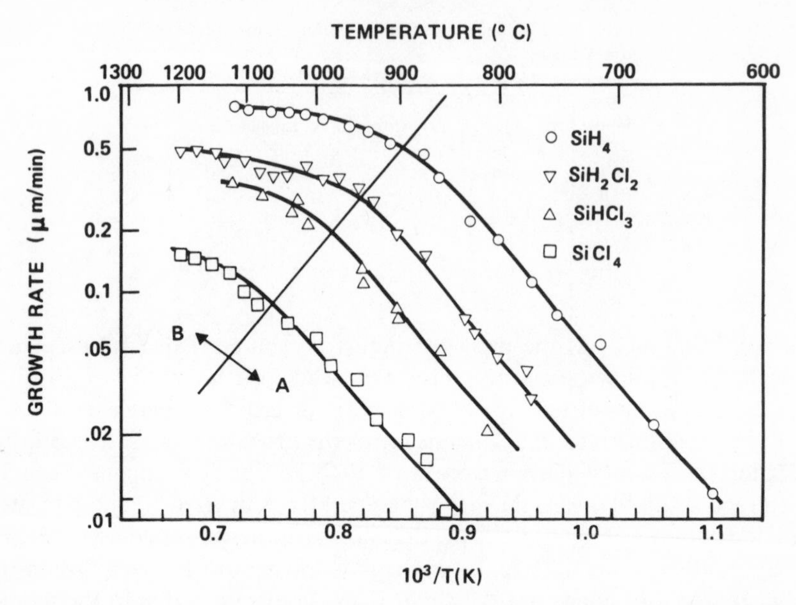

Figure 5.1 Temperature dependence of silicon growth rates for various precursors. After (17)

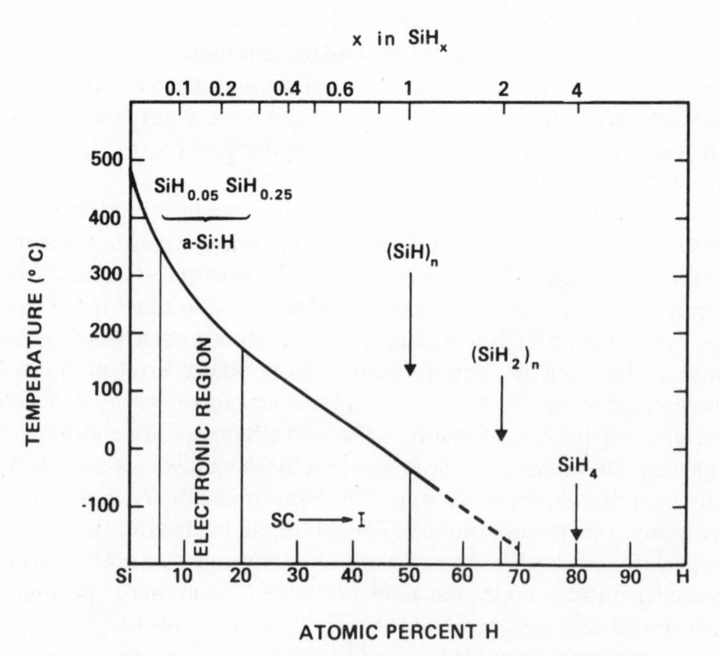

Figure 5.2 Temperature–chemical composition diagram. After (25)

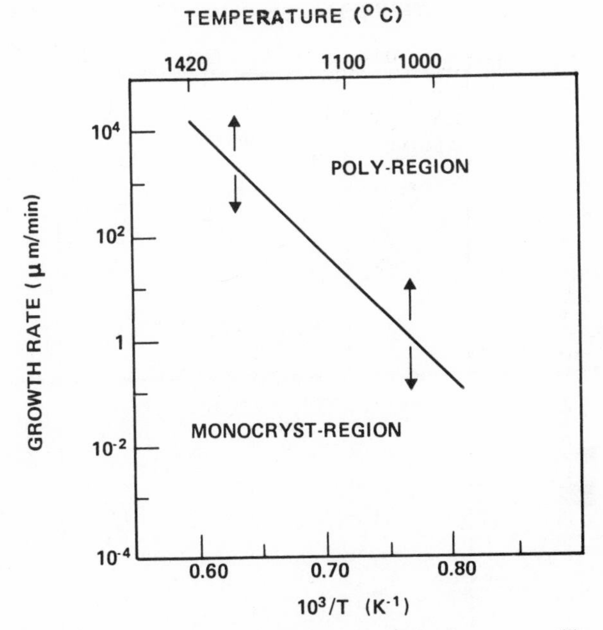

Figure 5.3 Illustration of the boundary between the growth of monocrystalline and polycrystalline silicon. After (21)

Post-deposition thermal annealing yields evidence as to the microstructural changes in the bulk material at the crystalline–polycrystalline boundary. On thermal annealing polysilicon, grain enlargement occurs almost exclusively in the $\langle 110 \rangle$ orientation (22). The grain boundaries are parallel to each other but are not orthogonal to the substrate surface.

As the temperature is lowered, the degree of crystallinity is reduced until the amorphous state is attained. The conditions for this phase change are shown in Figure 5.4. Amorphous silicon possesses no long-range structural order (23). The translational periodicity that characterizes a crystalline lattice is absent. X-ray and electron diffraction patterns of amorphous solids are broad, without distinct rings or spots. Of course, the same effect can be caused by the material comprising small (< 50 Å) crystallites embedded in an amorphous network. There is little ambiguity in the case of silicon (24) since the (220) and (311) Bragg reflections are completely absent. These reflections are replaced by a halo roughly midway between the crystalline peaks. At even lower temperatures, the major structural feature is the incorporation of covalently-bound hydrogen in the silicon network (25). This has a profound effect on the mechanical, optical and electrical properties of the material (26). The CVD of hydrogenated amorphous silicon, a-Si:H, is almost entirely the province of plasma techniques. Thermal CVD yields growth rates at these temperatures

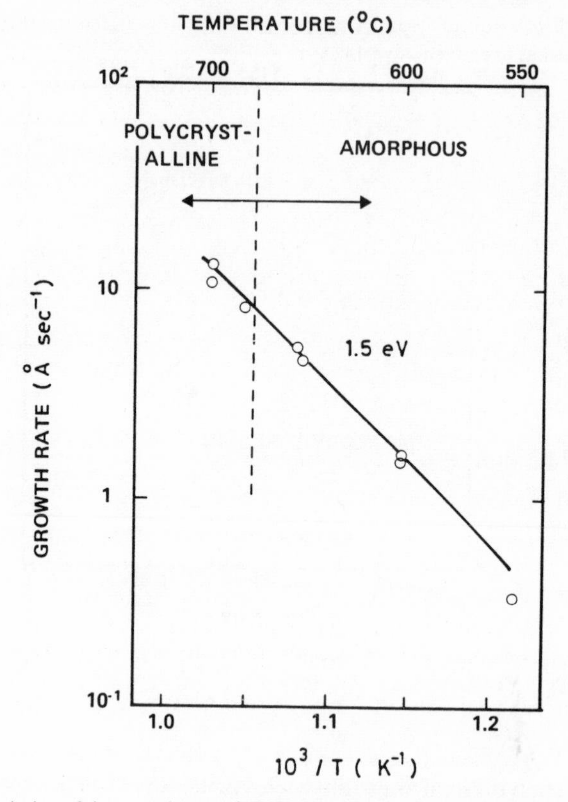

Figure 5.4 Variation of the growth rate of CVD amorphous silicon with temperature. After (81)

which are too low for commercial exploitation even using higher silanes (27). However, Scott and co-workers (25) have elucidated the composition of a-Si:H films deposited at low temperatures using homogeneous chemical vapour deposition. By employing substrate temperatures which differ from the gas-phase temperature, the substrate temperature can be maintained at ambient or sub-ambient. Under these conditions, films have been deposited with H-contents aproaching that of polysilane ($-(SiH_2)_n-$) chains (28). Similarly high H-contents have been observed in PECVD a-Si:H films prepared at low substrate temperatures.

It is at the lower temperatures that gas-phase reactions play a more dominant role and the growth rates are kinetically controlled rather than transport-limited.

Polycrystalline silicon

Polycrystalline silicon, or polysilicon, is employed as a gate electrode in MOS devices, as conducting material for multilevel metallization and as a contact material for shallow device junctions. Dopants can be introduced by diffusion from a dopant-rich layer, ion-implantation or *in situ* through the use of volatile precursors (29–34) such as B_2H_6 or PH_3. Heavily doped *n*-type polysilicon has been widely used as a gate electrode because of its ability to withstand high-temperature processing. In FET devices, the polysilicon gate is deposited over a thin (250–500 Å) silicon dioxide layer, and therefore the polysilicon is doped after deposition.

Polysilicon is universally deposited from the thermal decomposition of monosilane, SiH_4, in the temperature range 600–650° C. Neat silane is employed for LPCVD at 0.2–1 Torr pressure or as a 20–30% mixture in an inert gas such as N_2 at the same total pressure. Both variants produce 100–200 wafers per hour with deposition rates of 100–200 Å min^{-1} with a radial thickness uniformity better than 5%.

Polysilicon comprises single crystallites (22, 35, 36), of varying grain size (<0.1–$10\,\mu m$) separated by irregular grain boundaries. Films grow with a preferred ⟨110⟩ grain orientation at a substrate temperatures $>800°$ C with lesser preference below this temperature. Thick polysilicon films exhibit columnar structure (cf. PECVD a-Si:H films) orientated in the ⟨100⟩ direction. Thin films deposited at lower temperatures have smaller grain sizes which are more randomly orientated.

Amorphous silicon

Silicon deposited from APCVD or in a LPCVD system below *c.* 575° C is amorphous with non-discernible structure (36, 37) at a resolution of *c.* 50 Å. Annealing amorphous or polysilicon at $\sim 1050°$ C results in crystallization. The resulting grain size after annealing is a function of the annealing time and temperature and of dopant concentration (1). Ultimately the grain size reaches between 1 and 3 μm.

The transition from the truly amorphous state to a polysilicon structure is influenced by the experimental parameters of the deposition system, namely, growth rate, partial pressure of H_2, total pressure, presence and nature of dopants and impurities. In a LPCVD reactor this transition occurs between 575–625° C, whereas for as-prepared films grown at 600° C, annealing at 675–690° C is required.

Janai *et al.* (24) have reviewed a-Si:H films prepared by APCVD and provide a comprehensive bibliography. Kaplan (37) has reviewed the physical properties of CVD a-Si:H in comparison with the material produced by PECVD (38).

Hydrogenated amorphous silicon

Deposition of a-Si:H at low temperatures is confined to PECVD, homogeneous CVD, laser-assisted CVD or the CVD of higher homologues of SiH_4, e.g. Si_2H_6 (27). In this material significant amounts of hydrogen are present as SiH, SiH_2 or bound within $-(SiH_2)_n-$ chains and has been observed by secondary ion mass spectroscopy (SIMS) (39), nuclear reactions (20, 40), nuclear magnetic resonance (20) and infrared spectroscopy (20). The ensuing chemical complexity alters the structure and properties of the material (37). The composition, morphology and structure of a-Si:H, especially with regard to PECVD-produced material, has been extensively studied (38). Material produced by the HOMOCVD technique has revealed structural similarities (25) with a-Si:H produced by PECVD (38). Although some differences in the composition of films prepared by the two methods have been noted, the gross features are similar. This has led Kaplan (37) to suggest that hydrogen microstructures are intrinsic to amorphous structures derived from tetrahedrally-bonded materials.

Fabrication techniques

Industrial CVD equipment comprises a reactor plus an automatic gas delivery and control system. In the case of APCVD, many different types of reactor design have been described, whilst LPCVD systems have adopted a more standardized configuration.

Reactors designed to operate at close to atmospheric pressure are subdivided into high- and low-temperature varieties. High-temperature APCVD reactors are further classified into hot-wall or cold-wall types. Virtually all the high-temperature CVD reactors, including VPE reactors, employ the cold-wall variety, as shown in Figure 5.5. The substrate lies in thermal contact with a graphite or SiC-coated graphite susceptor which is heated by RF induction. Alternatively, a polysilicon or quartz susceptor can be heated with radiant heaters. In this way the susceptor and its immediate surroundings are the only parts which are at the elevated temperature.

Low-temperature reactors, for polysilicon deposition, normally employ a hot-wall tubular arrangement, shown in Figure 5.6, in which the gas flow can either be horizontal or vertical. Continuous flow reactors employ the geometrical arrangements shown in Figure 5.6. In each case resistively-heated elements are used to produce a particular axial temperature profile. Commercial APCVD reactors normally adopt a three-stage furnace which produces good axial uniformity in the deposited film thickness.

Low-pressure reactor designs are illustrated in Figure 5.7. The large packing densities achievable in LPCVD reactors are possible since at the low pressures adopted (0.1–0.5 Torr) the diffusion rate is sufficiently rapid that the deposition is not mass-transport-limited. The pressure dependence of the diffusion coefficient, D_T, is

$$D_{T,R} = D_{T,A} p_A / p_R$$

where the subscripts refer to atmospheric (A) and reduced (R) pressure

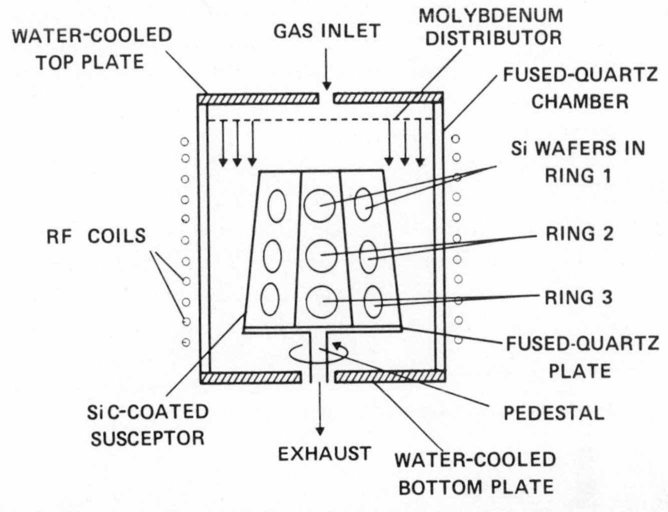

Figure 5.5 Schematic diagram of an RF-heated high-temperature CVD reactor. After (2)

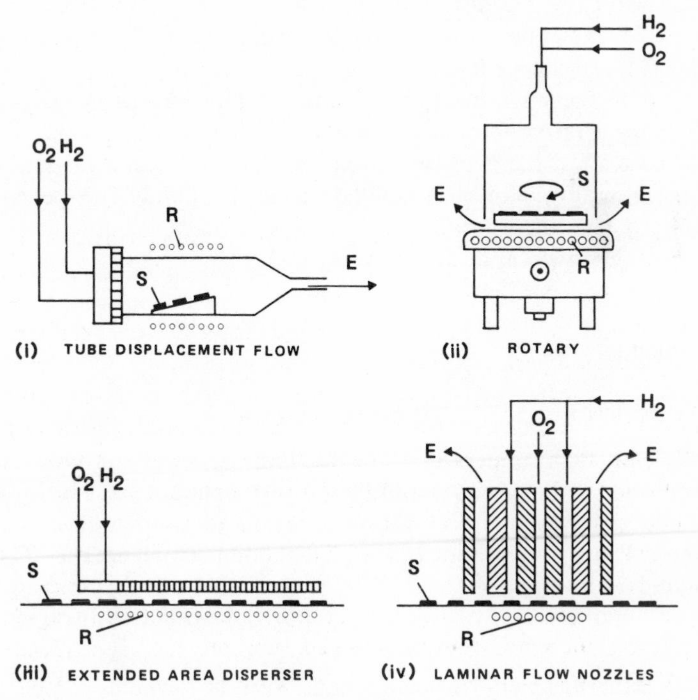

Figure 5.6 Schematic diagrams of various low-temperature APCVD reactors. (i) Horizontal tube-flow reactor; (ii) rotary vertical reactor; (iii) continuous-flow reactor; (iv) continuous-flow reactor with laminar flow nozzles. O_2, H_2, gas inlets; E, exhaust gases; R, resistance heater; S, substrate. After (2)

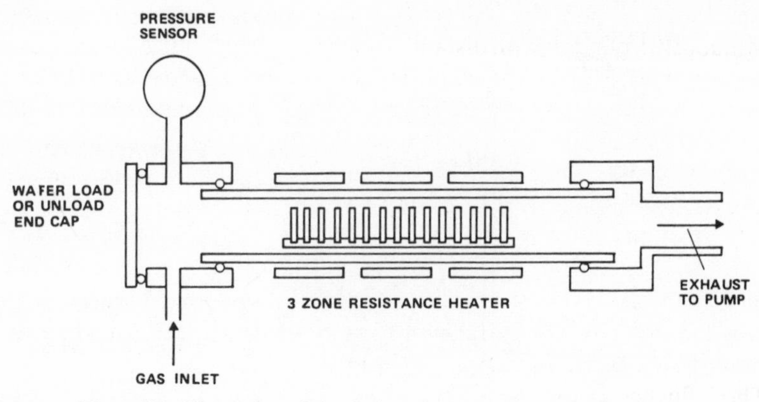

Figure 5.7 Illustration of the design of a low-pressure CVD reactor

conditions. Temperature control is more critical in the LPCVD process, to ensure good wafer-to-wafer uniformity. In most systems the diffusion-type hot-wall furnace is of horizontal geometry, as shown in Figure 5.7. The wafers are held vertically in a quartz carrier. Modifications to the design shown in Figure 5.7 are essential for the deposition of doped polysilicon (41) and silicon dioxide (41, 42) which entail incorporation of a carrier sleeve. This provision alters the furnace: wafer diameter ratio, which improves the radial uniformity. Once again axial uniformity is maintained using multiple-zone heaters to compensate for silane depletion along the reactor. Special gas inlet designs have beem employed for silicon dioxide deposition (2, 43). The specialized gas handling systems incorporating mass flow controllers and automatic throttle valves are described in detail elsewhere (44).

Deposition kinetics

Mass transport

At sufficiently high temperatures, the rate-limiting step of the overall reaction is the diffusion of the species to and away from the surface of the substrate. This transfer occurs via diffusion across a region of laminar flow above the susceptor. The region is termed the stagnant boundary layer. The thickness of the boundary layer, y, is

$$y = \sqrt{(D_r x / Re)}$$

where D_r is the diameter of the reactor tube, Re is the Reynolds number and x is the distance along the reactor. Most models of film growth rely on the diffusion of species across the interface between the boundary layer and the gas phase.

In the former region, the gas velocity and species concentrations increase continuously from zero with distance from the susceptor. The gas temperature decreases linearly over the same region. Above the boundary layer it is assumed that the parameters are constant. The products of the reaction diffuse away from the susceptor into the laminar flow region and are ultimately swept out of the reactor. The uniformity of film growth along the reaction zone is improved by varying the gas velocity and hence *Re*, since

$$Re = D_r v\sigma/\mu$$

where v is the gas velocity, σ the density and μ the viscosity. Values of D_r and v are of the order of 10^{-2} m and 10^{-1} m s^{-1} respectively, which results in laminar flow conditions since $Re < 2000$.

The influence of the chemistry on the deposition kinetics is minimal, the growth rates being determined by mass transport considerations. For the case of APCVD of silicon, the majority of models (45–64) assume that surface reactions predominate, viz.

$$SiH_4(g) + \text{surface site} \rightarrow SiH_4(ads)$$

$$SiH_4(ads) \rightarrow SiH_2(ads) + H_2(g)$$

$$SiH_2(ads) \rightarrow Si(ads) + H_2(g)$$

Silane depletion in the gas phase is compensated in practice by increasing the temperature at the rear of the furnace by 5–15° C. To maintain laminar flow, the susceptor is tilted with respect to the gas flow. The geometry is optimized to produce film thickness uniformity both radially and from wafer to wafer.

Kinetic control

For a horizontal reactor geometry, the diminution in growth rate as the temperature is lowered exhibits non-Arrhenius behaviour. The apparent activation energy varies over the temperature range encompassing mass transport and kinetic control. Even at temperatures where mass transport effects are minimal, contributions to the overall growth rate from homogeneous and heterogeneous reactions are present. The apparent activation energies have been attributed (65–73) to the kinetics of surface reactions. Data arising from HOMOCVD experiments (25) in which the substrate temperatures is sufficiently low to prevent heterogenous reactions occurring at the substrate delineates the two components. The experimental arrangement is shown schematically in Fig. 5.8. The variation in silicon growth rate with the temperature of the substrate, T_s, and furnace, T_g, is shown in Figure 5.9. A maximum in the deposition rate occurs at $T_g \sim 650°$ C and is due to the competition between homogeneous nucleation and decomposition to silicon. Homogeneous nucleation gives rise to particulates of high H-content which

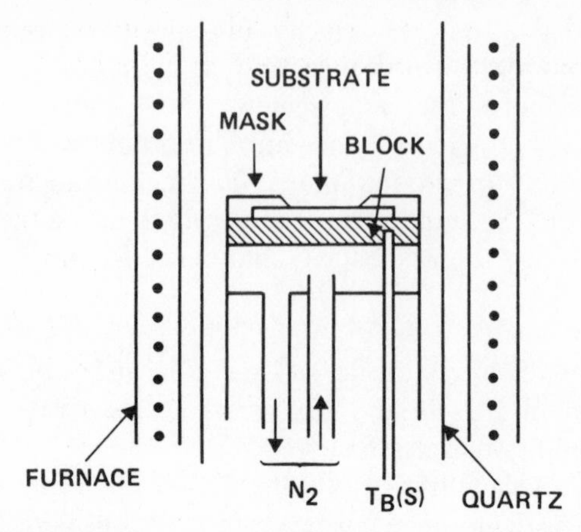

Figure 5.8 Schematic diagram of the design of an HOMOCVD reactor. After (25)

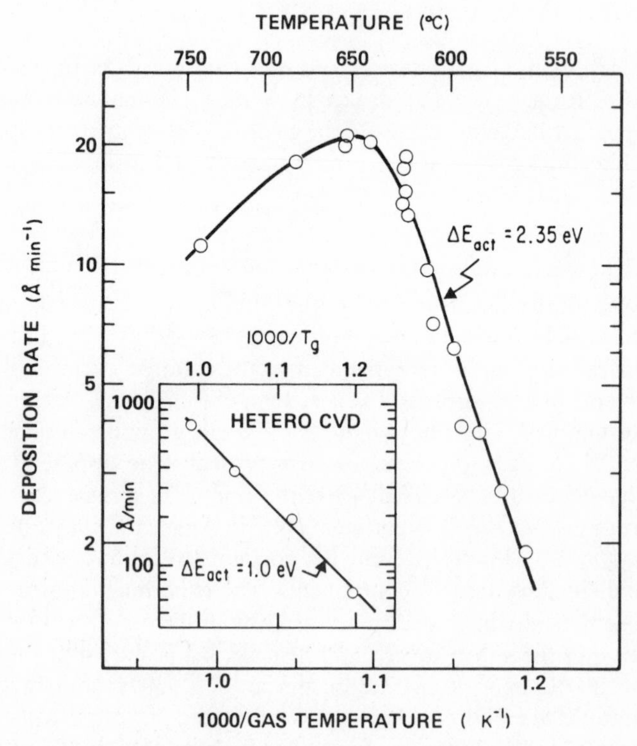

Figure 5.9 Arrhenius plot of the HOMOCVD growth rate. The insert shows the dependence of growth rate on temperature when $T_s = T_g$. After (25)

are swept out of the chamber. The detailed chemistry is uncertain but is thought to proceed by the insertion reactions (74) of SiH_2:

$$SiH_4 \rightarrow SiH_2 + H_2$$
$$SiH_2 + SiH_4 \rightarrow Si_2H_6$$
$$SiH_2 + Si_2H_6 \rightarrow Si_3H_8$$

and possibly by chain growth (75), or particle formation (76, 77) via the SiH_2 intermediate. Particulate formation in commercial CVD reactors is circumvented by lowering the SiH_4 partial pressure (LPCVD), by dilution in H_2 or other inert gas (APCVD) or by the addition of HCl (VPE).

The deposition rates in an HOMOCVD reactor are independent of T_s below $\sim 350°$ C; below this critical temperature the heterogeneous contribution to the growth rate is suppressed. In the insert to Figure 5.9, the relative growth rates of the homogeneous ($T_s \ll T_g$) and heterogeneous ($T_s = T_g$) reactions are compared. Clearly, in the temperature range employed for the CVD of polysilicon, a difference of some two orders of magnitude in favour of the heterogeneous pathway is favoured.

The homogeneous activation energy (2.35 eV) obtained from the HOMOCVD experiments (25) is satisfyingly close to the value of 2.35 eV obtained by gas-phase pyrolysis (78, 79). In comparison, the activation energy for heterogeneous reaction is significantly lower at 1–1.5 eV (80, 81). As a final point, the diminution in rate at T_g above 650° C is due to silane depletion at the reactor walls and nucleation of particulates in the gas phase.

Detailed gas dynamic and kinetic modelling of the processes occurring within a standard CVD reactor have been presented (82). The reaction pathways invoke silylene, SiH_2, as an important (if not the predominant) intermediate. The results of the calculations are shown in Figure 5.10 as temperature, pressure and flow-rate contours in an APCVD reactor of horizontal geometry. Monosilane decomposes to silylene as it diffuses across the boundary layer to the susceptor. The gas-phase reactions produce a steep concentration gradient, and at the higher temperatures the silane concentration at the surface is zero. The concomitant variation in the SiH_2 flux is shown in Figure 5.11. These results concur, coincidentally, with the presumptions of earlier models of the instantaneous decomposition of SiH_4 at the surface to give Si(s). The details of the complex reaction mechanism are given elsewhere (82) and are not reproduced here. The model relies heavily on gas-phase kinetics which have hitherto largely been ignored. The success in modelling the experimental rate data for silicon deposition in both APCVD and LPCVD reactors is impressive, as can be judged from Figure 5.12. Further confirmation of these conclusions arises from the direct laser-induced fluorescence observation (83) of Si_2. With the use of this technique to detect other important species such as Si (84), SiH (85) and SiH_2 (86–88) progress in understanding the reactor chemistry will be rapid.

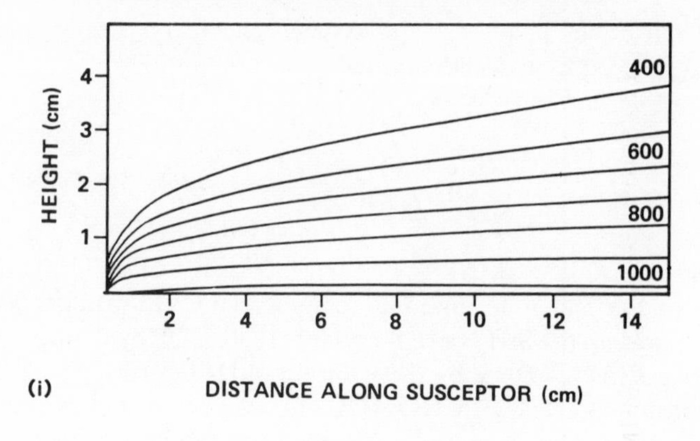

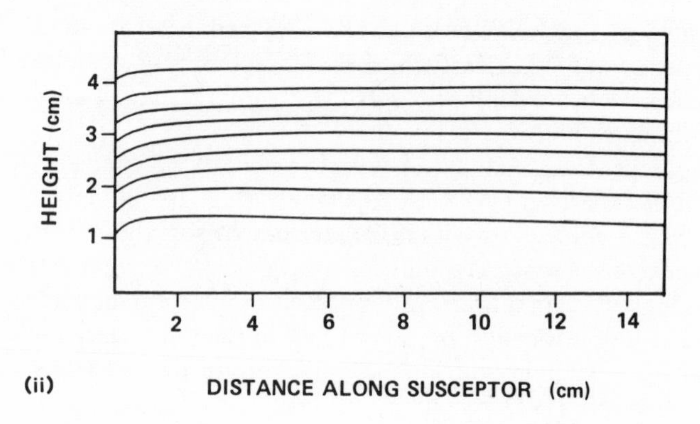

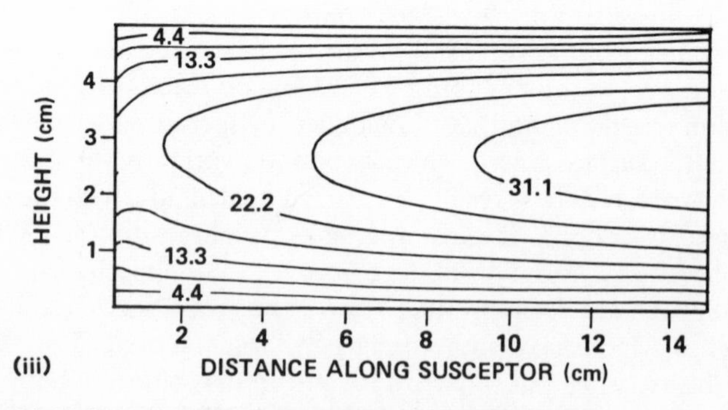

Figure 5.10 Fluid flow properties calculated for an APCVD horizontal reactor. Susceptor temperature = 1018 K; inlet gas temperature 300 K; gas velocity 15.3 cm s^{-1}; pressure 0.6 Torr SiH_4 in 640 Torr He. (i) Temperature contours (K); (ii) flow streamlines; (iii) gas velocities contours (cm s^{-1}). After (82)

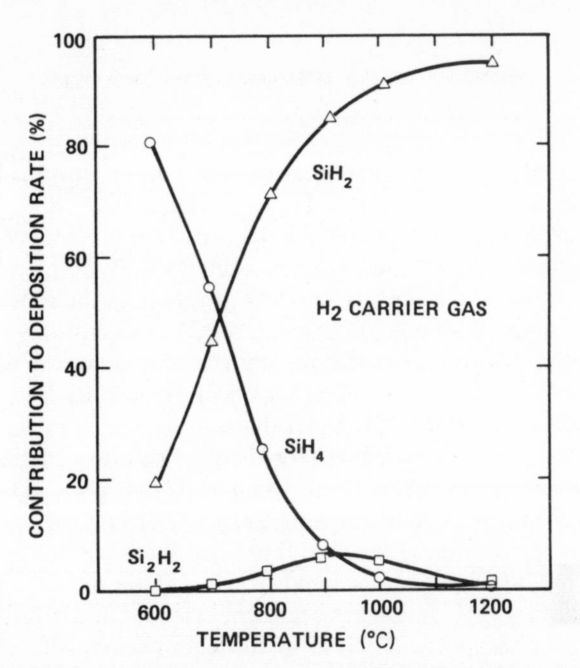

Figure 5.11 Contributions to the growth rate calculated from numerical CVD model as a function of temperature. After (82)

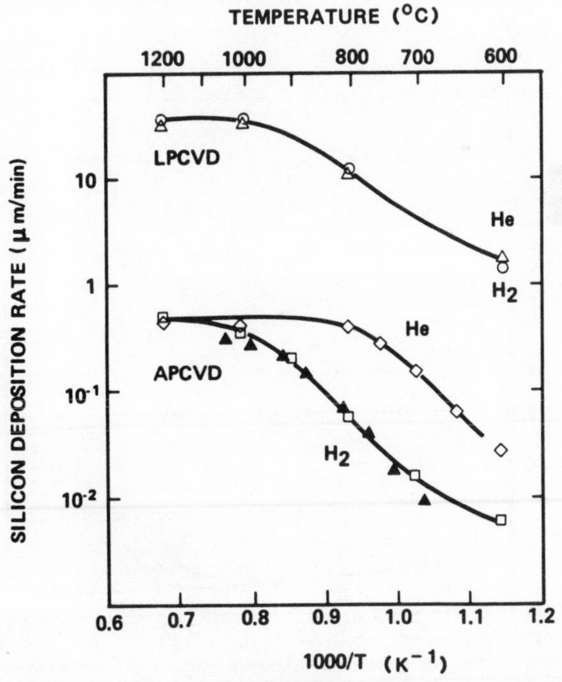

Figure 5.12 Comparison of the predicted growth rates from the CVD model with the experimental data (▲), from C. H. J. van den Bretzel, PhD thesis, University of Nijmegen, 1978 (unpubl.). After (82)

In-situ doping

The LPCVD of doped polysilicon is carried out by introducing a small percentage of dopant, e.g. AsH_3, PH_3 or B_2H_6, into the SiH_4 gas stream. Figure 5.13 demonstrates the marked influence of the presence of dopants on the deposition rate (29–31). The presence of diborane, B_2H_6, increases the rate significantly, whereas adding PH_3 or AsH_3 suppresses the growth rate (16). Similar phenomena have been observed in APCVD reactors (29). In addition to altering the growth rates, the presence of dopants also degrades the radial uniformity. This effect can be significantly reduced by changing the effective furnace diameter to the wafer diameter by the insertion of a sleeve around the wafer boat. The reasons for the growth rate variations for both dopants remain unclear (89). There is no reason *a priori* why the same explanation need apply to both dopants. Suggestions have been made that the activation energy differences result because the dopant atoms occupy active sites (30) or modify the electronic band structure in the vicinity of the bound atom (31). The first explanation has been supported (16) for PH_3 doping, from the finding that phosphine interferes with the heterogeneous decomposition of silane by preferentially occupying surface sites. Once adsorbed, phosphine effectively passivates the surface, leading to a slower heterogeneous SiH_4 decomposition and a diminution in the growth rate.

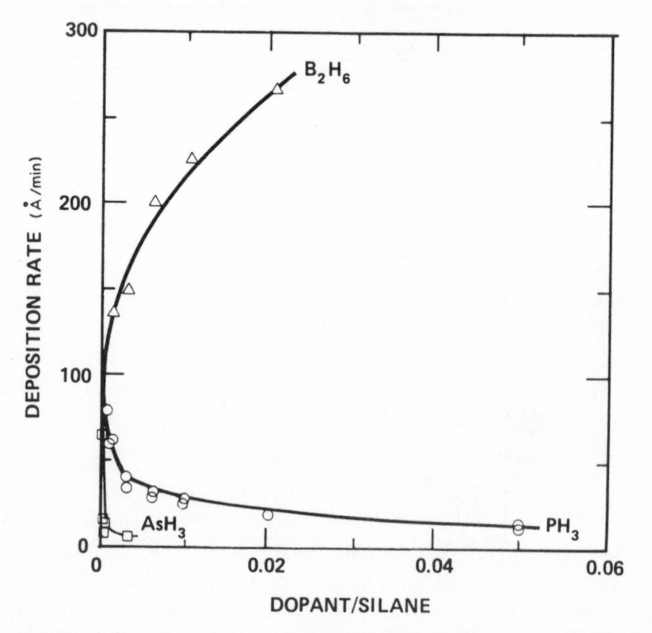

Figure 5.13 The effect of dopants on polysilicon deposition rate at 610° C. After (1)

The parallel homogeneous reaction will naturally play a more prominent role as the rate of the surface decomposition reaction is diminished. Under these conditions the transport of SiH_2 to the surface may become the rate-limiting step. The fast insertion reactions of SiH_2 into Si–H groups on the surface is described in more detail in the next section. In this context, however, the shift to a transport-limiting regime implies that the concentration of SiH_2 will be a strong function of the position in the reactor. In addition to an overall reduction in the growth rate, the inhomogeneous concentration profile of the film-forming intermediate produces relatively faster growth rates at the periphery of the wafer. The SiH_2 concentration at the wafer edge can be reduced by inserting a sleeve around the wafer boat. This acts as a radical trap, and, with suitable adjustments to the sleeve-to-wafer diameters, a uniform radial film thickness can be achieved.

The reasons for the enhanced growth rates in the presence of B_2H_6 are more problematical. The limited gas-phase kinetic data on the reactions of SiH_2 with Group III or V hydrides makes assessment of the gas-phase contributions difficult. It has been reported (90) that SiH_2 reacts rapidly with PH_3 in disagreement with earlier conclusions (91) about the relative rate of insertion into the P–H bonds in CH_3PH_2. There are no data on the reactions of SiH_2 with B_2H_6, although chemical arguments would suggest a similarly high rate. An investigation of the gas-phase chemistry might suggest a mechanism which increases the flux of SiH_2 to the surface or, conversely, surface reactions might be influenced by the presence of boron or boron-containing species. Evidence has been presented (95) which shows a more facile H_2 loss in B-doped a-Si:H on thermal annealing compared to either intrinsic or P-doped material.

Surface growth mechanisms

The proposed mechanism for the growth of amorphous silicon films at low temperatures draws heavily on reactions which have been studied in the gas phase. Under conditions where the flux of SiH_2 to the surface is responsible for film growth, the steps illustrated in Figure 5.14 have been independently suggested by Kampas (92) and Scott (25).

Scott (25) favours a scheme in which H_2 or SiH_4 are eliminated during the surface reconstruction. Polysilane chains readily undergo dehydrogenation (93, 94) at temperatures around 250–300° C. Evidence from the thermal dehydrogenation of plasma-produced a-Si:H films demonstrates (38) a facile loss of hydrogen at *c.* 350° C, whilst a second distinct dehydrogenation process occurs at 500° C. The presence of dopants affects the thermal stability of the a-Si:H films towards dehydrogenation (95). Annealing of B-doped films leads to a pronounced shift in both H_2 evolution peaks to lower temperatures. This effect is not observed in P-doped a-Si:H films. The relationship between loss of H_2 at the growing surface and the presence of boron hydride groupings

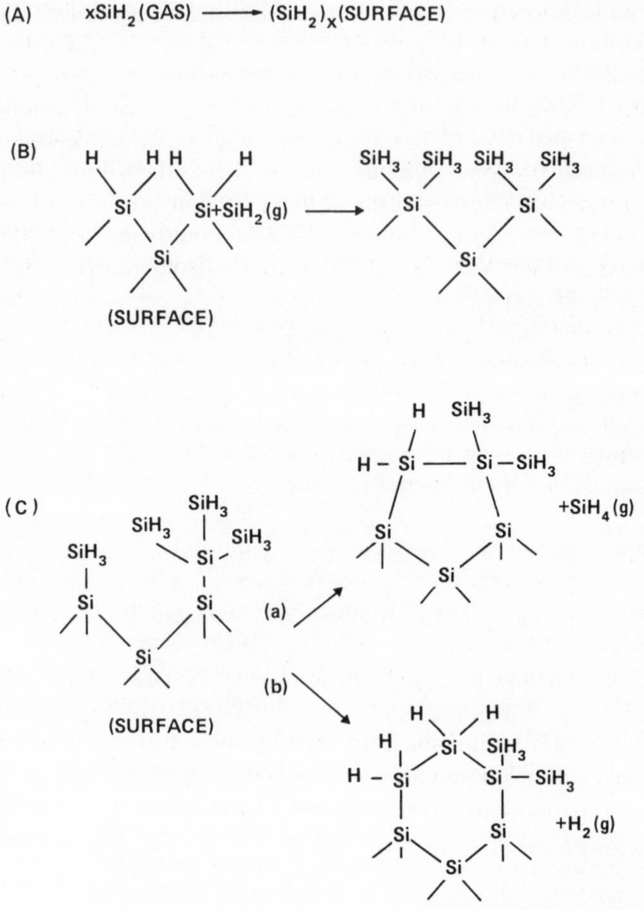

(A) $xSiH_2(GAS) \longrightarrow (SiH_2)_x(SURFACE)$

(B) $Si + SiH_2(g) \longrightarrow$ (SURFACE)

(C) (SURFACE) (a) $+SiH_4(g)$ (b) $+H_2(g)$

Figure 5.14 Proposed sequence of surface reactions leading to the growth of a-Si:H. After (25)

needs to be explored before the reasons for the enhanced growth rates can be elucidated.

CVD of electronic materials

Metals

Physical vapour deposition, namely electron-beam or RF induction-source evaporation or magnetron sputtering, has provided the most common means of forming thin metal or metal alloy films. These line-of-sight methods result in poor step coverage, and consequently reliability is impaired. This problem will be exacerbated as the trend towards smaller feature sizes continues. Most silicon MOS devices and bipolar integrated circuits are metallized with Al or one of its alloys. The advantages of CVD (96, 97) for the metallization process lie in the conformal nature of the coatings, film purity and the ability to deposit self-aligned layers selectively on to microelectronic device structures. Several studies of the LPCVD of Al have been reported using trimethylaluminium (98) or more recently tri-isobutylaluminium (99–102) as the source. For the latter, good conformal coverage, chemical purity, adhesion and electrical conductivity were noted. The disproportionation of aluminium monochloride has been investigated for the CVD of Al (103). In comparison with Al–0.5% Cu alloy evaporation, inferior electromigration properties are likely to impede the adoption of this technique.

Whilst CVD techniques for other metals (97) such as Ti (104), Mo (105) and W (106–108) have been reported, the selective deposition of W on to single or polycrystalline silicon has been of major interest (109–113). The reduction of tungsten hexafluoride, WF_6, with silicon, viz.

$$2WF_6 + 3Si \rightarrow 2W + 3SiF_4$$

is spatially selective and consumes the substrate, leading to a recessed W layer (108). The CVD of both the latter compound and WCl_6 (114) produces low-resistivity, high-purity polycrystalline W films at low temperatures (300–475° C).

This metal can also be deposited via the H_2 reduction of WF_6 (115) in the gas phase or via the thermal decomposition (116) of $W(CO)_6$. Defects caused by metallization failure arising from 'spiking' or electromigration (1) have been ameliorated with evaporated Si or Cu alloys respectively. A practical CVD process would need to produce films with resistances similar to that of competing phenomena. Incorporation of Si into LPCVD-deposited Al has been achieved (101) using SiH_4 as a co-reactant.

Insulators and dielectrics

A number of different methods are employed in the electronics industry to deposit insulating films such as SiO_2 or Si_3N_4. These methods are described in the relevant texts (1, 117) and have been reviewed for oxide growth (118). Silicon dioxide, SiO_2, is the pre-eminent insulator used in the microelectronics industry. The material formed by CVD is inferior in quality to thermally-grown material and has not replaced it as the gate insulator in FET devices. CVD-formed oxides are used for masking, as a diffusion source, as an insulator between multilayer metallization or over a polysilicon gate, and as a protective coating over an entire circuit. Deposition methods are categorized by the chemical source and furnace temperature.

Low-temperature oxide. Silicon dioxide, SiO_2, can be deposited (118–120) using SiH_4 in the presence of oxygen in a inert diluent gas, usually N_2, between

300–500° C. To prevent uncontrolled oxidation the silane is diluted (1%) in N_2. Scrupulous attention to process control is required for the fabrication of pinhole-free ($< 50\,cm^{-2}$) oxide layers, since the process is susceptible to particulate formation. The chemistry is complex and little understood. Growth rates are maximized at dilutions of *c.* 0.5% O_2 (121). At higher concentrations, particulate nucleation in the gas phase limits the film growth rate and the observed rates diminish.

Intermediate-temperature oxide. The pyrolysis (122, 123) of tetraethylorthosilicate (TEOS), or tetraethoxysilane $(C_2H_5O)_4Si$ in a N_2 carrier at temperatures between 500–850° C produces SiO_2 in a LPCVD reactor. The reaction yields a large number of volatile products and points to the complex chemistry that is involved in this process.

High-temperature oxide. Silicon dioxide is deposited under LPCVD conditions (120, 124, 125) employing the oxidation of dichlorosilane, SiH_2Cl_2, using nitrous oxide, N_2O, according to the stoichiometry

$$SiH_2Cl_2 + 2N_2O \rightarrow SiO_2 + 2N_2 + HCl$$

at around 900° C.

Other processes in the range 850–1100° C include the oxidation of SiH_4, diluted to *c.* 1% in N_2, using CO_2 and H_2 carrier gas (117). More emphasis is being placed on the development of LPCVD techniques at lower deposition temperatures. For example, the TEOS process (126, 127) can be operated at pressures between 15–60 Torr in a horizontal hot wall furnace at 750° C to achieve deposition rates of $300\,Å\,min^{-1}$.

Oxygen-doped polycrystalline silicon. Semi-insulating polysilicon (SIPOS) is the term given to oxygenated films of silicon (128–130) grown by CVD from a N_2- or Ar-diluted gas mix of SiH_4 and N_2O at *c.* 600–700° C. Oxygen incorporation reduces the leakage current in polysilicon layers used to replace thermally-grown oxide for high-voltage integrated circuits. The oxygen content in the films can be controlled by varying the ratio of N_2O to SiH_4 in the gas mixture. This is at the expense of decreasing the growth rates from approximately $500\,Å\,min^{-1}$ exhibited in the absence of oxidant. Resistivities vary from *c.* 3×10^6 ohm cm for undoped polysilicon to *c.* 1×10^9 ohm cm for films doped with 15–35% oxygen. Combined HREM imaging and selected area optical diffraction (131) demonstrates that SIPOS formed at 625° C is amorphous, but annealing in N_2 at 900° C induces crystallization.

Hitchman and Kane (42) have proposed that SiH_2 arising from the pyrolysis of SiH_4 is adsorbed on to O-containing surface sites. Reaction with adsorbed N_2O leads to oxygen incorporation into the material. The reaction model based on this scheme predicts accurately the radial non-uniformities in both SiO_2 (41) and SIPOS (42) deposition processes. The variation in growth rate across the wafer appears to be a common feature for low- and high-temperature stoichiometric oxides as well as silicon-rich oxide (SIROX) (41) and SIPOS (42).

Phosphosilicate glasses. Phosphosilicate glasses, PSG, are used for diffusion sources (5–15 wt% P), passivation layers or interlevel insulation (2–8 wt% P) or for use in the P-glass flow process (6–8 wt% P). The sharp angles produced on deposition of P-doped silicon dioxide on non-planar structures can be corrected by annealing at low temperatures until the oxide softens and flows to produce a conformal coating (P-glass flow). This improves device reliability. Further information on the CVD of PSG and other glasses and the reflow characteristics is available (132). The addition of boron lowers the flow temperature. The resulting borophosphosilicate glass (BPSG) can be deposited by APCVD (133) or LPCVD (134).

Silicon nitride. Stoichiometric silicon nitride, Si_3N_4, is employed as a passivating layer since the material is essentially impermeable to water and the transport of alkali ions. It is also employed as a mask for the selective oxidation of silicon (117).

Silicon nitride is deposited by reacting silane, or $SiCl_4$ diluted to 1–3% in N_2, and ammonia, NH_3, under APCVD conditions, at 700–900° C, or via the reaction between SiH_4 or SiH_2Cl_2 (135) and ammonia, using LPCVD, between 700–800° C (1, 117).

Photolytic CVD

Two basic photolytic CVD processes involve either Hg-sensitization or direct photolysis. Whilst laser CVD falls within the latter category, the potential novel applications warrant a separate section.

Incoherent sources

The reactor system employed for Hg-sensitized CVD is similar in design to LPCVD, except that the reactor chamber incorporates a quartz window above the heated substrate. Generally, low-pressure Hg lamps provide the irradiation source at 253.7 nm. A thermostat-controlled reservoir supplies the Hg vapour entrained within the reactant gas flow.

Two examples of photochemical deposition which demonstrate the potential of this technique for low-temperature processing are the PHOTOX process (136) for the deposition of SiO_2 films, and the direct deposition of a-Si:H (37).

In the former, the initial steps are

$$Hg(^1S_0) + h\nu \rightarrow Hg(^3P_J)$$
$$Hg(^3P_J) + N_2O \rightarrow Hg(^1S_0) + N_2 + O(^3P_J)$$

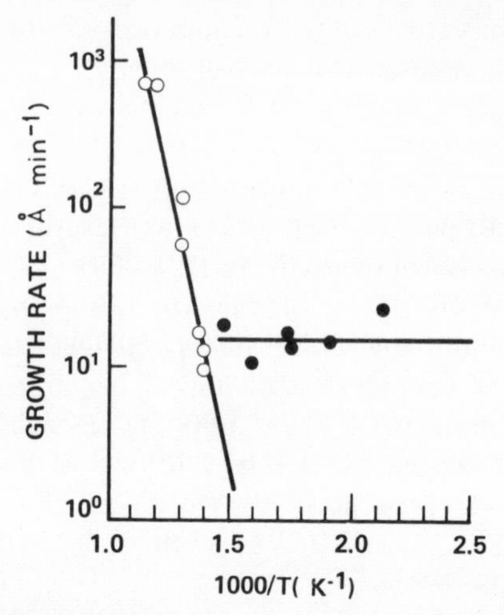

Figure 5.15 Growth rates of photo-CVD and thermal CVD silicon films produced from Si_2H_6. ○, thermal; ●, photochemical. After (137)

involves the excitation of the ground state Hg atom followed by reactive quenching with the liberation of O which leads to the silane oxidation. In the absence of N_2O it is possible to deposit a-Si:H. The stoichiometry of the resulting oxide can be altered by varying the SiH_4/N_2O ratio. The deposition rate varies directly with the light intensity and only slightly with the substrate temperature. Whilst the deposition rate varies with the reactor geometry, pressure and gas velocity, rates of *c.* 150 Å min^{-1} can be achieved over large areas (500 cm^2). This technique can be extended to other insulators such as silicon nitride and aluminium oxide.

Recently, device-quality a-Si:H has been deposited (137) from the direct photochemical CVD of disilane, Si_2H_6, at substrate temperatures below 300° C. Experiments were performed at atmospheric pressure employing 1% Si_2H_6 in He (400 sccm) in N_2 carrier. A comparison of the growth rates relative to thermal CVD from the same material is shown in Figure 5.15. The rates, *c.* 15 Å min^{-1} for an illumination power of 0.08 W cm^{-2} at 253.7 nm, are low but are higher than those for conventional CVD at the same temperature.

Laser-enhanced CVD

Laser-assisted CVD has emerged as a powerful technique for 'writing' micron-scale lines and for depositing large-area thin films. The scope of the method has been adequately described in a number of reviews (138–148). In this section, the salient features will be described briefly and the major applications, real and potential, of photothermal and photochemical CVD described. For a more comprehensive account of the laser processing of materials the reader is referred to the appropriate conference proceedings and reviews. This material should be read in the context of related topics which include laser annealing, laser alloying, laser doping and laser desorption (149). In addition, laser oxidation and laser etching could, in principle, be contained within a scheme for device fabrication without wet chemical processing (1).

Laser-induced CVD of thin films of lines can occur by either direct photodissociation or pyrolysis near the gas–surface interface. Photochemistry can be initiated in the gas phase, at the surface, or as a combination of both. As detailed elsewhere, photolytic CVD has been demonstrated to deposit metals, elemental and compound semiconductors and insulators. The majority of the precursors absorb in the ultraviolet or visible. As a consequence, excimer lasers or frequency-doubled Ar^+ lasers are commonly employed.

Deposition of materials with micron-scale resolution can be carried out, in the presence of a suitable source, by heating a surface with the focused beam of a pulsed or cw laser. The application of laser 'writing' to VLSI has been described (148). Whilst essentially thermal in character, the observed linewidths may be smaller than the laser beam waist. This feature reflects the additional complexities in the microchemistry occurring at the surface. The radial dependence of the surface temperature can be estimated. If the deposition is not transport-limited, the heterogeneous kinetics will display an activation energy. Thus the growth rates will not be linearly related to the laser beam contour. Line narrowing is not possible using this mechanism when diffusive transport is dominant. Other explanations include non-linearities in the absorption, nucleation or deposition processes.

Highlights in this area include the fabrication of NMOS transistors through a series of laser thermochemical reactions, the deposition of doped polysilicon links for customized gate arrays, and photo- and x-ray lithographic mask manufacture and repair.

Summary

The role of chemistry in semiconductor technology becomes apparent only when the manufacturing processes are examined in detail. Chemical vapour deposition in all its guises will remain at the heart of future developments. The recent progress made in elucidating the chemical reactions in CVD processes coupled with the development of laser diagnostic methods for identifying the reactive intermediates is a signpost for the direction of further research. If the spatial resolution can be improved then the prospect of characterizing the species within the important boundary layer is an exciting one. The information might then be used to improve the efficiency of current CVD reactors.

The trend towards low-temperature processing implies that photochemical techniques will become more prominent. Associated developments in laser fabrication of customized circuits are on the horizon.

A major challenge to chemists of all disciplines lies in the fabrication of compound semiconductor devices. Improvements in the chemical design of precursors, their purity and stability, the imaginative deployment of deposition techniques and characterization of new materials all lie within the province of chemistry. Applications to the growing areas of opto- and molecular electronics will naturally ensue.

References

1. S. M. Sze (ed.) *VLSI Technology*, McGraw-Hill, New York (1983).
2. Landolt-Bornstein, *Numerical Data and Functional Relationships in Science and Technology*, New series, Vol. **17**, *Semiconductors*, Springer Verlag, Berlin (1984).
3. H. K. Pulker, *Coatings on Glass*, Elsevier, Amsterdam (1984).
4. McD. Robinson, C. H. J. van der Brekel, G. W. Cullen, J. M. Blocher Jr. and P. P. Rai-Choudhury (eds.), *Ninth Int. Conf. on Chemical Vapour Deposition*, The Electrochemical Society (1984), and preceding conferences.
5. D. T. Hawkins (ed.) *Chemical Vapour Deposition: 1960–1980*, Plenum, New York, (1981).
6. See bibliography in ref. 2.
7. A. E. van Arkel and J. H. de Boer, *Z. Anorg. Allg. Chem.* **148** (1925) 345.
8. R. Holbing, *Z. Angew. Chem.* **40** (1927) 655.
9. R. Emeis and H. Henker, patent 1961, DBP 975158.
10. W. Dietze, W. Keller and A. Muhlbauer, in *Crystals: Growth, Properties and Applications*, Vol. 5, *Silicon*, Springer Verlag, Berlin (1981).
11. See Chapter 7.
12. B. A. Scott, R. M. Plecenik and E. E. Simonyi, *Appl. Phys. Lett.* **39** (1981) 73.
13. V. S. Ban, *Proc. VIth Int. Conf. on Chemical Vapour Deposition 1977*, Electrochemical Society (1977) 66.
14. C. W. Manke and L. F. Donaghey, *ibid*, (1977) 151.
15. K. F. Roenigk and K. F. Jensen, *J. Electrochem. Soc.* **132** (1985) 448.
16. B. S. Meyerson and W. Olbricht, *J. Electrochem. Soc.* **131** (1984) 2361.
17. F. C. Eversteyn, *Phillips Res. Rep.* **29** (1974) 45.
18. H. Basseches, R. C. Manz, C. O. Thomas and S. K. Tung, *AIME Semiconductor Metallurgy Conference*, Interscience, New York (1961).
19. S. P. Weeks, *Solid State Technol.* **24** (1981) 111.
20. B. A. Scott, J. A. Reimer, R. M. Plecenik, E. E. Simonyi and W. Reuter, *Appl. Phys. Lett.* **40** (1982) 973.
21. J. Bloem, *J. Cryst. Growth* **50** (1980) 581.
22. T. I. Kamins, *J. Electrochem. Soc.* **127** (1980) 686.
23. S. R. Elliot, *The Physics of Amorphous Materials*, Longman, London (1984).
24. M. Janai, D. D. Allred, D. C. Booth and B. O. Seraphin, *Solar Energy Mat.* **1**, (1979) 11.
25. B. A. Scott, in *Semiconductors and Semimetals*, J. I. Pankove (ed.), Academic Press (1984) 123.
26. F. Evangelisti and J. Stuke (eds.), *Proc. 11th Int. Conf. on Amorphous and Liquid Semiconductors*, Rome, 1985; published in *J. Non-Cryst. Solids* **77**, **78** (1985).
27. S. C. Gau, B. R. Weinberger, M. Akhtar, Z. Kiss and A. G. McDiarmid, *Appl. Phys. Lett.* **39** (1981) 436.
28. P. John, I. M. Odeh, M. J. K. Thomas, M. J. K. Tricker and J. I. B. Wilson, *Phys. Stat. Sol. (b)* **105** (1981) 499.
29. F. C. Eversteyn and B. H. Put, *J. Electrochem. Soc.* **120** (1973) 106.
30. R. F. C. Farrow, *ibid*, **121** (1974) 899.
31. C.-A. Chang, *ibid.* **123** (1976) 1245.
32. M. L. Yu and B. S. Meyerson, *J. Vac. Sci. Tech.* **A2** (1984) 446.
33. H. Kurokawa, *J. Electrochem. Soc.* **129** (1982) 2620.
34. A. Baudrant and M. Sacilotti, *ibid.* **129** (1982) 1109.
35. T. I. Kamins, M. M. Mandurak and K. C. Saraswat, *ibid.* **125** (1978) 927.
36. Y. Wada and S. Nishimatsu, *ibid.* **125** (1978) 1499.
37. D. Kaplan, in *Topics in Applied Physics* **55** (1984) 177.

38. J. C. Knights, *ibid.* **55** (1984) 5.
39. N. Sol, D. Kaplan, D. Dieumegard and D. Dubreuil, *J. Non-Cryst. Solids* **35**, **36** (1980) 291.
40. E. Bustarret, J. C. Bruyere, A. Deneuville, J. F. Currie, P. Depelsenaire and R. Groleau, *Proc. 8th Int. Conf. on Chemical Vapour Deposition*, eds. J. M. Blocher and G. E. Vuillard, Electrochemical Society (1981).
41. A. J. Learn, *J. Electrochem. Soc.* **132** (1985) 390; A. J. Learn and R. B. Jackson, *ibid.* **132** (1985) 2975.
42. M. L. Hitchman and J. Kane, *J. Cryst. Growth* **55** (1981) 485.
43. P. J. Tobin, J. B. Price and L. M. Campbell, *J. Electrochem. Soc.* **127** (1980) 2222.
44. W. R. Clark and J. J. Sullivan, *Solid State Technol* **25**, (1982) 105.
45. S. E. Bradshaw, *Int. J. Electronics* **21** (1966) 205.
46. S. E. Bradshaw, *ibid.* **23** (1967) 381.
47. F. C. Eversteyn, P. J. W. Severin, C. H. J. van der Brekel and H. L. Peek, *J. Electrochem.* **117** (1970) 925.
48. F. C. Eversteyn and H. L. Peek, *Philips Res. Report* **25** (1970) 472.
49. F. C. Eversteyn, *ibid.* **29** (1974) 45.
50. J. Bloem, *J. Electrochem. Soc.* **117** (1970) 1397.
51. P. van der Putte, L. J. Giling and J. Bloem, *J. Cryst. Growth* **31** (1975) 299.
52. S. Berkman, V. S. Ban and N. Goldsmith, in *Heteroepitaxial Semiconductor Electronic Devices*, eds. G. W. Cullen and C. C. Wang, Springer Verlag, New York (1978)
53. W. H. Shepherd, *J. Electrochem. Soc.* **112** (1965) 988.
54. R. W. Andrews, D. M. Rynne and E. G. Wright, *Solid State Technol.* **12** (1969) 61.
55. P. C. Rundle, *Int J. Electronics* **24** (1968) 405.
56. P. C. Rundle, *J. Cryst. Growth* **11** (1971) 6.
57. R. Takahashi, S. Sugawara, Y. Nakazawa and Y. Koga, in *Chemical Vapour Deposition*, eds. J. M. Blocher and J. C. Withers, Proc. Electrochem. Soc., New York (1970) 695.
58. K. Sugawara, H. Tochikubo, R. Takahashi and Y. Koga, *ibid.*, 73.
59. R. Takahashi, Y. Koga and K. Sugawara, *J. Electrochem. Soc.* **119** (1972) 1406.
60. K. Sugawara, *ibid.* **119** (1972) 1749.
61. E. Fujii, H. Nakamura, K. Haruna and Y. Koga, *ibid.* **119** (1972) 1106.
62. C. W. Manke and L. F. Donaghey, in *Chemical Vapour Deposition*, eds. L. F. Donaghey, P. Rai-Choudhury and R. N. Tauber, Proc. Electrochem. Soc., Princeton, (1977) 151.
63. C. W. Manke and L. F. Donaghey, *J. Electrochem. Soc* **124** (1977) 561.
64. M. L. Hitchman, *J. Cryst. Growth* **48** (1980) 394.
65. B. A. Joyce and R. R. Bradley, *J. Electrochem. Soc.* **110** (1963) 1235.
66. M. J. Duchemin, M. M. Bonnet and M. F. Koelsch, *ibid.* **125** (1978) 637.
67. J. Bloem and L. J. Giling, in *Current Topics in Materials Science*, ed. E. Kaldis, North-Holland, Amsterdam (1978).
68. M. L. Hitchman, in *Chemical Vapour Deposition*, eds. T. O. Sedgwick and H. Lydtin, Proc. Electrochem. Soc., Princeton, (1979) 59.
69. M. L. Hitchman, J. Kane and A. E. Widmer, *Thin Solid Films* **59** (1979) 231.
70. F. Hottier and R. Cadoret, *J. Cryst. Growth* **52** (1981) 199.
71. J. Holleman and T. Aarnink, in *Chemical Vapour Deposition*, eds. J. M. Blocher, G. E. Vuillard and G Wahl, Proc. Electrochem. Soc., Pennington (1981).
72. W. A. P. Claassen and J. Bloem, *J. Cryst. Growth* **51** (1981) 443.
73. W. A. P. Claassen, J. Bloem, W. G. J. N. Valkenburg and C. H. J. van der Brekel, *ibid.* **57** (1982) 259.
74. P. John and J. H. Purnell, *J. C. S. Faraday Trans. I* **69** (1973) 1455.
75. J. Dzarnoski, S. F. Rickborn, H. E. O'Neil and M. A. Ring, *Organometallics* **1** (1982) 1217.
76. R. M. Roth, K. G. Spears and G. Wong, *Appl. Phys. Lett.* **45** (1984) 28.
77. K. G. Spears and R. M. Roth, *Mat. Res. Soc. Symp. Proc.* Vol. 38, Materials Research Society, (1985).
78. J. H. Purnell and R. Walsh, *Proc. Roy. Soc. London Ser. A*, **293** (1966) 543.
79. R. Walsh, *Acc. Chem. Res.* **14** (1981) 246.
80. W. A. Bryant, *Thin Solid Films*, **60** (1979) 19.
81. M. Hirose, *J. Phys. Suppl.* **10** (1981) C4-705.
82. M. E. Coltrin, R. J. Kee and J. A. Miller, *J. Electrochem. Soc.* **131** (1984) 425.
83. P. Ho and W. G. Brieland, *Appl. Phys. Lett.* **44** (1984) 51.

84. D. M. Roth, K. G. Spears and G. Wong, *ibid.* **45** (1984) 28.
85. W. Bauer, K. H. Becker, R. Duren, C. Hubrich and R. Meuser, *Chem. Phys. Lett.* **108** (1984) 560.
86. G. Inoue and M. Suzuki, *Chem. Phys. Lett.* **105** (1984) 641.
87. D. M. Rayner, R. P. Steer, P. A. Hackett, P. John and C. L. Wilson, *Chem. Phys. Lett.* **123** (1986) 449.
88. J. W. Thoman Jr. and J. I. Steinfeld, *Chem. Phys. Lett.* **124** (1986) 35.
89. K. F. Roenigk and K. F. Jensen, *J. Electrochem. Soc.* **132** (1985) 448.
90. J. Blazejowski and F. W. Lampe, *J. Photochem.* **20** (1982) 9.
91. M. D. Sefcik and M. A. Ring, *J. Organometall. Chem.* **59** (1973) 167.
92. F. J. Kampas, in *Semiconductors and Semimetals*, Vol. 21, ed. J. I. Pankove, Academic Press, New York (1984).
93. D. K. Biegelsen, R. A. Street, C. C. Tsai and J. C. Knights, *Phys. Rev. B* **20** (1979) 4839.
94. P. John, I. M. Odeh, M. J. K. Thomas, M. J. Tricker, F. Riddoch and J. I. B. Wilson, *Phil. Mag. B* **42** (1980) 671.
95. W. Beyer and H. Wagner, *J. Physique* **42** (1981) C4-783.
96 W. Kern and V. S. Ban, 'Chemical vapour deposition of inorganic thin films', in *Thin Film Processes*, eds. J. L. Vossen and W. Kern, Academic Press, New York (1978).
97. C. F. Powell, Chemically deposited metals, in *Vapour Deposition*, eds. C. F. Powell, J. H. Oxley and J. M. Blocher Wiley, New York (1966).
98. H. O. Pearson, *Thin Solid Films* **45** (1977) 257.
99. R. A. Levy, M. L. Green and P. K. Gallagher, *J. Electrochem. Soc* **131** (1984) 2175.
100. M. L. Green, R. A. Levy, R. G. Nuzzo and E. Coleman, *Thin Solid Films* **114** (1984) 367.
101. M. J. Cooke, R. A. Heinecke, R. C. Stern and J. W. C. Maes, *Solid State Technol.* **25** (1982) 62.
102. A. Malazgirt and J. W. Evans, *Metall. Trans.* **11B** (1980) 225.
103. R. A. Levy, P. K. Gallagher, R. Contolini and F. Schrey, *J. Electrochem. Soc.* **132** (1985) 457.
104. M. J. H. Kemper, S. W. Koo and F. Huizinga, *Electrochem. Soc. Extended Abstr.* **84** (1984) 533.
105. J. K. Chu, C.-C. Tang and D. W. Haas *Appl. Phys. Lett.* **41** (1982) 75.
106. J. M. Shaw and J. A. Amick, *RCA Review* **31** (1970) 306.
107. N. Miller and R. Herring, *Electrochem. Soc. Extended Abstr.* **81** (1981) 712.
108. E. K. Broadbent and W. T. Stacey, *Solid State Technol.* **28** (1985) 51.
109. C. Crowell, J. Savace and S. Sze, *Tr. Metal Soc. AIME* **233** (1965) 478.
110. J. M. Shaw and J. A. Amick, *RCA Review* **31** (1970) 306.
111. C. M. Melliar-Smith, A. C. Adams, R. H. Kaiser and R. A. Kushner, *J. Electrochem. Soc.* **121** (1974) 298.
112. C. E. Morosanu and V. Soltuz, *Thin Solid Films* **52** (1978) 181.
113. W. A. Bryant and G. H. Meir, *J. Electrochem. Soc.* **120** (1973) 559.
114. C. F. Powell, J. H. Oxley and J. M. Blocher, in *Chemical Vapour Deposition*, Wiley, New York (1966).
115. E. K. Broadbent and C. L. Ramiller, *J. Electrochem. Soc.* **131** (1984) 1427.
116. H. Peter, W. Hey and J. N. Fordemwalt, *Workshop on Tungsten for VLSI Applications*, Sandia National Laboratories, Albuquerque (1984).
117. E. H. Nicollian and J. R. Brews, *MOS Physics and Technology*, Wiley, New York (1982).
118. W. Kern and R. S. Rosler, *J. Vac. Sci. Technol.* **14** (1977) 1082.
119. N. Goldsmith and W. Kern, *RCA Rev.* **28** (1967) 153.
120. R. S. Rosler, *Solid State Technol.* **20** (1977) 63.
121. Bell Telephone Laboratories, *Silicon Oxidation Guide.*
122. E. L. Jordan, *J. Electrochem. Soc.* **108** (1961) 478.
123. J. Klerer, *ibid.* **108** (1965) 503.
124. W. A. Brown and T. I. Kamins, *Solid State Technol.* **22**, (1979) 51.
125. K. Watanabe, T. Tanigaki and S. Wakayama, *J. Electrochem. Soc.* **128** (1981) 2630.
126. H. Hupertz and W. L. Engl., *IEEE Trans Electron. Devices* **ED–26**, (1979) 658.
127. A. C. Adams and C. D. Capie, *J. Electrochem. Soc.* **126** (1979) 1042.
128. T. Aoki, T. Matsusshita, H. Yamoto, H. Hayashi, M. Okayama and Y. Kawana, *Electrochem. Soc. Extended Abstr.* **75** (1975) 352.
129. T. Matsushita, T. Aoki, T. Ohtsu, H. Yamoto, H. Hayashi, M. Okayama and Y. Kawana, *IEEE Trans. Electron. Devices* **ED–23** (1976) 826.

130. H. Mochizuki, T. Aoki, H. Yamoto, M. Okayama, M. Abe and T. Ando, *Jap. J. Appl. Phys. Suppl.* **15** (1976) 35.
131. J. Wong. D. A. Jefferson, T. G. Sparrow, J. M. Thomas, R. H. Milne, A. Howie and E. F. Koch, *Appl. Phys. Lett* **48** (1986) 65.
132. R. A. Bowling and G. B. Larrabee, *J. Electrochem. Soc.* **132** (1985) 141.
133. R. M. Levin and K. Evans-Lutterodt, *J. Vac. Sci. Technol.* **B1** (1983) 54.
134. T. Foster, G. Hoeye and J. Goldman, *J. Electrochem. Soc.* **132** (1985) 505.
135. P. Pan and W. Berry, *J. Electrochem. Soc.* **132** (1985) 3001.
136. J. Y.-T. Chen, R. C. Henderson, J. T. Hall and J. W. Peters, *J. Electrochem. Soc.* **131** (1984) 2146.
137. Y. Mishima, M. Hirose, Y. Osaka, K. Nagame, Y. Ashida, N. Kitagawa and K. Isogaya, *Jap. J. Appl. Phys.* **22** (1983) L46.
138. D. J. Ehrlich, R. M. Osgood and T. F. Deutsch, *IEEE J. Quant. Electronics* **QE-16** (1980) 1233.
139. D. J. Ehrlich and J. Y. Tsao, *VLSI Electronics: Microstructure Sci.* **7**, (1983) 129.
140. I. P. Herman, *Laser Processing and Diagnostics*, Springer Ser. in Chem. Physics, **39** (1984) 396.
141. D. Bauerle, *Surface Studies with Lasers*, Springer Ser. in Chem. Physics **33** (1983) 179.
142. R. M. Osgood, *Ann. Rev. Phys. Chem.* **34** (1983) 77.
143. T. F. Deutsch, Laser Proc. and Diagnostics, Springer Ser. in Chem. Physics **39** (1984) 239.
144. Y. Rytz-Froidevaux, R. P. Salathe and H. H. Gilgen, *Appl. Phys. A* **37** (1985) 121.
145. R. Solanki, C. A. Moore and G. J. Collins, *Solid State Technol.* **28** (1985) 220.
146. R. M. Osgood and T. F. Deutsch, *Science* **227** (1985) 709.
147. J. T. Yardley, in *Laser Handbook*, Vol. 5 eds. M. Bass and M. L. Stitch, Elsevier, Amsterdam (1985).
148. D. J. Ehrlich, *Solid State Technol.* **28** (1985) 81.
149. T. J. Chuang, *Surface Sci. Rep.* **3** (1983) 1.

Figures 5.10, 5.11 and 5.12 appear by permission of the publisher, The Electrochemical Society, Inc.

6 Plasma-enhanced chemical vapour deposition

P. JOHN and B. L. JONES

Introduction

There is considerable interest in the deposition of inorganic materials at low temperatures for three-dimensional VLSI, solar cells, silicon-on-insulator (SOI) techniques and for quantum-well technology. Low processing temperatures ($< 600°$ C) are essential to minimize the effects of diffusion on junction profiles, metal migration and structural changes. It is for this reason that plasma-enhanced chemical vapour deposition (PECVD) has played a prominent role in thin film research of late.

Plasmas used for these applications are usually produced by applying a high-frequency electric field across a gas in which the discharge is generated between two parallel plate electrodes. Such plasmas are typical of 'cold' plasmas in which the ions and electrons are not in thermal equilibrium (see Chapter 15). Typically the ion temperature is a few hundred degrees Kelvin whilst the electron temperature may be greater than a few thousand degrees.

The composition, physical, chemical and electrical properties of the deposited films depend in a complex fashion on many deposition parameters. These include gas pressure, temperature, flow rate and dwell time, chamber geometry, RF power, frequency and method of power coupling, the substrate electrical and thermal bias, nature of the substrate and the type of reactive and diluent gases. An empirical approach has been adopted in the past and it is only of late that details of the chemistry have been revealed.

This chapter will consider the basic principles of the plasma process using the plasma chemistry of silane as a specific example. The plasma deposition of hydrogenated amorphous silicon, a-Si:H, and related alloys has been one of the major breakthroughs in semiconductor materials science in recent years and many studies have been made of this complex system (1, 2). The properties of a-Si:H as a semiconductor material depend on the fraction, chemical configuration and location of hydrogen within the silicon framework. The flux of species to the growing film is essentially controlled by the electron energy distribution within the plasma and the resulting ionization and radical formation processes through electron impact. These in turn are governed by the macroscopic reactor design and operating parameters. The next section of this chapter briefly describes the most commonly used RF reactors.

Other sections cover the chemistry of the plasma processes, the use of plasma diagnostic tools, and the deposition of materials using gases other than silane (including non-hydride gases) and provides some insight into the versatility of this growth technique. Some thought is given to possible future developments in the final section.

RF glow discharge reactors

PECVD has been performed using various types of plasma reactors. Reactor design has continually improved with emphasis on the electrode configuration, gas handling and substrate geometry.

Inductively-coupled tube reactor

This type of reactor system was used extensively in the pioneering days of glow discharge research by virtue of its simple design (Figure 6.1*a*). The RF power is coupled to the gas by an inductive coil wound around the quartz chamber. The substrate holder can be placed either in or below the plasma. Because of the contamination problems associated with the reactor wall and the difficulty of scale-up, this geometry has been superseded by planar electrodes.

Parallel-plate capacitatively-coupled reactor

This reactor type is used for pilot production work primarily because of the ease of scaling up the reactor size (Figure 6.1*b*). The use of planar electrodes allows various sizes, shapes and configurations of substrates. The substrate holder doubles as the heater and earthed electrode; the substrate is earthed to reduce the influence of ion bombardment. Since the grounded electrode is at a negative potential with respect to the plasma, the substrate does, however, suffer positive ion bombardment.

Hotwall tube reactor

The third reactor is based on the design of a conventional tubular CVD furnace (Figure 6.1*c*). A series of RF-driven electrodes creates the plasma, and support the substrates. Heat is supplied from the surrounding furnace to maintain the substrate at a nominal temperature.

For all three system types the gaseous precursors are admitted to the chamber by mass flow controllers. Typically, flow rates of between 20 sccm to 3 slm are employed depending on reactor size. The pumping speed of the system is controlled by a baffle or throttle valve and the process pumps are a Roots blower and rotary combination to give the pressure required (several hundred mTorr). A schematic of the system layout is shown in Figure 6.2, and typical conditions for plasma deposition of a-Si:H are shown in Table 6.1.

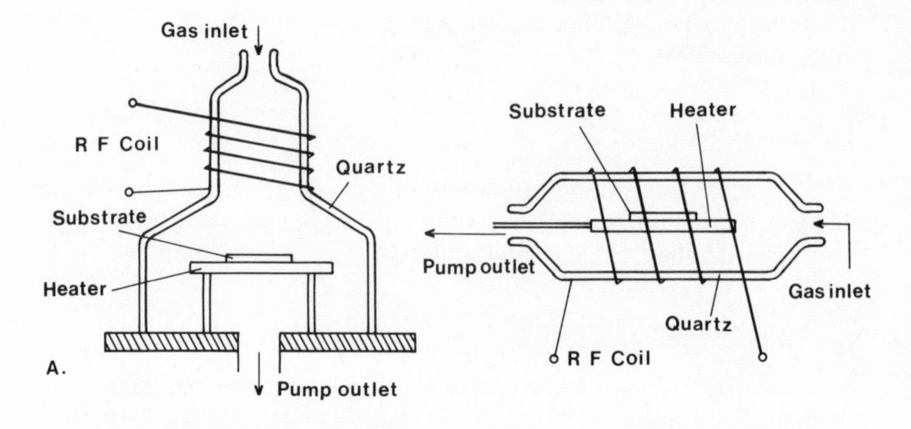

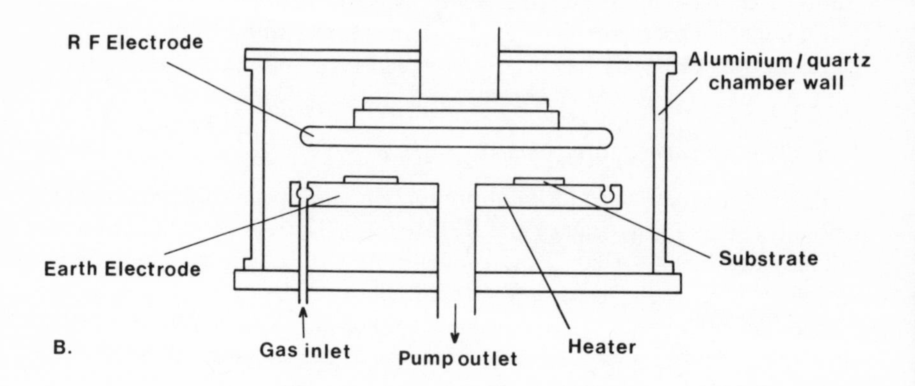

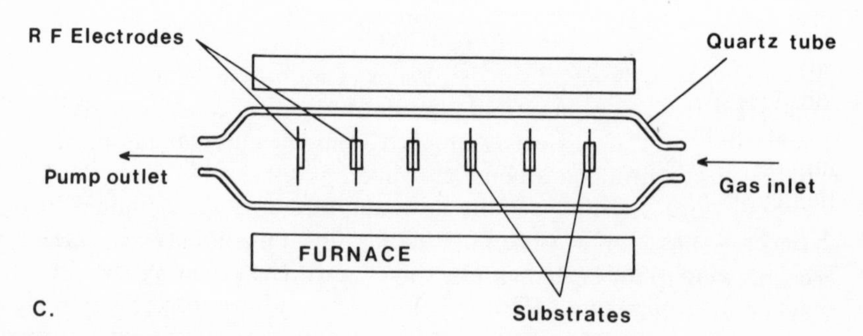

Figure 6.1 RF reactor configurations. *A*, inductive tube; *B*, parallel plate capacitor; *C*, hotwall tube

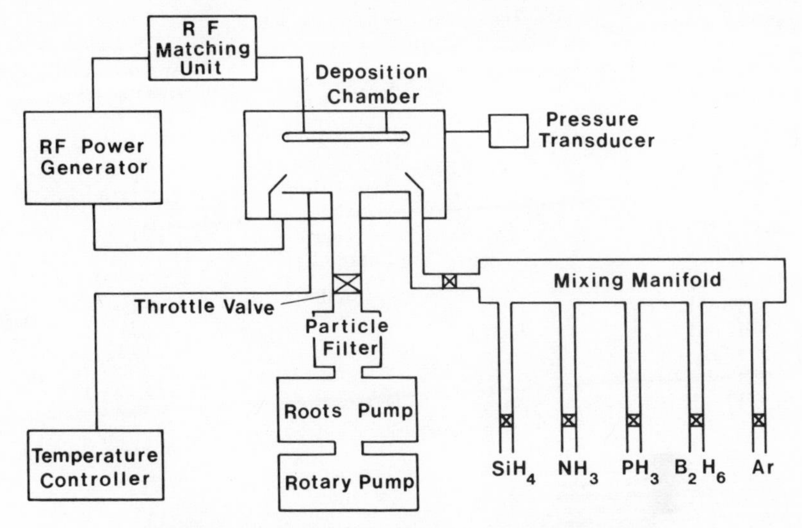

Figure 6.2 Glow discharge RF plasma deposition system

Table 6.1 Discharge conditions

Parameter	Range
SiH_4 concentration	10—100% in H_2 or inert gas
Total pressure	0.05–2.0 Torr
Flow rate	20–200 sscm
RF power	1–100 W
Substrate temperature	200–300° C
Bias	Normally zero

Plasma processes

The plasma deposition process consists of four main steps (3, 4):

(i) Plasma excitation/dissociation of the gaseous precursors
(ii) Transport of the chemically active species (ions, atoms and radicals) to the substrate surface
(iii) Heterogeneous reactions of adsorbed molecules on the substrate surface
(iv) Desorption of species from the substrate surface

It is the combination of these processes, in conjunction with a variety of secondary reactions in the plasma, which result in film formation.

Primary reactions

The plasma consists of partially ionized gas. The degree of ionization, i.e. the ratio of electron to gas molecule density, is low, at most around 10^{-4}. The

Table 6.2 Primary electron-impact reactions in hydrogen and silane

Reaction products	Process type	Appearance potential (eV)
$H_2 + e^-$		
$2H + e^-$	Dissociation	4.5
$H^+ + H + e^-$	Dissociative excitation	>14.7
$H^+ + H + 2e^-$	Dissociative ionization	18.1
$SiH_4 + e^-$		
$SiH_2 + H_2 + e^-$	Dissociation	2.2
$SiH_3 + H + e^-$		4.0
$Si + 2H_2 + e^-$		4.2
$SiH + H_2 + H + e^-$		5.7
$SiH^+ + H_2 + H + e^-$	Dissociative excitation	> 8.9
$Si^+ + 2H_2 + e^-$		9.5
$SiH_2^+ + H_2 + 2e^-$	Dissociative ionization	11.9
$SiH_3^+ + H + 2e^-$		12.3
$Si^+ + 2H_2 + 2e^-$		11.7
$SiH^+ + H_2 + H + 2e^-$		16.1

electrons gain energy in the RF field and undergo both elastic and inelastic collisions with the neutral atoms or molecules present. At RF frequencies in excess of approximately 1 MHz the ions, due to their low mobility, do not instantaneously respond to the ac field.

Table 6.2 lists the important primary reactions between electrons and hydrogen and silane respectively together with the appearance potentials (5, 6), updated, where appropriate, from the original work of Potzinger and Lampe (5).

Electron attachment occurs only for the low-energy electrons. Since their fraction is small in these plasmas, the process is of minor significance. At high electron energies, > 11eV, dissociative ionization occurs, resulting in the formation of radical ions SiH_n^+, $n = 0$ to 3.

From fundamental discharge physics, the reaction rates are dependent on macroscopic plasma or reactor parameters. The electron energy distribution function, which is approximately Maxwellian at high energies (> 1 eV), is determined by the plasma species and the voltage and frequency of the applied RF field. Typical plasmas use RF power densities of up to $1\ W\,cm^{-2}$ at 13.56 MHz and operate in the 100–1000 m Torr range. The excitation frequency is greater than the collision frequency to ensure that the electrons undergo primarily inelastic collisions with the atoms and molecules present. Under these conditions the mean electron energy is usually several electron volts with a density of 10^{10}–10^{11} electrons cm^{-3}.

Mass transport

Very little information is available as to the identity and energy of the species transported to the substrates. By using a grid to separate the discharge in which the species are generated from the transport region (7), the deposition rate can be measured as a function of grid–substrate spacing. It is then possible to use a one-dimensional transport model to determine the lifetimes and nature of the species. The important difference between SiH_4 and Si_2H_6 plasmas lies in the nature of the neutral fragments responsible for film growth. Diffusion of a single species towards the substrate in SiH_4 plasmas, in contrast to two species in Si_2H_6 plasmas, provides a possible explanation of the increased growth rates of the latter. The proposal (7) that one species is common to both systems suggests this may be either SiH_2 or SiH_3.

There is a consensus that, apart from cathodic films and plasma deposition at very low pressures, a-Si:H films arise primarily from the flux of neutral radicals to the gas–solid interface. The steady-state concentration of ions in silane plasmas is relatively low, and the measured ion flux onto the surface is insufficient to account for the observed growth rates (8). The importance of the ionization processes and ion–molecule reactions lies instead in their influence in establishing the levels of neutral species in the plasma. Mass spectroscopy has, to date, been the major technique for determining the free radical concentrations. Weakliem has pointed out the difficulties in quantifying such measurements (9).

Under low-power conditions, which give rise to electronic-quality a-Si:H, Gallagher and Scott (10) determined that SiH_3 is the most abundant radical and that the relative concentrations (11) are $SiH_3 > SiH > Si > SiH_2$. At higher RF powers the order differs in that Si atoms predominate. Earlier work by Bourquard (12) suggested, albeit by indirect methods, that the relative proportions were $SiH_2 > SiH_3 > Si > SiH$. This underlines the need for direct spectroscopic investigation of the temporal and spatial variations of radicals within the plasma. Progress in this area is reviewed in a subsequent section.

In a series of experiments aimed at identifying the species responsible for film growth, Longeway (13) has measured the initial rates of silane depletion and formation of stable products, e.g. H_2, Si_2H_6 and Si_3H_8 by selective-ion mass spectroscopy. A dramatic decrease in the rates of formation of products and the virtual elimination of film growth was encountered in the presence of NO. The latter is a well-known radical scavenger in hydrocarbon chemistry, but its reaction with silyl radicals is less well understood (14). As a result, the observation does not necessarily discriminate between SiH_2 and SiH_3 as the species responsible for film growth. Recent kinetic measurements (15) by Jasinski have revealed that SiH_2, as well as SiH_3, reacts rapidly with NO. Notwithstanding this observation, the plasma chemistry initially may not be representative of conditions at steady state. Summarizing, the question as to the identity of key species in film growth of a-Si:H remains unanswered.

Secondary and heterogeneous reactions

Irrespective of the relative flux of neutral radicals, including Si_2, to the substrate, there is little known about the reactions of these species with the growing film. Much of the speculation (16, 17) has been based on the established gas-phase chemistry. Attempts have been made recently to understand the nucleation phenomena by *in-situ* ellipsometry (18) but the detailed chemistry at the solid–gas interface is obscure.

Plasma diagnostics

The principal traditional techniques used to measure plasma properties are emission spectroscopy in the visible/UV spectral range, mass spectrometry and electrical measurements. In all cases the discharge must be analysed with zero or only minor disturbance to the plasma.

In optical spectroscopy the excited states of species in the plasma glow can be monitored by conventional spectroscopic techniques, and emissions from Si (288 nm), SiH (414 nm), H(656 nm) and H_2 have been characterized (19, 32). In the wavelength range 180–650 nm, emission due to excited electronic states of Si, SiH, H and H_2 is observed plus emission from N_2 and Cl impurities. Attempts at distinguishing the routes to the excited states have been reported (16). While the technique is non-intrusive, the intensity of emission does not, in general, provide information on the concentration of the ground-state species. It can be employed as a diagnostic technique, and empirical relationships (2) have been claimed for the dependence of optical emission intensity on RF power, gas flow rate, pressure and magnetic field. This is not surprising since the electron energy distribution and density varies with the reactor parameters. The disadvantage of optical emission is that, whilst it is useful for process control, it does not directly provide information on the species that are responsible for film growth.

Mass spectroscopic analysis (9) is perhaps the most common technique used to provide information about the chemical reactions that occur in the plasma. The discharge is sampled through an aperture small with respect to the plasma sheath thickness, although sampling without undue disturbance of the plasma is a major difficulty. Neutral species are detected by magnetically deflecting the ions in the extracted effusive beam and subjecting the neutrals to an external ionizer. A detailed discussion of ion–molecule reactions occurring in silane plasmas and further evidence as to the neutral radical SiH_x density has been given elsewhere (9, 16, 17). The findings can be summarized by stating that optical emission is observed only from the silicon species SiH and SiH^+, and that SiH_2^+ and SiH_3^+ are the dominant ions in a SiH_4 discharge.

The use of a Langmuir probe of insulated wire is a simple technique. The probe is biased by a battery and the probe current is measured, with respect to a reference electrode, in contact with the plasma. The resulting I–V characteristic can, with correct interpretation, reveal the plasma electron temperature, plasma potential and plasma density. In all three cases the probe stands a high probability of disturbing the plasma potential and hence the plasma itself (20, 33). In addition, film deposition on to the exposed probe constantly occurs, thus changing the probe characteristics. This appears to be less important in pure silane plasmas than in silane plasmas diluted with B_2H_6. Thus conclusions based on probe diagnostics are to be treated with a certain amount of suspicion and the data require care in interpretation (9).

High resolution infrared absorption and emission spectra of SiH_4/H_2 plasmas have been measured (21) in the 1800–2300 cm^{-1} range corresponding to the silicon–hydrogen stretching frequencies. Emission from the SiH radical was observed using a Connes-type interferometer with a 1 m optical path (21). Higher spatial resolution can be achieved (22) using coherent anti-Stokes Raman spectroscopy (CARS) using the ν_1 Q-branch of SiH_4. Preliminary results (23) have indicated that SiH_2 may be detected in SiH_4 plasmas, but this result awaits confirmation. Similarly, laser diode spectroscopy has demonstrated the presence of SiH and, possibly, other neutral radicals (24).

Long pathlength frequency modulation absorption spectroscopy, using a cw dye laser, has been employed (25) for detecting SiH_2 in both SiH_4 and Si_2H_6 RF discharge. There is growing interest in laser-induced fluorescence (LIF) for the *in-situ* measurement of Si atoms, Si_2, SiH and SiH_2 (26). The technique is not necessarily non-intrusive. Laser-induced photochemistry of the species within the plasma may lead to incorrect interpretation of the data. Work has also been reported (27) on the LIF detection of halogenated radicals, e.g. SiF within the corresponding plasma.

Plasma deposition of electronic materials

The versatility of the PECVD process is such that, by selection of suitable gaseous precursors, a variety of materials and alloys can be deposited.

Amorphous silicon

Most device-quality material is currently produced using silane as the reactive gas, with the option of hydrogen or helium as a diluent. Although amorphous silicon solar cells are produced industrially (28), the deposition rate is a crucial factor in determining the commercial viability.

A number of ways of increasing this have been explored. One possibility is to use higher silanes, such as Si_2H_6 or Si_3H_8, for which there is a rate enhancement (29, 30) of a factor of over 20 compared to silane. The resulting films exhibit comparable photoluminescence properties compared to SiH_4 films but enhanced AMI photoconductivity. At lower substrate temperatures, the method suffers from the disadvantage of increasing the proportion of H in the SiH_2 configuration and a concomitant increase in the defect state density.

Table 6.3 Amorphous silicon deposition conditions

	Low frequency (85 kHz)	High frequency (13 MHz)
Flow rate (sccm)	106	100
Pressure (mTorr)	130	120
Power (W)	25	5
Temperature (°C)	300	300
Deposition rate (Å min^{-1})	100	75
Bandgap (eV)	1.68	1.74
H content (at %)	22	8

Other halogens, especially F, have been advocated as passivating atoms for amorphous silicon; however, the larger and heavier the atom, the less efficacious the compensation. The use of SiF_4 with H_2 or SiH_4 also increases the deposition rate (31). *P*-type or *n*-type material can be grown *in situ* using the preferred dopant gases of diborane or phosphine respectively. Typical a-Si:H characteristics for low-frequency (85 kHz) and high-frequency (13 MHz) deposition conditions are shown in Table 6.3 (32).

Silicon–carbon alloys

The interest in silicon-carbon alloys stems from their applications in C-doped silicon films as wide bandgap windows for heterojunction solar cells (28) to SiC as an insulator and for hard coatings. Additionally this material is of interest for dielectric insulation where superior physical properties such as stress and cracking resistance are expected compared to silicon nitride or silicon oxide. Commonly the films are deposited from mixtures of silane and methane (33) or ethylene (34). One of the biggest problems is that of forming defect states within the band gap associated with $SiCH_3$ and similar groupings. Recently there has been interest in using freon-based gases to prevent polymerization (31), although these films prove to be less photoconductive than films prepared from hydrides.

Carbon

Carbon layers can be deposited using CH_4, C_2H_2 or C_6H_6 as the precursors. Typically the carbon films are insulating (1012 ohm cm), with a high dielectric strength (106 V cm^{-1}) and are extremely hard 4000 HV (Vickers hardness $\sim$9000 kg mm^{-2}) (32). Applications for these films include glass coatings, antireflection coatings for Ge-window IR detectors, anti-abrasion coatings and heat sinks.

Amongst the different phases of carbon, the production of diamond is an intriguing possibility. In order to preserve the diamond phase the process must be carried out at low pressures and temperatures such that the phase transition of diamond into graphite is retarded. Higher power is necessary to initiate the discharge, but suffers from the possibility of contamination with silica impurities (36). Diamond growth occurs in the presence of H_2 diluent compared to Ar or He which promote the formation of graphite (36). To create 'diamond-like' carbon films a high proportion of tetrahedrally bonded carbon is required. A recent model has shown that the films are more usually threefold co-ordinated graphitic sheets linked by tetrahedrally bonded carbon atoms (37). Under conditions investigated to date, the tetrahedrally bonded carbon is invariably attached to at least one H atom. There is convincing evidence (38), however, for the growth of a true diamond phase in thermal (38) and microwave plasma-enhanced CVD systems (39).

Silicon–germanium alloys

At the other end of the solar spectrum the red response is improved by reducing the bandgap by the incorporation of germanium. Common preparations (40) use GeF_4 in conjunction with SiF_4 and SiH_4 respectively. In some cases germane is used in conjunction with disilane in photolytic deposition or ultraviolet-enhanced CVD.

Silicon nitride

PECVD silicon nitride is perhaps the most popular low-temperature passivation dielectric. The nitrogen source is ammonia, nitrogen or a combination of both. The macroscopic reactor parameters determine whether the covalently-bound hydrogen is located at Si or N sites, and this in turn determines the electronic characteristics of the material (41). By judicious selection of operating parameters it is possible to fabricate a material suitable for gate dielectric applications, with a breakdown strength in excess of $8\,MV\,cm^{-1}$. Recent work involves the use of NF_3 instead of NH_3 for higher thermal stability.

Silicon oxide

The plasma production of silicon oxide is one of the most natural directions for low-temperature insulator work to proceed. The high performance of thermally-grown oxide layers indicates that the properties of silicon oxides could be superior to nitrides. Silicon oxide can be fabricated by the plasma oxidation of SiH_4 using O_2 (42), N_2O (43) or CO_2 (44) and via the plasma decomposition of tetraethoxysilane (TEOS) (45).

Nitrous oxide, N_2O, is the most commonly used oxidant as it is readily available, non-toxic and does not leave carbon contamination. Plasma reactions using oxygen suffer from the problem of producing comparatively large quantities of silica dust and the process is more hazardous. Low-temperature oxide films usually contain traps (formed by strained bonds) throughout the deposited layer, which can be removed by high-temperature annealing. In some instances, however, it is possible to fabricate an amorphous stoichiometric SiO_2, when the SiH_4 and O_2 flow rates are approximately equal (46). A very large excess of N_2O is required ($N_2O/SiH_4 \sim 50$) to produce stoichiometric SiO_2, although the exact ratio proves to be dependent on reactor parameters and in particular the RF power.

Silicides

Aluminium and polysilicon for VLSI interconnections suffer from the severe limitations of electromigration and high sheet resistivity respectively. The versatility of the PECVD process offers the possibility of producing refractory metal silicides. The source gases typically used are the corresponding metal halides or carbonyls due to the high vapour pressures of these compounds (45). The ratio of metal to silicon incorporated in the silicide is controlled by the ratio of the source gases and it is possible to deposit films with only a few percent silicon.

In the case of tungsten silicide (WSi_2), the reactions between SiH_4 and WF_6 or $W(CO)_6$ have been employed, although the latter process may lead to carbon contamination. The step coverage of these films is superior to the competing physical vapour techniques. The conductivity of as-deposited films can be improved by post-deposition anneals ($\sim 600°$ C), driving out fluorine or hydrogen from the film. In the case of $TiSi_x$, resistivities of < 20 μohm cm have been achieved (46).

Future considerations

The most important question to be asked in PECVD processes concerns the identification of the species responsible for film growth. Many arguments have been presented in support of a particular SiH_n ($n = 0$ to 3) radical in silane plasmas but definitive conclusions have not yet been reached. The role of ions in the growth mechanism is yet to be understood and the effect of secondary decomposition on the growing film is unknown. Despite the extensive work performed to date more questions are now posed, in the field of plasma chemistry and kinetics, than have been answered. Plasma diagnostics will play an increasingly important role in helping to address these questions. Especially important will be the development of non-intrusive techniques which allow both the temporal and spatial concentrations of species responsible for film growth to be measured.

References

1. J. Stuke, *Ann. Rev. Mater. Sci.* **15** (1985) 79.
2. M. Hirose, in *Semiconductors and Semimetals*, Vol. 21A, *Hydrogenated Amorphous Silicon*, ed. J. I. Pankove, Academic Press, New York (1984).

3. S. Hotta, N. Nishimoto, Y. Towada, H. Okamoto and Y. Hamakawa, *Jap. J. Appl. Phys.* **21** (1982) Suppl. 21, 289.
4. S. Hotta, Y. Towada, H. Okamoto and Y. Hamakawa, *J. Phys.* **42** (1981) C4.
5. P. Potzinger and F. W. Lampe, *J. Phys. Chem.* **73** (1969) 3912.
6. G. Turban, Y. Catherine and B. Grolleau, *Thin Solid Films* **67** (1980) 309.
7. A. Matsuda, T. Kaga, H. Tanaka and K. Tanaka, *J. Non-Cryst. Solids* **59/60** (1983) 687.
8. G. Turban, Y. Catherine and B. Grolleau, *Thin Solid Films* **77** (1981) 287.
9. H. A. Weakliem, in *Semiconductors and Semimetals*, Vol. 21A, *Hydrogenated Amorphous Silicon*, ed. J. I. Pankove, Academic Press, New York (1984).
10. A. Gallagher and J. Scott, Ann. Rep. to SERI for contract nos. XB-2-02085-1 and DB-2-02189-1, Solar Research Institute, Golden, Colorado (1983).
11. R. Robertson, D. Hills, H. Chatham and A. Gallagher, *Appl. Phys. Lett.* **43** (1983) 554.
12. S. Bourquard, D. Erni and J. M. Mayor, in *Proc. V Int. Symp. on Plasma Chemistry*, Heriot-Watt University, eds. B. Waldie and G. A. Farnell (1981) 664.
13. P. A. Longeway, R. D. Estes, H. A. Wiekliem, *J. Phys. Chem.* **88** (1984) 73.
14. I. M. T. Davidson, *RSC Ann. Rep.*, Section C (1986) 47.
15. J. M. Jasinski, private communication.
16. F. J. Kampas, in *Semiconductors and Semimetals*, Vol. 21A, *Hydrogenated Amorphous Silicon*, ed. J. I. Pankove, Academic Press, New York (1984).
17. P. A. Longeway, *idem.*
18. R. W. Collins and A. Pawlowski, *J. Appl. Phys.* **59** (1986) 1160.
19. C. C. Tang, J. K. Chu, D. W. Hess, *Solid State Technol.* (1983) 125.
20. R. S. Rosler and G. M. Engle, *J. Vac. Sci. Technol. (B)*, **4**, (1984) 733.
21. J. C. Knights, J. P. M. Schmitt, J. Perrin and G. Geulachvili, *J. Chem. Phys.*, **76** (1982) 3414.
22. N. Nata, A. Matsuda, K. Tanaka, K. Kajiyama, N. Moro and K. Sajiki, *Jap. J. Appl. Phys.* **22** (1983) L1.
23. N. Hata, A. Matsuda and K. Tanaka, *J. Non-Cryst. Solids* **59/60** (1983) 667.
24. P. B. Davies, N. A. Isaacs, S. A. Johnson and D. K. Russell, *J. Chem. Phys.* **83** (1985) 2060.
25. J. M. Jasinski, E. A. Whittaker, G. C. Bjorklund, R. W. Dreyfus, R. D. Estes and R. E. Walkup, *Appl. Phys. Lett.* **44** (1984) 115.
26. For references on the LIF of the individual species see: Si, (*a*) R. M. Roth, K. G. Spears and G. Wong, *Appl. Phys. Lett.* **45** (1984) 28; (*b*) R. M. Roth, K. G. Spears and G. Wong, *Mat. Res. Soc. Symp. Proc.* **38** (1985) 117: SiH, (*c*) J. P. M. Schmitt, P. Gressier, M. Krishnan, G. deRosny and J. Perrin, *Chem. Phys.* **84** (1984) 281; (*d*) W. Bauer, K. H. Becker, R. Duren, C. Hubrich and R. Meuser, *Chem. Phys. Lett.* **108** (1984) 560 and references within: SiH_2, (*e*) G. Inoue and M. Suzuki, *Chem. Phys. Lett.* **105** (1984) 641; (*f*) *ibid.*, **122** (1985) 361; (*g*) D. M. Rayner, R. P. Steer, P. A. Hackett, C. L. Wilson and P. John, *Chem. Phys. Lett.* **123** (1986) 449; (*h*) J. W. Thoman and J. I. Steinfeld, *Chem. Phys. Lett.* **124** (1986) 35: Si_2, (*i*) P. Ho and W. G. Brieland, *Appl. Phys. Lett.* **44** (1984) 51.
27. H. U. Lee, J. P. Deneufuille and S. R. Ovshinsky, *J. Non-Cryst. Solids* **59/60** (1983) 671.
28. (*a*) K. Takahashi and M. Konagai, *Amorphous Silicon Solar Cells*, North Oxford, London (1986); (*b*) Y. Hamakawa (ed.) JARECT, *Amorphous Semiconductor Technologies and Devices*, North-Holland, Amsterdam (1984).
29. B. A. Scott, R. M. Plecenik and E. E. Simonyi, *Appl. Phys. Lett.* **39** (1981) 73.
30. A. Matsuda, T. Kaga, H. Tanaka and K. Tomaka, *J. Non-Cryst. Solids* **59/60** (1983) 687.
31. H. Matsumura, T. Uesugi and H. Ihara, *MRS Symp. Proc.* **49**, (1985) 175.
32. J. Janoa, M. Drsticka and R. Covelik, *Scr. Fac. Sci. Nat. Univ. Pirkynianae Brunensis Phys.* (Czechoslovakia) **15** (1985) 251.
33. Y. Tawada, K. Tsuge, M. Kondo, H. Okamoto and Y. Hamakawa, *J. Appl. Phys.* **53** (1982) 5273.
34. D. A. Anderson and W. E. Spear, *Phil. Mag.* **35** (1976), 1.
35. V. P. Varmin, I. G. Terenktskaya, D. V. Fedoseev and B. V. Deryagin, *Sov. Phys. Dokl* **29** (5) (1984) 419.
36. O. Matsumoto, H. Toshima and Y. Kanzaki, *Thin Solid Films* **128** (1985) 341.
37. D. R. McKenzie, R. C. McPhedran and N. Savvides, *Thin Solid Films* **108** (1983) 247.
38. S. Matsumoto, Y. Sato, M. Tsutsumi and N. Setaka, *J. Mat. Sci.* **17** (1982) 3106.
39. M. Kamo, Y. Sato, S. Matsumoto and N. Setaka, *J. Cryst. Growth* **62** (1983) 642.
40. D. Slobodin, S. Aljiski and S. Wagner, *MRS. Symp. Proc.* **49** (1985) 153.

41. B. L. Jones, *J. Non-Cryst. Solids* **77/78** (1985) 957.
42. C. Pavelsecu and C. Cobianu, *J. Electrochem. Soc.* **130** (1983) 975.
43. B. Gorowitz, T. B. Gorczyca and R. J. Saia, *Solid State. Technol.* **29** (1985) 197.
44. S. P. Mukharjee and P. E. Evans, *Thin Solid Films* **14** (1972) 105.
45. K. Strater, *RCA Rev.* **29** (1968) 618.
46. A. Chevenas-Paule, private communication.

7 Metal-organic vapour phase epitaxy

N. J. MASON

Introduction

Ever since the days of the cat's whisker and the crystal radio set, the production of single-crystal semiconductor material has been an important aspect of research and development in the electronics industry. The current state of the growth of bulk single-crystal materials has been discussed in Chapters 2 and 3. This chapter deals with a technique for the growth of films on single-crystal substrate wafers which are treated chemically to act as a crystal 'template' for the continued growth of the crystal. Thus, to a first approximation, there is no change in the lattice of the semiconductor in moving through the interface. Other parameters, e.g. dopant concentration or type, may change at the interface, but the crystal structure is continuous. This effect is called *epitaxy*, from the Greek *epi* (upon) and *taxis* (ordered).

This aspect of semiconductor growth leads to the first necessity of any growth technique: it must be capable of growing good-quality single-crystal material, with an interface epitaxially congruent with the substrate. For an elemental semiconductor with a diamond-like lattice, for example silicon (see Chapter 2), this involves growing the crystal without atoms missing from lattice sites (vacancies), or atoms in between their correct location in the lattice (interstitials), or with two planes of atoms not meeting correctly (dislocations). Compound semiconductors like gallium arsenide (GaAs) have a zinc blende structure, i.e. a diamond lattice with every alternate atom either gallium or arsenic. Growth of compound semiconductors may thus pose the additional problem of the crystal growing with gallium on the arsenic sub-lattice and vice-versa (antisite defects). With the addition of a third element to make a ternary alloy like aluminium gallium arsenide (AlGaAs) there are further complications to growing good-quality crystals which will be discussed below.

The nomenclature of the film-growth technique outlined in this chapter is sometimes confused. The term *metal-organic chemical vapour deposition* (MOCVD) covers both the growth of epitaxial material and the growth of polycrystalline or metallic layers. On the other hand, *metal-organic vapour phase epitaxy* (MOVPE) is specific to the growth of single-crystal material. It is this type of growth that predominates in the UK semiconductor industry, and the international conference on the subject has adopted ICMOVPE as its initials. However, most of the chapter is also relevant to polycrystalline growth.

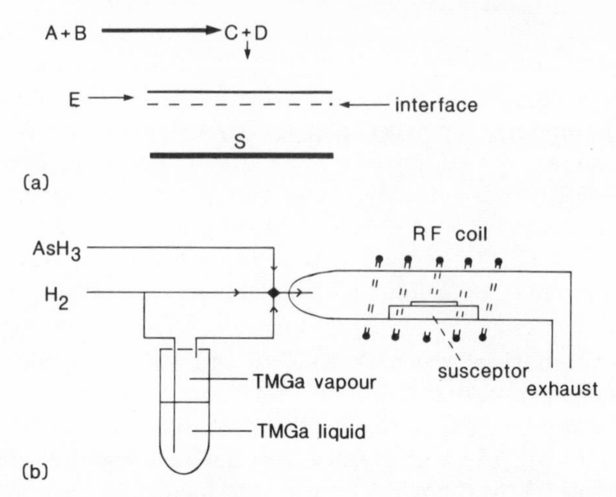

Figure 7.1 (*a*) Schematic of MOVPE process; (*b*) schematic of MOVPE equipment

MOVPE is part of a wider family of VPE (vapour-phase epitaxy) growth techniques which make use of the reactions occurring between the vapours of volatile chemical compounds *A* and *B* when they are heated together (see Chapter 5). These reactions produce chemically active species *C* and *D* that interact, either in the vapour phase or on a solid surface of the substrate *S* to produce a corresponding epitaxial layer *E*. This is illustrated in Figure 7.1(*a*).

MOVPE specifically utilizes vapours from metal-organic compounds. (The term 'metal-organic' tends to be used to include organometallic compounds like metal alkyls, for instance trimethylgallium, and, more rarely, the metal alkoxides which have no metal–carbon bond and are therefore not strictly organometallic. Each alkyl is usually referred to as a four-letter acronym which includes the chemical symbol of the metal. Typical starting materials used to grow GaAs are trimethylgallium (TMGa) and arsine (AsH_3). These two compounds can react at 700° C under hydrogen to produce epitaxial GaAs.

$$(CH_3)_3Ga + AsH_3 \xrightarrow{H_2} GaAs + 3CH_4 \tag{7.1}$$

This reaction was first reported in 1968 (1) and can be used as an illustration of a simple MOVPE reactor. Trimethylgallium is liquid at STP, and is held in a stainless steel bubbler where the bubbles of hydrogen become saturated with TMGa vapour; this vapour is then transported to the reaction cell. Gaseous AsH_3 is supplied either neat or diluted with hydrogen in a steel cylinder. The pressure is reduced to the reactor pressure (usually atmospheric but sometimes 77 Torr). A typical schematic set-up is shown in Figure 7.1(*b*).

The two gas streams are intimately mixed and passed over a heated graphite block called a susceptor, which holds a GaAs single-crystal wafer.

Energy for heating the susceptor is usually provided by a high-power radiofrequency generator operating at about 450 kHz. TMGa and AsH_3 are heated in the reaction zone and react to deposit GaAs. This simple scheme ignores all the temperature-controlled baths, flow controllers and valves that make up a real reactor, but it does give the basis of MOVPE. The growth of indium phosphide is similar, save that the carrier gas is sometimes helium or nitrogen or a combination of these with hydrogen.

Probably the most important aspect of material preparation in the semiconductor industry is obtaining material of known (usually high) purity. In traditional areas of chemistry, materials or compounds of sufficient purity are readily obtainable by standard methods, but semiconductors are difficult to obtain with an acceptable purity. Dopants may be added in minute, controlled, amounts in order to impart desired electrical properties to the semiconductor; but this needs pure starting materials to be successful.

In real semiconductor materials, total purity is not possible: a crystal of 'pure' GaAs contains about 10^{23} atoms cm^{-3} of which approximately 10^{13} would be impurities. A useful growth technique should be able to achieve this level of purity (less than 1 part per billion), which places stringent constraints on the purity of the starting materials.

Group VI elements in GaAs (e.g. S, Se, Te) replace As in the lattice, release electrons in the conduction band, and are *n*-type. Group II elements (Be, Mg, Zn, Cd) replace Ga, accept electrons, and are *p*-type. When such elements are unavoidably present, they are known as *n*-type and *p*-type impurities; if added deliberately, they are dopants. Oxygen and transition metals behave as deep centres able to trap either electrons or holes.

After crystal quality and purity, the final important parameter in any growth technique is control of growth rate of the crystal in order to produce layers of uniform and reproducible thickness and to produce interfaces of suitable abruptness for the device in question. In the case of vapour-phase epitaxy this implies close control of the concentration of the reactants in the vapour phase. These three parameters of crystal growth (crystal quality, purity and growth rate) will be discussed with respect to MOVPE.

MOVPE is used to grow compound semiconductors. These are more complicated to grow than elemental semiconductors (like silicon or germanium) but they have special properties that make them an increasingly important aspect of industrial semiconductor growth. The group of binary III–V semiconductors includes GaAs, InP and other combinations of III and V elements, e.g. AlSb, InSb, AlP, GaSb, AlAs and GaP. It is also possible to have different Group III elements on the same sub-lattice, e.g. aluminium gallium arsenide (AlGaAs), or to combine different Group V elements, e.g. gallium arsenide antimonide (GaAsSb), thus growing a ternary alloy. For some applications it is necessary to combine four elements into a quaternary semiconductor, e.g. indium gallium arsenide phosphide (InGaAsP). For ternary and quaternary III–V materials, the ability to vary the physical

Table 7.1 Compound semiconductors and their applications

Compound	Lattice constant (Å)	Bandgap (eV)	Wavelength (μm)	Reference	Industrial application
GaP	5.450	2.24(I)	—	(2)	See InGaP
AlAs	5.66	2.16(I)	—	(3)	See AlGaAs
AsSb	6.136	1.6(I)	—	(4)	Solar cells
GaAs	5.653	1.43(D)	0.92	(5)	Waveguides
InP	5.87	1.36(D)	1.05	(6)	MESFET
GaSb	6.10	0.72(D)	1.83	(7)	See GaAsSb
InAs	6.06	0.35(D)	3.76	(8)	See InGaAs
InSb	6.479	0.18(D)	7.32	(9)	IR photodetectors
$GaAs_{1-x}P_x$	5.65	1.9	—	(9)	Lasers, $x = 0.39$
$GaAs_{1-x}Sb_x$	5.65–6.10	—	—	(11)	Solar cells
$InAs_{1-x}Sb_x$	—	—	—	(12)	IR photodetectors
$Al_xGa_{1-x}As$	5.65	1.43–2.16	—	(13)	Lasers, transistors $x = 0.3$
$In_xGa_{1-x}As$	5.87	—	—	(14)	Lasers, $x = 0.47$
$In_xGa_{1-x}P$	5.65	1.9	—	(8)	$x = 0.5$ (LED)
$Al_xIn_{1-x}As$	5.87	—	—	(15)	$x = 0.48$

I, indirect; D, direct bandgap. These terms refer to the electrical efficiency of the transition responsible for the emission of radiation. In an indirect transition, momentum is not conserved and the radiative recombination has a much lower probability than a direct transition. The quantum efficiency of indirect transitions can be 2 orders of magnitude less than direct ones.

The x values for InGaAs and AlInAs are those for lattice matching to an InP substrate, those for InGaP and GaAsP, to a GaAs substrate.

properties with alloy compositions is one of the factor which render these systems potentially useful in solid state electronics. The III–V bandgaps range from 0.172 eV (InSb) to 2.45 eV (AlP), corresponding to light of wavelength 7.30 μm to 0.51 μm. If an alloy of two binaries is formed, the ternary or quaternary compound has a bandgap and a lattice constant that can be varied over the entire range between the corresponding values of the two binaries. Thus it is possible to 'engineer' the bandgap and hence the optical wavelength in a semiconductor. However, not all the material systems needed can yet be grown, and this is the driving force for much of the research in MOVPE.

The role of impurities and dopants discussed above is the same in all types of III–V semiconductor materials. Some of the compound semiconductors that can be grown by MOVPE, together with their industrial application, are given in Table 7.1. Because of the importance of the emission and adsorption of light by III–V semiconductors, the table includes the wavelength of radiation equivalent to the energy gap.

The Group II–VI semiconductors, for example ZnS and ZnSe, are also grown by MOVPE. These have important optical characteristics not yet available to III–V semiconductors, including use in image intensifiers for night vision systems (see Chapter 4 for a fuller description of II–VI semiconductor materials).

This chapter concentrates on the growth of III–V semiconductors; most of the discussion will be relevant to growth of II–VI semiconductors, although some of these materials, such as CdHgTe, are more difficult to grow by MOVPE than III–V material. Such problems are discussed in a recent review (16)

Starting materials for MOVPE

As well as TMGa and AsH_3, there are a number of other compounds used either to grow the III–V host lattices or to dope the lattice *n*- or *p*-type. The compounds are listed in Table 7.2, together with their use in each compound system and an indication of their volatility. Sometimes two compounds (trimethyl and triethyl) are available to transport a given element, and these alternatives will be discussed below in the relevant section.

The compounds listed in Table 7.2 do not exhaust the possible sources of the elements used in the growth and doping of III–V semiconductors; however, with the exception of adducts mentioned below, they are those found to be of practical value in MOVPE. Higher homologues, e.g. *n*-propyl or butyl alkyls are too involatile, at the temperatures easily obtained in MOVPE apparatus, to be transported in reasonable concentration by the carrier gas. The compounds must also be capable of purification to below the ppb level.

Table 7.2 III–MOVPE starting materials and their volatility

Compound	Acronym	Role of element	B.p. (1 atm.) °C	Reference
Trimethylgallium	TMGa	III sub-lattice	55.7	(17)
Triethylgallium	TEGa	III sub-lattice	143	(17)
Trimethylaluminium	TMAl	III sub-lattice	126	(18)
Triethylaluminium	TEAl	III sub-lattice	194	(18)
Trimethylindium	TMIn	III sub-lattice	133.8	(19)
Triethylindium	TEIn	III sub-lattice	184	(20)
Trimethylantimony	TMSb	III sub-lattice	80.6	(7)
Triethylantimony	TESb	V sub-lattice	160	(12)
Arsine	AsH_3	V sub-lattice	gas	(21)
Phosphine	PH_3	V sub-lattice	gas	(22)
Hydrogen selenide	H_2Se	*n*-dopant	gas	(23)
Hydrogen sulphide	H_2S	*n*-dopant	gas	(24)
Silane	SiH_4	*n*-dopant	gas	(25)
Disilane	Si_2H_6	*n*-dopant	gas	(26)
Tetramethyltin	TMSn	*n*-dopant	78	(27)
Tetraethyltin	TESn	*n*-dopant	181	(28)
Dimethylzinc	DMZn	*p*-dopant	46	(29)
Diethylzinc	DEZn	*p*-dopant	118	(29)
Diethylberyllium	DEBe	*p*-dopant	194	(30)
Dimethylcadmium	DMCd	*p*-dopant	105.5	(31)
Bis-cyclopentadienyl magnesium	BCp_2Mg	*p*-dopant	solid	(32)

Synthesis of III–V alkyls

Until recently the routes to metal alkyl compounds could be broadly divided into two groups–those which start from the pure metal and those which start from the metal halide. Recently (33) Epichem has pioneered the preparation of high-purity TMGa from gallium–magnesium alloy and MeI (Table 7.3).

The provision of pure metal-organics is sufficiently important to have led to a number of studies of the gas-phase purity of alkyl compounds. A variety of commercial alkyl compounds were investigated by distilling them and analysing the distillate before it condenses (34). A molecular jet separator inlet system was used in conjunction with a quadrupole mass spectrometer. All TMGa peaks were identified together with some major impurities; the main

Table 7.3 Synthetic routes to TMGa

Gallium source	Alkyl source	Reaction type	Solvent	Reaction time (h)	Yield (%)	Purity of starting materials
Ga	MeMgI	Electrolysis	R_2O	360	90	Very good
Ga	MeMgI	Electrolysis	Et_2O	180	66	Very good
$GaCl_3$	MeMgI	Chemical	R_2O	19	70	good
$GaCl_3$	MeMgI	Chemical	Et_2O	36	60	good
Ga_2Mg_5	MeI	Chemical	R_2O	42	50	good

impurity was MeI, other possible contaminants being ethers and low-molecular-weight hydrocarbons. One serious problem encountered was the difficulty of separating TMGa and MeI, even by a distillation column equivalent to 100 theoretical plates. This has worrying implications for the purification of TMGa by distillation.

When grown layers are analysed, the dominant residual acceptor impurity is found to be carbon (35, 36) or silicon (37). The incorporation of carbon has been extensively investigated, especially with regard to the influence of the alkyl type and the growth pressure. All the Group III metals have at least two, and sometimes three, metal alkyls suitable for use as agents to transport the metal species to the reaction zone. An alternative source of gallium is triethylgallium (TEGa), although in this case it is only possible to use the compounds in a reactor where the ambient pressure is 77 Torr or thereabouts; this is also true of TEIn. Initially MOVPE growth took place at or slightly above atmospheric pressure using TMGa and TMIn. This is still the case in the majority of industrial and academic growth facilities in the UK. However, overseas growers, especially in the USA (37) and France (38), have championed growth at 77 Torr or below. They claim that undoped GaAs grown using TEGa as a starting material has considerably fewer impurities than that grown from TMGa (it is only possible to use TEGa at reduced pressure, usually 77 Torr). It is thought that this effect is due to the relative stabilities of TMGa and TEGa (39), and two studies (40,41) appear to support this view. The first (40) involved the investigation of the diffusivity and thermal cracking rate of various alkyls by chromatography. The alkyls TMGa, TEGa, TMAl and TEAl were found to have diffusivities of the same magnitude as the common hydrocarbons with the same carbon number. From the thermal cracking data, it was concluded that trimethyl compounds are more stable than the corresponding triethyl compounds. The second (41) is a mass spectrometric study ot TMGa and TEGa decomposition in H_2 and N_2. TMGa is found to decompose to methane, but TEGa to ethylene. However, perhaps because few of the industrial devices grown by MOVPE in the UK need undoped material of very high purity, the technique of growth at 77 Torr has not established itself in the UK.

The choice between TMIn and TEIn is also complicated by their respective side reactions with PH_3. Much research has been carried out on the reaction of the indium alkyls with AsH_3 and PH_3. The alkyls are planar molecules with an empty *p*-orbital above and below the plane of the molecule. The Group V species is pyramidal, with a lone pair occupying the apical position. This lone pair can be donated on to the vacant *p*-orbital on the alkyl. TMGa and AsH_3 are not observed to undergo this reaction, but in the growth of InP or related materials the reaction

$$Me_3In + PH_3 \rightarrow InMe_3.PH_3 \rightarrow (\text{—}InMePH\text{—})_n + 2nCH_4 \qquad (7.2)$$

takes place in the gas phase after the reactants have been mixed but before they reach the heated zone of the cell (42). The polymeric compound is an involatile solid, and, depending on the location of the reaction (in a narrow tube or in the cell) the consequences can be either disastrous or merely inconvenient. A variety of ingenious and elegant solutions to this problem has been suggested, including pre-cracking the PH_3 to P_4 (by passing the PH_3 through a chamber at 500° C), avoiding mixing the species in question until just before the reaction zone, and the chemical alteration of the alkyl to block the vacant *p*-orbital. It is this latter route that has led to the important work on adducts discussed below.

This reaction has been investigated by many groups, and recently it has been reported (43) that the problem is no longer observable in recent batches of TMIn. This raises the possibility that the reaction was caused by impurities that are now removed by the improved synthetic and purification techniques used in alkyl manufacture.

III–V hydrides

Arsine and phosphine are toxic and colourless gases with a garlic-like and decaying fish-like odour respectively. Although the Group V elements P and As can be transported as their hydrides, it is not possible to use stibine (SbH_3) to transport antimony because of instability of the gas.

The synthesis and purification of the hydrides, arsine (AsH_3) and phosphine (PH_3), is relatively straightforward. In practice, the purity of the gases which actually enter the reactor leaves more to be desired than the purity of the alkyls. Unlike the alkyl manufacturers, the UK's manufacturers of semi-conductor gases and MOVPE growers have no interaction, so this problem has yet to be resolved.

The main drawback of the hydrides is contamination by water and oxygen (44). Whilst this is not especially critical in the growth of GaAs, it has serious consequences in the growth of AlGaAs, which are of considerable commercial significance. The problem lies in the reaction of the oxygen or water with the aluminium alkyl in the vapour phase. The reactions are (45):

$$2R_3Al + O_2 \rightarrow 2AlOR.R_2 \tag{7.3}$$

$$2R_3Al + 3H_2O \rightarrow Al(OH)_3 + 3RH \tag{7.4}$$

In the first reaction, the alkyl forms a volatile alkoxide which is transported to the reaction zone and results in the inclusion of oxygen in the grown layer. This forms an electron trap which, for some applications, ruins the optical and electrical properties of the material. In the second reaction the alkyl forms the involatile oxide and, whilst this will deplete the available alkyl from the gas stream and alter the composition of the ternary compound, it is not as important as the first reaction. A number of attempts have been made to circumvent the problem of water and oxygen in the gas stream entering the reactor and these are detailed below.

Many contaminant gases can be removed from AsH_3 or PH_3 by the use of a simple molecular sieve of suitable size (46). However, the sieve soon becomes saturated with water and this is not easily monitored except by the deterioration in the semiconductor crystal quality. A liquid-nitrogen cold trap has also been employed to remove water and oxygen from reactor gas streams (47). Whilst this has the advantage of monitoring the contamination present in the trap, care must be exercised in warming the trap to room temperature.

A third alternative for removing water and oxygen from the gas stream is the use of an Al-In-Ga eutectic (48). Elemental aluminium is usually rendered passive to the attack of water or oxygen by a layer of oxide that grows on the surface of the metal. By bubbling the gas stream through the eutectic, the contaminants are able to come into contact with elemental aluminium. The Al_2O_3 is precipitated to the surface of the eutectic and no longer prevents the continuing oxidation of the remaining aluminium. The solid oxide indicates when the eutectic needs replacing.

Three other methods have been developed to deal with the gas stream contaminants, and with residual contaminants in the cell caused by insufficient purging of the cell after opening and loading the reactor. Ideally, the loading of samples into a MOVPE reactor should be followed by extensive purging of the reactor and leak checking to ensure good integrity of joints and seals that have been disturbed during loading. Commercial pressures do not always allow this, especially in a production environment, and the two methods below allow a compromise solution to the problem.

The first (49) involves the use of graphite baffles in the reaction cell (upstream of the susceptor). These adsorb TMAl or TEAl from the reactant gas stream; this adsorbed TMAl reacts as outlined in (7.3) and (7.4) above, but being adsorbed it also leaves the $Al(OH)_3$ and alkoxide adsorbed and unable to be incorporated in the growth of AlGaAs. This solution is only suitable if abrupt interfaces are not needed in the structures or devices being grown, because the adsorption of material on to the graphite smears out any changes in gas composition. This leaves the interface with a graded composition rather than an abrupt one.

The incorporation of the oxygen into the growing layer is not a problem for GaAs material or AlGaAs with a small mole fraction ($x < 0.15$) of aluminium. Thus if a layer of AlGaAs is grown with the aluminium content increasing from $x = 0$ to $x = 0.3$ before the device or active part of the structure is grown, the oxygen will be 'gettered' out of the reaction cell into material that will not be damaged by it. Subsequently the device can be grown on top of this 'getter buffer' layer (50). This method is clearly more appropriate for contamination from insufficient purging than from gas-stream contamination.

The final alternative for growth of AlGaAs uncontaminated with oxygen is to grow it at a temperature 100–200° C higher (51) than that typically used for GaAs. This increases the volatility of the aluminium alkoxides, removing them from the reaction zone. The serious disadvantage is the increase in carbon and silicon (17) incorporation at the high temperature.

Dopant alkyls

The choice of dopant is usually made on the basis of the level of doping needed, the device or structure to be grown and the limitations of the apparatus. A typical MOVPE apparatus will have between two and six temperature-controlled baths. If all of these are already in use, then the choice of dopant is restricted to a gas source. Like the III–V compounds, the alkyl sources are easier to obtain free from water and oxygen contamination and do not need the elaborate drying precautions of a gas source. It is perhaps for this reason that after initial monopoly of Si and Se (as SiH_4 and H_2Se), Sn doping using TMSn is becoming a viable alternative. All the *p*-type dopants (Be, Cd, Zn and Mg) are available as alkyls. The choice between the methyl or ethyl source is usually made on the basis of the concentration range needed. It is more satisfactory to choose the less volatile ethyl compound if low doping levels are needed than to use very small flows of the more volatile methyl, with the alkyl having to be held at inconveniently lower temperatures.

Dopant hydrides

The hydrides of Si and Se have almost completely monopolized *n*-doping until recently. Each has drawbacks that prevent either one becoming the predominant compound, and it remains to be seen whether the Sn compounds are able to avoid these problems.

The problem with Se doping is mainly related to the molecule of transport (H_2Se) which has a water-type structure in the gas phase. This structure lends the molecule a strong dipole and makes it very prone to adsorption on the walls of the reactor lines and reaction cell (52). When abrupt changes of doping level or type are needed, this adsorbed material creates a delay in the change of the vapour concentration as it desorbs off the walls (53). It was for this reason that non-polar SiH_4 was preferred, being much less prone to adsorption. Whilst it had the advantage of achieving abrupt changes in doping level, SiH_4 suffers from a number of drawbacks. One is that the incorporation of Si is very temperature-dependent. The same vapour concentration can dope 10^{17} *n*-type at 650° C and 10^{18} at 800° C. The temperature variation over the susceptor is enough to make uniform doping with silane very difficult to achieve. Si_2H_6 has been suggested as an alternative source of Si, as the decomposition kinetics of this compound are different from those of SiH_4, eliminating the temperature effect (26).

A more serious drawback is the site of incorporation of the Si atom. Group IV elements (C, Si, Ge and Sn) are amphoteric (they can be incorporated on to either sub-lattice). Si and Sn are usually incorporated on to the III sub-lattice where, having one electron more than the host, they act as *n*-type dopants. This is only true for Si up to a certain concentration (about $2 \times 10^{18}\,cm^{-3}$). Above this concentration the Si is incorporated as nearest neighbour pairs (Si_{Ga}–Si_{As}) (54). This causes the material and electrical properties of the semiconductor to be severely degraded, so the material of choice for high levels of doping [$> 5 \times 10^{18}$] is either Se or, more recently, Sn.

Adducts

The unwanted side reaction between TMIn and PH_3 which was mentioned above has been reported since the early days of indium phosphide growth by MOVPE. Along with the engineering and physical solutions already described, a chemical solution was investigated.

The intention was to block off the vacant *p*-orbital on the TMIn, as this was thought to be the source of the problem, especially as it had been demonstrated that trimethylarsine did not lead to the same difficulties as AsH_3 when growing InAs. Thus an adduct was formed between an electron acceptor (TMIn) and an electron donor (usually from Group V).

$$InR_3 + XR_3 \rightarrow InR_3.XR_3 \tag{7.5}$$

where X = a Group V element, usually N or P, and R = alkyl or aryl. The donor bond is weaker than the metal–carbon bond, its strength depending on the organic groups attached to the Group III and V elements. This reaction can either take place *in situ* (in the gas stream before entering the reactor), or in a separate stage, after which the adduct can be stored and purified like a normal alkyl. One advantage over a normal alkyl is the ease of handling of adducts, glass containers being quite suitable storage media.The Group V element need not be that which will be used in the actual growth; e.g. $InMe_3.PEt_3 + AsH_3$ results in the growth of an InAs layer without any appreciable incorporation of phosphorus (55).

Recently the purification of TMIn by reaction with diphos(bis 1, 2-diphenylphosphinoethane) as a step in the synthesis and subsequent resubliming of the TMIn from the adduct has produced InP of very high purity.

It has also been an aim of adduct researchers to provide adducts which can act as single sources of both the Group III and V elements (56). Such compounds as (C_6F_5) $Me_2Ga.AsEt_3$ and $(ClEt_2Ga.AsEt_2)_2CH_2$ have been used to grow GaAs successfully. However, the level of impurities in the grown layer is such that this technique is some distance away from commercial exploitation.

Vapour-phase concentration of reactants

Table 7.4 shows some typical concentrations, hydrogen and reactant flows and mole fractions for a particular MOVPE reactor. The ratio of the hydrogen

Table 7.4 Typical characteristics of an MOVPE reactor

Source	Hydrogen flow ($cm^3 min^{-1}$)	Source flow ($cm^3 min^{-1}$)	Molar flow (molecules min^{-1})	Mole fraction
H_2	21 000		5.63×10^{23}	1
TMGa (0° C)	30	2.08	5.59×10^{19}	9.94×10^{-5}
TMAl (20° C)	75	0.88	2.36×10^{19}	4.2×10^{-5}
AsH_3		350	5.63×10^{21}	1.0×10^{-2}
SiH_4(4 ppm)		0.237	6.46×10^{15}	1.2×10^{-8}
DEZn (20° C)	60	0.82	2.20×10^{19}	3.9×10^{-5}

At a growth temperature of 760° C these flows give the following:

	Solid composition	$= Al_{0.4}Ga_{0.6}As$
	Growth rate	= 500 Å min^{-1}
either	SIH_4	$= 10^{18} cm^{-3}$ *n*-type
or	DEZn	$= 10^{18} cm^{-3}$ *p*-type

flow through the bubblers to the flow of alkyl material depends on the temperature of the bubbler.

MOVPE requires control of the concentration of species in the reactant gas stream to control the growth rate and composition of the semiconductor layer. This concentration control is achieved by two main features in a MOVPE reactor; firstly accurate control of the temperature of the metal-organic bubbler baths which directly control the vapour pressure of the alkyl sources, and secondly accurate control of the flow of the vapour to the reaction zone using mass flow controllers.

As well as being able to change the concentration of the reactants rapidly, it is also important to be able to maintain a constant concentration when needed. This in turn requires strict control of the vapour pressure of the metal-organics by control of the bath temperature. A slight variation in the bath temperature is not a problem in the growth of binary compounds (the vapour pressure only affects the growth rate). In the growth of ternaries and quaternaries, however, any slight change in the vapour pressure will alter the composition of the ternary, which in turn will alter the lattice constant and other properties of the crystal.

The way in which reactants enter, and the type of valve used, are of crucial importance to the interface quality of the grown layer. Unfortunately the subject is too complex to discuss here, but references (53) and (57) deal with the problems and possible solutions.

Film growth

The mechanism of MOVPE single-crystal growth is a matter for continuing conjecture. The main stages in the mechanism consist of the thermal decomposition, followed by the diffusion of the atoms and molecular nuclei to the substrate or codeposition of the atomic vapours on the substrate and decay of the organo-function to the saturated hydrocarbon and hydrogen species. Intermediate processes, involving the adduction of a constituent organometallic with the atomic species forming as a result of premature decomposition of the other constituent metal-organic or hydride, have been found to occur. The primary process occurring in the metal-organic decomposition is determined by the volatility, thermal stability, Lewis acid and base strength and steric influence of the alkyl groups. Thus it is possible that the detailed mechanism of growth may be quite different for different alkyls.

From a kinetic study of the TMGa-AsH_3 reaction (58) it has been shown that the reaction is first-order in TMGa with an apparent activation energy of $54\,kJ\,mol^{-1}$, and heterogeneous because of very high bond dissociation energy of the third Ga–Me bond, $-322\,kJ\,mol^{-1}$, lending support for a decomposition occurring by the successive loss of methyl radicals. However, observation of $(GaAs)_n$ ($n = 5$–10) aggregates (59) forming in the gas phase and diffusing to the substrate support the mechanism of initial formation of a 1:1 TMGa–AsH_3 adduct and its successive demethanation analogous to that observed for TMIn and PH_3. It has been shown that both models of growth can explain the observed growth behaviour (60).

The site of the gas phase reaction is determined by the type of cell used. One design of cell, the horizontal Bass cell, was shown in the introduction to this chapter. There are alternatives to a horizontal arrangement of cell and susceptor, the most common being a horizontal susceptor inside a bell-jar where the gas flow is from top to bottom. Other designs have been investigated, but since the introduction of the Bass cell in 1975 (61) most UK reactors are of the horizontal type or modifications of it. The following discussion on cell design is restricted to this type of cell.

Having been mixed with the carrier gas, the reactant vapours enter the reaction cell and then form part of non-turbulent flow over the susceptor. The cell must be designed to allow a smooth path for the reactants from the inlet over the susceptor and out of the exhaust if rapid changes of gas composition are to be achieved.

The concentration of reactants in the vapour phase in the reaction zone is influenced both by the chemical activity of the species and the complicated gas dynamics over the susceptor. These gas dynamics can be described in a simple model of a boundary layer of stagnant gas over the susceptor. This is illustrated in Figure 7.2.

In order to participate in the crystal growth, an element or compound must pass through this boundary layer. The reaction equations for the growth of GaAs are as follows:

$$(CH_3)_3Ga(v) + \tfrac{3}{2}H_2(v) \rightarrow Ga(v) + 3CH_4(v) \qquad (7.6)$$

$$AsH_3 \rightarrow \tfrac{1}{4}As_4(v) + \tfrac{3}{2}H_2(v) \qquad (7.7)$$

$$Ga(v) + \tfrac{1}{4}As_4(v) \rightarrow GaAs(s) \qquad (7.8)$$

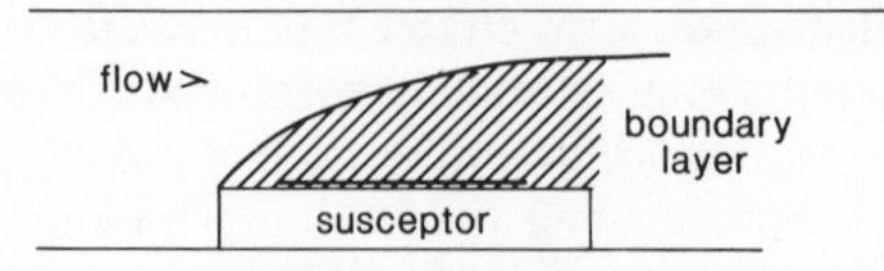

Figure 7.2 Cross-section through cell showing boundary layer (not to scale)

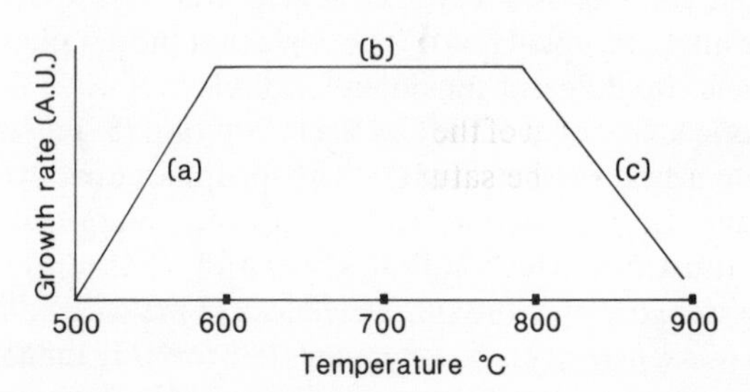

Figure 7.3 Growth rate *v.* substrate temperature. (*a*) Kinetic-controlled region; (*b*) mass-transfer controlled region; (*c*) desorption-controlled region.

The gas composition over the boundary layer has been investigated by IR spectroscopy (62) and atomic absorption spectrometry (63). There is disagreement on whether the alkyl decomposes before transfer through the boundary layer or on the semiconductor crystal surface; it is agreed that the rate-determining step is transfer of the alkyl species through the boundary layer. This is reflected in the temperature-dependence of the growth rate. The effect of temperature on the growth rate can be split into three regions (64) (Figure 7.3).

(i) *Low-temperature kinetic-controlled region.* As the temperature of the susceptor is raised, the growth rate also increases, because the rate-determining step is the cracking of AsH_3 to elemental arsenic. (reaction 7.7).

(ii) *Medium-temperature mass-transfer-controlled region.* In this region there is sufficient arsenic to be in excess, and the rate-determining step is the transfer of the gallium species across the boundary layer. Thus between 600–850° C the growth rate is controlled solely by the concentration of TMGa and is unaffected by the temperature or the AsH_3 concentration. It is this temperature region that is used for MOVPE growth. The temperature is usually kept as low as possible for high-purity (undoped) material, because the impurity (mainly carbon or silicon) concentration increases with temperature.

(iii) *High-temperature desorption-controlled region.* In this region an increase in temperature leads to an increase in the desorption of reactant species from the growth surface and consequent lowering of the growth rate.

No matter which, or how many, Group III species are involved in growth, they are always solely responsible for their own growth rate, and varying one alkyl concentration does not affect the growth of another.

Epitaxy

The provision of good-quality epitaxial material is hampered by a crystal's ability to grow in ways other than that needed to produce a perfect lattice. These 'side reactions' include growing a lattice with Group III atoms on a Group V lattice (antisite defect), a lattice with atoms sitting between the lattices rather than on them (interstitials), and lattices with atoms missing from their correct positions (vacancies).

Chemists are usually conversant with crystals that grow on many different facets at once, gradually increasing in volume. Semiconductor crystal growth only takes place on one facet, usually the (100) crystal plane. It is this face that has been found to give epitaxial material of the best quality. The crystal thus grows in increasing thickness on one face only.

The first requirement for good epitaxy is to ensure that the initial stages of growth have a clean crystal face on to which to 'key'. If this does not happen, the growth will not recover, no matter how thick a layer is grown. In order to ensure that growth starts with good crystal quality, the substrate is prepared in a scrupulously clean environment. Wafers 400 μm thick and 2″ in diameter are cut from ingots, and polished to a mirror-like finish. Before growth, about 20 μm of material is etched off the surface of the wafer to remove regions of the crystal that were damaged during the cutting and polishing (65, 66). If growth is conducted on substrates that have not been etched, the material grown is full of defects that can act as electron or hole traps and spoil the electrical properties of the layer. The etch is usually a combination of oxidant and complexing agent. The latter removes the oxidized species from the liquid–solid interface and prevents it from precipitating. Examples of etches are H_2O_2/H_2SO_4 for GaAs and Br_2/HBr for InP substrates. These are the two substrates mainly used. Other etches have been investigated, including NaOH and NH_4OH as the complexing agents in combination with H_2O_2 (67).

The etching leaves the substrate covered in a thin layer of oxide on the surface. This oxide is removed in the first stage of the MOVPE process. However, if the surface of the substrate is left stained by chemicals or is dirty in any way, the surface of the grown crystal will be full of defects which extend up through the layer from the substrate. This will seriously impair some applications of semiconductor material.

To ensure good epitaxy, the lattice constant of the epitaxial layer must closely match that of the substrate. This is straightforward in homoepitaxy (GaAs on GaAs or InP on InP) or even some types of heteroepitaxy (AlGaAs on GaAs), because AlAs and GaAs have nearly the same lattice constant (5.66A) thus allowing any composition of AlGaAs from $Al_{0.01}Ga_{0.99}$ As to

$Al_{0.99}Ga_{0.01}As$. However, other types of heteroepitaxy are not so simple. InGaAs varies its lattice constant from InAs (6.06) to GaAs (5.66) depending on the composition, and only one composition corresponds to the lattice constant of InP (5.87), namely $In_{0.47}Ga_{0.53}As$. The lattice matching when growing quaternaries is even more susceptible to changes in the composition. In these systems special care must be taken to ensure that there is the minimum variation in the gas composition over the slice, as any such variation will alter the alloy composition away from that required for lattice matching. It has been shown (69) that even a mismatch of 0.1% in lattice constant leads to about 10^8 dislocations cm^{-3}. This mismatch also leads to strained layers that are sometimes bent to minimize the strain, causing difficulties in handling material that will readily delaminate from the substrate.

It is also necessary to control the ratio of Group III to Group V species in the gas phase. Table 7.4 shows an excess of Group V over Group III, by about 100:1, although ratios as low as 5:1 are common. This excess of Group V is needed to overcome the higher volatility of the Group V element. However, if there is too much overpressure of the Group V material, the Group III sub-lattice will develop vacancies or As_{Ga} i.e. arsenic on gallium sites.

A more important problem, if the III/V ratio is not correct, is the lattice site of the amphoteric impurities silicon and carbon, since both of these impurities are capable of doping either *n*- or *p*-type, depending on which sub-lattice they occupy. If the III/V ratio is too lean in Group V, the C or Si atoms will be incorporated on to the V sub-lattice as acceptors, making the material *p*-type. For most devices it is more useful to have unintentionally doped material with a background doping of donors. This is achieved by making the III/V ratio high so as to incorporate the amphoteric impurities on the III lattice, thus making the material *n*-type.

The exhaust material from a reactor consists mainly of unreacted Group V hydride and elemental Group V species either in the vapour phase or as dust. The excess of the Group V species leads to a large build-up of elemental phosphorus or arsenic in the exhaust pipework of an MOVPE reactor, as well as a proportion of the hydride passing through the reactor unchanged. There are a variety of ways of dealing successfully with the solid material, providing that any filter arrangement is placed far enough downstream of the reactor to allow all the elemental Group V material to condense (both elements have an appreciable vapour pressure at the typical reactor temperatures, 600–850° C). Large quantities of phosphorus present a fire as well as a toxic hazard. The disposal of the hydrides poses more of a problem; any solution must not cause pressure fluctuations upstream in the reactor cell, as this could affect growth rate and alloy composition.

One solution is to burn the gases and scrub the exhaust with alkali; alternatively, they can be adsorbed on to large quantities of activated charcoal. Both of these solutions merely postpone the problem of disposal, effectively concentrating an already toxic compound into a major hazard. The solid charcoal is perhaps preferable to the liquid alkali in this respect.

Future trends

The industrial future of MOVPE seems assured, especially as there is an increasing trend towards large-scale wafer production. It is possible to scale up MOVPE growth with greater facility than other growth techniques, especially as this scale-up has already been achieved in the field of silicon epitaxy.

The promotion of the gas-phase reactions above the substrate by laser or UV radiation is becoming increasingly important, especially in the growth of II–VI material (69). It is likely soon to become of industrial importance.

There is an active research interest, both academic and industrial, towards understanding and improving the technique. This includes not only the growth itself but also the provision of high-purity alkyls that can be relied on from batch to batch. Adducts continue to be investigated and their purity is gradually reaching that of alkyls. New starting materials are being investigated, using fluorocarbons in an attempt to solve the problem of carbon incorporation. Finally, there is an increasing trend towards lower pressures in MOVPE at one extreme and molecular-beam epitaxy (MBE) using either AsH_3 or alkyl sources at the other. This hybrid technique is known as metal-organic molecular beam epitaxy (MOMBE) and might have the advantages of both techniques (70).

Acknowledgements

I would like to thank Peter Walker of the Clarendon Laboratory for helpful criticism, and John Roberts of the University of Sheffield for introducing me to MOVPE.

References

1. H. M. Manasevit, *Appl. Phys. Lett.* **12** (1968) 156.
2. H. M. Manasevit and W. I. Simpson, *J. Electrochem. Soc.* **116** (1969) 1725.
3. H. M. Manasevit, *J. Electrochem. Soc.* **118** (1971) 647.
4. A. Tromson-Carli, P. Gibrat and C. Bernard, *J. Cryst. Growth* **55** (1981) 125.
5. R. G. Walker and R. C. Goodfellow, *Electron. Lett.* **19** (1983) 590.
6. J. M. Woodal, *Science* **208** (1980) 908.
7. C. B. Cooper III, R. R. Saxena and M. J. Ludowise, *Electron. Lett.* **16** (1980) 892.
8. H. M. Manasevit and W. I. Simpson, *J. Electrochem. Soc.* **120** (1971) 135.
9. P. K. Chiang and S. M. Bedair, *J. Electrochem. Soc.* **131** (1984) 2422.
10. M. D. Camras, J. M. Brown, N. Holonyak, Jr, M. A. Nixon, R. W. Kaliski, M. J. Ludowise, W. T. Dietze and C. R. Lewis, *J. Appl. Phys.* **54** (1983) 6183.
11. T. Fukui and Y. Horikoshi, *Jap. J. Appl. Phys.* **20** (1981) 587.
12. T. Fukui and Y. Horikoshi, *Jap. J. Appl. Phys.* **19** (1980) L53.
13. R. W. Glew and M. S. Frost, *J. Cryst. Growth* **68** (1984) 450.
14. H. M. Manasevit and W. I. Simpson, *J. Electrochem. Soc.* **118** (1971) C291.
15. C. C. Hsu, J. S. Yuan, R. M. Cohen and G. B. Stringfellow, *J. Appl. Phys.* **59** (1986) 395.
16. J. B. Mullin, S. J. Irvine, J. Giess and A. Royle, *J. Cryst. Growth* **72** (1985) 1.

17. T. F. Keuch and R. Potemski, *App. Phys. Lett.* **47** (1985) 821.
18. M. Mizuta, T. Iwamoto, F. Moriyama, S. Kawata and H. Kukimoto, *J. Cryst. Growth* **68** (1984) 142.
19. S. J. Bass and M. L. Young, *J. Cryst. Growth* **68** (1984) 311.
20. T. Fukui and Y. Horikoshi, *Jap. J. Appl. Phys.* **19** (1980) L395.
21. L. M. Fraas, J. A. Cape, P. S. McLeod and L. D. Partain, *J. Vac. Sci. Technol.* **B3** (1985) 921.
22. G. B. Stringfellow, *J. Cryst. Growth* **62** (1983) 225.
23. C. R. Lewis, W. T. Dietze, and M. J. Ludowise, *J. Electron. Mater.* **12** (1983) 507.
24. S. J. Bass, *J. Cryst. Growth* **31** (1975) 172.
25. S. J. Bass, *J. Cryst. Growth* **47** (1979) 613.
26. T. F. Keuch, E. Veuhoff and B. S. Meyerson, *J. Cryst. Growth* **68** (1984) 48.
27. A. P. Roth, R. Yakimova and V. S. Sundaram, *Electron. Lett.* **19** (1983) 1062.
28. C. Y. Chang, M. K. Lee, Y. K. Su and W. C. Hsu, *J. Appl. Phys.* **54** (1983) 5464.
29. R. W. Glew, *J. Cryst. Growth* **68** (1984) 44.
30. N. Bottka, R. S. Sillmon and W. F. Tseng, *J. Cryst. Growth* **68** (1984) 54.
31. A. W. Nelson and L. D. Westbrook, *J. Cryst. Growth* **68** (1984) 102.
32. C. R. Lewis, W. T. Dietze, and M. J. Ludowise, *Electron. Lett.* **18** (1982) 569.
33. A. C. Jones, A. K. Holliday, D. J. Cole-Hamilton, *J. Cryst. Growth* **68** (1984) 1.
34. J. I. Davies, R. C. Goodfellow and J. O. Williams *J. Cryst. Growth* **68** (1984) 10.
35. L. Samuelson, P. Omling, H. Titze and H. G. Grimmeiss, *J. Cryst. Growth* **55** (1981) 164.
36. K. L. Hess, P. D. Dapkus, H. M. Manasevit, T. S. Low, B. J. Skromme and G. E. Stillman, *J. Electron. Mater.* **11** (1982) 1115.
37. T. F. Keuch and E. Veuhoff, *J. Cryst. Growth* **68** (1984) 148.
38. M. Razeghi and J. P. Duchemin, *J. Cryst. Growth* **64** (1983) 76.
39. Y. Seki, K. Tanno, K. Iida and E. Ichiki, *J. Electrochem. Soc.* **122** (1975) 1108.
40. M. Suzuki and M. Sato, *J. Electochem. Soc.* **132** (1985) 1684.
41. M. Yoshida, H. Watanabe and F. Uesugi, *J. Electrochem. Soc.* **132** (1985) 677.
42. R. H. Moss, *Chem. Brit.* **19** (1983) 733.
43. G. B. Stringfellow, *J. Cryst. Growth* **68** (1984) 111.
44. W. Kroll, *Solid State Technol.* **27** (1984) 220.
45. H. Terao and H. Sunakawa, *J. Cryst. Growth* **68** (1984) 157.
46. E. J. Thrush and J. E. A. Whiteway, *Gallium Arsenide and Related Cpds 1980*, IOP (1981) 337.
47. L. M. Fraas, J. A. Cape, P. S. McLeod and L. D. Partain, *J. Vac. Sci. Tech.* **B3** (1985) 921.
48. J. R. Shealy, V. G. Kreismanis, D. K. Wagner and J. M. Woodall, *Appl. Phys. Lett.* **42** (1983) 83.
49. D. W. Kwisker, J. N. Miller and G. B. Stringfellow, *J. Appl. Phys.* **40** (1982) 614.
50. S. D. Hersee, M. A. Di Forte-Poisson, M. Baldy and J. P. Duchemin, *J. Cryst. Growth* **55** (1981) 53.
51. M. J. Tsai, M. M. Tashima and R. L. Moon, *J. Electron. Mater.* **13** (1984) 437.
52. C. R. Lewis, W. T. Dietze, and M. J. Ludowise, *J. Electron. Mater.* **13** (1984) 447.
53. J. S. Roberts, N. J. Mason and M. Robinson, *J. Cryst. Growth* **68** (1984) 422.
54. M. Druminski, H.-D. Wolf, K.-H. Zschauer, *J. Cryst. Growth* **57** (1982) 318.
55. R. H. Moss, *J. Cryst. Growth* **68** (1984) 78.
56. F. Maury, A. El Hammadi and G. Constant, *J. Cryst. Growth* **68** (1984) 88.
57. E. J. Thrush, J. E. A. Whiteway, G. Wale-Evans, D. R. Wight and A. G. Cullis, *J. Cryst. Growth* **68** (1984) 412.
58. W. H. Petzke, V. Gottschalch and E. Butler, *Krist. Tech.* **9** (1974) 763.
59. I. A. Frolov, P. B. Boldysevskii, B. L. Druz and E. B. Sokolov, *Inorg.* Mater. **13** (1977) 632.
60. N. Putz, J. Korec, M. Heyen and P. Balk, *Proc. 4th Eur. Conf. on CVD* (1983) 103.
61. S. J. Bass, *J. Cryst. Growth* **31** (1975) 172.
62. M. R. Leys and H. Veenvliet, *J. Cryst. Growth* **55** (1981) 145.
63. J. Haigh, *J. Mater. Sci.* **18** (1983) 1072.
64. D. H. Reep and S. K. Ghandi, *J. Electrochem. Soc.* **130** (1983) 675.
65. P. J. Caldwell, W. D. Laidig, Y. F. Lin and C. K. Peng, *J. Appl. Phys.* **57** (1985) 984.
66. F. D. Auret, A. W. R. Leitch and J. S. Vermaak, *J. Appl. Phys.* **59** (1986) 158.
67. J. C. Dyment and G. A. Rozgonyi, *J. Electrochem. Soc.* **118** (1971) 1346.
68. H. F. Matare, *Solid State Technol.* 25 (1976).
69. J. Y. Tsao and D. J. Ehrlich, *J. Cryst. Growth* **68** (1984) 176.
70. L. M. Fraas, P. S. McLeod, J. A. Cape and L. D. Partain, *J. Cryst. Growth* **68** (1984) 490.

8 Liquid-phase epitaxy of III–V compounds

P. D. GREENE

Introduction

The process known as liquid-phase epitaxy (LPE) is a technique for the deposition of epitaxial layers from supersaturated solutions. For growth of semiconductors, the chosen solvent is always a metal with a low melting point and a low vapour pressure, implying an extensive liquid range. The most suitable metals are gallium (liquid from 30–2070° C), indium (157–2050° C) and tin (232–2720° C). A seed crystal, known as the substrate, is provided in the form of a flat and polished wafer cut from a large single crystal in a specific direction with respect to the crystal lattice. The orientation usually preferred is with the surface parallel to the (100) plane of the crystal lattice. The term epitaxy implies that the crystal lattice of the material deposited from the solution is aligned with the lattice of the substrate. The epitaxial material and the substrate have the same crystal structure and closely matched crystal lattice parameters so that they form a continuous single crystal, even though there may be a composition change at the interface.

Although the LPE method is used occasionally for the growth of elemental semiconductors such as silicon, atoms of the solvent are inevitably incorporated into the deposited solid and only highly doped material can be grown. On the other hand, LPE is particularly suited to the growth of III–V compound semiconductors such as GaAs and InP because one of the elements in these compounds, the Group III metal, can be used as the solvent. This allows compound semiconductor layers with a high purity to be produced.

Because the use of tin as a solvent leads to highly doped layers, this metal is rarely used for LPE growth of III–V compounds, but their greater solubility in tin allows lower temperatures to be used, and this can be an advantage in special circumstances.

A distinction is made between homoepitaxy and heteroepitaxy. The former term applies when the substrate and the epilayer are the same chemical compound although with a change in the doping, for example, *p*-GaAs on *n*-GaAs. Heteroepitaxy involves a gross change in chemical composition, for example, the deposition of the ternary alloy $Ga_{1-x}Al_xAs$ on GaAs.

To achieve satisfactory heteroepitaxy it is necessary for the lattice parameter (the length of the side of the cubic unit cell) to be similar for the materials on both sides of the junction, though some mismatch can be tolerated. Because of different thermal expansion coefficients, the mismatch at the growth temperature is usually slightly different from the measured mismatch at room temperature. With small mismatches (up to about one part in 10^3) the discrepancy may be accommodated by elastic strain, virtually all of which is in the epitaxial layer because it is much thinner than the substrate. The layer adopts the lattice parameter of the substrate in the two directions parallel to the surface, but is not constrained in the perpendicular direction. The stretching or contraction of the lattice in this direction is approximately twice as much as it would be without the constraint in the other directions. The exact factor is given by $(1 + \sigma)/(1 - \sigma)$ where σ is Poisson's ratio for the material. At some critical degree of mismatch (1), which decreases as the layer thickness becomes greater, the elastic strain becomes excessive and is relieved by the generation of dislocations parallel to $\langle 011 \rangle$ directions in the epitaxial layer. Such dislocations can have an adverse influence on performance and reliability of electronic devices. Ultimately, if the mismatch is too great, no epitaxial growth occurs.

By a fortunate coincidence, the lattice parameters of GaAs and AlAs are identical at about 950° C and differ by only 1.2 in 10^3 at room temperature. On GaAs substrates it is therefore possible to grow epitaxial layers of the alloy compositions $Ga_{1-x}Al_xAs$ with any value of x from 0 to 1 inclusive. For the analogous antimonides GaSb and AlSb, the lattice match is conspicuously worse, about 6.5 parts in 10^3, and addition of small amounts of arsenic as a fourth component is usually necessary to obtain satisfactory lattice matching to a GaSb substrate. However, the antimonide solid solutions are not of such great technological importance. To maintain lattice matching in alloys to an accuracy of 1 part in 10^3, the extra degree of freedom provided by a fourth component is usually essential. The most widely used system of this type is the quaternary alloy $In_{1-x}Ga_xAs_yP_{1-y}$ which can be grown to match InP substrates by setting y equal to $2.1x$, or to match GaAs substrates by setting $(1 - y)$ equal to $2.1\ (1 - x)$. Figure 8.1 shows how the lattice parameter and the bandgap of this quaternary alloy depend on the composition parameters x and y. The realization of some compositions is hindered by solid state immiscibility, but this is not a major technological problem.

Within the alternative quaternary system $In_{1-x}Ga_xAs_{1-y}Sb_y$, alloys can be grown with lattice parameters to match either GaSb or InAs substrates. However, an extensive solid-state miscibility gap occurs in this system and only a very limited range of alloys can be produced. A discussion of the growth of quaternary alloys is presented in a later section.

In the LPE process, the interface at which growth is occurring is not far from equilibrium. Consequently surface energy effects can have a marked influence on the growth rate. Concave surfaces in substrates in which grooves have been etched generally encourage faster growth, so that the groove fills quickly and a planar configuration is eventually restored. Slower growth occurs on convex surfaces. In addition to these general effects of curvature, different growth rates occur in different directions with respect to the crystal lattice. The most

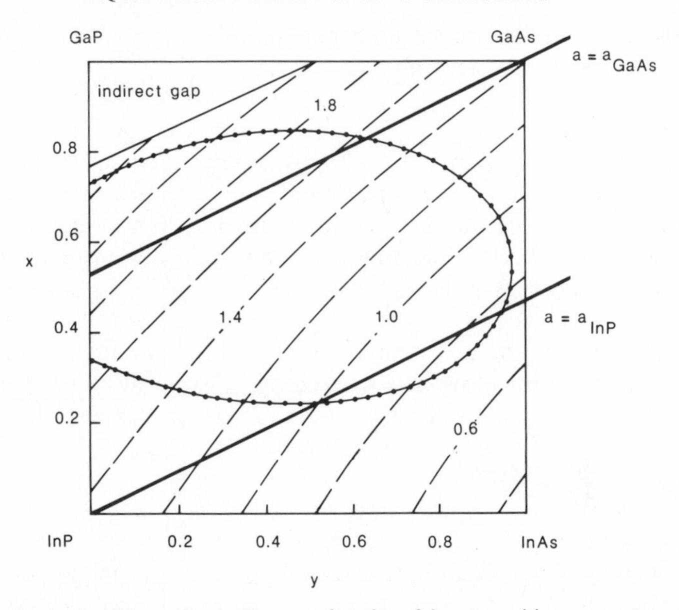

Figure 8.1 Properties of $In_{1-x}Ga_xAs_yP_{1-y}$ as a function of the composition parameters x and y. The energy-gap contours (in eV) are shown by broken lines. Compositions with the same lattice constant as InP or GaAs are shown by full lines. The calculated region of immiscibility at 600° C is contained within the dotted line

remarkable difference is found when ridges or channels present (111)A faces to the liquid (2). Growth of the binary semiconductors and some of the ternary and quaternary alloys is very slow on such faces. An understanding of these surface energy effects enables the semiconductor technologist to produce structures in which the composition of the material is tailored in more than one dimension.

Growth of binary compounds

The aluminium binary compounds are of limited importance in semiconductor device technology because of their high reactivity in a normal atmospheric environment, although protection can be afforded by a native oxide film if the preparation and the environment are controlled appropriately. The binary compounds of gallium and indium have no such stability problem at room temperature. All the compounds of these two elements with phosphorus, arsenic or antimony are of interest, but two binary compounds, GaAs and InP, are much more widely exploited than the remainder and consequently this section will concentrate on this pair of materials.

Extensive and thorough reviews of the solubility of and the thermodynamic data for GaP, GaAs, InP and InAs have been published (3–6). Perea and Fonstad (4) give the following expressions to describe the solubilities (as mole fractions) as a function of temperature (K):

$$\begin{aligned} &\text{GaP:} && x = 2700 \exp(-15123/T) \\ &\text{GaAs:} && x = 4779 \exp(-13112/T) \\ &\text{InP:} && x = 2059 \exp(-11570/T) \\ &\text{InAs:} && x = 404 \exp(-7877/T) \end{aligned}$$

The fundamental chemical process on which LPE is based is the reversible dissolution of the III–V compound in the corresponding Group III liquid metal. Using InP as an example:

$$\mathrm{InP(c) \rightleftharpoons In(l) + P(l\ in\ In)}$$

The description of this process as chemical can be justified by considering the enthalpy change associated with it. Using the classical thermodynamic relationship

$$\Delta H^{\ominus}_{\mathrm{sol}} = -R\frac{\mathrm{d}(\ln x)}{\mathrm{d}(1/T)}$$

the standard enthalpy of solution $\Delta H^{\circ}_{\mathrm{sol}}$ is calculated to be about $96\,\mathrm{kJ\,mol^{-1}}$.

This result can be compared with the sum of three processes for each of which the enthalpy change is known. The first is the reverse of the standard formation reaction and the others are simply the melting of the constituent elements.

$$\begin{aligned} &\mathrm{InP(c) \rightleftharpoons In(c) + P(c)} \\ &\mathrm{In(c) \rightleftharpoons In(l)} \\ &\mathrm{P(c) \rightleftharpoons P(l)} \end{aligned}$$

The standard enthalpies (5, 7) of these three separate processes are $+61.5$, $+3.3$ and $+18.0\,\mathrm{kJ\,mol^{-1}}$, giving a total of $82.8\,\mathrm{kJ\,mol^{-1}}$. Within the limits of experimental error this is similar to or slightly smaller than the standard enthalpy of the dissolution process, indicating that chemical decomposition of the semiconductor occurs on dissolution.

The process of liquid phase epitaxy involves the following steps.

(i) Producing a solution with a solute concentration corresponding to a saturation temperature T_s
(ii) Cooling this solution to a lower temperature T_g to induce supersaturation
(iii) Bringing the supersaturated solution into contact with the substrate for the time needed to deposit material to the required thickness (in some cases cooling is continued throughout the growth)
(iv) Separating the liquid from the solid to terminate growth.

The saturation temperature is usually between 1 and 20° C higher than the growth temperature. The growth temperature for InP is usually in the range

600–660° C and for GaAs in the range 700–800° C. The solute content of the solution would normally be in the range 0.3–3%.

Two methods of defining the saturation temperature T_s can be distinguished. In the single-phase method, the precise quantity of solute for the saturation temperature T_s is provided, allowing the dissolution to occur at any temperature above T_s. In the two-phase method an excess of the solid semiconductor is provided, and the amount dissolved is determined by maintaining the temperature precisely at T_s. The time required to achieve a homogeneous liquid depends on the dimensions, the amount of convection (which may be solutal as well as thermal) and the diffusion coefficient (3) (usually in the range 1×10^{-5} to $5 \times 10^{-5}\,\text{cm}^2\text{s}^{-1}$). Under typical circumstances saturation requires a few hours.

Ideally the cooling of the solution to the growth temperature would be instantaneous, but in practice the cooling rate is limited by the mass of the furnace and its contents. During the cooling the amount of solute in the liquid is reduced by deposition on the excess solid present if the two-phase method is adopted. However since the cooling time (minutes) is short compared to the time required to approach equilibrium (hours), a useful amount of supersaturation is retained. Alternatively, it is possible to divide the solution immediately before cooling, thereby separating the part of the solution to be used for the epitaxy process from the remainder which retains the excess floating solid. Even with a single-phase growth solution a premature deposition can occur if the supersaturation produced by cooling is excessive, and spontaneous nucleation and growth is provoked.

Growth is initiated by moving either the solution or the substrate to bring them into contact with each other. A typical experimental arrangement is the graphite boat in Figure 8.2, in which there is linear movement of a slider. Another movement in the same direction is used to terminate growth. Most electronic devices require several layers which are produced from a sequence of solutions. As an alternative to linear movement, various rotary systems have been devised. The use of centrifugal force (8) provides rapid and efficient removal of the liquid.

In most circumstances the growth rate on flat substrates is diffusion-limited, and the thickness d of the deposited layer depends on the growth time t

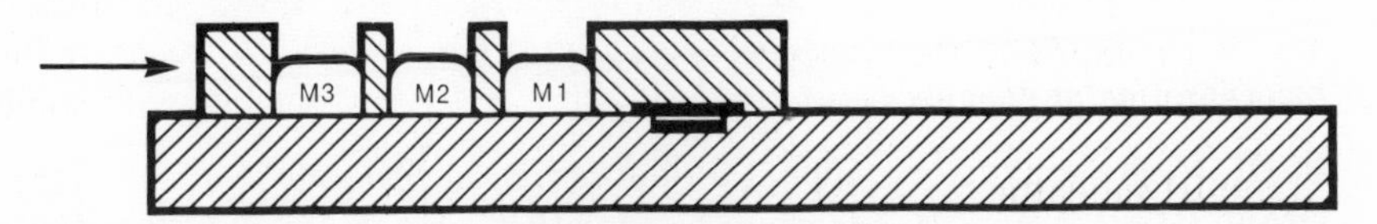

Figure 8.2 Graphite slider board for LPE growth. The slider is in the initial position, with the substrate protected by a cover slice. By pushing the slider, solutions M1, M2 and M3 are successively moved over the substrate for a time appropriate to the layer thickness required

according to a relationship (3) of the form

$$d = C\left(\Delta T t^{1/2} - \frac{2}{3}\frac{\mathrm{d}T}{\mathrm{d}t} t^{3/2}\right)$$

where $\Delta T (= T_s - V_g)$ is the supersaturation at the start of growth and $\mathrm{d}T/\mathrm{d}t$ the rate of cooling, assumed here to be kept constant while growth continues. The term C incorporates the diffusion coefficient of the Group V atoms and therefore has some dependence on temperature. For short growth times the first term in the brackets is dominant. Layers of thickness in the range 10–100 nm can be achieved using low supersaturations and contact times below one second. At the other extreme, long growth times, up to an hour, accompanied by cooling rates of perhaps 1° C per minute, allow thicknesses in the range 5–50 μm to be produced, the second term in the brackets being dominant.

Efficient separation of the solution and the solid at the termination of growth remains one of the most difficult parts of the technology, but a detailed discussion of the techniques is not appropriate in a text primarily concerned with the chemistry of LPE.

The electrical properties of the semiconductor are extremely sensitive to the presence of impurities. Measurement of the concentration of free electrons or holes and their mobility down to liquid-nitrogen temperature provides a convenient method of estimating the donor and acceptor concentrations in the material. To identify the chemical species present, photoluminescence or photothermal ionization spectroscopy at liquid helium temperature is required (9–12). Nominally pure GaAs and InP are always found to be *n*-type, though with fastidious attention to purity and baking of melts under appropriate conditions, material with ionized impurity concentrations well below 10^{15} cm^{-3} can be produced. In LPE material the usual residual donors are S and Si. The main residual acceptors are Cd, Zn and C in InP and Si, Ge and C in GaAs. The incorporation of dopants is considered in greater detail in the next section, and the effect of furnace atmosphere on contamination in a later section.

Acceptors are normally present in GaSb without acceptor impurities, because the semiconductor contains native defects usually considered (13) to be complex centres, each consisting of a Ga atom on an Sb site and a vacancy on an adjacent Ga site. The concentration of this defect is generally of the order of 10^{16} cm^{-3}. Nominally undoped GaSb is therefore generally *p*-type.

Incorporation of dopants

In III–V semiconductors, dopant atoms can be incorporated on either the Group III sub-lattice or the Group V sub-lattice. Dopants from Group II of the Periodic Table (Be, Mg, Zn and Cd) only substitute for the Group III elements and function as acceptors. Transition metals such as Mn usually behave in the same way. Dopants from Group VI (S, Se and Te) only occupy sites on the Group V sub-lattice and function as donors. Dopants from Group IV may occupy positions on either or both sub-lattices. They function as donors when replacing Group III elements and as acceptors when on the other sub-lattice. An understanding of the different characteristics of dopant incorporation on the two sub-lattices provides the basis for interpreting the amphoteric behaviour of the Group IV dopants.

The energy level of common donors is only a few meV below the conduction band. The energy level of most acceptors lies a little further above the valence band, but acceptors formed by transition metal atoms on Group III sites are significantly more remote from the valence band. As a general rule, the further the energy level is from the valence band, the harder it is to incorporate the transition metal dopant. Although Cr and Fe are commonly used to produce high resistivity in bulk melt-grown crystals of GaAs and InP, at the lower temperatures used for LPE growth their solubility in the semiconductor is generally too low for technological exploitation. However, Mn can be

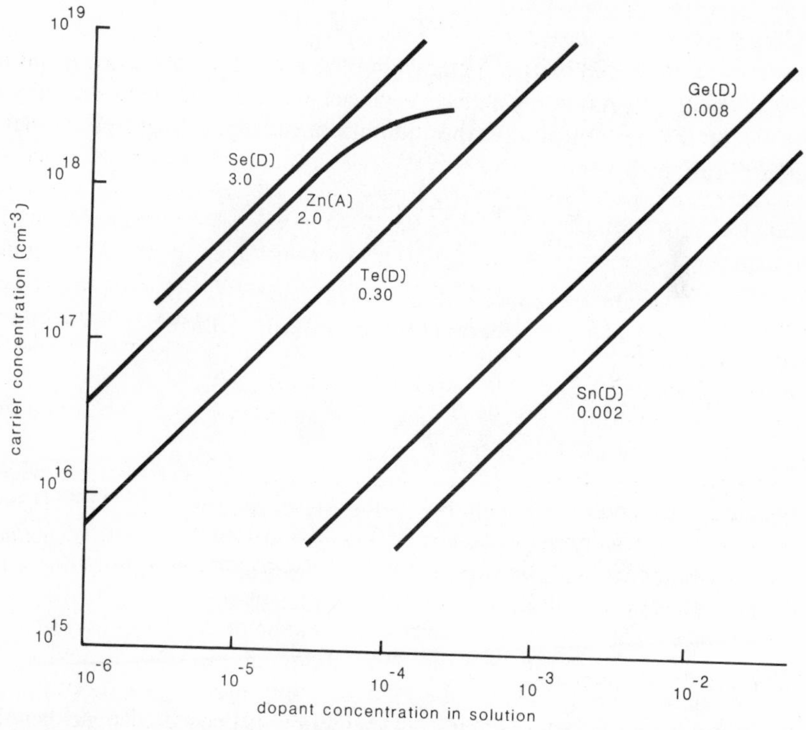

Figure 8.3 Relationship between dopant concentration in the solution and the carrier concentration in LPE InP grown at about 630° C. The distribution coefficients are shown for each element. The data are derived from Astles (23) (Zn, Te, Ge, Sn); Rosztoczy (24) (Te, Ge, Sn); and Greene (22, 26) (Se, Zn, Te)

incorporated more readily as it forms an acceptor rather closer to the valence band (14, 15, 16): 0.11 eV above in GaAs, 0.23 eV in InP and only 0.05 eV above in $In_{0.53}Ga_{0.47}As$.

Within one group of the Periodic Table, the smaller the atom the more readily it is incorporated into the semiconductor. Two examples of this can be seen in Figure 8.3. Fortunately carbon is an exception to the rule–graphite is the most commonly-used material for LPE boats, but carbon contamination of the solution and the epitaxial semiconductor is usually negligible although traces are a characteristic of LPE material of the highest purity (11).

It is generally satisfactory to consider the incorporation of dopants during growth by LPE as a chemical reaction involving direct competition between the components of the liquid phase for the lattice sites in the solid semiconductor. For the purpose of the present discussion the primary constituents will be considered to be Ga and As, competing with two unidentified dopants, Dm and Dn, with preference for the Group III sites and the Group V sites respectively.

$$\mathrm{Ga_{Ga} + Dm(l) \rightleftharpoons Ga(l) + Dm_{Ga}}$$
$$\mathrm{As_{As} + Dn(l) \rightleftharpoons As(l) + Dn_{As}}$$

Since Ga_{Ga}, As_{As} and Ga(l) are almost in their standard states their activities are all close to 1. Assuming activity coefficients of unity for the other species present at low concentrations, the equilibrium constants K and K' relate the concentrations by

$$K = [\mathrm{Dm_{Ga}}]/[\mathrm{Dm(l)}]$$
$$K' = [\mathrm{As(l)}][\mathrm{Dn_{As}}]/[\mathrm{Dn(l)}]$$

The ratio of the dopant concentration in the solid to the concentration in the liquid is known as the distribution (or segregation) coefficient, here written as K_m and K_n.

Therefore

$$K_m = K$$
$$K_n = K'/[\mathrm{As(l)}]$$

Because the concentration of As in the liquid is low but increases with temperature, wheareas the concentration of Ga remains close to unity, a marked difference is to be expected in the magnitude and the temperature-dependence of K_m and K_n when K and K' are similar.

The linear relationship holds over a wide range of dopant concentrations, but there is a limit to the solid solubility of dopants, and deviation from the linear relationship is always observed at sufficiently high concentrations, typically around 10^{18} to 10^{19} cm^{-3}. The limit of the linearity depends both on the semiconductor and the dopant.

The maximum free hole concentration in GaAs exceeds the maximum free electron concentration, but in InP the reverse occurs (Figure 8.3). Attempts to introduce higher doping than the semiconductor will tolerate produce increasing concentrations of compensating point defects, such as vacancies on the Ga sub-lattice in highly Se-doped GaAs or Zn in interstitial positions in InP doped with large amounts of Zn.

It is noteworthy that the incorporation of dopants can be represented satisfactorily by purely chemical equations, even though the dopant is ionized in the semiconductor. When this electrical behaviour is considered, the equations take the form

$$\mathrm{Ga_{Ga}} + \mathrm{Dd(l)} \rightleftharpoons \mathrm{Ga(l)} + \mathrm{Dd_{Ga}^{+}} + e^{-}$$

$$\mathrm{Ga_{Ga}} + \mathrm{Da(l)} \rightleftharpoons \mathrm{Ga(l)} + \mathrm{Da_{Ga}^{-}} + h^{+}$$

depending on whether the dopant is a donor Dd or an acceptor Da. For GaAs at a typical LPE growth temperature of 1050 K the intrinsic carrier concentration is $6.5 \times 10^{16}\,\mathrm{cm^{-3}}$. For InP, with a smaller bandgap and a lower growth temperature, the relevant intrinsic carrier concentration is of similar magnitude. For dopant concentrations higher than the intrinsic carrier concentration, the free electron (or hole) concentration is equal to the dopant concentration, suggesting that the square of the dopant concentration in the solid would be proportional to the concentration in the solution, instead of the observed linear relationship. The explanation (17) is thought to be that the Fermi level at the surface of the semiconductor is determined by the characteristics of the junction with the liquid metal and not by the bulk electrical properties of the semiconductor layer. An equilibrium between the liquid phase and the epitaxial semiconductor further from the interface is only approached when there is an exceptional combination of low growth rate, a highly doped and narrow junction and a rapidly diffusing dopant.

For the majority of dopants, the standard free energy of the incorporation reaction is positive, i.e. energy has to be expended to place a dopant atom on the lattice site instead of the usual occupant. Since this free energy is related to the equilibrium constant by a relationship of the form

$$\Delta G^{\ominus} = -RT \ln K$$

the distribution coefficient of most dopants incorporated on the Group III sub-lattice increases with rising temperature, but remains less than one. A well established example (18, 19) is the behaviour of Sn in the LPE growth of GaAs, presented in Figure 8.4. Because the distribution coefficient is less than one for most impurities, LPE is a purification as well as a crystal-growth process. On the other hand a dopant such as Be, which has a very small atomic radius, behaves in the opposite way (22). The standard free energy of the incorporation reaction is negative and energy is released when Be replaces a Group III element, so that the distribution coefficient is greater than 1.

The relative size of the atoms is crucial in determining the behaviour of dopants, including the predominance of donor or acceptor characteristics with

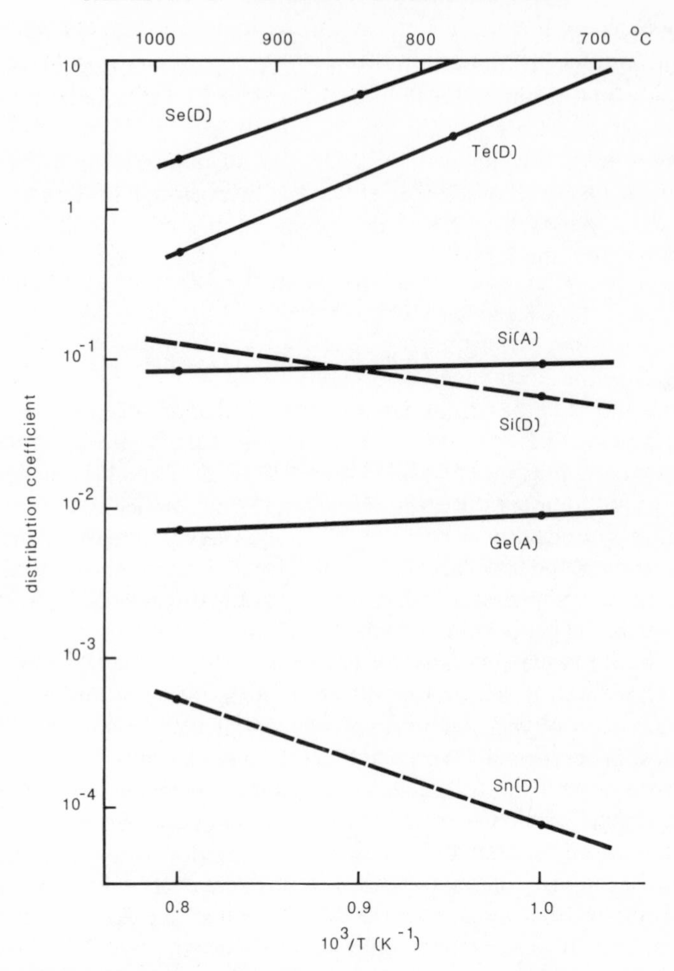

Figure 8.4 Distribution coefficient for various dopants (18–22) in LPE GaAs as a function of temperature. Full lines are used for dopants on Group V sites and broken lines for dopants on Group III sites. *D*, donor; *A*, acceptor

Group IV dopants. Since generally it is easier for a dopant atom to replace a larger atom than a smaller one, Ge substitutes for In in InP and is a donor, but replaces Sb in GaSb to form an acceptor. In GaAs the sizes of the atoms are all similar, so that during growth of GaAs at its melting point (20) the incorporation of Ge is almost exactly shared between the two sub-lattices. However, with decreasing growth temperature the As concentration drops, so that Ge competes more successfully against As and is primarily an acceptor (21). The distribution coefficient (Figure 8.4) is almost independent of temperature because of the effect of temperature on the As concentration.

In the growth of InP, all Group IV dopants behave as donors, but with

GaAs some subtle chemical effects can be found in the comparison of the behaviour of Si, Ge and Sn, illustrated in Figure 8.4. At all realistic growth temperatures Sn has a strong preference for Group III sites and is a donor. The distribution coefficient (20) for Si as a donor has the same type of temperature dependence as that of Sn, whereas the distribution coefficient for Si as an acceptor is almost independent of temperature, because of the temperature dependence of the As content of the solution. The two distribution coefficients for Si are equal at about 860° C, with donor behaviour dominant above this temperature and acceptor behaviour below. By growing over an appropriate temperature range a single Si-doped melt can produce GaAs containing a graded *p–n* junction. The precise temperature of the change-over is sensitive to the direction of growth with respect to the crystal lattice and to the Si concentration (25).

The behaviour of the Group VI donors (26, 27) such as Se and Te is also illustrated in Figures 8.3 and 8.4. The effect of competing with a temperature-dependent amount of As instead of Ga has already been discussed. In addition there is a considerable deviation from ideal solution behaviour in the liquid phase as well as in the solid. This is evident from the binary phase diagrams of Se or Te with Ga or In, which contain extensive regions of liquid–liquid immiscibility. The deviations from ideal solution behaviour are of the same sign in both the solid semiconductor and the solution from which it is grown. In other words, the reluctance of the solid semiconductor to incorporate the dopant is countered by a tendency of the liquid phase to reject the dopant also. Taking into account the As concentration effects as well, the overall result is that these dopants have distribution coefficients which are relatively large in typical LPE conditions and decrease as the growth temperature rises.

The availability of two types of donor dopants with distribution coefficients possessing opposite types of temperature dependence makes it possible for thick epitaxial layers grown over a wide temperature range to maintain greater uniformity of free electron concentration. However, use of this method must be limited to total dopant concentrations low enough for the two species to remain independent.

Distribution coefficient calculations can be based on different conventions for the definition of concentration in the liquid of the solid phase. All LPE distribution coefficients in Figures 8.3 and 8.4 have been unified to a single convention, requiring in some cases a change by a factor of two from the number in the original publication.

Ternary and quaternary alloys

The existence of solid solutions such as $Ga_{1-x}Al_x$ As and $In_{1-x}Ga_xAs_yP_{1-y}$ with lattice parameters equal to those of GaAs and InP was mentioned in the Introduction. The composition of the ternary or the quarternary alloy can be chosen to provide the required value of some parameter which is not adjustable in simpler semiconductors. Usually the composition is chosen to provide an appropriate bandgap, but sometimes another property (such as refractive index in opto-electronic devices) can be of equal or greater significance.

Controllable bandgap profiles within a single crystal provide the semiconductor device engineer with options that are not available in elemental or binary-compound crystals. Furthermore, the tailoring of the structure need not be limited to only one dimension. The combination of multilayer growth with photolithography and selective chemical etching enables complex three-dimensional structures to be produced.

Although the bond lengths in AlAs and GaAs are almost the same, the bond strength is greater in AlAs than in GaAs. This is apparent from a comparison of the melting points (1740° C for AlAs and 1240° C for GaAs) and from the compositions of solutions for the growth of the ternary alloy $Ga_{1-x}Al_xAs$. The Al content of the solid, characterized by the value of x, is always much greater than that of the liquid from which the solid is grown[28,29]. When x is small enough to allow $(1-x)$ to be close to unity, the incorporation of Al can be treated in a similar way to the incorporation of a dopant such as a Be on a Group III site, and the ratio of concentrations in the two phases can be characterized by a distribution coefficient which is always much greater than unity.

Such a distribution coefficient is temperature-dependent, and a constant ternary alloy composition is therefore not compatible with the growth of epitaxial material over a wide temperature range. Furthermore, for a fixed temperature, an increase in the Al content of the liquid decreases the amount of As required for saturation. Thus equivalent growth times yield thinner layers as the Al content is increased. In practice the required melt compositions are produced by weighing appropriate quantities of Ga, Al and GaAs.

One important additional difficulty arises from the high (negative) free energy of formation of Al_2O_3. If the growth is interrupted, leaving an exposed layer of $Ga_{1-x}Al_xAs$, the formation of an oxide film on the surface can prevent the deposition of further layers. Multilayer structures containing Al-rich ternary alloys need to be grown without any delay between the removal of one solution and the arrival of the next.

For the growth of quaternary alloys in the $In_{1-x}Ga_xAs_yP_{1-y}$ system the required melt compositions are normally produced by dissolving appropriate amounts of InAs, GaAs and InP in In metal. Quantitative data for a wide range of compositions and growth conditions have been provided by Nakajima (30, 31) and by Kuphal (32). The same size effects occur in quaternary alloy growth as in the growth of $Ga_{1-x}Al_xAs$ or in the doping of binary materials. With two atoms of the same group, the smaller is always more readily incorporated in the solid, this effect being less marked as the growth temperature is raised. LPE growth of layers of quaternary material needs to be started and completed within a narrow temperature range if uniform composition is to be achieved.

Supersaturated solutions are normally produced in the same way as for the growth of binary layers. However, saturation followed by cooling is not the only possible method when ternary or quaternary layers are to be grown. A reaction converting an appropriate mixture of binary III–V compounds into a ternary or quaternary alloy is accompanied by an appreciable gain in configurational entropy. The formation of the ternary alloy $In_{1-x}Ga_xAs$ is a simple example:

$$(1-x)\mathrm{InAs(c)} + x\mathrm{GaAs(c)} \rightarrow \mathrm{In}_{1-x}\mathrm{Ga}_x\mathrm{As(c)}$$

Such a reaction is normally not reversible at the growth temperature, so that a solvent such as metallic In at a *constant* temperature can dissolve InAs and GaAs to an extent sufficient to render it supersaturated with respect to the deposition of a ternary alloy. Such an entropy-driven reaction allows alloys to be growth under conditions which are isothermal in space and time.

A similar process involving only a single III–V compound as source material has been demonstrated by Benchimol and Quillec (33).

$$(1-x)\mathrm{In(l)} + \mathrm{GaAs(c)} \rightarrow \mathrm{In}_{1-x}\mathrm{Ga}_x\mathrm{As(c)} + (1-x)\mathrm{Ga(l\,in\,In)}$$

In this case the chemical reaction involving replacement of Ga in GaAs by In is actually endothermic, but the gain in configurational entropy is sufficient to drive the reaction continuously in the direction shown. The composition (30, 33) of the ternary alloy required for lattice-matching to an InP substrate requires a fixed temperature around 510° C. The experimental arrangement is attractively simple, requiring only a source wafer of GaAs above the liquid metal and an InP substrate below.

One important consideration peculiar to the alloy systems is whether the components are always miscible in the solid phase. In alloys containing antimony, extensive miscibility gaps are encountered and only a very limited range of alloys in the quaternary system $In_{1-x}Ga_xAs_{1-y}Sb_y$ can be produced (34). In the quaternary system $In_{1-x}Ga_xAs_yP_{1-y}$, which is of far greater importance in the semiconductor industry, the effects are much less drastic, and will be considered in greater detail.

From the standard enthalpies of formation (5, 6) of the four binary semiconductors involved, the standard enthalpy of the following reaction can be calculated:

$$\begin{array}{lcccc} & \mathrm{GaAs(c)} + & \mathrm{InP(c)} \rightarrow & \mathrm{GaP(c)} & + \mathrm{InAs(c)} \\ \Delta H^{\ominus}_{\mathrm{f}298} & -81{\cdot}76 & -61{\cdot}63 & -100{\cdot}12 & -59{\cdot}79\,\mathrm{kJ\,mol^{-1}} \end{array}$$

$$\Delta H^{\ominus}_{\mathrm{R}298} = -16{\cdot}53\,\mathrm{kJ\,mol^{-1}}$$

The result indicates a preference for the pairings on the right-hand side of the equation as far as the enthalpy is concerned. The tendency for the atoms to pair in this way is opposed by the entropy gain associated with random pairings that would occur in an ideal solid solution. The extent to which pairing and phase separation occur has been the subject of both theoretical and practical investigations (32, 34–37). In a limited alloy composition range, with y around 0.75 (see Figure 8.1) transmission electron microscopy has revealed contrast patterns ascribed to spinodal decomposition of the quarternary alloy into regions of different chemical composition on a scale of the order of 50–100 nm. However, it appears that the presence of the substrate stabilizes the solid solution, and phase separation on a macroscopic scale does not occur if the layer is sufficiently thin. Keeping the overall composition constant, a higher growth temperature favours a more random alloy, as expected, and this improves the carrier mobilities by reducing the alloy scattering.

The hydrogen environment

A hydrogen environment at about one atmosphere pressure is almost always used for the LPE process. Hydrogen is generally purified by diffusion through a Pd or Pd–Ag alloy membrane, enabling a total impurity content below one part per million to be achieved. The use of hydrogen prevents the formation of an oxide scum on the melt which would adversely affect the contact between liquid and substrate. Hydrogen, however, also reacts with materials containing elements from Groups V and VI and with silica, which is nearly always used for the furnace tube and the linkage to the slider. These reactions will be considered in some detail.

The loss of arsenic from GaAs (or InAs) substrates occurs to some extent at temperatures used for LPE growth (700–850° C for GaAs), but this is not a technologically significant problem. In contrast, precautions to prevent loss of phosphorus from InP substrates in any type of epitaxy furnace during the pre-growth heating to around 600–650° C are very important (38–40). The erosion or decomposition is frequently described as the incongruent evaporation of phosphorus, leaving indium-rich metallic droplets on a pitted surface:

$$\mathrm{InP\,(c)} \rightleftharpoons \mathrm{In\,(1)} + \tfrac{1}{2}\mathrm{P_2(g)}$$

However, consideration of the equilibria

$$2\mathrm{P_2(g)} \rightleftharpoons \mathrm{P_4(g)}$$
$$\mathrm{P_2(g)} + 3\mathrm{H_2(g)} \rightleftharpoons 2\mathrm{PH_3(g)}$$

shows that for the normal conditions in an LPE furnace, PH_3 is the dominant P-containing species in the gas phase and a more realistic representation of the substrate decomposition involves the hydrogen atmosphere (39).

$$\mathrm{InP(c)} + \tfrac{3}{2}\mathrm{H_2(g)} \rightleftharpoons \mathrm{In(1)} + \mathrm{PH_3(g)}$$

To suppress the substrate decomposition it is necessary to introduce P-containing species into the gas phase. One method involves the direct addition of PH_3 to the hydrogen introduced into the furnace.

At a typical working temperature of 900 K the addition of PH_3 at a concentration of 3.5×10^{-5} is enough to suppress erosion. However, even lower concentrations of PH_3 are sufficient if the partial pressure of hydrogen in the system is reduced. If 90% of the hydrogen is replaced by a gas such as argon or nitrogen, the PH_3 requirement is reduced by an order of magnitude. A continuing reduction of the hydrogen content yields diminishing benefit because P_2 and P_4 become the dominant P-containing species in the gas instead of PH_3.

Addition of PH_3 to the entire furnace gas stream increases both the cost and the danger of the process, so that alternative methods of substrate protection are generally preferred. The simplest method (Figure 8.2) uses a larger covering wafer, usually of the same material as the substrate, to limit the exposure of the substrate to the gas ambient before epitaxial growth is started. The mechanical movement that brings the supersaturated solution into contact also removes the covering wafer. However, this method is never completely effective in preventing erosion.

To produce higher concentrations of the required P-containing molecules in the gas phase near the substrate, an anti-erosion melt based on solutions of InP in liquid Sn is often placed nearby (40). The liquid In produced when InP decomposes has a much reduced activity in the anti-erosion melt because the In is in a dilute solution in Sn, and such solutions do not deviate greatly from ideal behaviour, as the two liquid metals have cohesive energy densities of the same magnitude. The reduced activity of the In in the Sn means that a higher partial pressure of PH_3 is required for equilibrium at the surface of the anti-erosion melt. It is often found in practice that anti-erosion melts made only from Sn and InP produce overpressures of PH_3 that are too high, affecting the composition of adjacent growth melts and also making nucleation on the substrate difficult. A mixture of Sn, In and InP appropriate for the boat and furnace in question usually has to be established by experiment.

Hydrogen is also important in the transport of sulphur, one of the two major donors controlling background purity (11, 12) in an LPE furnace. Consideration of thermodynamic data for reactions such as

$$2H_2(g) + S_2(g) \rightleftharpoons 2H_2S(g)$$

shows that H_2S is the dominant S-containing species in the gas in an LPE furnace at all normal operating temperatures. The removal of sulphur, by prolonged baking of the melt, is therefore achieved more rapidly in hydrogen than in an inert atmosphere. Transfer of sulphur from highly doped dislocation-free InP substrates or hot stainless steel to nominally pure growth solutions can occur by the same chemical mechanism.

Another problem occurs when an InP substrate does not have the usual simple planar form, but has a more complicated shape containing ridges or channels. Since vapour pressures are greater for convex surfaces than for concave ones, material can be transported through the vapour phase to produce a smoother and less angular profile. The mechanism by which the phosphorus is transported is the same as that involved in the erosion process, but the transport of the indium is less well understood. It is known that the transport of InP can be enhanced by the presence of halides, a process which is exploited in the production of MTBH (mass-transported buried heterostructure) lasers (41), but some transport occurs even when precautions are taken to exclude halides. A plausible explanation is that the metal is transported through the vapour phase as InS, sulphur being a common contaminant of epitaxial systems.

Although the mechanism for the metal transport remains debatable, it has been found empirically (42) that the shape of non-planar InP substrates undergoes less change when the covering slice consists of GaAs instead of InP. Auger studies (22) of the surface of an InP slice protected by a GaAs wafer, with a separation of 50–60 μm, show that Ga as well as As is transported across the gap. It can be argued that the stronger bonding between Ga and P atoms is responsible for the increased stability of the surface of the InP substrate.

Hydrogen plays an important part in determining the silicon contamination of solutions and the epitaxial layers grown from them. An exhaustive treatment of both the kinetics and the thermodynamics of the various possible reactions was given by Weiner (43), who considered a wide range of gaseous species including SiO and Ga_2O. However the analysis can be greatly simplified (44) by considering only the predominant equilibrium:

$$SiO_2(\text{vitreous}) + 2H_2(g) \rightleftharpoons Si(\text{l in Ga or In}) + 2H_2O(g)$$

Since both the hydrogen and the silica have activities always close to unity, the equilibrium constant for this reaction relates the equilibrium concentration of dissolved silicon to the partial pressure of water in the gaseous environment by the simple equation

$$K_R(T) = [Si][H_2O]^2$$

K_R can be calculated for any required temperature from published thermodynamic data (7) for the substances involved in the reaction.

As an example, Figure 8.5 presents results of calculations of the silicon contamination of initially pure gallium held at 800° C in a hydrogen flow containing concentrations of water from 1 in 10^5 to 1 in 10^7. The equilibrium silicon concentrations can be read from the flat regions on the right of the diagram. These depend only on the temperature and the water content of the input gas. However, the time taken to reach the equilibrium depends on other parameters. In this case it is assumed that the liquid metal is in contact with vitreous silica, that diffusion within the liquid metal is not the rate-limiting process, and that the ratio of the gas flow rate to the mass of gallium is 40 ml min^{-1}g^{-1}. For a water content of 1 in 10^6, equilibrium is achieved

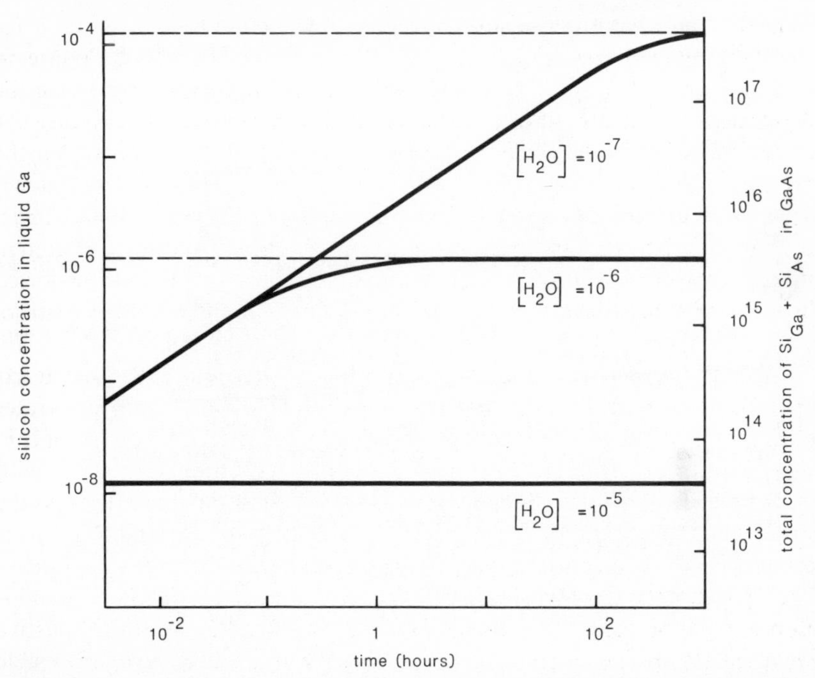

Figure 8.5 Development of Si contamination is an initially pure Ga + GaAs solution in SiO_2 at 800° C showing dependence on H_2O content of the H_2 environment. The scale on the right indicates the corresponding ionized impurity concentration in the epitaxial GaAs grown from such a solution at the same temperature

within the time required for saturation of the liquid with the III–V semiconductor solute. The concentration of silicon (just over 1×10^{-6}) in the liquid is sufficient to dope the epitaxial layer at a concentration of around 10^{-7}, equivalent to an ionized impurity concentration of $4 \times 10^{15}\,cm^{-3}$. To achieve LPE GaAs of the highest purity, a higher water content and a lower temperature are required. When the water content is lower than 1 in 10^6, not only does the equilibrium silicon contamination rise, but the time taken to reach equilibrium usually also exceeds the process time. Therefore changes in the purity of the original gallium metal and changes in the duration of the pre-growth saturation can cause variable results.

When aluminium is present in the melt so that $Ga_{1-x}Al_x$ As epitaxial layers can be grown, low water contents are required to prevent oxidation of aluminium and this then becomes the dominant consideration. For this reason the high purities achieved in the LPE growth of GaAs and other materials without aluminium have not been reproduced with $Ga_{1-x}Al_x$ As.

References

1. K. Nakajima, S. Komiya, K. Akita, T. Yamaoka and O. Ryuzan, *J. Electrochem. Soc.* **127** (1980) 1568.

2. S. E. H. Turley and P. D. Greene, *J. Crystal Growth* **58** (1982) 409.
3. J. J. Hsieh, *Proc. 6th Int. Symp on GaAs and Related Compounds*, St. Louis 1976, Inst. Phys. Conf. Ser. 33b, 74.
4. E. H. Perea and C. G. Fonstad, *J. Electrochem. Soc.* **127** (1980) 313.
5. M. Tmar, A. Gabriel, C. Chatillon and I. Ansara, *J. Crystal Growth* **68** (1984) 557.
6. M. Tmar, A. Gabriel, C. Chatillon and I. Ansara, *J. Crystal Growth* **69** (1984) 421.
7. O. Kubaschewski and C. B. Alcock, *Metallurgical Thermochemistry*, 5th edn., Pergamon, Oxford (1979).
8. E. Bauser, L. Schmidt, K. S. Loechner and E. Raabe, *Jap. J. Appl. Phys.* **16** [supp. 1] (1977) 457.
9. L. W. Cook, M. M. Tashima, N. Tabatabaie, T. S. Low and G. E. Stillman, *J. Crystal Growth* **56** (1982) 475.
10. C. J. Armistead, P. Knowles, S. P. Najda and R. A. Stradling, *J. Phys. C* **17** (1984) 6415.
11. B. J. Skromme, T. S. Low and G. E. Stillman, *Proc. 10th Int. Symp. GaAs and Related Compounds*, Albuquerque 1982, Inst. Phys. Conf. Ser. **65**, 45, 515.
12. M. S. Skolnick, P. J. Dean, S. H. Groves and E. Kuphal, *Appl. Phys. Lett.* **45** (1984) 962.
13. C. Woelk and K. W. Benz, *J. Crystal Growth* **27** (1974) 177.
14. R. A. Chapman and W. G. Hutchinson, *Phys. Rev. Lett.* **18** (1967) 433.
15. A. W. Smith, L. G. Shantharama, L. Eaves, P. D. Greene, J. R. Hayes and A. R. Adams, *J. Phys. D* **16** (1983) 679.
16. N. Chand, P. A. Houston and P. N. Robson, *Electron. Lett.* **20** (1981) 726.
17. H. C. Casey, M. B. Panish and K. B. Wolfstirn, *J. Phys. Chem. Solids* **32** (1971) 571.
18. C. S. Kang and P. E. Greene, *Proc. 2nd Int. Symp. GaAs*, Dallas 1968, Institute of Physics, 18.
19. J. Vilms and J. P. Garrett, *Solid State Electronics* **15** (1972) 443.
20. P. D. Greene, *J. Crystal Growth* **50** (1980) 612.
21. F. E. Rosztoczy, F. Ermanis, I. Hayashi and B. Schwartz, *J. Appl. Phys.* **41** (1970) 264.
22. P. D. Greene, unpublished.
23. M. G. Astles, F. G. H. Smith and E. W. Williams, *J. Electrochem. Soc.* **120** (1973) 1750.
24. F. E. Rosztoczy, G. A. Antypas and C. J. Casau, *Proc. 3rd Int. Symp. GaAs and Related Compounds*, Aachen 1970, Inst. Phys. Conf. Ser. **9**, 86.
25. B. H. Ahn, R. S. Shurtz and C. W. Trussel, *J. Appl. Phys.* **42** (1971) 4512.
26. P. D. Greene, *Solid State Comm.* **9** (1971) 1299.
27. V. M. Lakeenkov, L. M. Morgulis, M. G. Mil'vidskii and O. V. Pelevin, *Izv. Akad. Nauk SSSR, Neorg. Mat.* **13** (1977) 1373.
28. M. B. Panish and S. Sumski, *J. Phys. Chem. Solids* **30** (1969) 129.
29. W. G. Rado, W. J. Johnson and R. L. Crawley, *J. Electrochem. Soc.* **119** (1972) 652.
30. K. Nakajima, T. Kusunoki, K. Akita and T. Kotani, *J. Electrochem. Soc.* **125** (1978) 123.
31. K. Nakajima, in *GaInAsP Alloy Semiconductors*, ed. T. P. Pearsall, John Wiley (1982), Chapter 2.
32. E. Kuphal, *J. Crystal Growth* **67** (1983) 441.
33. J. L. Benchimol and M. Quillec, *Journal de Physique* **43-C5** (1982) 69.
34. G. B. Stringfellow, *J. Crystal Growth* **58** (1982) 194.
35. M. Quillec, H. Launois and M. C. Joncour, *J. Vac. Sci. Technol. B* **1** (1983) 238.
36. B. de Cremoux, P. Hirtz and J. Ricciardi, *Proc. 8th Int. Symp. GaAs and Related Compounds*, Inst. Phys. Conf. Ser. **56**, 115.
37. P. Henoc, A. Izrael, M. Quillec and H. Launois, *Appl. Phys. Lett.* **40** (1982) 963.
38. A. R. Clawson, W. Y Lum and G. E. McWilliams, *J. Crystal Growth* **46** (1979) 300.
39. P. D. Greene and E. J. Thrush, *J. Crystal Growth* **72** (1985) 563.
40. G. A Antypas, *Appl. Phys. Lett.* **37** (1980) 64.
41. A. Hasson, L. C. Chiu, T. R. Chen, U. Koren, Z. Rav-Noy, K. L. Yu, S. Margalit and A. Yariv, *Appl. Phys. Lett.* **43** (1983) 403.
42. J. Kinoshita, H. Okuda and Y. Uematsu, *Electron. Lett.* **19** (1983) 215.
43. M. E. Weiner, *J. Electrochem. Soc.* **119** (1972) 496.
44. P. D. Greene, *J. Phys. D.* **6** (1973) 1550.

9 Photoresists

A. LEDWITH

Introduction

Microcircuit fabrication requires the selective diffusion of tiny amounts of impurities into regions of a semiconductor substrate such as silicon, to produce the desired electrical characteristics of the circuit. These regions (together with the metal conductor paths that link the active circuit elements) are defined by lithographic processes in which the desired pattern is first generated in a resist layer (usually a polymeric film 0.5–1.0 μm thick which is spin-coated onto the substrate) and then transferred via processes such as etching into the underlying substrate. In silicon integrated-circuit manufacture, the latter is usually SiO_2 which is present as a thin layer on top of the silicon and which functions as the actual mask for the diffusion process. The definition of the pattern in the resist layer is achieved by exposing the resist to some suitable form of patterned radiation such as ultraviolet light, electrons, x-rays or ions. The resist contains radiation-sensitive groups which chemically respond to the incident radiation, forming a latent image of the circuit pattern which can subsequently be 'developed' by solvent or plasma treatment to produce a three-dimensional relief image. Radiation-induced chemical reactions cross-linking and/or chain degradation, for instance, enable exposed regions to be differentiated from unexposed regions either by solubility differences or by differences in plasma etch rate. The areas of resist remaining after development (the exposed or non-exposed areas as the case may be, depending on whether irradiation causes the exposed areas to become more soluble or less soluble than the unexposed areas) must now protect the underlying substrate during the variety of additive and/or subtractive processes encountered in semiconductor processing.

The basic steps of the lithographic process are shown schematically in Figure 9.1. The example shown corresponds to photolithography in which the photosensitive resist or photoresist is applied as a thin film to the substrate (SiO_2 on Si) and subsequently exposed in an image-wise fashion through a mask. The mask contains clear and opaque features that define the circuit pattern. The areas in the photoresist that are exposed to light are made either soluble or insoluble in a specific solvent termed a developer. In the first case, the irradiated regions are dissolved leaving a positive image of the mask in the resist. Therefore, the photoresist is termed a positive resist. In the second case, the unirradiated regions are dissolved leaving a negative image; hence, the

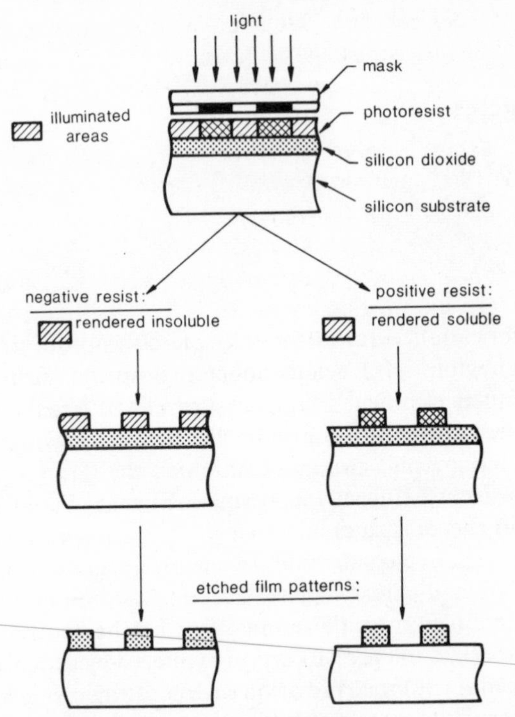

Figure 9.1 Schematic diagram of the photolithographic process. Adapted from (8)

resist is termed a negative resist. Following development, the exposed SiO_2 areas are removed by etching, thereby replicating the mask pattern in the oxide layer. The underlying silicon is exposed by this etching procedure, and impurities or dopants such as boron or arsenic can now be diffused into these exposed regions. Complex circuits are built into the silicon by a succession of such procedures. The polysilicon and/or metal conductor interconnections, together with the terminals that make the circuit accessible to the outside world, are also fabricated by lithographic techniques.

The role of the resist in this process is seen to be twofold. First, it must respond to the exposing radiation to form a latent image of the circuit pattern described in the mask which can subsequently be developed to produce a three-dimensional relief image. Second, the areas of resist remaining after development (the exposed or non-exposed areas as the case may be) must protect the underlying substrate during subsequent processing. Indeed, the generic name 'resist' no doubt evolved as a description of the ability of these materials to 'resist' etchants.

Types of radiation for microlithography

In the semiconductor industry, smaller line widths translate into higher speeds, less heat build-up, more devices per wafer, greater energy efficiency, and more powerful devices. The 1986 manufacturing state of the art involves 1.5 μm lines, and further shrinkage of device geometries will most probably require methods and materials unlike anything previously seen in microlithography.

Conventional exposure sources, however, approach their fundamental theoretical limit when submicrometre resolution is required. The exact resolution limit is inherently dependent on the wavelength of the exposing device. The Rayleigh diffraction limit for any electromagnetic radiation is equal to approximately twice the wavelength of the exposure light. The wavelength of a common exposure source is 436 nm, which translates to a diffraction-limited resolution of 0.87 μm. A potential short-term improvement in resolution can be obtained by decreasing the exposure wavelength to the 365, 313, or 254 nm lines of a mercury arc lamp. The latter two lines have been used loosely to define the mid- and deep UV exposure wavelengths, respectively. But it appears that increasingly stringent resolution requirement will stretch the limits of optical lithography, even with mid- and deep UV exposure, in the areas of diffraction and depth of focus. Several non-optical technologies involving electron beams, x-ray, or ion-beam exposures have the potential to achieve these high-resolution goals, and an excellent introduction to these techniques has been given by Thompson *et al.* (1) (see also Chapter 10). Additionally, excimer lasers can be used to expose resists at a variety of short wavelengths including 193 and 157 nm (2). Each type of exposure requires its own unique resist material. Of the four technologies, electron-beam lithography has received the most attention and already has a growing number of practical applications. X-ray lithography is emerging as the exposure method of choice for high-throughput submicrometre lithography. Advances in sources and masks have increased the potential feasibility of commercial x-ray integrated circuit production. Ion-beam lithography, a relatively new development, is based on using a beam of accelerated cations to induce chemical transformations in a polymer film.

Irrespective of what particular lithographic technology is being employed, whether it be conventional photolithography or the other high-resolution techniques mentioned above, a resist system is required which enables the exposed regions to be differentiated from the unexposed regions. As shown earlier, there are two types of resist, depending on the tone of the image following exposure and development. Positive resists produce positive-tone images in which the exposed areas are removed (either by solvent as mentioned earlier or by dry processes). Conversely, negative resists give rise to negative-tone images in which the non-irradiated areas are removed. Both types of resist are used in circuit manufacture. Nearly all resists are organic polymers which are designed to respond to the exposing radiation, although resists derived from inorganic materials, including chalcogenide glasses, have shown potential (3). Most recently, inorganic polymers and oligomers with silicon–silicon bonds in the main chain have become of interest, particularly as potentially self-developing resists (4).

One-component systems, as the name implies, are polymers that combine radiation sensitivity, etching resistance and film-forming characteristics within a single species. Two-component systems are most often used in photochemistry and here the resist is formulated from an inert matrix resin which serves as a binder and film-forming material, and which is usually monomeric in nature and undergoes the radiation-induced chemical transformations that are responsible for imaging. For example, poly(methyl methacrylate) (PMMA) is a classic one-component resist which has found wide utility as a deep UV, electron, x-ray and ion-beam resist. It is a single homogeneous material that combines the properties of excellent film-forming characteristics, resistance to chemical etchants, and (albeit low) intrinsic radiation sensitivity. Detailed descriptions of the two component systems (used in photolithography) together with their chemical structures and photochemistry will be given later.

$$\text{Ⓟ}-CH_2-C(CH_3)(C{=}O\,OCH_3)-CH_2-C(CH_3)(C{=}O\,OCH_3)-\text{Ⓟ} \xrightarrow[\alpha\ \text{scission}]{h\nu} \text{Ⓟ}-CH_2-\dot{C}(CH_3)-CH_2-C(CH_3)(C{=}O\,OCH_3)-\text{Ⓟ} + \underbrace{CO, CO_2, \bullet CH_3, \bullet OCH_3}_{\text{fragments}}$$

$$\xrightarrow{\beta\ \text{scission}} \text{Ⓟ}-CH_2-C(CH_3){=}CH_2 + \bullet C(CH_3)(C{=}O\,OCH_3)-\text{Ⓟ} \longrightarrow \text{further degradation}$$

The end objective of the lithographic process is the faithful replication of the pattern originally specified by the device designer. Just how successfully this will be accomplished is determined by the physics and chemistry of resist exposure and development, and of subsequent pattern transfer. For the purposes of this chapter the practical problems involved in fine-line lithography will not be considered further—they are adequately described in recent books (5) and a number of reviews (6). Rather it is the intention to outline the basic chemical transformations involved in commonly used photoresist technology, concentrating on pattern delineation where an understanding of the theory and chemistry of resist design and processing is essential. For more details of the design and function of materials useful in fine-line lithography the reader is referred to the excellent Symposium Report edited by the Thompson, Wilson and Fréchet (7) who, together with Bowden (6), have pioneered chemistry development in the field of microlithography.

Resist requirements

Resist requirements for microfabrication are similar, regardless of the exposure technology, and the general requirements have been discussed in many articles (1, 6, 7). It is useful to indicate the more important requirements, particularly in view of the special limitations imposed by the need to fabricate fine-line VLSI devices highlighted by Bowden (8).

Solubility

Since films are normally deposited on a substrate by spin-coating from solution, solubility in organic solvents is a necessary requirement. Solubility in aqueous media may be undesirable if wet etching is contemplated, since many of the etching solutions are aqueous. Although other coating techniques such as plasma deposition have been used, such resist systems generally employ dissolution techniques to differentiate between exposed and non-exposed regions. One goal of resist research is the design of material systems capable of all-dry processing in which the resist would be applied to the substrate by CVD or plasma deposition techniques, exposed, and then developed by plasma treatment. Such systems would remove solubility requirements in resist design.

Adhesion

A variety of substrates is encountered in semiconductor processing. These include metal films such as aluminium, chromium, and gold, insulators such as silicon dioxide and silicon nitride, and semiconductor materials. The resist must possess adequate adhesion to withstand all processing steps. Poor adhesion leads to marked undercutting, loss of resolution and, in the worst case, complete destruction of the pattern. Pattern transfer to the underlying substrate has been conventionally accomplished using liquid etching techniques which require tenacious adhesion between resist and substrate in order to minimize undercutting and maintain edge acuity and feature size control. In some cases, adhesion between the resist and substrate may be enhanced with adhesion promoters such as hexamethyldisilazane. As geometries are reduced, liquid etching is being replaced by dry-etching techniques which place less stringent adhesion requirements on the resist. It should be noted, however, that there still has to be sufficient adhesion to withstand development, and this becomes increasingly difficult as feature sizes decrease.

Etchant resistance

Etchant resistance refers to the ability of the resist to withstand the etching environment during the pattern transfer process. The most common method of pattern transfer is wet chemical etching which places emphasis on the adhesion and chemical stability of the resist. Etchant solutions may be either acidic or basic, depending on the type of substrate to be etched. For example, buffered hydrofluoric acid is used to etch SiO_2. However, lateral penetration of the chemical etchant is significant for thick substrate films and results in unacceptable line width control for feature sizes $< 2\,\mu m$. For thin substrate films, as found for example in chromium photomasks where the thickness of the chromium is approximately 50–100 nm, liquid etching is capable of delineating submicron features. The need to pattern fine features in thick substrates has led to the development of anisotropic etching methods such as reactive-ion etching, plasma etching, ion milling, and sputter etching. While not requiring a premium in terms of adhesion, these techniques place other very stringent requirements on the resist. Most dry-etching techniques rely on plasma-induced gas reactions in an environment of high radiation flux and temperatures in excess of 80° C (see Chapter 15). The physical and chemical properties of the resist therefore play an important part in determining dry-etchant resistance and may present a fundamental limitation. For example, dry-etchant resistance is generally observed in polymers that possess thermal and radiation stability such as obtain in polymers that exhibit high glass transition temperatures and contain radiation-stable groups, e.g. aromatic structures. High sensitivity in positive resists, on the other hand, usually requires polymers that are unstable to radiation. Thus the requirement of high sensitivity coupled with adequate resistance to dry-etching environments presents an apparent dichotomy to the resist designer.

Sensitivity

Sensitivity is conventionally defined as the input incident energy (measured in terms of energy or the number of photons or particles (fluence) per unit area) required to attain a certain degree of chemical response in the resist that results, after development, in the desired relief image. This represents an operational, lithographic definition of sensitivity (8).

Typically a plot is made of normalized film thickness remaining as a function of log (dose) as shown in Figure 9.2*a*, *b*. Note that sensitivity increases as the dose required to produce the lithographic image decreases. The sensitivity of a positive resist is the dose (D_c in Figure 9.2*a*) required to effect complete solubility of the exposed region under conditions where the unexposed region remains completely insoluble. In the case of a negative resist, sensitivity is defined as the dose at which a lithographically useful image is formed. Although gel (insoluble cross-linked resist) begins to form at the interface gel dose (D_g^i in Figure 9.2*b*), no lithographically useful image has been formed and additional dose is required to build the relief image to the desired thickness (usually to $D_g^{0.5}$ which is the dose at which 50% of the original film thickness has been retained after development).

Contrast (resolution)

The pattern resolution attainable with a given resist for a particular set of processing conditions is determined in large part by the resist contrast (γ).

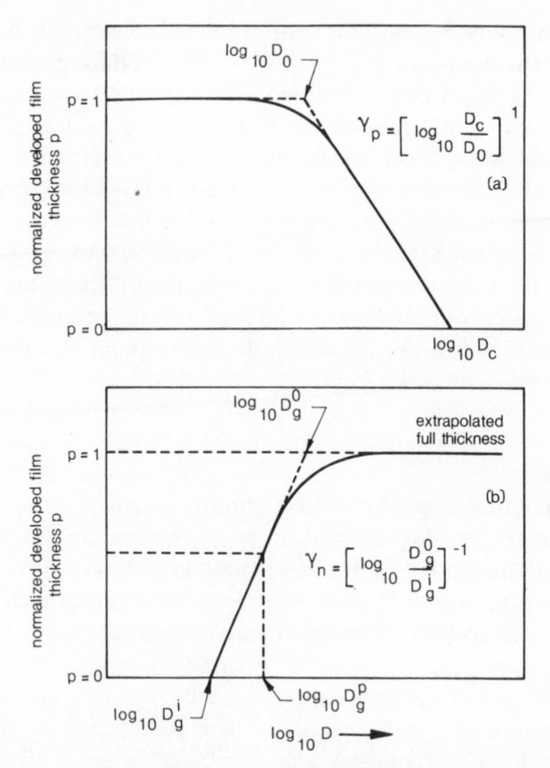

Figure 9.2 Typical lithographic response or contrast plots for (*a*) positive resists and (*b*) negative resists in terms of the developed thickness normalized to initial resist thickness (*p*) as a function of log (dose). As described by Bowden (8)—see also Chapter 10

Contrast in the case of a negative resist is related to the rate of cross-linked network (gel) formation and, for a positive resist, to both the rate of degradation and the rate of change of solubility with molecular weight with the latter being markedly solvent-dependent. The numerical value of γ is obtained from the slope of the linear portion of the response curves shown in Figures 9.2*a*, *b*. The contrast of a positive resist is expressed mathematically as

$$\gamma_p = 1/(\log D_c - \log D_0) = [\log(D_c/D_0)]^{-1}$$

where D_0 is the dose at which the developer begins to dissolve the irradiated film and is determined by extrapolating the linear portion of the film thickness remaining *v.* dose plot to a value of 1.0 normalized film thickness; D_c is the complete development dose (8).

The contrast of a negative resist is expressed as

$$\gamma_n = 1/(\log D_g^{100} - \log D_g^i) = \left(\log\left[\frac{D_g^{100}}{D_g^i}\right]\right)^{-1}$$

where D_g^{100} is the dose required to produce 100% initial film thickness and is obtained by extrapolating the linear portion of the thickness *v.* dose plot to a value of 1.0 normalized film thickness; D_g^i is the interface gel dose (8).

High contrast is important since it minimizes the deleterious effects due to scattering of radiation in the resist film. All exposure techniques result in some energy being deposited outside the primary image area. Diffraction of light in photolithography, scattering of primary and secondary electrons in electron-beam lithography and of photoelectrons following x-ray absorption or scattering of ions in ion-beam lithography all contribute to energy deposition in areas remote from the desired exposure area. These effects present fundamental physical limitations to resolution, although the situation may be improved by high-contrast resists which do not respond significantly to low levels of scattered radiation.

The chemistry of photoresists

Commercial microlithography relies almost exclusively on two well-established chemistries. Most positive resists involve the photofunctional change of naphthalene quinone diazides in novolac resins, and negative resists utilize the photodecomposition (followed by cross-linking) of bisazides in a cyclized polyisoprene matrix. Together, these two resist systems dominate the commercial semiconductor industry because they can produce high-resolution, high-quality patterns in a manufacturing environment. There is, however, a continuing trend toward positive resist systems because they appear capable of higher resolution than their negative counterparts. A useful and up-to-date review of the chemistry of photoresists has been given by Blevins, Daly and Turner (9). Bisazide/cyclized polyisoprene resists are by far the most popular negative-working materials for microelectronic and many macro-electronic applications. Cyclized polyisoprene resist materials are popular because they are inexpensive, have high sensitivity, adhere well to most surfaces, and have excellent resistance to wet chemical etching by acids or bases. They also have good plasma etch resistance, but their glass transition temperatures are too low for acceptable thermal image stability.

Cyclized polyisoprene is prepared by thermal degradation of the long, linear polymer chains of either natural rubber or synthetic *cis*-1,4-polyisoprene, followed by acid-catalysed isomerization. These two reactions create smaller segments of polymer with various degrees of unsaturation and cyclization. The reaction generally takes place in xylene or toluene at high temperatures in the presence of strong protonic acid (sulphuric, *p*-toluenesulphonic) or Lewis-acid catalysts ($AlCl_3$, $TiCl_4$, BF_3). The linear polyisoprene polymer cyclizes via carbocation intermediates which are formed by protonation of a double bond. This carbocation may either react with an adjacent double bond and subsequently deprotonate to form a monocyclic structure, or continue to react with neighbouring double bonds to yield various multicyclic structures. Thus, deprotonation competes with cyclization to produce a mixture of mono-, bi-, tri-, and multicyclic structures. On the average, microelectronic resist polyisoprenes contain 1–3 rings per cyclic unit, with 5–20% unreacted isoprene units remaining. Cyclized rubber has higher density, glass transition temperature, and refractive index than natural rubber but has lower solution viscosity. Since several other reactions are theoretically possible, the exact structure of cyclized polyisoprene has been the subject of considerable controversy and research (10). The overall cyclization processes are indicated in Figure 9.3.

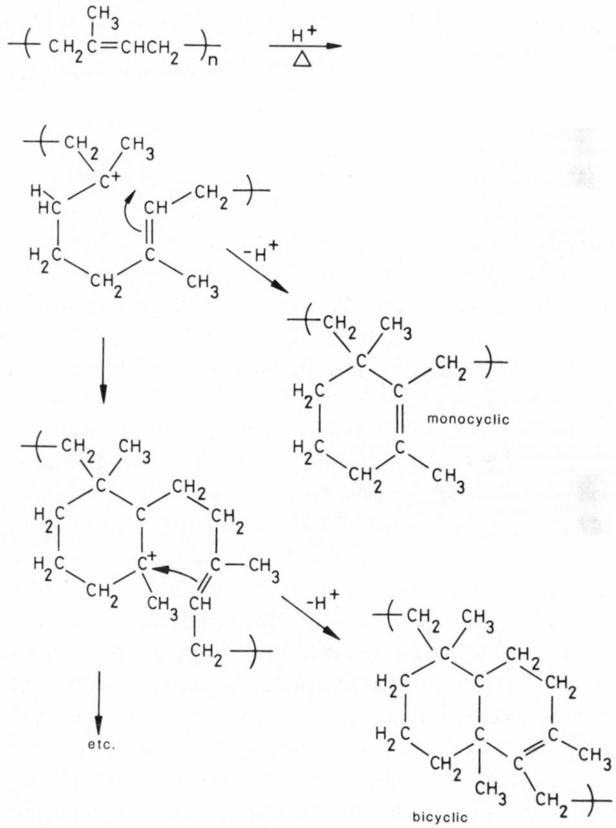

Figure 9.3 Acid-catalysed cyclization of polyisoprene. Taken from (9)

Aromatic azides readily decompose on irradiation to yield highly reactive chemical intermediates known as nitrenes. Nitrenes undergo insertion reactions with aliphatic carbon–hydrogen bonds to produce secondary amines, and undergo addition reactions with carbon–carbon double bonds to yield aziridines:

$$ArN_3 \xrightarrow{h\nu} Ar\ddot{N}: + N_2 \longrightarrow Ar\dot{N}\cdot$$

aromatic azide — singlet nitrene — triplet nitrene

$$Ar\ddot{N}: + RH \longrightarrow ArNHR$$

$$\cdot Ar\dot{N}\cdot + RH \longrightarrow Ar\dot{N}H + R\cdot \longrightarrow ArNHR$$

$$Ar\ddot{N}: + RCH = CHR' \longrightarrow RCH{-}CHR' \text{ (bridged by N–Ar)}$$

It is easy to understand therefore that bis-azido compounds (N_3ArN_3) when photolysed in a polymer matrix such as cyclized polyisoprene readily produce crosslinks.

N_3–C_6H_4–R–C_6H_4–N_3 + 2 ~CHR′–CH$_2$~ $\xrightarrow{h\nu}$ ~CHR′–CH(NH–C_6H_4–R–C_6H_4–NH)CH–CHR′~ + 2N_2

where R is >CO; –CH=CH–; –CH$_2$–; –NH–; or –CH= (2,6-cyclohexanone, 4-CH$_3$) =CH–

and ~CHR′–CH$_2$~ represents cyclized polyisoprene

The very high chemical reactivity of nitrene intermediates suggests that bis-azido compounds could be used to effect photochemical cross-linking of synthetic polymers other than cyclized polyisoprene. A number of polymers have been employed as resists for this purpose and recent work (11) highlights the value of poly(para-hydroxystyrene), alternatively called poly(4-vinylphenol).

~CH$_2$–CH(C$_6$H$_4$OH)–CH$_2$–CH(C$_6$H$_4$OH)~

As for the novolac resins to be discussed later, the phenol-containing polymer is soluble in alkaline solutions and hence unexposed or unaffected areas of the resist are readily removed. Photodecomposition of the azido compounds in poly(4-vinylphenol) produces secondary amino groupings and these cause the polymer to be insoluble in alkali, with the added advantage of absence of swelling. Detailed mechanisms of all the chemical transformations involved are not yet available, but by selection of chromophore associated with the azido group, the resists can be made sensitive to deep UV or the usual 350–400 nm radiation (11).

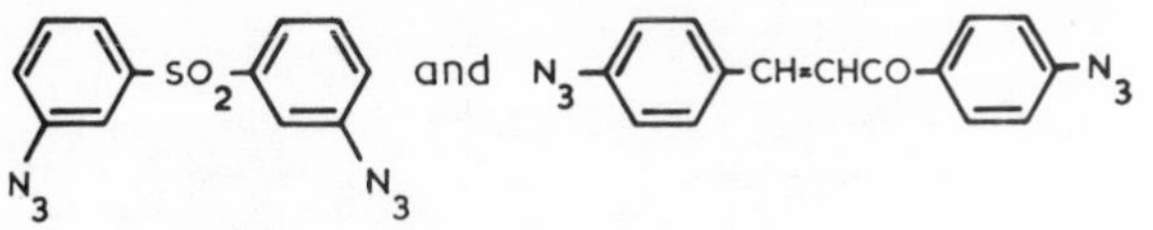

Conventional positive resist systems usually comprise an aqueous-base soluble novolac binder polymer and a quinone diazide photoactive component (PAC). Positive novolac-based resists are often favoured over polyisoprene-based negative-working materials because they exhibit less swelling and image distortion during the development of the resist image. Disadvantages include their higher cost and lower sensitivity compared to cyclized polyisoprene resists. Current formulations are capable of sub-micrometre resolution with excellent control of critical dimensions and edge profiles. The novolac polymer, often called the binder or resin in a resist, plays a key role in the performance of the total resist package. It has excellent film-forming properties, adhesion, and dry etch resistance, and is base-soluble at low molecular weights. Commercial positive resists generally contain an *m*-cresol resin synthesized from acid-catalysed condensation of *m*-cresol and formaldehyde, e.g.

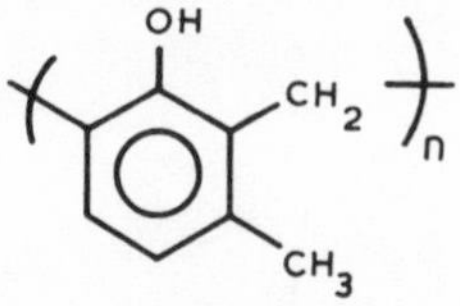

The *m*-cresol resins are preferred because of their good solubility in organic coating solvents and because they can be manufactured on a large scale at workable molecular weights. Mixtures of the three naturally-occurring cresol isomers form low-molecular-weight resins with broad polydispersities. Typically, novolac photoresist polymers made from *m*-cresol have molecular weights of 2000–12 000 with moderately broad polydispersities, and glass transition temperatures are in the range 90–145° C. Polyphenols such as novolacs are soluble in alkaline solutions. For resist applications the alkali solubility of the novolac is inhibited by the presence of a photoactive compound (PAC). This compound is invariably a derivative of a 1,2-naphthoquinone diazide (12). The latter readily decomposes on photolysis to yield ultimately indene carboxylic acids via successive keto-carbene and ketene intermediates (13).

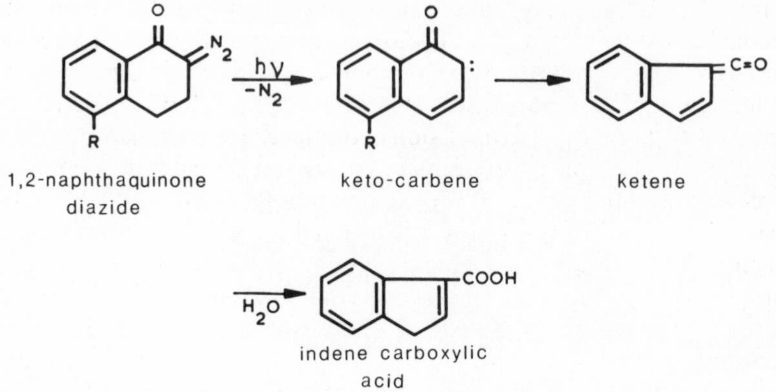

$h\nu$, $-N_2$

1,2-naphthaquinone diazide

keto-carbene

ketene

H_2O

COOH

indene carboxylic acid

It is frequently overlooked that water is an essential reactant in the resist chemistry and is presumably present as adventitious impurity in the ingredients, notably the novolac. Controlled humidity is also essential for reproducible positive working. In the absence of water the ketene intermediates will cause some degree of cross-linking by reactions with the phenolic groups (13). Where the image-wise exposure produces carboxylic acid, the area readily dissolves in solutions containing metallic or quaternary ammonium hydroxides. In the absence of photodecomposition to produce alkali-soluble carboxylic acids, the novolac is inhibited from dissolving in alkali by the hydrophobic character of the quinone diazide. Quinone diazides may constitute up to 40% by weight of a typical positive resist, and dissolution inhibition is ensured by using rather bulky polyfunctional materials as indicated below.

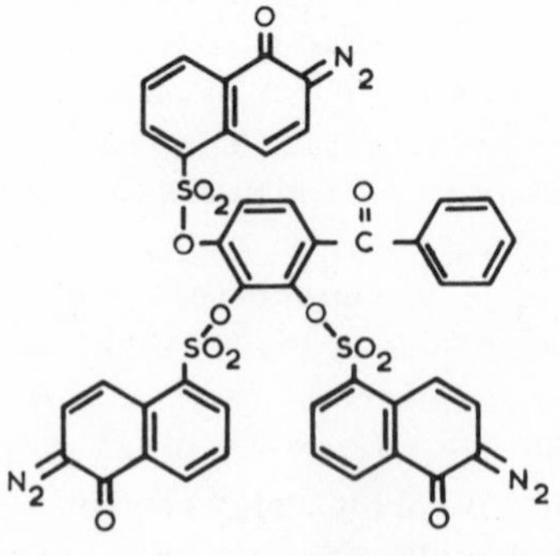

The photodecomposition of quinone diazides has been studied extensively and is known as the Süs reaction, a special form of the well-known Wölff rearrangement (12, 13). The mechanism of image formation and development as given above appears simple but has yet to be fully elucidated. The absorption of a photon of light causes three things to occur: (1) the PAC is destroyed, (2) nitrogen is given off, and (3) in the presence of water an indene carboxylic acid is formed. Historically, the first and third events have been considered the key to the developer discrimination. Recent studies suggest that the formation of acid may be less important than previously believed (9). It appears that the image discrimination of the developer is primarily a result of the inhibition of novolac dissolution by the unexposed hydrophobic PAC and increased dissolution caused by the gaseous nitrogen released into the exposed areas. The release of nitrogen in the exposed areas creates a microporous structure, which facilitates the rapid diffusion of the developer through the resist, thereby increasing its effectiveness (14). Regardless of how it occurs, an enormous change takes place in the dissolution kinetics of the irradiated resist (9). Several polymers are being considered as replacement for novolacs in an alkaline-developable photoresist package (9). Poly (*p*-hydroxystyrene) has received considerable attention as a resist binder. The different backbone and improved molecular-weight control yield a polymer with a glass transition temperature of about 180° C, versus 110–120° C for novolacs.

The well-known film-forming and thermal properties of novolacs have stimulated a search for techniques which would permit negative working modes in addition to, or instead of the usual positive working. Research workers at IBM Laboratories have reported a promising process (15). In positive working, the resists function, at least in part, because light transforms the dissolution inhibitor into a carboxylic acid soluble in the developer solution. If the carboxylic acid photo-product could itself be trapped in some way or decomposed to form a neutral molecule, the alkali developer would no longer be able to dissolve away the resist in the initially exposed areas. However, subsequent flood exposure of the previously unexposed areas, followed by normal development, would then yield overall a polymer resist in the initially image-wise exposed areas and the system would be properly classified as negative working. It is known that indene carboxylic acids may be decomposed, with loss of carbon dioxide, by heating in the presence of amines or bases at temperatures in excess of 100° C.

COOH heat / amine → + CO_2
R R

Thus if a typical positive working resist is formulated to include about 1% by weight of one of a number of amines known to cause decarboxylation, then baking at 100–120° C prior to development will cause the indene carboxylic acid to produce alkali-insoluble indenes. Flood exposure of the complete resist surface, followed by normal development would then yield a negative image of the original mask pattern. Positive and negative working is shown in Figure 9.4, and the negative process has been given the trivial name 'monazoline

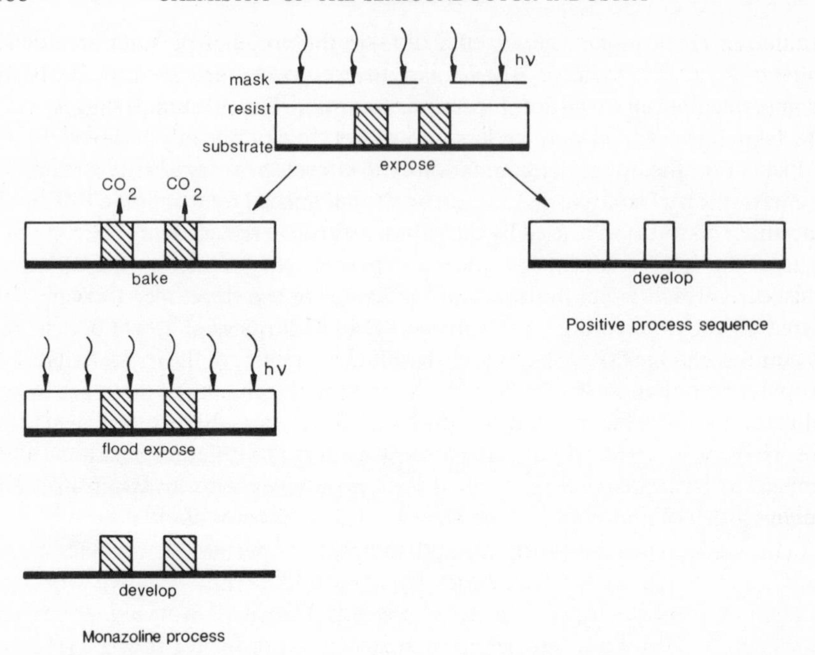

Figure 9.4 Sequence of processing steps that allow a doped positive photoresist to generate positive or negative tone images

process' because the monazolines are a range of commercially available detergents (1-hydroxyethyl-2-alkyl-imidazolines) which show useful catalytic activity in the required decarboxylation (15). So far it is not clear whether the presence of amines like monazolines have any effect on shelf life of optical resists or even more importantly, no speed and resolution.

In a useful modification of the Monazoline process, the positive resist is reversed in action by careful baking in an atmosphere of ammonia in order to induce the required decarboxylation (15*b*). Economic pressures for higher throughputs in wafer manufacture demand that certain stages utilize mask plates with transparent backgrounds. Image reversal, by changing exposure sequence, on a common mask plate, is a highly convenient expedient for this purpose and is likely to grow in importance. The de-carboxylation required for image reversal may also be accomplished by post-treatment with acids or by incorporation of a variety of dye molecules (15*c*). Combination of image reversal with acid-catalysed further condensation of the novolac resins has recently afforded (15*d*) novel ways of producing either positive or negative images which are stable to temperatures above 300° C.

During the last few years there have been many reports of new optical resists systems for microlithography (7, 16). Most of these remain as essentially in-house developments of individual companies active in manufacturing integrated circuits, and there has been little or no opportunity for wider evaluation and comparison. Overall the new materials offer either (or both) enhanced activity at shorter optical wavelengths (e.g., deep UV, 230–280 nm) or improved plasma resistance facilitating all-dry development. Two systems will be described to illustrate the types of approach currently in favour.

Plasma resistance in organic polymers is enhanced by the presence of aromatic and heteroaromatic groups. Taylor and his collaborators at the Bell Laboratories have utilized this result in a very elegant resist technique referred to as 'photo-locking', (Figure 9.5, see also Chapter 16). In the original work the resist consisted of a plasma-degradable acrylic polymer intimately mixed with a volatile but readily polymerized solid monomer such as N-vinylcarbazole or acenaphthylene. Highest sensitivity was achieved using poly(2, 3-dichloropropyl acrylate) as the host polymer, and the optical wavelength sensitivity was adjusted by incorporation of sensitizers such as phenanthraquinone which absorb in the 290–350 nm range. It is assumed that photolysis of N-vinylcarbazole/sensitizer combinations leads to cationic and free radical polymerization of the monomer with the resulting polymeric product 'locked' into the original host polymer matrix. Whether the chlorine substituents in the

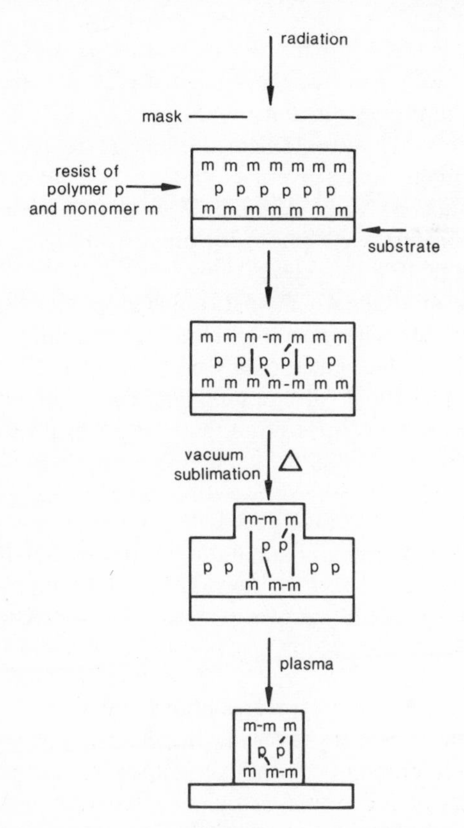

Figure 9.5 Schematic representation of photo-locking technique

host polymer play any significant role by facilitating chemical binding of the newly-formed polymer to the host remains unresolved. General principles of photo-locking are shown schematically in Figure 9.5. Essential features are vacuum sublimation of any remaining monomer (e.g. N-vinylcarbazole) and the high sensitivity to plasma attack of the acrylic host polymer. As originally conceived, and recently extended, the photo-locking technique may be used for optical, electron beam, and x-ray resists with only the first-named requiring sensitization. It is particularly applicable to x-ray lithography and in a significant development, Taylor and collaborators (18) have shown that much improvement may be obtained by replacement of the rather toxic N-vinylcarbazole with a number of polymerizable silicon compounds of which bis-acryloxy and methacryloxy derivatives are best:

$$CH_2{=}CHCOO(CH_2)_n{-}\underset{\underset{CH_3}{|}}{\overset{\overset{CH_3}{|}}{Si}}{-}O{-}\underset{\underset{CH_3}{|}}{\overset{\overset{CH_3}{|}}{Si}}{-}(CH_2)_nOCOCH{=}CH_2$$

Values of n may be varied to improve compatibility with the host polymer, and the monomer with $n=4$ appears most useful. A feature of the use of silicon-containing monomers is that resist treatment with an oxygen plasma produces a protective thin layer of silicon dioxide. These developments are at a very early stage and much improvement and amendment may be anticipated as regards type of host polymer, type of added polymerizable monomer, and the development of improved 'photo-locking' attachment sites in the host polymer.

In an alternative approach which offers much scope for future development, Willson and collaborators have demonstrated the feasibility of positive photo-lithography via light-induced acid-catalysed hydrolysis (19). It is well known (20) that aryl iodonium and sulphonium salts decompose on photolysis to yield a mixture of products including the appropriate protonic acid, e.g.

$$Ar_3S^+AsF_6^- \xrightarrow[RH]{h\nu} Ar_2S + Ar\cdot + R\cdot + H^+AsF_6^-$$

Onium salts of these types have maximum absorbance in the 250 nm region and hence offer ideal sensitizers for deep UV lithography, but may also be sensitized for photolysis at longer wavelengths. The polymer most useful for

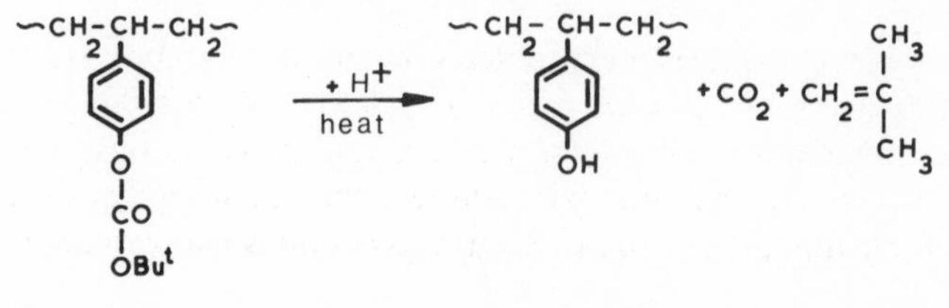

$$\sim CH_2{-}CH{-}CH_2\sim (C_6H_4{-}O{-}CO{-}OBu^t) \xrightarrow[heat]{+H^+} \sim CH_2{-}CH{-}CH_2\sim (C_6H_4{-}OH) + CO_2 + CH_2{=}C(CH_3)_2$$

this type of lithography is the tert-butoxycarbonyl ester of poly(4-vinylphenol). Tert-butoxycarbonyl esters of phenols are highly reactive towards acids, decomposing to give the appropriate phenol, carbon dioxide and gaseous isobutene.

In this way a neutral styrene-based polymer is converted into a poly(4-vinylphenol) which may in principle be extracted with alkali, as noted earlier, to give a positive resist. In practice it is preferable to extract with polar organic solvents, again yielding a positive resist, because of the enormous difference in solubility between neutral initial ester units and the product phenols. It should noted that complete acidolysis of the phenol esters requires post-illumination baking at about 100° C. The idea of using acid-sensitive units to protect phenolic groups in polymer resists is innovative and capable of wide utilization. Similar transformations would be expected of esterified novolac materials and polymers having acid-sensitive groups in the backbone, and the versatility and efficiency of onium salts as latent sources of protonic acids should stimulate further research in this area (15*d*, 21).

Finally, mention must be made of the most exciting prospects offered by the recent demonstration by Srinivasan (2, 22) of laser-induced ablative decomposition, at high resolution, of polymer films. Excimer laser light sources emitting at 193 nm are now readily available commercially and, because of the relatively large area of the laser beam, offer highly convenient radiation sources for lithography by projection printing through masks. Absorption by oxygen in the air at this short wavelength is easily overcome by nitrogen flushing. At 193 nm nearly all organic materials absorb very strongly and hence give rise to short-lived excited states. Photon energies are extremely high and result in extensive bond fission processes. Above a certain threshold energy the laser illumination produces a mini-explosion in organic materials (e.g. polymer films). Polymers are converted into a large number of volatile fragments with a concomitant increase in volume. This causes the molecular fragments to explode away from the surface with essentially no heat loss to the adjoining films. As a result extremely sharp-edged ablative etching may be accomplished. Results for ablative imaging in polyethyleneterephthalate are quite spectacular and, since no wet or dry development is necessary, offer potential advantage over other current lithographic processes (2, 22). It remains to establish whether the excimer laser sources can be interfaced with integrated circuit design and production requirements.

Multilayer resists

Two of the most troublesome problems in microlithography are the reflectivity of the electronic substrate and irregular resist coating thickness over previous geometries. During processing, the original silicon substrate is etched, added to, and otherwise altered; the resulting device surface has considerable topography (up to 3 μm) and contains many different materials.

When new resist layers are applied over such a topography, they tend to planarize rather than conforming to the surface, and this results in a resist coating with variable thickness. It is extremely difficult to develop precise geometries from a resist with large fluctuations in coating thickness—not least because the various materials in the device have different properties, and variations in reflectivity cause the resist on top to receive an uneven exposure. The combined effects of reflectance and uneven coating thickness can create unacceptable line-width deviations for manufacture of high-resolution devices.

Attempts to utilize a single resist coating to produce submicrometre images while overcoming the problems of reflectivity, topography, and chemical or physical processing frequently results in unsatisfactory compromises. Use of several resist layers, each of which has been optimized to address one particular concern such as planarization, etch resistance, or improved lithographic resolution offers one solution to the problem. Unfortunately, considerable process complexity is introduced by multiple coatings and their interfaces, but the materials specifications for the individual layers are much more flexible. A multilayer approach offers the opportunity to separate the lithographic imaging function from the fabrication function (23, 24). Frequently, thermal stability or etch resistance is achieved only at the expense of lithographic speed or resolution. When the two functions can be carried out by separate layers, each can be optimized.

Uneven topography can be minimized with a planarizing or smoothing layer, over which a resist can then be coated with greater uniformity, resulting in a more accurate developed image (Figure 9.6). Since the planarizing layer is not necessarily imageable, the transfer of the resist image onto the device topography now must involve the selective removal of not only the imaged resist but also the underlying planarizing layer. Two different methods for this have been proposed, each of which requires a unique polymer. The portable conformable mask approach uses a positive-working deep-UV (250 nm) absorbing resist as the planarizing layer (25). Appropriate polymers for this approach are poly(methyl methacrylate) and poly(methyl isopropenyl ketone). Since the usual quinone diazide/novolac resists absorb strongly at 250 nm, the exposed and developed top resist layer acts as a contact mask for the exposure of the planarizing layer. Difficulties with this approach arise if there is any intermixing at the interface between resist and planarizing layer. In an alternative approach, the planarizing layer is developed by oxygen reactive-ion etching (RIE). However, for most polymer resists the ratio of etch rates is small and it is difficult to accomplish the required level of image transfer. Silicon-containing resists offer special advantages in this respect.

Silicon-containing resists have the potential to simplify the device manufacturing procedure by combining both lithographic performance and plasma resistance into the same layer. If a resist material is used for image formation that either contains silicon initially or may be modified to accept it later, then

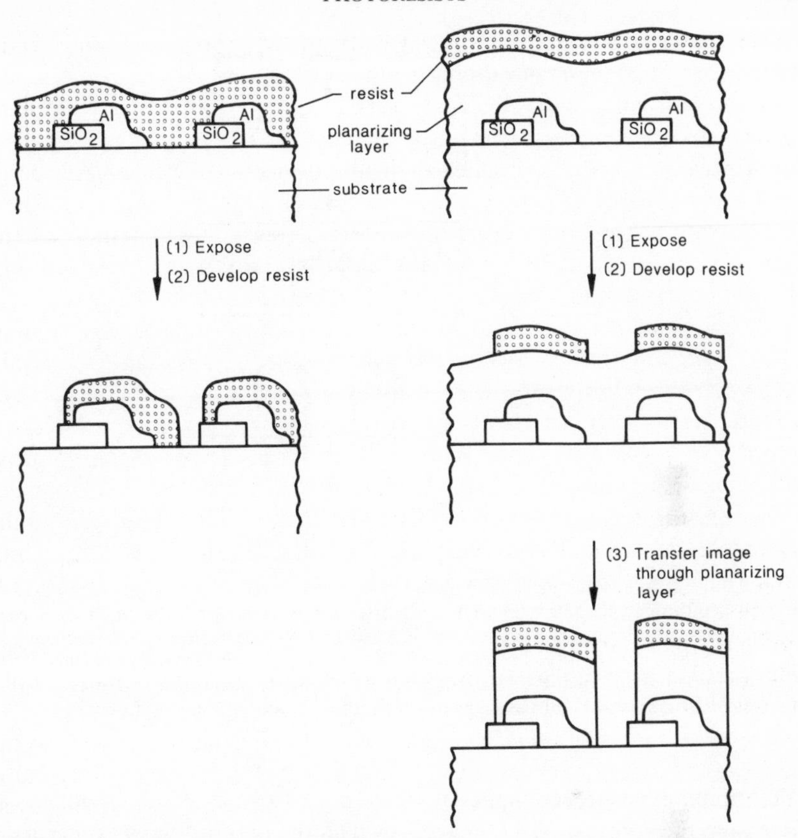

Figure 9.6 Use of planarizing layer. Typically the resist could be of the quinone diazide/novolac type, and the planarizing layer PMMA. Taken from (9)

the image can act as an extremely effective reactive-ion-etch mask. Organic silicon compounds are converted into SiO_2 when subjected to an oxygen plasma etch, and this SiO_2 rubble is resistant to further oxygen RIE etching. Sufficient oxygen etch resistance to form an etch-stop can be obtained at about 10–12 wt% of silicon in the resist coating (26, 27). The earliest studies used commercially available siloxanes with pendant vinyl groups that cross-link on exposure (28), and high-resolution images with excellent etch characteristics were reported. Other examples of this type are reviewed by Blevins *et al.* (9).

Contrast enhancement layers

The apparent contrast response of commercial photoresists can be dramatically improved by imaging through contrast enhancement layers. Contrast enhancement layers consist of films of photobleachable materials, layered on top of the photoresist, which exhibit large non-linear transmission character-

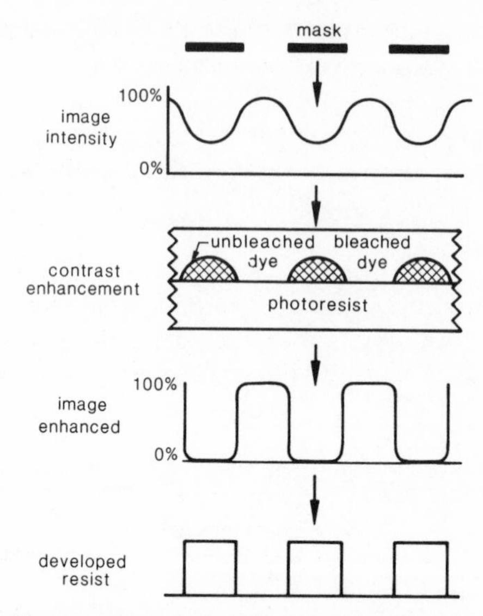

Figure 9.7 This figure illustrates the contrast enhancement process. The image is degraded from transmission through the mask and lens system so that there is significant intensity in areas corresponding to opaque portions of the mask. A partially bleached contrast enhancement layer can increase the contrast of the image because the unbleached, light-absorbing portions of the layer correspond to the opaque areas of the mask. The figure shows an enhanced image to which the photoresist is exposed and the sharper resist profiles expected. Adapted from (30)

istics upon exposure to light of different intensities. They function by increasing the contrast of the image to which the photoresist is exposed and thus extending the limits of the optical resolution system. Typically, the contrast enhancement layer is made up of a polymer doped with a dye which bleaches at the wavelength of light used for the photoresist exposure. In order to be effective, the exposure time of the resist must be shorter than the time required for the dye to bleach. During resist exposure, the unbleached portion of the contrast enhancement layer acts as a portable conformable mask (Figure 9.7). This results in an increase in the contrast of the image exposing the photoresist and leads to much sharper resist profiles. The idea was first proposed by Griffing and West (29), who utilized the loss of absorption occurring on photoinduced cyclization of diaryl nitrones (30).

$$\mathrm{Ar{-}CH{=}\overset{+}{N}(O^-){-}Ar'} \xrightarrow{h\nu} \mathrm{Ar{-}CH{-}N{-}Ar'}\ \text{(with O bridging CH and N)}$$

By varying substituents in the aryl groups, the absorption profile of the initial nitrone can be adjusted to match the light wavelengths required for activating the photoresist. Similarly, photo-induced decomposition of diphenylamine-*p*-diazonium sulphate in water-containing polymers provides a useful contrast enhancing bleaching process (30).

$$C_6H_5-NH-C_6H_4-N_2^+\ HSO_4^- \xrightarrow[H_2O]{h\nu} C_6H_5-NH-C_6H_4-OH + N_2 + H_2SO_4$$

Multilayer techniques, including anti-reflection coatings (31), generally yield much more vertical profiles in photolithography and, although many problems associated with their use remain to be solved, they are likely to find increased application because of the ever-increasing requirements for smaller feature sizes and more complex circuit geometries.

References

1. L. F. Thompson, C. G. Willson, and M. J. Bowden (eds.) *Introduction to Microlithography.* ACS Symp. Ser. **219** (1983).
2. R. Srinavisan, *J. Vac. Sci. Technol.* **B1** (1983) 923; D. J. Ehrlich, *Polym. Eng. Sci.* **26** (1986) 1146.
3. D. A. Doane and A. Heller (eds.) Inorganic resist systems, *Electro-Chemical Soc. Proc.* **82–9** (1982).
4. J. M. Ziegler, L. A. Harrah, and A. Wayne Johnson, in *Advances in Resist Technology and Processing, Proc. SPIE* **539** (1985) 166. R. D. Miller, D. Hofer, G. N. Fickes, D. Trefonas and R. West, *Polym. Eng. Sci.* **26** (1986) 1129.
5. W. S. DeForest, *Photoresist-Materials and Processes,* McGraw-Hill, New York (1975); D. J. Elliott. *Integrated Circuit Fabrication Technology,* McGraw-Hill, New York (1982); H. Ahmed and W. C. Nixon (eds.) *Microcircuit Engineering,* Cambridge University Press (1980).
6. M. J. Bowden, *Solid State Technol.* **73** (1981) (June); M. J. Bowden, *CRC Critical Reviews in Solid State Sciences* **223** (1979).
7. L. F. Thompson, C. G. Willson and J. M. J. Fréchet (eds.) *Materials for Microlithography.* ACS Symp. Ser. **266** (1984).
8. M. J. Bowden, Chapter 3 of ref. 7.
9. R. W. Blevins, R. C. Daly and S. R. Turner, 'Lithographic photopolymers', *Encyclopedic of Polymer Science and Engineering,* John Wiley, New York (in press).
10. M. A. Golub, Cyclised and isomerised rubber, in *Polymer Chemistry of Synthetic Elastomers,* eds. J. P. Kennedy and E. G. M. Tornqvist, John Wiley, New York (1969) 939.
11. S. Koibuchi, A. Isobe, D. Makino, T. Iwayanagi, M. Hashimoto and S. Nonogaki, in *Advances in Resist Technology 11, SPIE Proc.* **539** (1985) 182; S. Nonogaki, M. Hashimoto, T. Iwayanagi, and H. Shiraishi, in *Advances in Resist Technology II, SPIE Proc.* **539** (1980) 189; M. Hashimoto, T. Iwayanazi, H. Shiraishi and S. Nonogaki, *Polym. Eng. Sci.* **26** (1983) 1090.
12. V. V. Ershov, G. A. Nikiforov, and C. R. H. De Jong, *Quinone Diazides.* Elsevier, Amsterdam (1981).
13. J. Pacansky, *Polym. Eng. Sci.* **20** (1980) 1049.
14. A. C. Ouano, *Polymers in Electronics,* ACS Symp. Ser. **242** (1984) 79.
15. (*a*) S. A. MacDonald, R. D. Miller, C. Grant Willson, G. M. Feinberg, R. T. Gleason, R. M. Halverson, M. W. MacIntyre, and W. T. Motsiff, *IBM Res. Rept* RJ 3624 (42420) 10th January 1982; H. Moritz and G. Paal, US Patent (IBM) 4, 104, 070 (1978); M. Watanabe, US Patent (Fuji) 4, 196, 003 (1980). (*b*) E. Alling and C. Slauffer, *Image Reversal and Positive Photoresists,* IMTEC Pub. 2001-2M-28S, Sunnyvale, California (1985). (*c*) L. H. Kaplan and S. M. Zimmerman, US Patent (IBM), 4, 007, 047 (1977); Y. Takahashi, H. Tachikawa, F. Shinozaki, T. Ikeda, US Patent (Fuji) 4,356,254 (1982); T. Chonan and A. Morishige, US Patent (Fujitsu), 4,444,869 (1984). (*d*) W.E. Feely, J. C. Imhof, and C. M. Stein, *Polym. Eng. Sci.* **26** (1986) 1101.
16. B. D. Grant, N. J. Clecak, R. J. Twieg, and C. Grant Willson, *IEEE Trans. Electron. Devices*

ED–28 (1981) 1300; C. W. Wilkins, Jnr., E. Reichmanis, and E. A. Chandross, *J. Electrochem. Soc.* **127** (1980) 1510; R. L. Hartless and E. A. Chandross, *J. Vac. Sci. Technol.* **19** (1981) 1333; E. Reichmanis, C. W. Wilkins, Jnr., and E. A. Chandross, *J. Vac. Sci. Technol.* **19** (1981) 1338.

17. G. N. Taylor, T. M. Wolf, and M. R. Goldrick; *J. Electrochem. Soc., Solid State Sci. and Technol.* **128** (1981) 361.
18. G. N. Taylor, T. M. Wolf, and J. M. Moran, *J. Vac. Sci. Technol.* **19** (1981) 872.
19. J. M. J. Fréchet, T. G. Tessier, C. G. Willson and H. Ito, *Macromolecules* **18** (1985) 317; J. M. J. Frechet, E. Eichler, H. Ito and C. G. Willson, *Polymer* **24** (1983) 955.
20. J. V. Crivello, ACS Symp. Ser. **14** (1979) 1.
21. J. M. J. Fréchet, F. M. Houlihan, and C. G. Willson, *Polym. Mater. Sci. Eng.* **53** (1985) 268; J. M. J. Fréchet, F. Bouchard, F. M. Houliham, B. Kryczka and E. Eichler, *J. Imag. Sci.* **30** (1986) 59.
22. R. Srinivasan, B. Braren, D. E. Seeger and R. W. Dreyfus, *Macromolecules* **19** (1986) 919; R. Srinivasan and W. J. Leigh, *J. Amer. Chem. Soc.* **104** (1982) 6784; R. Srinivasan, *Polymer* **23** (1982) 1863.
23. B. J. Lin, P. 287 in ref. 1.
24. M. Hatzakis, Solid State Technol. **24** (1981) 74.
25. B. J. Lin, E. Bassovs, V. W. Chao, and K. E. Petrillo, *J. Vac. Sci. Technol.* **19** (1981) 1313; B. J. Lin, V. W. Chao, K. E. Petrillo and B. J. L. Yang, *Polym. Eng. Sci.* **26** (1986) 1112.
26. M. Suzuki, K. Saigo, H. Gokan, and Y. Ohnishi, *J. Electrochem. Soc.* **130** (1983) 1962; see also ref. 4.
27. C. W. Wilkins, Jnr., E. Reichmanis, T. M. Wolfe and B. C. Smith, *J. Vac. Sci. Technol.* **B3** (1985) 306.
28. M. Hatzakis, J. Paraszczak, J. Liutkus and E. Babich, *Polym. Eng. Sci.* **23** (1983) 1054.
29. B. F. Griffing and P. R. West, *Polym. Eng. Sci.* **23** (1983) 947; B. F. Griffing and P. R. West, *Electron Device Lett.* **EDL-4** (1983) 14; P. R. West, G. C. Davis, and B. F. Griffing, *J. Imag. Sci.* **30** (1986) 65.
30. L. F. Halle, *J. Vac. Sci. Technol.* **B3** (1985) 323.
31. Y. C. Lin, V. Marriot, K. Orvek and G. Fuller, in *Advances in Resist Technology*, SPIE Proc. **469** (1984) 30; K. Barlett, G. Hillis, M. Chen, R. Trutna and M. Watts, in *Optical Microlithography, Proc. SPIE* **394** (1983) 49.

10 Electron beam and x-ray resists

E. D. ROBERTS

In Chapter 9 the use and chemistry of standard photoresists is described. Since the inception of planar semiconductor technology, which has been described fully by Elliott (1), photolithography has been the standard method of defining the shaped areas required for making diffused regions, insulator and conductor patterns, etc., which are used in suitable combinations to form discrete devices and integrated circuits. The shaped details needed are initially formed in a photoresist film on the planar substrate by the action of radiation in the wavelength range 365–430 nm, followed by a suitable development process, when a pattern of windows corresponding to the irradiation pattern is formed in the resist film. The pattern is transferred by subsequent operations on to the surface of the planar substrate or into its bulk.

As the technology has improved, it has become possible to produce progressively smaller details by photolithography. This has enabled devices to be made smaller, with consequent benefits of reduced power consumption and increased frequency of operation. More complex integrated circuits containing larger numbers of components in a given area of substrate material can be produced, which has led to substantial reductions in cost per component, and hence to complete systems having improved capability, smaller sizes and lower prices. This is amply demonstrated by the availability at minimal cost of consumer goods such as hand calculators, watches, personal radio sets, home computers, etc., and similar changes have taken place with professional electronic equipment. There is a desire among makers and users of integrated circuits to reduce device dimensions further, in the expectation of still further improvements.

However, as the sizes of exposed details are reduced and approach the wavelength of the exposing radiation, diffraction effects at the edges of patterns become increasingly troublesome. They spoil the edge definition, and make dimensional control and accuracy of alignment between successive stages in the device fabrication process difficult or impossible to achieve. In factories, the smallest details defined by photolithography at present are about 1.5 μm wide. In laboratories, where more time and care can be taken in pattern alignment procedures, it has been possible to make devices with details as small as 1 μm or even a little less. Nevertheless, diffraction effects do impose a fundamental limit upon the size of details which can be successfully exposed by conventional photolithography, and this limit is rapidly being approached.

Table 10.1 Wavelength and photon energy of radiation used in various lithographic techniques

Lithographic technique	Wavelength range (nm)	Photon energy range (eV)	Mask needed	Minimum pattern dimensions achieved (nm)
* Photo	365–430	3.4–2.9	Yes	~ 1000
* Deep UV	240	5.2	Yes	700
* Excimer laser	249	5.0	Yes	500
	193	6.4	Yes	500
* X-ray	0.4–4	3×10^3–3×10^2	Yes	20
** Electron beam				
Scanning	0.25–0.01	5×10^3–1×10^5	No	5
Projection	0.06–0.03	2×10^4–4×10^4	Yes	250
** Ion beam				
Scanning	0.024–$<$0.006	5×10^4–$>2 \times 10^5$	No	$<$100
Projection	0.024–$<$0.006	5×10^4–$>2 \times 10^5$	Yes	20

* The radiation energy is usually specified as its wavelength.
** The radiation energy is usually specified as the voltage through which the electrons or ions are accelerated, or as the photon energy.

That this situation would occur eventually has been recognized for at least twenty years. Techniques using deep ultraviolet light, electron beams, ion beams and x-rays for exposing the patterns have been developed to overcome the limitations experienced in the practice of photolithography. All these methods use radiation of higher photon energy than that used in conventional photolithography, as shown in Table 10.1. Their effective wavelengths are lower, so diffraction effects are correspondingly smaller, and it is possible to obtain good definition of details well into the submicron region. The decrease in effective wavelength in deep ultraviolet and excimer laser exposure systems is not very large, so these will allow only a modest decrease in detail sizes, perhaps to 0.3–0.5 μm. Electron beams, ion beams and x-rays, however, have effective wavelengths in the nanometre range, so diffraction effects are negligible, which enables details with dimensions of a few nm to be defined successfully.

Electron beam exposure equipment

There are two basic types of electron beam exposure equipment—scanning systems (also called pattern generators) and projection systems.

The scanning system is the only method known by which submicron patterns can be generated. In these, the electron beam is focused to a fine spot at the surface of a resist-coated substrate, and is scanned electronically to trace out the pattern it is required to produce in the resist film. Electron beams can already be focused to diameters as small as 1 nm (2), though at present beams about 100 nm in diameter are generally used in pattern generators for device fabrication. Many systems have been built and the main types have been reviewed by Pfeiffer (3).

Devices and integrated circuits are already being made on silicon slices four inches in diameter, and it is expected that soon slices five and six inches across will be in use. Drawing patterns over substrates with such a large area with a beam which may be only 0.1 μm or less in diameter can take many hours or even days, so scanning beam systems are not particularly suited to mass production of circuits in factories. Electron image projection has been investigated to cater for this need.

A projection system cannot generate patterns—it can only reproduce them from a suitable mask which has first to be fabricated by some other means. If the mask is to contain submicron details, it must be made by a scanning electron beam system. An electron projection system was described first by O'Keeffe (4) and subsequently developed by Scott (5) and later by Ward (6). It enables the whole area of the slice to be exposed simultaneously, so in principle a slice of any size can be exposed in a few seconds, and the method is suitable for mass production. In passing, it may also be noted that an image projector does not necessarily need a very sensitive resist for its operation to be economically viable. This allows more freedom to select a resist on the basis of properties other than high sensitivity.

Ion scanning systems (7), which can also generate submicron features, and ion projection systems (8) have been constructed. They differ in detail from the corresponding electron systems, but the general principles are similar. However, ion beam systems have not yet been developed to the same extent as electron beam methods.

Resists for submicron lithography

Standard photoresists are exposed by radiation having photon energy about 3 eV (Table 10.1). Most chemical bond energies lie within the range 200–1000 kJ mol^{-1}, equivalent to 2.2–10.8 eV. Thus it is only weaker chemical bonds which are broken in conventional photolithography, and photoresists depend for their action upon the presence of a suitable sensitizer containing such weaker bonds. The sensitizers are frequently organic azides. Light acts selectively upon the azide group in the sensitizer to produce an intermediate which initiates some secondary chemical change, e.g. polymerization, or a change of polarity, which results in the difference in solubility needed for pattern development in the resist material (see Chapter 9).

If the photon energy is greater than the energy of any bond in an irradiated compound, as occurs, for example, in a mass spectrometer using electrons of 100 eV energy, it is possible to detect the presence of fragments showing that every bond in the molecule can be broken. In submicron lithography, electron beams, ion beams or x-rays are used to expose the resist and these radiations have even higher photon energies (Table 10.1). Thus, there is sufficient energy available to break every bond in an irradiated molecule and, at least in principle, anything may happen! However, there seems still to be some selectivity when high-energy beams interact with matter, presumably because weaker bonds are still broken preferentially.

Most electron and x-ray resists in use at present are organic materials. Sensitizers are not required, for the high-energy radiation itself can break bonds in the resist material. The active centres thus formed initiate further reactions, leading to changes in chemical structure and physical properties which can be used in the practice of lithography. Most of the development work has involved selecting or preparing materials having the optimum combination of properties required in both lithography and in subsequent processing such as etching.

Properties required in resists

Any type of resist is required to perform satisfactorily two distinct functions. First, it must be possible to form in it, by the lithographic process, the pattern of windows needed. Second, the film pattern defined by the lithographic process must act as a mask during subsequent processing in which the pattern is transferred on to the surface of the substrate, or even into its bulk. It is desirable that both negative- and positive-working forms of resist are available for each type of exposure tool.

The first of these functions requires that the resist have the following properties:

(i) It must be capable of easy application by a simple process, to give films of uniform and reproducible thickness, free from pinholes or other defects
(ii) It must be sufficiently sensitive to the exposing radiation to allow economic production of semiconductor devices
(iii) The pattern development process should be as simple as possible and should not need critical control of the conditions
(iv) It must allow adequate resolution of fine patterns to be obtained
(v) Preferably, electron and x-ray resists should not be photosensitive, so that special safe lighting need not be provided.

The second function needs the following properties in the resist:

(i) It must retain its integrity during any process used to transfer the pattern into the substrate, at least for the time needed to complete the transfer
(ii) Its adhesion to the substrate must be sufficiently strong to protect masked parts from the effects of the pattern transfer process
(iii) Its glass transition temperature, T_g, should be higher than any

temperature to which it may be subjected during pattern transfer

(iv) It must be possible to remove the resist residue completely in a simple way after its pattern transfer function has been performed.

The most commonly used method of forming resist films is to apply a solution of the resist polymer to the substrate, which is then spun rapidly. In recent years, there have been a few examples of resist film formation by other methods such as vacuum evaporation (9), plasma polymerization (10) and monomolecular layer deposition by the Langmuir–Blodgett technique (11), though these processes are generally slower than spinning from a solution. Similarly, there have been attempts to evolve dry development techniques using plasma etching processes, though none of these appears yet to have developed sufficiently to challenge, at least in factory practice, wet development methods involving immersion in, or spraying with, a solvent.

The pattern transfer stage usually involves an etching process. Traditionally, wet etching methods have been used and the main requirement has been that the resist should withstand the effects of the aggressive reagents (hydrofluoric acid, phosphoric acid, etc.) used in etching. One of the main difficulties encountered has been that areas of the substrate which should be masked by the resist are etched partially or even completely. This is ascribed to poor adhesion between the resist and the substrate, allowing the etchant to be drawn by capillarity into the interface between them. The use of adhesion promoters, and baking the resist after developing the patterns, have usually enabled the problem to be confined within reasonable bounds. Lack of fidelity in pattern transfer by wet etching methods can also occur, even when resist adhesion to the substrate is adequate, because lateral etching occurs as well as the vertical etching required. In large area details, such undercutting may not matter or it may be compensated for. However, as details become smaller and more closely spaced, undercutting becomes more troublesome and may even destroy narrow features it is required to produce. Dry etching methods using gaseous plasmas, in which resist adhesion need not be so strong, are coming into use. The forms known as reactive ion etching and ion milling are anisotropic processes in which undercutting can be eliminated entirely, so faithful pattern transfer can be obtained. However, in dry etching, the resist is subjected to an environment containing charged particles and free radicals, which is very different from that encountered in wet etching. New types of resists have been needed to withstand the effects of dry etching media.

A high glass transition temperature in the resist polymer is of more consequence in positive-working resists, for the masking film consists of the original polymer. It must not become distorted or flow if it becomes heated during the pattern transfer stage. Negative resists become cross-linked during exposure, which automatically raises the glass transition temperature substantially, so the initial value is of less interest in this case. Indeed, silicone fluids have been successfully used as negative electron resists (12).

Electron resists

Mechanism of action of electron resists

Most electron resists are organic polymers. The smooth, pinhole-free films required in lithography can be formed very easily and rapidly by spinning a substrate coated with a solution of a polymer.

It is generally agreed that the first event which occurs when an electron strikes organic matter is that an electron is expelled from some bond within the target molecule to produce an ion-radical. There has been some speculation upon, though little experimental work to substantiate, what succeeding reactions occur, and the details of reaction mechanisms are not understood in most cases. It can be inferred, however, from the nature of the final reaction products that the reaction may take two possible courses after it is initiated. If the ion-radicals or their subsequent derivaties join together, larger molecules are formed and ultimately this process leads to a three-dimensional cross-linked network of polymer. Alternatively, the polymer main chain may break to give fragments smaller than the original molecules. Usually, both effects occur simultaneously, though in most materials one predominates largely over the other. Thus, in a given material there is either a net increase or a net decrease in molecular size (and molecular weight) after irradiation.

The changes in molecular size result in changes in physical properties, and generally use has been made of the change in solubility which occurs to develop patterns in irradiated resist films. When the molecular size is increased upon irradiation, solubility is reduced; more usually, in practical resist systems, the polymer actually becomes cross-linked and insoluble upon irradiation. The development process will then leave a deposit of cross-linked film corresponding to the irradiation pattern. Such resists are called negative-working. Conversely, when the molecular size is reduced during exposure, the irradiated material becomes more soluble and can be removed selectively by immersion in a suitable solvent (or mixture of solvents). These materials form the basis of positive-working resists.

Almost all organic compounds become cross-linked upon irradiation and can act as negative electron resists (though their other properties may not be ideal), but relatively few are degraded and show positive action. In fact, except for a few 'self-developing' resists (see later), all positive-working electron resists become negative-working at higher exposures and there is only a limited range of exposures within which positive action is observed. This occurs because the rate of degradation, although initially higher than the rate of cross-linking, decreases with increasing exposure as more of the degradable bonds are broken, and eventually becomes zero. The simultaneous cross-linking which occurs at other sites in the molecules continues unabated, and ultimately unites even the degraded fragments into an insoluble network.

Side-effects of electron irradiation

In addition to the degradation or cross-linking reactions used to obtain lithographic action, electron irradiation can produce other less desirable effects.

It causes the deposition of insulating films upon any surface which is struck by electrons, by cross-linking molecules within the system, such as those of pump oil, which happen to be adsorbed on the surface. Over a period of time, heavy deposits are formed by this mechanism in parts of the equipment, the electron beam column being particularly vulnerable. The deposits can store charge which distorts the beam, so the equipment must be dismantled periodically to remove the deposits.

The resist film itself is generally insulating, so it can retain charge which deflects the beam, displacing exposed patterns. This effect can be nullified by applying a very thin film of aluminium over the resist film before exposure or by introducing a small concentration of ions of an element such as caesium into the resist. Both methods provide means for the trapped charge to leak away.

During their passage through the film, some of the electrons will be scattered, both forwards and backwards, and they are also back-scattered from within the substrate. Scattered electrons also expose the resist, so the pattern profile may be trapezoidal, the sloping edges changing detail sizes and giving some loss of resolution at the substrate.

The sensitivity of electron resists

Polymers of different chemical structures may, of course, show large differences in sensitivity to electron irradiation. Even in a given type of resist polymer, the apparent sensitivity depends upon a number of factors.

(i) The electrons must always be sufficiently energetic to pass through the resist film and into the substrate. Otherwise, that part of the film in contact with the substrate will remain unexposed. Beyond this, the greater the electron energy the higher the exposure needed to allow development of patterns. Nevertheless, as the electron energy is increased, the effects of scattering become less pronounced—wall profiles are more nearly vertical and finer patterns in thick resist films may be made (13)

(ii) The nature of the substrate influences sensitivity—the heavier the atoms of the substrate, the greater the back-scattering, and the higher the apparent sensitivity

(iii) The higher the molecular weight of the resist polymer, the greater the sensitivity. For positive resists, Ku and Scala (14) showed that when the difference in solubility needed to allow pattern development results from a large number of chain scissions per molecule, sensitivity is independent of the original molecular weight. However, when the initial molecular weight is very high (a million or more), only one or two chain scissions per molecule may be sufficient to alter the solubility, and a significant increase in sensitivity may then be observed.

(iv) The characteristics of resists are usually expressed as sensitivity curves, constructed from exposure measurements made on large features, several tens of microns wide. They have the forms shown in Figures 10.1 and 10.2.

In any real polymer, there is always a distribution of molecular weights, so the exposure of a real film does not occur suddenly at a specific exposure—a range of exposures is needed to affect all the molecules in the film. Generally, the wider the molecular weight distribution the flatter the exposure curve.

The slope of the exposure curve is often used to quantify the contrast, γ, of a resist (15), which is defined as

$$\gamma = \frac{1}{\log(D_{100}/D_0)}$$

As previously outlined in Chapter 9, the contrast may give only a rough

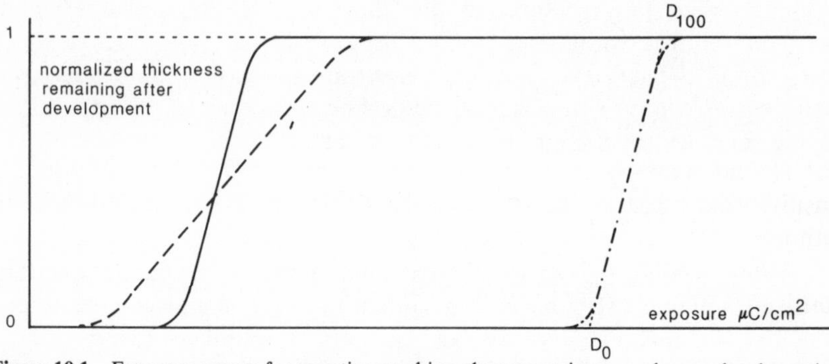

Figure 10.1 Exposure curves for negative-working electron resists: –·–·, low-molecular-weight resist; ——, high-molecular-weight resist; – – –, high-molecular-weight resist with wide molecular weight distribution (see also Chapter 9)

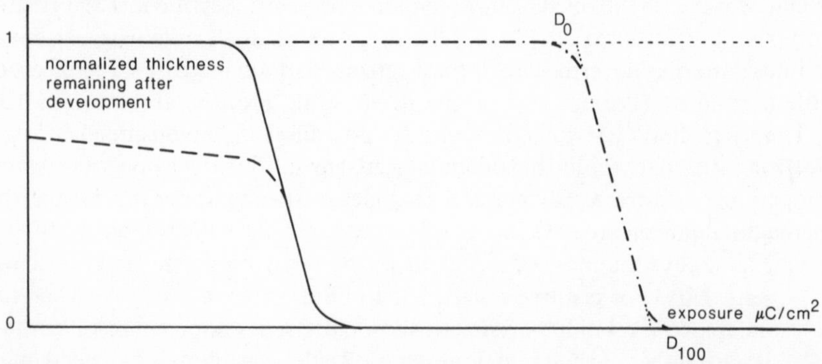

Figure 10.2 Exposure curves for positive-working electron resists: –·–·, low-molecular-weight resist; ——, high-molecular-weight resist; – – –, high-molecular-weight resist with wide molecular weight distribution (see also Chapter 9)

guide to the resolution capability of a resist, for the usual method of calculation ignores the presence of 'tails' on the sensitivity curve and these can have a marked influence upon the lithography.

(v) The lithographic exposure needed depends upon the size of the feature exposed and upon its proximity to other features. An isolated narrow line will require the highest exposure in a given resist. The exposure needed falls when the effect of electrons scattered during the exposure of adjacent features manifests itself (at detail separations around 1 μm) and becomes a minimum when the lines overlap to expose effectively a large area feature, the situation which prevails when sensitivity curves are determined.

(vi) The sensitivity may also depend upon the development conditions, at least in positive-working resists in which the development process must distinguish between areas of film differing usually only in molecular weight. The developer is generally a mixture of a solvent for the resist polymer and a non-solvent, in proportions such that the higher molecular weight (unexposed) regions are insoluble (or less soluble) but degraded (exposed) regions are readily soluble. The measured sensitivity can depend upon the composition of the developer, its temperature and upon the time of development. A higher proportion of solvent in the mixture results in apparently higher sensitivity, though this may be accompanied by some dissolution of unirradiated regions of the film and possibly also in different pattern profiles.

It should now be apparent that great care is needed in interpreting sensitivity data, particularly when comparing the results obtained by different authors.

Positive-working electron resists

Solvent-developed systems

Until relatively recently, only two main classes of polymers were known which showed positive action when irradiated by electrons. These were poly(methacrylates) and poly(sulphones). A few other materials, such as cellulose esters, have been shown to act as positive resists but they have been little used in practice.

The characteristic which is common to most of the positive-working electron resist materials is that when they are heated above the polymer ceiling temperature, they depolymerize completely to monomer. Poly(methyl methacrylate) (PMMA) (**1**), the first example, is still widely used by litho-

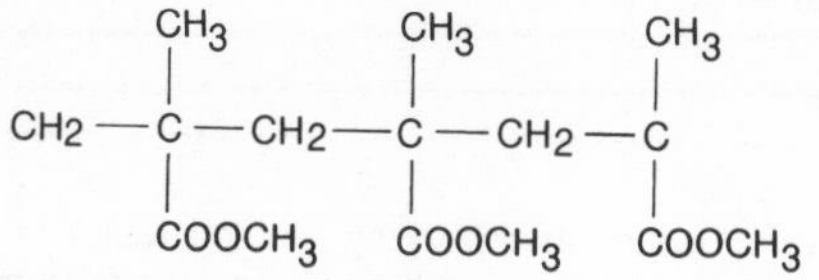

```
        CH3          CH3          CH3
        |            |            |
CH2 —  C — CH2 —  C — CH2 —  C
        |            |            |
     COOCH3       COOCH3       COOCH3
```

(1)

graphers. Its main disadvantage is that about 50 μC/cm^2 is the lowest practicable exposure value, which makes economic operation difficult.

The relatively easy thermal degradation of PMMA—and its positive action upon electron irradiation—is due to the presence of monomer units in which at least one of the main chain carbon atoms is not directly bonded to hydrogen. This realization led to the preparation (16–32) of many polymers and copolymers having the same structural feature in attempts to obtain resists having higher sensitivity than PMMA. Some products of these developments are shown in Table 10.2, though the list is by no means exhaustive.

It can be seen that the incorporation of other groups, particularly halogenated or other polar ones, both as substituents on the main chain quaternary carbon atom and in the ester side chain, can increase the sensitivity by large factors. It seems that the more sensitive materials often lack the latitude in exposure and development conditions shown by less sensitive materials. Small variations in exposure, developer composition and developing temperature can result in large changes in apparent sensitivity, giving substantial errors in developed detail sizes. This is particularly evident in poly(hexafluorobutyl methacrylate) (**2**), for which a computer-controlled development process has been devised to overcome the difficulty (33).

Table 10.2 Positive-working electron resists

Polymer type	Sensitivity (μC/cm^2)	Reference
α-Methyl styrene	100	16
Isobutylene	50	16
Methyl methacrylate	50	16
Methyl methacrylate-co-methacrylic acid	20	17
Methyl isopropenyl ketone	12	18
Isobutylene oxide	8	19
Methacrylonitrile	6	20
Methyl methacrylate-co-itaconic acid	5	21
Methyl methacrylate-co-isobutylene	4.5	22
Methyl methacrylate-co-acrylonitrile	4	23
2,2-propane-bis-4-phenyl carbonate	3	24
Methyl α-chloroacrylate	2.5	25
Isopropenyl trifluoroacetate	2	26
2,2,2-trichloroethyl methacrylate	1.25	27
Trifluoroethyl α-chloroacrylate	0.75	28
Butyl α-cyanoacrylate	0.6	29
Ethyl α-cyanoacrylate	0.5	30
Hexafluorobutyl methacrylate	0.3	31
n-butyl methacrylate	0.1	32
Styrene sulphone	10	34
Methyl cyclopentene sulphone	2.6	35,36
Butylene sulphone	0.8	37

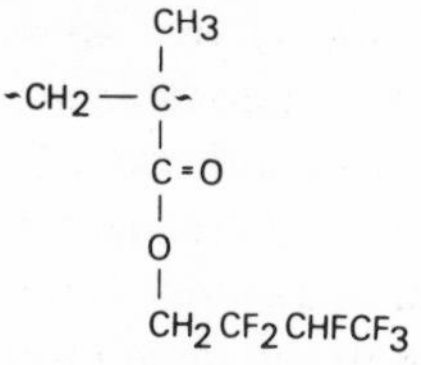

$$
\begin{array}{c}
\quad CH_3 \\
\quad | \\
-CH_2-C- \\
\quad | \\
\quad C=O \\
\quad | \\
\quad O \\
\quad | \\
\quad CH_2CF_2CHFCF_3
\end{array}
$$

(2)

A large number of poly(sulphones) have been investigated (34–37), but only poly(butene-1-sulphone) (37) has come into general use. The sulphones depend for their action upon easy rupture of the main chain at the carbon–sulphur linkage. Poly (butene-1-sulphone) (**3**) serves well as a sensitive electron resist for making optical masks economically by electron lithography. A wet etching process is used for transferring the developed resist pattern into the chromium film on a glass substrate, and the resist is still much used in this application. However, users have experienced difficulty in defining smaller details on other types of substrates where thicker resist films have been required, and its performance in the more modern dry etching processes leaves much to be desired.

An early system which did not rely upon main chain scission for its operation was based upon a methyl methacrylate polymer cross-linked by anhydride groups (38), which were formed after applying to the substrate a film of mixed copolymers, by the following reaction (MMA = methyl methacrylate):

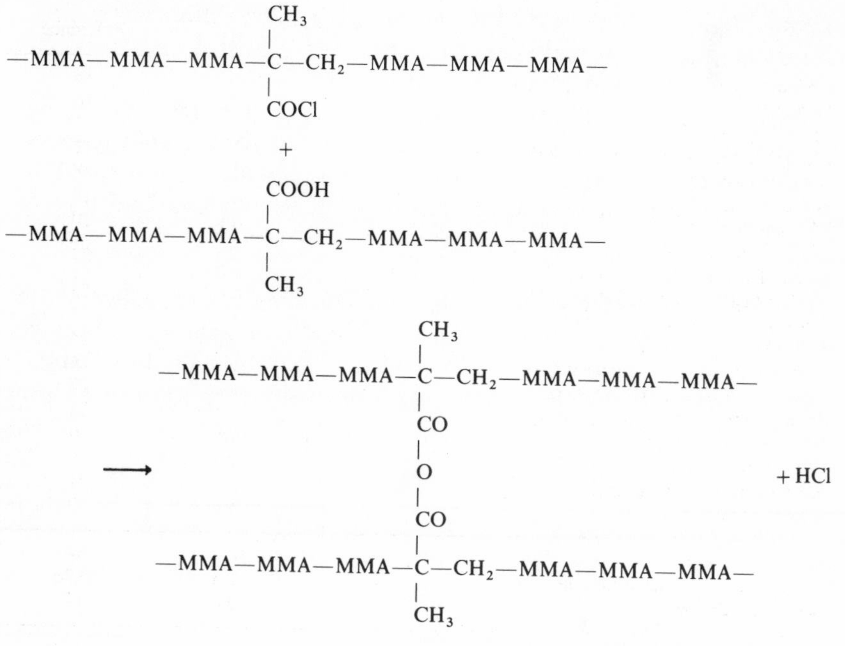

$$
\begin{array}{c}
CH_3 \\
| \\
-MMA-MMA-MMA-C-CH_2-MMA-MMA-MMA- \\
| \\
COCl \\
+ \\
COOH \\
| \\
-MMA-MMA-MMA-C-CH_2-MMA-MMA-MMA- \\
| \\
CH_3
\end{array}
$$

$$
\longrightarrow
\begin{array}{c}
CH_3 \\
| \\
-MMA-MMA-MMA-C-CH_2-MMA-MMA-MMA- \\
| \\
CO \\
| \\
O \\
| \\
CO \\
| \\
-MMA-MMA-MMA-C-CH_2-MMA-MMA-MMA- \\
| \\
CH_3
\end{array}
+ HCl
$$

The anhydride cross-links confer complete insolubility upon the resist film. They are ruptured upon irradiation so the exposed parts of the film can be dissolved selectively to achieve development of a positive pattern. Such resists can be formulated to cover a range of sensitivities, 2–40 $\mu C/cm^2$ at 10 kV, by changing the density of cross-links (39). Some forms of this resist are also more tolerant of variations in development conditions because the development process need distinguish only between soluble and completely insoluble regions of the resist film. The developed film, being cross-linked, is also very resistant to thermal deformation.

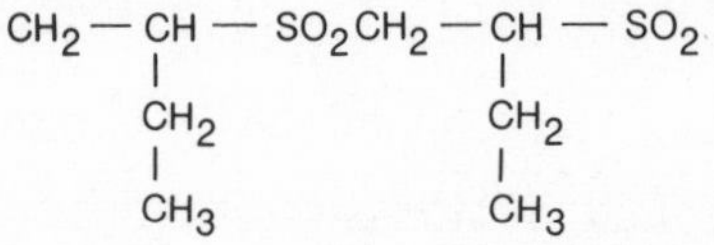

$$CH_2-\underset{\underset{\underset{CH_3}{|}}{\underset{CH_2}{|}}}{CH}-SO_2CH_2-\underset{\underset{\underset{CH_3}{|}}{\underset{CH_2}{|}}}{CH}-SO_2$$

(3)

Self-developing systems

Self-development was first observed in poly(butene-1-sulphone) (34, 37). When this resist was exposed at 300 $\mu C/cm^2$ the main chain rupture was so extensive that the fragments produced volatilized during exposure and no subsequent development was needed. This mechanism can only produce positive working, of course. It was also found that if the substrate was heated to 140° C during irradiation, self-development could be achieved at an exposure of only 6 $\mu C/cm^2$. Here, the temperature during exposure is above the ceiling temperature of the polymer, so once the main chain is ruptured it unzips spontaneously, which accounts for the high sensitivity.

Other systems in which unzipping results in very sensitive resists have been described more recently. Certain aliphatic aldehyde copolymers (40) required exposures between 0.3 and 20 $\mu C/cm^2$, depending upon the resist composition and the film thickness at which it was used. In this material and in poly (butene-1-sulphone) it is reported that very slight traces of residue remain in self-developed patterns.

Poly(phthalaldehyde) (**4**) also has a low ceiling temperature and will unzip even at ambient temperatures. Polymers stable at room temperature can be made by capping the polymer with an alkyl or acyl group and the capped polymer can be cast into films. Unzipping is initiated by irradiation (41) which

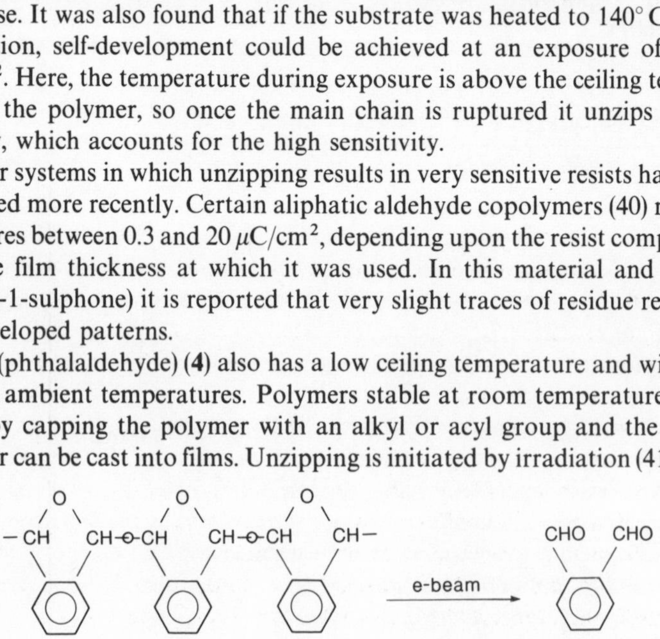

–CH–O–CH–O–CH–O–CH–O–CH–O–CH– (each O-bridged CH pair fused to a benzene ring) $\xrightarrow{\text{e-beam}}$ CHO CHO (o-phthalaldehyde, benzene ring)

(4) (5)

ruptures the main chain acetal linkages, and the resulting monomer (5) (*o*-phthalaldehyde) is volatile in vacuum. However, in initial experiments with this material, depolymerization was only partial about 60% of the film being removed during exposure.

'Onium' salt-sensitized resists

A dramatic difference was observed when about 10% of an 'onium' salt was mixed with poly(phthalaldehyde). Suitable salts were triphenylsulphonium hexafluoroarsenate, diphenyliodonium hexafluoroarsenate and *p*-hexyloxybenzene-diazonium tetrafluoroborate (41). Upon irradiation, each of these salts is converted into the corresponding strong acid which catalyses the degradation of poly (phthalaldehyde). Complete self-development was obtained, and lines as narrow as 0.25 μm were made at a exposure of only 1 $\mu C/cm^2$.

An interesting system described by the same authors (41) which relied upon a change of polarity for its operation, comprised a mixture of poly(4-*t*-butoxycarbonyl oxystyrene) and an 'onium' salt. In this, the acid produced during irradiation catalysed the following reaction when the exposed film was subsequently baked at 100° C.

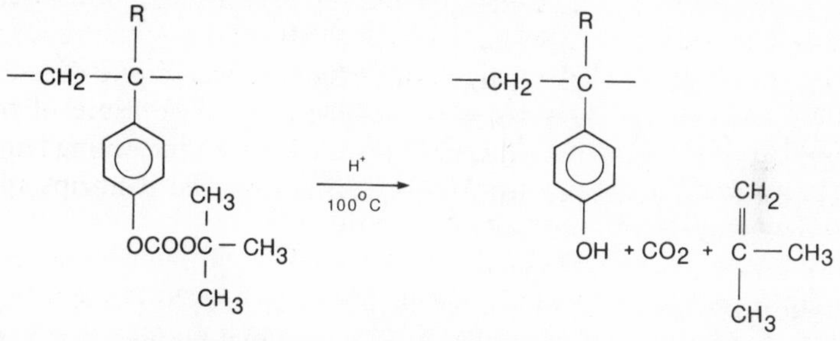

$$-CH_2-\underset{\underset{\displaystyle OCOOC(CH_3)_3}{|}}{\overset{R}{\overset{|}{C}}}(C_6H_4)- \xrightarrow[100^\circ C]{H^+} -CH_2-\overset{R}{\overset{|}{C}}-(C_6H_4)-OH + CO_2 + CH_2{=}C(CH_3)-CH_3$$

The product, being a phenol, could be dissolved in polar solvents such as aqueous alkaline solutions, in which the original polymer is insoluble, so a positive image resulted. In a non-polar developer the solubilities were reversed and a negative image could be produced from the same exposure pattern. The exposure needed was about 1 $\mu C/cm^2$.

Yamada and his coworkers have described copolymers of methyl methacrylate, prepared by plasma polymerization directly on to the substrate. Copolymers with tetramethyl tin were self-developed at 20 $\mu C/cm^2$ (42). Higher sensitivity (1 $\mu C/cm^2$) was obtained in an analogous system prepared from hexafluorobutyl methacrylate with tetramethyl tin (43). Copolymers of hexafluorobutyl methacrylate with styrene could also be self-developed, though these were considerably less sensitive (100 $\mu C/cm^2$) (43).

Green and his co-workers have recently made a preliminary study of the use of anhydrous cadmium chloride films as self-developing resists (44). In these, irradiation produces electron–hole pairs resulting in dissociation of the resist into its elements. Chlorine is pumped away and if the exposure is conducted at elevated temperature (> 150° C) the cadmium is also volatilized. At present, this resist needs several thousand $\mu C/cm^2$ to remove about 100 nm of film, but it is a new and simple system which holds the promise of increased sensitivity after further development work.

Plasma etch-resistant materials

The edge irregularity produced in features by under-etching and undercutting in wet etching processes can often be tolerated or compensated for in large area details. But when detail sizes are reduced below 1 μm, it becomes increasingly difficult to transfer the resist window pattern faithfully into the substrate by wet etching methods. Dry etching processes are expected to overcome such difficulties.

In these, substrate material is removed as volatile matter formed by reaction between the substrate and excited species found in gaseous plasmas, which may be free radicals, ions, electrons or excited molecules. Sputtering may also occur in some dry etching processes. The forms known as ion milling and reactive ion etching employ an electric field which directs ions normally towards the substrate being etched. Etching can then be anisotropic with no undercutting and faithful pattern transfer is obtained.

Positive-working electron resists have been designed to degrade readily when they are exposed to electrons, or indeed, to any kind of ionizing radiation. It is not surprising that when they are used as etch masks in dry etching processes, they are degraded quite rapidly by the species present in the plasma. Generally, the more sensitive the positive resist, the more rapidly it is eroded and, of course, the self-developing resists are eroded most rapidly of all.

In recent years, one of the most pressing problems in positive electron resist technology has been the provision of a resist having both high sensitivity to irradiation and high resistance to erosion during dry etching processes. Much needs still to be done to elucidate the mechanisms of dry etching processes, but it has been observed that compounds containing aromatic groups generally have much greater resistance to erosion in dry etching than aliphatic compounds have. All the sensitive positive-working electron resists known at the time dry etching was introduced were aliphatic. Invariably, the introduction of aromatic groups into the structure reduces the sensitivity. Initially, when dry etching was to be used, electron lithographers often reverted to using positive photoresists which, being mixtures of aromatic compounds, had the required erosion resistance. Their lower sensitivity was tolerated pending the appearance of more sensitive dry etch-resistant electron resists.

An early attempt to reconcile in positive-working electron resists the opposing requirements of high sensitivity and high erosion resistance involved adding free radical scavengers or plastics antioxidants to the standard resists

PMMA and poly(hexafluorobutyl methacrylate) (45), though in most cases sensitivity was reduced to some extent. The improvement in plasma etch resistance was said to be due to stabilization, by the additive, of the radicals produced in the resist polymer by the plasma, so that chain scission was reduced. However, the additives were all aromatic compounds which in themselves may have contributed to the improved performance in dry etching.

Bowden described an electron analogue of positive photoresists, consisting of a novolac resin with about 10% of the radiation-degradable poly(2-methyl-1-pentene sulphone) (46). The latter is volatilized and removed from the film during irradiation i.e. it is a self-developing resist in its own right. Positive images could be formed in films of the mixture by developing in aqueous alkaline solutions, the sulphone remaining in unirradiated regions acting as a dissolution inhibitor. An exposure of 16 $\mu C/cm^2$ was required, though 3 $\mu C/cm^2$ was adequate if the resist film was heated after exposure, a process which enhanced the degradation of the polysulphone. The residual novolac resin film provided the dry etch resistance.

A styrene analogue of the anhydride cross-linked methacrylate described above, in which methyl methacrylate units were replaced by styrene, showed

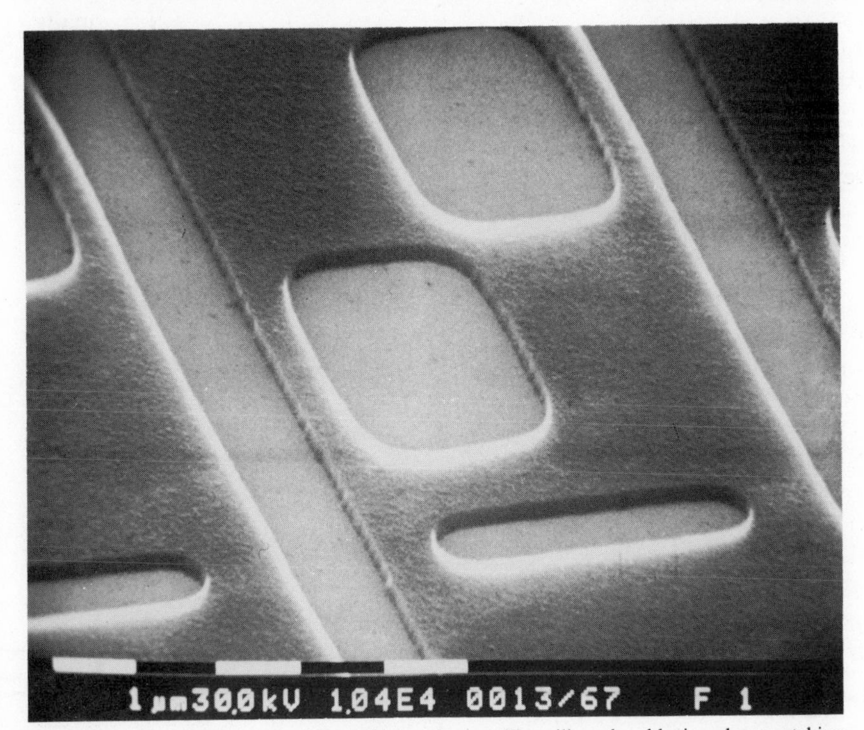

Figure 10.3 Pattern defined in 0.2 μm thick polycrystalline silicon by chlorine plasma etching, using mask of styrene-based anhydride cross-linked positive-working electron resist (47). The pattern was exposed in an electron image projector (5, 6)

very good erosion resistance in chlorine-based plasmas (47). However, the exposure required, which can range from 30 $\mu C/cm^2$ upwards, was higher than is desirable for use in scanning systems, though it is very suitable for use in electron image projectors. An etched pattern produced with this resist is shown in Figure 10.3.

More recently, attention has turned to the possibility of modifying the properties of the resist film after the pattern in it has been developed. Roberts used poly(methacryloyl chloride) (**6**) as the sensitive resist (48) which can be exposed at doses less than 10 $\mu C/cm^2$ with solvent development. The film was treated subsequently with a compound which could react with the chlorocarbonyl group in the polymer, to introduce into its structure an etch-resistant group. Aniline was a particularly effective reagent, the etch resistance being conferred by the aromatic group introduced when the acid chloride polymer was converted to its anilide (**7**). This system was used for making, by a dry etching process, mask patterns on silica like that shown in Figure 10.4.

Anderson and his co-workers used a similar idea in which barium ions were introduced into poly(methyl methacrylate-co-itaconic acid) resist (**8**) after pattern definition (49). In this case, erosion resistance was given by the metal which formed refractory oxide or non-volatile chloride or fluoride.

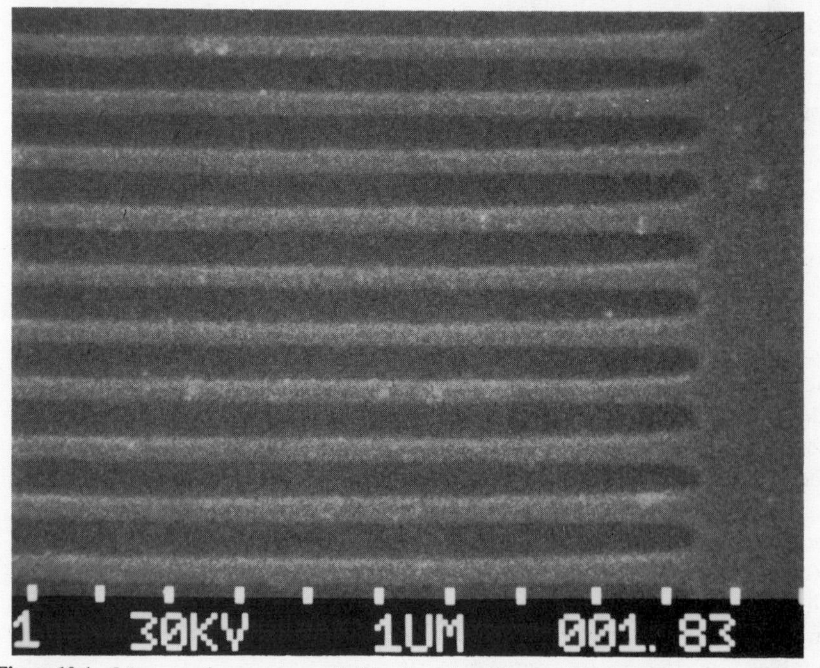

Figure 10.4 0.4 μm grating pattern etched in chromium film on silica by carbon tetrachloride-based plasma etching process. Etch mask was poly(methacryloyl chloride) resist treated after development, with aniline (48)

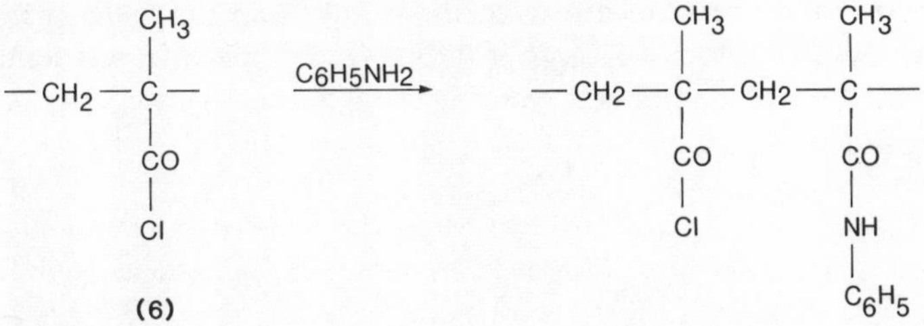

$$-CH_2-\underset{\underset{Cl}{|}}{\underset{CO}{|}}{\overset{\overset{CH_3}{|}}{C}}- \xrightarrow{C_6H_5NH_2} -CH_2-\underset{\underset{Cl}{|}}{\underset{CO}{|}}{\overset{\overset{CH_3}{|}}{C}}-CH_2-\underset{\underset{\underset{C_6H_5}{|}}{NH}}{\underset{CO}{|}}{\overset{\overset{CH_3}{|}}{C}}-$$

(6) (7)

$$-CH_2-\underset{\underset{COOCH_3}{|}}{\overset{\overset{CH_3}{|}}{C}}-CH_2-\underset{\underset{COOH}{|}}{\overset{\overset{CH_2COOH}{|}}{C}}-$$

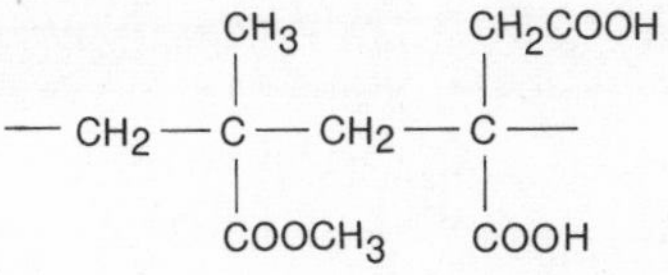

(8)

A similar approach was used by Moreau to improve the dry-etch resistance of positive photoresists (50), in which sodium, potassium or magnesium ions provided the erosion resistance.

Negative-working electron resists

Solvent-developed systems

In general, negative resists have been favoured less by lithographers because their resolution capability is inferior. The disadvantages become potentially more serious as detail sizes are reduced. Nevertheless, it is convenient to have available negative resists for each exposure technique and in scanning systems they can be essential to economic operation.

In negative resists, the effects of scattered electrons result in some residue outside the boundary of the irradiated area at exposures approaching those needed to cross-link the film fully. Extensive residues may be left between closely-adjacent features so it has become customary to expose negative resists so that only 50–70% of the initial thickness of resist remains after development. However, this strategy does not eliminate the problem entirely and it is usually necessary to remove thin unwanted residues from negative resist images by a brief treatment in oxygen plasma, a process called descumming. Reducing the exposure in this way can introduce a different problem in that the exposed resist is less extensively cross-linked and is susceptible to swelling in the developer, as shown in Figure 10.5*a*. Careful selection of the developer enables this effect to be minimized (Figure 10.5*b*).

Most polymers become cross-linked upon electron irradiation and so are potential candidates for use as negative electron resists, though not all have the

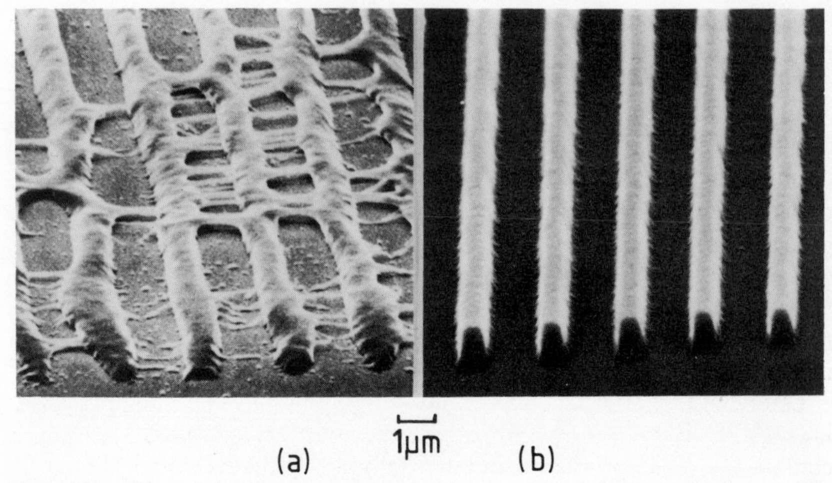

Figure 10.5 Effect of developer on styrene copolymer negative electron resist images. The exposure is the same in each case. (*a*) Development in toluene results in swelling and image distortion; (*b*) development in methyl isobutyl ketone avoids swelling and produces an acceptable image

Table 10.3 Negative-working electron resists

Polymer type	Sensitivity ($\mu C/cm^2$)	Reference
Methylcyclosiloxane + silylene phosphate	800	51
Styrene	100–400	52
Methylcyclosiloxane	130	53
Vinylcyclosiloxane	4	53
Vinylcyclosiloxane + silylene phosphate	4	51
Diallyl orthophthalate	2–6	54
Styrene-co-glycidyl methacrylate	1–150	55
Glycidyl methacrylate-co-3-chlorostyrene	1–4	56
Esterified acrylic terpolymer	1	57
ω-tricosenoic acid and its unsaturated esters used as monomolecular layers	0.5–50	58
Glycidyl methacrylate-co-ethyl acrylate	0.4	59
Epoxidized butadiene	0.1–1	60
Epoxidized isoprene	0.05	60
Glycidyl methacrylate	0.05	60
Glycidyl methacrylate	1–20	57
Half esters of maleic anhydride-co--vinyl ether	0.02–0.12	61
Vinyl carbazole	8	62
Styrene-co-4-chlorostyrene	5–7	55
Iodinated polystyrene	2	63
3-bromo-9-vinylcarbazole	1–2	62
Chloromethylated polystyrene	0.1–1	64

The ranges of sensitivity shown in some cases arise because the sensitivity depends upon the molecular weight of the polymer. Higher molecular weights result in higher sensitivity.

right combination of properties for lithographic applications. Some of the materials which have been used are shown in Table 10.3.

The earliest studies indicated that saturated molecules, especially if they contained aromatic groups (e.g. polystyrene), were rather insensitive. All early sensitive negative electron resists contained either olefinic linkages or epoxy groups, which may be at any site in the polymer molecule. Episulphide polymers were also described, though these have not been used much. All these groups can take part in cross-linking, and once initiated, the reaction can proceed spontaneously by a chain mechanism. This self-propagating reaction leads to very high sensitivity ($\sim 1\ \mu C/cm^2$) but there is no control over where the reaction spreads through the film. Most early sensitive negative electron resists actually showed very poor contrast and resolution of fine patterns.

Later developments showed that unsaturation was not essential to providing high sensitivity. The presence of halogen atoms, even in saturated aromatic compounds, could produce resists having very high sensitivities (Table 10.3) (55, 62–64). Being aromatic, these resists also have adequate erosion resistance in dry etching processes, and they have now largely displaced the earlier aliphatic types of resist. They can show much increased contrast values over those of the earlier materials, though the majority are still susceptible to the effects of stray electrons and swelling in the developer.

Examples of negative-working electron beam resists are poly(glycidyl methacrylate-co-ethylacrylate) (**9**), poly(styrene-co-4-chlorostyrene) (**10**), and poly(styrene-co-chloromethylstyrene) (**11**).

An interesting new type of negative electron resist consisted of tetra-

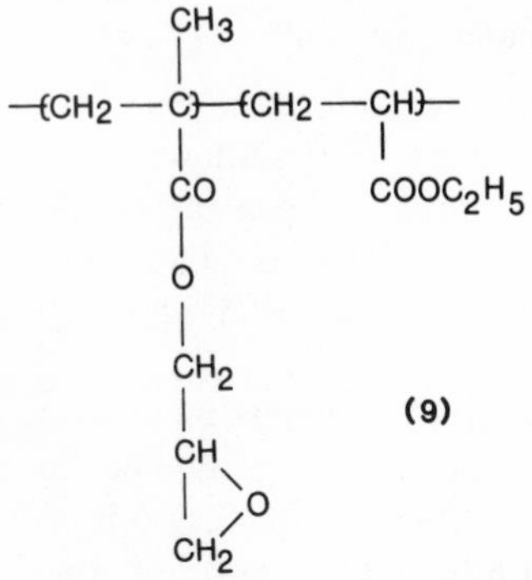

(9)

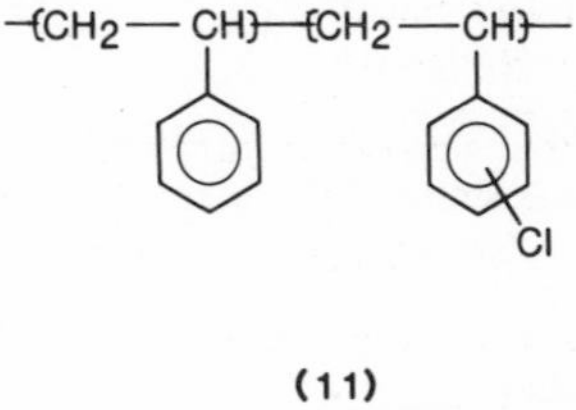

(11)

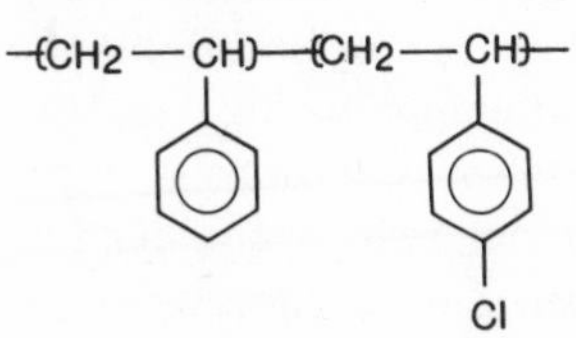

(10)

Figure 10.6 0.3 μm lines in polymethylcyclosiloxane electron resist (53) defined by a scanning electron beam exposure system

thiafulvalene covalently bonded to polystyrene and mixed with an electron acceptor such as carbon tetrabromide (65). Upon irradiation with electrons at about 10 $\mu C/cm^2$, charge transfer occurred to ionize the tetrathiafulvalene. A non-polar developer removed unirradiated material, but the ionized (irradiated) parts remained insoluble, and, being polar, they were not swollen by the developer.

The polysiloxane resists described by Roberts (53) were intended to allow direct definition by electron beams of siliceous film patterns. These could be converted by heating to silica film which could serve as diffusion barriers and as insulating layers in the semiconductor fabrication process. Figure 10.6 shows some narrow lines defined in this resist. Phosphosilicate glass patterns were made similarly by incorporating a suitable phosphorus compound into the resist (51). This was one of the earliest attempts to eliminate wet etching and the consequent undercutting from the fabrication process to obtain full benefit of the resolution capability of electron beam techniques. In recent years, with the advent of multilevel processing, interest in silicon-containing resists has revived, and many other materials of this type—including positive-working ones and those intended for use in other exposure techniques—have been described (66–69)

Dry-developable electron resists

Swelling of exposed resist patterns during development in organic solvents can have disastrous effects upon pattern definition. It occurs mainly in negative resists and is increasingly troublesome as detail dimensions are reduced, as in electron and x-ray exposure techniques.

A scheme for using a dry-developable negative photoresist was proposed first by Hughes and his colleagues (70), and it operated as shown schematically in Figure 10.7. The same principle has been used by other workers to make both positive- and negative-working dry-developable resists for each exposure technique. Generally, the negative resists have consisted of a host polymer mixed with a moderately volatile etch-resistant compound. During exposure, the etch-resistant compound is chemically bonded to the host polymer. The film is then baked in vacuum, when the etch-resistant compound is volatilized from unexposed areas where it is still uncombined. Subsequent treatment in an oxygen plasma (often in the reactive ion etching mode) erodes unexposed areas more rapidly, so a negative image is formed.

A difficulty in electron resists operating in this way is that exposure must be conducted in vacuum. The etch-resistant additive in its free state must not be volatile at this stage. Its subsequent removal from unirradiated areas can be assisted only by an increase in temperature without the decrease in pressure which can assist removal after exposure by other methods which are conducted at atmospheric pressure. This limits the choice of possible additives. Nevertheless, the method has been operated successfully, particularly by

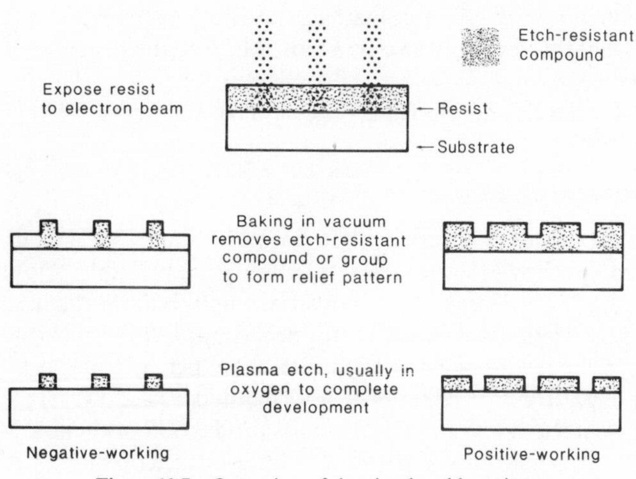

Figure 10.7 Operation of dry-developable resists

workers in Japan. Tsuda has used mixtures of poly(methylisopropenyl ketone) with various aromatic diazides to produce resist compositions for each exposure technique (71). The aromatic diazide residue is incorporated into the resist structure as cross-links during exposure. 0.3 μm lines have been defined in this type of resist by electron beams.

Yoneda used mixtures of standard electron resists with various silicon compounds to formulate both negative- and positive-working dry-developable resists (72). Although the positive version needs a final treatment with fluorine plasma or with hydrofluoric acid to remove small residues of silica from exposed regions, the published pictures of submicron grating patterns are very impressive.

An interesting positive-working system, developed by Meyer and coworkers as a deep ultraviolet resist (73), is a copolymer of methyl methacrylate with 4-(trimethylsilylmethylene)-acetophenoneoxime methacrylate (**12**). The silicon atom, which gives oxygen plasma resistance, is chemically

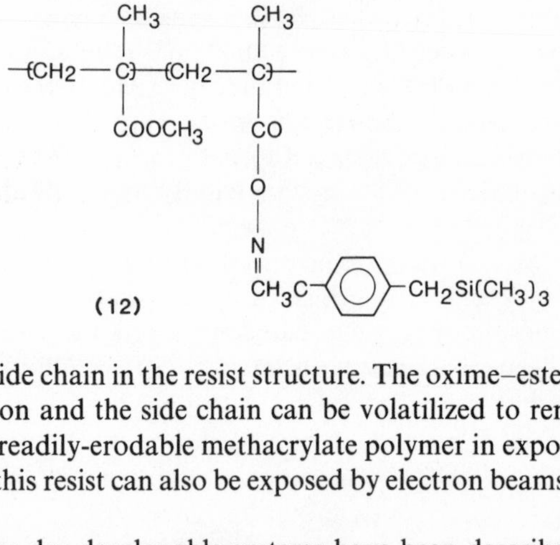

$$-\!\left[CH_2-\underset{COOCH_3}{\overset{CH_3}{C}}\right]\!-\!\left[CH_2-\underset{\underset{\underset{\underset{CH_3C-C_6H_4-CH_2Si(CH_3)_3}{\|}}{N}}{|}}{\underset{O}{\underset{|}{CO}}}}{\overset{CH_3}{C}}\right]\!-$$

(12)

bound into a side chain in the resist structure. The oxime–ester linkage breaks upon irradiation and the side chain can be volatilized to remove the silicon atom, leaving readily-erodable methacrylate polymer in exposed regions. It is reported that this resist can also be exposed by electron beams with sensitivity 1 $\mu C/cm^2$.

Several other dry-developable systems have been described (74–77).

X-ray exposure systems

An x-ray exposure system was first described by Smith (78) and has since been developed by many other workers. It is similar, in principle, to photographic contact printing, an x-ray beam instead of a light beam being used to expose the sensitive material. A mask is still needed, and this is held within a few microns of the resist-coated substrate to avoid penumbral effects which can spoil the edge definition of patterns. This method also allows large areas to be exposed, so it should be suitable for mass production of circuits. Masks have been made from gold films supported on (thinned) silicon slices or on materials such as Mylar film stretched on a frame or on thin membranes of boron nitride.

Again, if submicron details are required on the mask, they must be generated by a scanning electron beam system.

X-ray resists

It is believed that in x-ray exposure the structural changes in the resist which allow development of the exposed pattern are actually caused by secondary electrons produced by the primary radiation on its passage through the resist film. The secondary electrons have low energy and short ranges, so scattering within the resist is so small as to be negligible. Consequently, it is possible to produce by x-ray exposure pattern profiles with vertical walls and with much higher aspect ratios than can be achieved by electron exposure. The difference is illustrated by the micrographs in Figure 10.8, which are grating patterns in a positive electron resist made by each exposure method.

When the x-ray exposure technique was first practised, electron resists were used, PMMA being favoured. Most of the electron resists available at that time contained only carbon, hydrogen and oxygen. These elements are almost transparent to the x-rays used, so the resists were very insensitive in their reaction to x-rays. As was the case with electron resists, much of the research effort has been devoted to finding materials having higher sensitivities.

The first report of a resist designed specifically for x-ray exposure was by Brault (79), who used a water-soluble mixed barium/lead acrylate as a

Table 10.4 X-ray resists

Polymer type	Image tone	X-ray sensitivity (nm)	X-ray sensitivity (mJ/cm^2)	Reference
Methyl methacrylate	+	0.54	2400	31
Hexafluorobutyl methacrylate	+	0.54	45–80	31
Tetrafluoroethylene-co-vinyl ethers	–	0.54	11–260	82
Allyl maleate-co-vinyl ethers	–	0.54	2.4–10	82
Epoxidized butadiene	–	0.54	5	82
Methyl methacrylate	+	0.83	500	17
Methyl methacrylate-co-methacrylic acid	+	0.83	100–200	17
MMA-co-MAA, thallium salt	+	0.83	24	17
MMA-co-MAA, caesium salt	+	0.83	40	17
MMA-co-MAA, barium salt	+	0.83	400	80
MMA-co-MAA, lead salt	+	0.83	400	80
Barium/lead acrylate mixture	–	0.83	60	79
Butene-1-sulphone	+	0.437	94	81
Glycidyl methacrylate-co-ethyl acrylate	–	0.437	159	81
Glycidyl methacrylate	–	0.437	67	81
1,3-dichloro-2-propyl acrylate	–	0.437	39	81
2-chloroethyl acrylate	–	0.437	27	81
2,3-dichloro-1-propyl acrylate	–	0.437	13	81
2,3-dichloro-1-propyl acrylate-co-glycidyl acrylate	–	0.437	13	81

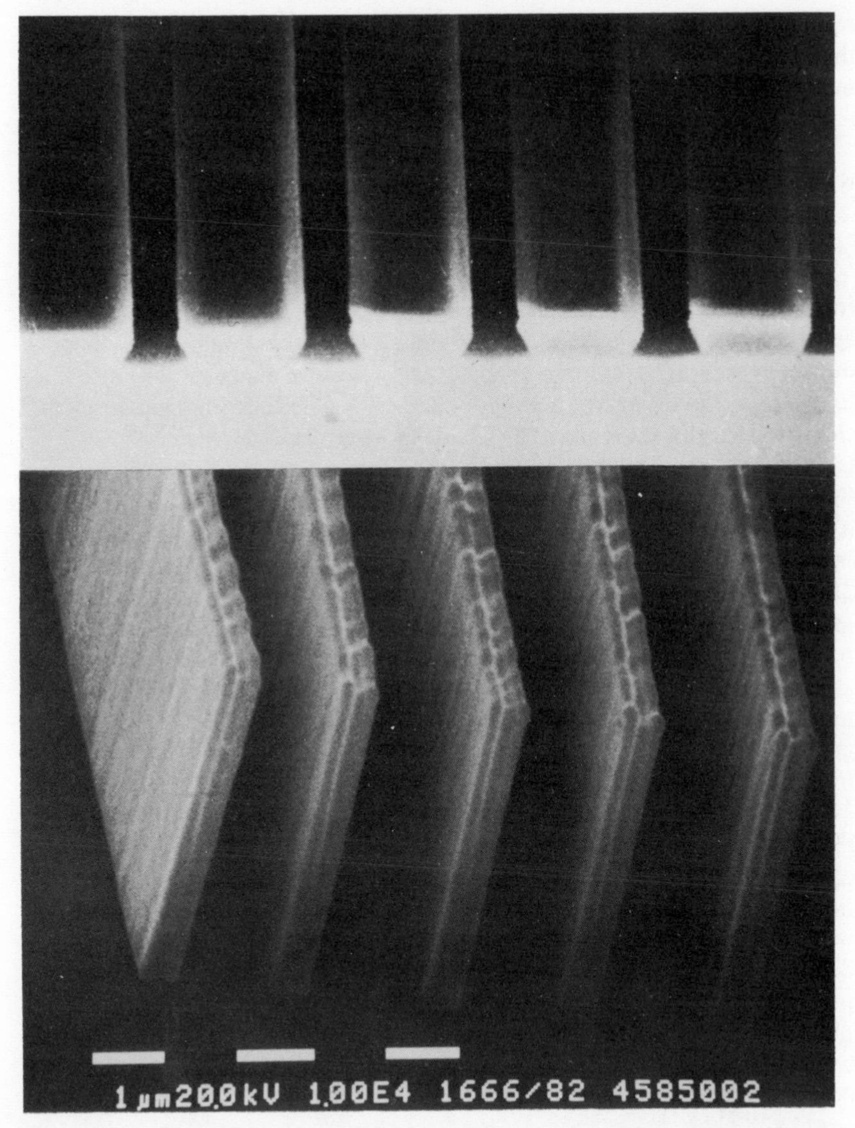

Figure 10.8 Grating patterns in cross-linked methacrylate positive-working electron resist (38) exposed by (top) 20 kV electron beam and (bottom) synchrotron x-radiation

negative-working x-ray resist. This was soon followed by the use of thallium, caesium and other metal salts of poly(methyl methacrylate-co-methacrylic acid) as positive resists (17, 80). All these materials relied upon the heavy metal atom to increase the absorption of x-rays, increasing the sensitivity by factors of ten to twenty over that of PMMA.

Taylor, working with a system employing x-rays of shorter wavelength, realized that chlorine has an absorption edge at a wavelength slightly higher than that of his exposure system. He obtained much improved sensitivity in a number of chlorine-substituted polymers and copolymers synthesized specially for the purpose (81).

A list of the main compounds which have been used as x-ray resists, together with a summary of their properties, is given in Table 10.4.

Future developments

Improvements in optical exposure techniques have, in the past, resulted in an approximate halving every five years in the size of the smallest details defined during the fabrication of semiconductor devices. At present, the ultimate limit to which detail sizes can be reduced without altering the performance characteristics of the resulting devices is not known, though some calculations have been made in attempts to answer this question (83).

A few years ago, it was thought that diffraction effects would preclude the use of photolithographic methods in making details less than about 1 μm wide. Improvements to equipment and to resists and the use of shorter wavelength ultraviolet light have, in fact, allowed remarkable successes to be achieved in reproducing smaller details, and even now, in laboratories, 0.7 μm details can be properly exposed. But these successes have depended crucially upon the ability of electron beam exposure methods and electron resists to make the masks containing such small details, which are necessary to the photolithographic technique. Scanning electron lithography is the best method known by which submicron features can actually be generated. (Scanning ion lithography can also do this, though it is as yet less well developed.) In its ability to make the masks needed, electron lithography has enabled the limits of photolithography to be extended to dimensions hitherto unexpected, and ironically, has delayed its own adoption as the method of fabricating devices with details in the range 0.5–1 μm, with future potential for even smaller details. Synchrotron x-ray sources are now available, at least for experimental purposes. Patterns with dimensions in the 0.2–0.5 μm range are readily obtained, and existing UV and e-beam resists may be employed (84). Japanese work with chloromethylated polystyrene resists is particularly promising (85).

Electron and x-ray resist development is likely to be concentrated upon positive-working resists with high sensitivity and with improved erosion resistance in dry etching. Probably more attention will be given to methods of changing the properties of resist films by processes applied after the pattern development stage. The study of this possibility has only just begun. It also seems probable that improved resists capable of dry development after exposure will be introduced. This will help not only in dimensional control (by avoiding swelling problems); it should give considerable environmental and economic advantages.

Self-developing resists are attractive at first sight, for processing is even more simple. It seems likely, though, that they will not be readily accepted in electron or ion lithography because the degradation products will cause relatively rapid accumulation of insulating deposits inside the equipment. They may, however, find some use in x-ray or deep ultraviolet exposure systems, where deposits, even if they were formed, would not cause pattern distortion.

At present, multilevel processing is being investigated actively. Many workers believe this to be the only way to achieve adequate dimensional control of patterns exposed on the contoured surfaces which arise in the later stages of integrated circuit manufacture. Device fabricators, however, regard it as a last resort and would prefer to retain single resist level processing as far as possible. To this end, there will be developments in other methods of application of resist films–plasma polymerization and deposition of organic films, evaporation, sputtering, etc., of inorganic materials. Such methods could enable thin films to be applied uniformly over contoured surfaces, thereby eliminating the need for multilevel processing. Better-quality thin films may be obtainable by such methods, thereby circumventing the other objection which device fabricators have to using very thin spun films, the greater likelihood of pinhole formation. The improved erosion resistance available from suitable inorganic compounds is also consistent with the use of thinner films. Only a few inorganic systems have been examined so far and, in general, they are too insensitive to allow their economic use. There seems little doubt, though, that new inorganic systems and methods of increasing the sensitivity of those known already, will be discovered, and as their molecular sizes will be small relative to organic polymers, they may prove ultimately to be the most suitable materials to provide the highest resolution of fine patterns.

The published proceedings (84) of a recent review symposium on materials for lithography give an excellent overview of the scope of materials development likely in the next few years.

Acknowledgements

The author thanks the Director of Philips Research Laboratories, Redhill, Surrey, UK, for permission to publish the photographs in this chapter. He is also grateful to Dr H. Lühtje of Philips GmbH Forschungslaboratorium, Hamburg, West Germany, who produced the x-ray exposed pattern in Figure 10.8*b* and supplied the scanning electron micrograph and to Mr P. W. Whipps who supplied the micrographs in Figure 10.5.

References

1. D. J. Elliott., *Integrated Circuit Fabrication Technology*, McGraw-Hill, New York (1982).
2. K. L. Lee and H. Ahmed, *J. Vac. Sci. Technol.* **19** (1981) 946–949.
3. H. C. Pfeiffer, in *Microcircuit Engineering 83*, eds. H. Ahmed, J. R. A. Cleaver and G. A. C. Jones, Academic Press, London (1983) 3–14.
4. T. W. O'Keeffe, J. Vine and R. M. Handy, *Solid State Electronics* **12** (1969) 841–848.

5. J. P. Scott, in *6th Int. Conf. on Electron and Ion Beam Science & Technology*, ed. R. Bakish, Electrochem. Soc., Princeton, N. J. (1974) 123–136.
6. R. Ward, A. R. Franklin, P. Gould and M. J. Plummer, *J. Vac. Sci. Technol.* **19** (1981) 966–970.
7. P. D. Prewett, *Proc. SPIE* **537** (1985) 126–137.
8. G. Stengl, H. Löschner, W. Maurer and P. Wolf, *Proc. SPIE.* **537** (1985) 138–145.
9. H. Nagai, A. Yoshikawa, Y. Toyoshima, O.Ochi and Y. Mizushima, *Appl. Phys. Lett.* **28** (1976) 145–147.
10. S. Hattori, J. Tamano, M. Yamada, M. Ieda, S, Morita, K. Yoneda and S. Ishibashi, *Thin Solid Films* **83** (1981) 189–194.
11. A. Barraud, C. Rosilio and A. Ruaudel-Teixier, *Thin Solid Films* **68** (1980) 91–98.
12. Y. Yatsui, T. Nakata and K. Umehara, *J. Electrochem. Soc.* **116** (1969) 94–97.
13. T. R. Neill and C. J. Bull, *Electronics Lett.* **16** (1980) 621–622.
14. H. Y. Ku and L. C. Scala, *J. Electrochem. Soc.* **116** (1969) 980–985.
15. E. D. Feit, M. E. Wurtz and G. W. Kammlott, *J. Vac. Sci. Technol.* **15** (1978) 944–947.
16. I. Haller, M. Hatzakis and R. Srinivasan, *IBM J. Res.*, **12** (1968) 251–256.
17. I. Haller, R. Feder, M. Hatzakis and E. Spiller, *J. Electrochem. Soc.* **126** (1979) 154–161.
18. A. W. Levine, M. Kaplan and E. S. Poliniak, *Polymer Eng. & Sci.* **14** (1974) 518–524.
19. Jpn. Kokai Tokkyo Koho, 78, 81/113; *Chem. Abstr.* **90**, 64500d.
20. C. U. Pittman, Jr., M. Ueda, C. Y. Chen, J. H. Kwiatkowski, C. F. Cook, Jr. and J. N. Helbert, *J. Electrochem. Soc.* **128** (1981) 1758–1762.
21. Y. M. N. Namasté, S. K. Obendorf, C. C. Anderson, P. D. Krasicky, F. Rodriguez and R. Tiberio, *J. Vac. Sci. Technol.* **B1** (1983) 1160–1165.
22. E. Gipstein, O. Need and W. Moreau, *ACS Coatings & Plastics Preprints* **35**, 246–51, American Chemical Society (1975).
23. S. Nonogaki, ACS 170th Meeting, Chicago, 1975, Extended Abstracts **78–1**, 946–9, American Chemical Society (1975).
24. British Patent 1,481,887.
25. J. H. Lai, J. N. Helbert, C. F. Cook, Jr. and C. U. Pittman, Jr., *J. Vac. Sci. Technol.* **16** (1979) 1992–1995.
26. Jpn. Kokai Tokkyo Koho, 78/100, 224; *Chem. Abstr.* **90**, 64502f.
27. T. Tada, *J. Electrochem. Soc.* **126** (1979) 1635–1636.
28. T. Tada, *J. Electrochem. Soc.* **130** (1983) 912–917.
29. Jpn. Tokkyo Koho, 80/21, 332; *Chem. Abstr.* **93**, 213410r.
30. US Patent 4,279,984; *Chem. Abstr.* **95**, 142038d.
31. M. Kakuchi, S. Sugawara, K. Murase and K. Matsuyama, *J. Electrochem. Soc.* **124** (1977) 1648–1651.
32. T. Nakayama, K. Kotake, T. Sato, S. Ichikawa, S. Asaumi, A. Yokota and H. Nakane, *Denki Kagaku oyobi Kogyo Butsuri Kagaku* **49** (1981) 245–248;.*Chem. Abstr.* **95**, 33359p.
33. K. Harada, K. Miyoshi, H. Namatsu and S. Moriya in *Microcircuit Engineering 83*, eds. H. Ahmed, J. R. A. Cleaver and G. A. C. Jones, Academic Press (1983) London, 313–320.
34. L. F. Thompson and M. J. Bowden, *J. Electrochem. Soc.* **121** (1974) 1620–1623; **120** (1973) 1722–1726.
35. A. S. Gozdz, H. G. Craighead and M. J. Bowden, *Polym. Eng. Sci.* **26** (1986) 1123–1128.
36. R. J. Himics, N. V. Desai, M. Kaplan E. S. Poliniak, *ACS Coatings & Plastics Preprints* **35**, 266–272, 273–280, American Chemical Society (1975).
37. M. J. Bowden and L. F. Thompson, in *Applied Polymer Symp. No. 23*, ed. L. H. Princen, John Wiley, New York (1974) 99–106.
38. E. D. Roberts *Applied Polymer Symp. No. 23*, ed. L. H. Princen, John Wiley, New York (1974) 87–98.
39. E. D. Roberts, ACS Symp. Ser. 184, eds. E. D. Feit and C. W. Wilkins, Jr., American Chemical Society (1982) 1–17.
40. K. Hatada, T. Kitiyama, S. Danjo, H. Yuki, H. Aritome, S. Namba, K. Nate and H. Yokono, *Polymer Bull.* **8** (1982) 469–472.
41. H. Ito and C. G. Willson, *Polymer Eng. & Sci.* **23** (1983) 1012–1018.
42. M. Yamada, J. Tamano, K. Yoneda, S. Morita and S. Hattori, *Jap. J. Appl. Phys.* **21** (1982) 768–771.
43. S. Morita, K. Yoneda, S. Ishibashi, J. Tamano, Y. Yamada, S. Hattori and M. Ieda, in *5th Int. Symp. on Plasma Chem.* **1** (1981) 259–264.

44. M. Green, C. J. Aidinis and O. A. Fakolujo, *J. Appl. Phys.* **57** (1985) 631–633.
45. K. Harada *J. Electrochem. Soc.* **127** (1980) 491–497.
46. M. J. Bowden, L. F. Thompson, S. R. Fahrenholtz and E. M. Doerries, *J. Electrochem. Soc.* **128** (1981) 1304–1313.
47. E. D. Roberts, *Polymer Eng. & Sci.* **23** (1983) 968–974.
48. E. D. Roberts, *Proc. SPIE* **539** (1985) 124–130.
49. C. C. Anderson and F. Rodriguez, *J. Vac. Sci. Technol.* **B3** (1985) 347–352.
50. W. M. Moreau, in *Microcircuit Engineering 83*, eds. H. Ahmed, J. R. A. Cleaver and G. A. C. Jones, Academic Press, London (1983) 321–327.
51. E. D. Roberts, in *5th Int. Conf. on Electron and Ion Beam Science and Technology*, ed. R. Bakish, Electrochem Soc., Princeton, N. J. (1972) 102–111.
52. T. L. Brewer, *Polymer Eng. & Sci.* **14** (1974) 534–537.
53. E. D. Roberts, *J. Electrochem. Soc.* **120** (1973) 1716–1721.
54. J. L. Bartelt, in *Applied Polymer Symp. 23*, ed. L. H. Princen, John Wiley, New York (1974) 139–146.
55. J. C. Jagt and P. W. Whipps, *Philips Tech. Rev.* **39** (1980 346–352.
56. L. F. Thompson, E. M. Doerries and L. Yau, *J. Electrochem. Soc.* **126** (1979) 1699–1702, 1703–8.
57. J. L. Bartelt and E. D. Feit, *J. Electrochem. Soc.* **122** (1975) 541–544.
58. A. Barraud, C. Rosilio and A. Ruaudel-Teixier, *Solid State Technol.* **22** (8) (1975) 120–124.
59. E. D. Feit, L. F. Thompson and L. E. Stillwagon, *ACS Coatings and Plastics Preprints* **35** (1975) 287–293, American Chemical Society.
60. T. Hirai, Y. Hatano and S. Nonogaki, *J. Electrochem. Soc.* **118** (1971) 669–672.
61. H. S. Cole, D. W. Skelly and B. C. Wagner, *IEEE Trans. Electron Devices* **ED-22** (1975) 417–420.
62. J. C. Jagt and A. P. G. Sevriens, *Polymer Eng. & Sci.* **20** (1980) 1082–1086.
63. H. Shiraishi, Y. Taniguchi, S. Horigome and S. Nonogaki, *Polymer Eng. & Sci.* **20** (1980) 1054–1057.
64. S. Imamura, *J. Electrochem. Soc.* **126** (1979) 1628–1630.
65. Y. Tomkiewicz, E. M. Engler, J. D. Kuptsis, R. G. Schad, V. V. Patel and M. Hatzakis, *Appl. Phys. Lett.* **37** (1980) 314–316.
66. Y. Ohnishi, M. Suzuki, K. Saigo, Y. Saotome and H. Gokan, *Proc. SPIE* **539** (1985) 62–69.
67. R. G. Brault, R. L. Kubena and R. A. Metzger, *Proc. SPIE* **539** (1985) 70–73.
68. M. A. Hartney and A. E. Novembre, *Proc. SPIE* **539** (1985) 90–96.
69. J. M. Zeigler, L. A. Harrah and A. W. Johnson, *Proc. SPIE* **539** (1985) 166–174.
70. H. G. Hughes, W. R. Goodner, T. E. Wood, J. N. Smith and J. V. Keller, *Polymer Eng. & Sci.* **20** (1980) 1093–1096.
71. M. Tsuda, S. Oikawa, W. Kanai, K. Hashimoto, A. Yokota, K. Nuino, I. Hijikata, A. Uehara and H. Nakane, *J. Vac. Sci. Technol.* **19** (1981) 1351–1359.
72. Y. Yoneda, M. Miyagawa, S. Fukuyama, T. Narusawa and H. Okuyama, *Proc. SPIE* **539** (1985) 145–150
73. W. H. Meyer, B. J. Curtis and H. R. Brunner, *Microelectronic Eng.* **1** (1983) 29–40.
74. M. S. Chang, T. W, Hou, J. T. Chen, K. D. Kolwicz and J. N. Zemel, *J. Vac. Sci.* Technol. **16** (1980) 1973–1976.
75. S. Morita, J. Tamano, S. Hattori and M. Ieda, *J. Appl. Phys.* **51** (1980) 3938–3941.
76. H. Hiraoka, *J. Electrochem. Soc.* **128** (1981) 1065–1071.
77. V. P. Korchov, T. N. Martynova and V. S. Danilovich, *Thin Solid Films* **101** (1983) 369–376.
78. D. L. Spears and H. I. Smith, *Electronics Lett.* **8** (1972) 102–104.
79. R. G. Brault, in *6th Int. Conf. on Electron and Ion Beam Sci. and Technol.*, ed. R. Bakish, Electrochem. Soc., Princeton (1974) 63–70.
80. D. J. Webb and M. Hatzakis, *J. Vac. Sci. Technol.* **16** (1979) 2008–2013.
81. G. N. Taylor, G. A. Coquin and S. Somekh, *Polymer Eng. & Sci.* **17** (1977) 420–429.
82. S. Imamura, S. Sugawara and K. Murase, *J. Electrochem. Soc.* **124** (1977) 1139–1140.
83. R. W. Keyes *Proc. IEEE* **63** (1975) 740–767.
84. M. Tsuda, S. Oikawa, M. Yabuta, A. Yokota, H. Nakane, N. Atoda, K. Hoh, K. Gamo, and S. Namba, *Polym. Sci. Eng.* **26** (1986) 1148–1152; H.-L. Huber, H. Betz, and A. Heuberger, *Polym. Sci. Eng.* **26** (1986) 1153–1157.
85. K. Shibayama and T. Kato, *Polym. Sci. Eng.* **26** (1986) 1140–1145.
86. L. F. Thompson, C. F. Wilson and J. M. J. Fréchet (eds.) ACS Symp. Ser. **266** (1984).

11 Wet etching

W. KERN and G. L. SCHNABLE

Introduction

Etching, wet etching, and dissolution

This chapter is principally concerned with wet-chemical etching as used in high-volume fabrication of integrated circuits. Many other applications of wet etching, in semiconductor device research and in fabrication of developmental devices, have appeared in the literature, and others are being considered for fabrication of future devices.

Etching can be defined as removal of solid material from a substrate by means of a chemical reaction. The reactant to effect etching can be a liquid or gaseous chemical agent. Dissolution, in contrast, is a physical process and, strictly speaking, should not be considered etching unless some chemical reaction is involved that changes the chemical composition of the material, which is sometimes the case. In semiconductor technology, all wet chemical processes are referred to as 'wet processes', of which wet-chemical etching is a sub-category.

In this chapter, wet etching is considered to be a process in which a chemical reaction occurs at a liquid–solid interface, and results in dissolution of the surface of the solid. The process may result in material removal of all surfaces of a substrate, or may be localized to regions not protected by a mask, as in lithography. Unless otherwise specified, the concentration of the acids, alkalis, and solutions described in this chapter are as in Table 11.1. Mixtures of liquids are by volume unless otherwise stated.

Dry etching

Dry etching processes use gaseous or vaporized chemical reactants, at high temperature. They are also used at low temperature and low pressure in combination with RF-induced plasma discharges that can generate from the reactant gas chemically extremely active species, such as free radicals and excited neutral species. The volatile reaction products of these non-directional (isotropic) chemical dry etching processes are pumped off and scrubbed or exhausted (see Chapter 15).

Another type of dry etching is strictly physical and directional (anisotropic) in its effect. Cations from an inert gas are created in a plasma and impinge on

Table 11.1 Concentration of common reagent-grade chemicals

Acids		wt%
Acetic acid	CH_3CO_2H	99.7 (minimum)
Hydrochloric acid	HCl	36.5–38.0
Nitric acid	HNO_3	69.0–71.0
Perchloric acid	$HClO_4$	69.0–72.0
Phosphoric acid	H_3PO_4	85.0 (minimum)
Sulphuric acid	H_2SO_4	95.0–98.0
Alkalis		wt%
Ammonium hydroxide	NH_4OH	28.0–30.0 (as NH_3)
Hydrazine monohydrate	$NH_2NH_2 \cdot H_2O$	99.0 (minimum)
Potassium hydroxide	KOH	85.0 (minimum)
Sodium hydroxide	$NaOH$	90.0 (minimum)
Other chemicals		wt%
Ammonium fluoride sol.	NH_4F	40–41
Hydrogen peroxide sol.	H_2O_2	30.0–32.0
Methanol	CH_3OH	99.8
2-Propanol	$(CH_3)_2$ CHOH	99.5 (typical)
Sodium peroxide	Na_2O_2	96.0 (minimum)
Mixtures*		vol
Aqua regia	HNO_3–HCl	1:3
Buffered HF (BHF)	HF–NH_4F	1:x (typically 1:6)
P-etch	HNO_3–HF–H_2O	1:3:60

*All mixtures in this chapter are by volume unless stated otherwise.

the substrate surface to remove substrate atoms and molecules physically and non-selectively by bombardment impact. Energetic processes of this type include sputter or ion etching, ion-beam etching or milling, and magnetic-confinement ion etching.

In chemical/physical dry etching, chemical and physical processes are combined, resulting in synergistic effects that can substantially enhance both the selective and directional etching reactions. This type of process includes the techniques of reactive ion or reactive sputter etching, reactive ion-beam etching, chemically assisted ion-beam etching, magnetic-confinement reactive ion etching, and triode etching. A special case of chemical dry etching is chemical transformation of a substrate surface, such as oxidation, siliconization, nitridation, carburization, and other similar reactions.

Objectives and scope

The main purpose of this chapter is to survey and describe wet chemical etching in its broadest sense and as it applies to the manufacture of integrated circuits. The vast majority of integrated circuits and other solid-state devices are based on silicon technology, although gallium arsenide and other compound semiconductors are rapidly gaining in importance for the fabrication of opto-electronic and very-high-speed solid-state devices and integrated circuits for special applications. Accordingly, we are placing primary emphasis on silicon technology, which comprises wet etching of silicon materials, dielectrics, insulators, silicide conductors, and metals. Associated materials used in the packaging of integrated circuit and discrete component chips include additional metals and alloys. Important compound semiconductors, especially gallium arsenide, will also be covered. We will describe and review basic chemical processes underlying wet etching, the technology to implement the processes, uses of wet etching in integrated circuit fabrication, and industrially important wet-etch processes for specific electronic materials. A separate section is devoted to important analytical applications of wet etching.

Representative references to the literature up to early 1986 have been cited, with emphasis on review papers, papers concerned with etching mechanisms, and papers which have been published recently. Numerous treatises and reviews have been published over the years on various aspects associated with chemical etching; these are listed chronologically in Part A of the references (1–126). They describe the chemistry, electrochemistry, technology and practical applications. Frequent reference is made to a chapter published in 1978 on chemical etching of electronic materials (82); we have attempted to complement and extend this previous work by emphasizing the most recent information.

Wet etching v. dry etching in microelectronics

There are three application categories of wet etching: (i) etching for the removal of thin films from a substrate, and layers from the substrate material itself; (ii) etching for analytical purposes; and (iii) etching for pattern transfer. The first type includes many important uses throughout microcircuit fabrication: removal of work damage, chemomechanical polishing of semiconductor slices, production of optically flat surfaces, precision thinning of semiconductor wafers for electro-optical devices, chemical cleaning and decontamination of surfaces, removal of oxide layers from silicon, stripping of insulator layers from the backside of wafers, removal of glass layers after impurity gettering steps, selective stripping of thin oxide films from conductor surfaces prior to metal deposition, development and stripping of certain photoresists, and removal of photoresist residues after the numerous photolithographic processing steps needed in device manufacture. Wet-chemical etching for most of these isotropic etching applications is superior to dry etching in terms of effectiveness, reliability, simplicity, selectivity, and cost effectiveness. Wet-chemical etching for analytical applications for structural studies, compositional tests, and defect detection and characterization is unique and obvious and needs no further explanation.

The third type of application, etching for pattern transfer, is the area where dry etching is most likely to displace wet etching. With the exception of crystallographic anisotropic etches, defect decoration etches, and junction delineation etches, most wet-chemical etchants are isotropic in nature, resulting in roughly equal removal of material in all directions. The upper portions of etched patterns are therefore smaller in dimension than the masking resist pattern and a tapered or concavely-shaped profile results. For linewidths of 3 μm and larger this shrinking effect of conventional isotropic wet etching is acceptable. Since a large percentage of present-day integrated circuits are still being manufactured within this dimensional range, wet-etch processes are generally adequate and advantageous. The critical problem arises with the linewidths and feature sizes of less than 3 μm required for VLSI and ULSI devices to meet the demands for high-density patterns with conservation of device area. Here the width and pitch of the finished delineated patterns and lines must be very close to the masked area, and this can only be attained by an anisotropic etch process. Physical or chemical–physical dry-etch processes meet these demands and are generally the only acceptable choice at present.

However, there are numerous intrinsic and technical difficulties associated with these relatively new dry-etch technologies. Perfect pattern transfer implies vertical walls of the patterns, which leads to non-uniformity of thickness and thinning over the top edges in subsequently deposited films. A reliably controllable taper angle is required, but is not easy to attain for all materials by dry etching on a production scale. Furthermore, the etch process must be highly selective, which is not always possible at present by dry etching for certain important structures, such as SiO_2 on silicon. Sophisticated improved techniques of highly selective wet etching, such as semi-anisotropic spray etching, may actually be superior for these applications.

A major drawback of dry etching, especially the physical and chemical–physical directional processes, concerns various types of damage inflicted on the material being etched (127). Ion bombardment and plasma environment can modify the structural, chemical and electrical properties of insulators and semiconductors due to intrinsic bonding damage, permeation of impurities, and the formation of polymer by-products and impurity films (128–133). The electrical characteristics of subsequently fabricated devices may therefore be adversely affected or even destroyed, as, for example, in the case of dielectric breakdown of gate insulators due to reactive ion etching (131). In the case of plasma etching of aluminium and its alloys, the highly reactive aluminium trichloride in the etch gas can cause photoresist failure, poor selectivity towards oxide, particle generation in the reactor and corrosion of the etched aluminium pattern and corrosion of the metal leads after packaging due to absorbed residual $AlCl_3$ (134). Particle contamination associated with plasma etching processes and equipment is another problem area (135). Poor etch ratios of the films to be etched over the substrates not to be attacked to an appreciable extent are a general problem in dry etching, in contrast to wet etching. In its most sophisticated forms, the much more mature wet-etching technology may be a viable alternative with fewer problems for situations where dry etching fails.

Wet-etching applications in the semiconductor industry

Preparation of semiconductor wafers

Semiconductor wafers are prepared by the sequence of process steps of sawing of single-crystal boules, mechanical polishing, chemical polishing and chemical–mechanical polishing. Chemical polishing is used to remove the work damage from the preceding sawing and lapping operations (136–138). High-quality final surfaces are achieved by chemical–mechanical polishing (139–146). Wet etching is also extensively used to reclaim semiconductor wafers (145) which were used as process monitors, or which did not produce satisfactory yields.

Semiconductor wafer processing

Wafer processing to prepare silicon integrated circuits involves hundreds of steps, and typically includes more than twenty wet-chemical etching steps. Wet etches are used to pattern and to strip oxides and nitrides, to pattern metal, to develop photoresists, and to pattern and strip organic layers. Wet etches are also used as anisotropic etchants in special devices, and to pattern polycrystalline silicon in silicon-gate MOS devices.

The number of etching steps depends upon the type of device being fabricated and the number of photomasks used for the process. Wafer processing techniques for silicon devices (53, 59, 93, 103, 111, 147–164), and to a lesser extent gallium arsenide devices (103, 165-169), have been reviewed in detail in a number of publications. An advanced 2-μm geometry CMOS wafer processing line, using positive resists and dry processing for each patterning step, will use more than ten wet-chemical etch steps. Wet processes are also extensively used in preparation of various types of silicon discrete devices and integrated circuits, and in fabrication of devices based on gallium arsenide and other compound semiconductors.

In dielectric-isolated integrated circuits, the active device regions are completely surrounded by a layer of SiO_2. To accomplish the desired geometry, (100)-plane Si wafers are used, and a crystallographically preferential anisotropic etchant is used to achieve (111) etched planes at the edges of isolated single-crystal regions (170). These (111) planes intersect the (100) surface at an angle of 55°. Dielectric-isolated integrated circuits are used for radiation-hardened bipolar integrated circuits, and for specialized high-voltage applications.

Wet-chemical etchants are used to strip various doped silicon dioxide films which served as diffusion or ion implant masks for boron, phosphorus, arsenic or antimony, and to develop positive photoresists, which are widely used for high-resolution ($< 3\ \mu m$) applications.

Wet etching is extensively applied to pattern oxides and metals in fabrication of semiconductor devices. The patterned structures must have high resolution and very low density of defects. An additional requirement of considerable importance is the contour of the edge of the patterned films. Edge contours which are too steep can result in problems in the continuity and edge coverage of subsequently deposited layers, and thus can have large effects on yield and/or reliability of completed devices (47, 51). Dissolution processes are used to remove certain forms of contamination from surfaces.

Etching of organic materials

Several dissolution processes, while not strictly wet etching, are of considerable importance in integrated circuit fabrication technology. One of the most important of the dissolution processes is the development of negative photoresists where unpolymerized resist areas are dissolved in a liquid developer. Development of negative photoresists involves simple dissolution of a cyclized rubber matrix in a non-polar organic solvent such as toluene or xylene.

Polyimides have been used as an inter-aluminium dielectric in multilevel metallized integrated circuits (see Chapter 12). In this application, wet-chemical etching of partly polymerized films is employed to pattern films (171). Some integrated circuits employ a layer of polyimide as the passivation layer, either singly or in combination with an inorganic layer such as a phosphosilicate glass. Wet etchants are used to remove the polyimide over bonding pad areas of the aluminium metallization.

Analytical applications

Wet etching is a very important tool in assessing semiconductor crystal quality, both in starting single-crystal semiconductor wafers and in wafers processed through the various steps in fabrication of completed integrated circuit wafers and devices (34, 35, 54, 65). It provides a convenient, although generally destructive, technique for determining dislocations, stacking faults, surface work damage and other crystallographic defects in single-crystal semiconductor wafers. Wet etching also enables delineation of microprecipitates and of metallurgical and/or electronic junctions in semiconductor device structures, and is frequently used to determine junction depth and lateral junction spreading.

Wet etchants are widely used to determine the factors which limit integrated circuit yield or performance, or to ascertain the fundamental physical mechanisms of failure of devices. They can reveal structural defects in dielectric layers, diffusion pipes in bipolar integrated circuits, and localized crystal dislocations in MOS integrated circuits. Wet etchants have been widely used to study solid-state aluminium–silicon interaction in contact areas during sintering of aluminium metallization (formation of etch pits in the silicon, particularly at the edges of contact cuts).

Semiconductor device packaging

Wet etching is extensively utilized in the preparation of packages for integrated circuits (172–174). Etching may be used to remove oxides from metal surfaces, to polish metal, to pattern lead frames, or to remove moulding flash from epoxy-encapsulated integrated circuits. The etchants for these purposes are similar to those employed elsewhere in the electronics industry, and include acids for removal of oxides and for patterning metal.

Chemistry of etching

Electrochemistry of etching

The electrochemistry of etching of germanium, silicon and other semiconductors has been extensively studied (15, 23, 30, 46, 77, 84, 88, 104, 120). Electrochemical techniques were widely used in preparation of early semiconductor devices, but are no longer commonly used in high-volume fabrication of integrated circuits and other microelectronic devices. The principal applications of electrochemistry in present-day solid-state technology are for characterizing semiconductor materials, delineating *p–n* junctions, and minimizing corrosion effects (87) during device fabrication and during usage in equipment. Electrochemical principles are important in understanding processes for etching of semiconductors (15, 84) and metallization systems (86), and for designing products which are less susceptible to corrosion (87).

Electrochemistry plays an important role in understanding the results obtained when semiconductor structures containing regions of several doping types and concentrations are etched. For example, when semiconductors with exposed *p–n* junctions are etched under illumination, the photons create hole–electron pairs in the semiconductor and thus generate photocurrents which cause anodic etching of some regions, while cathodically protecting others.

Kinetics and diffusion processes

The rate at which a solid surface is chemically etched by a liquid can depend on a number of factors, including concentration of reactants in the liquid, viscosity of the liquid, temperature, rate of agitation of the system, and solubility of the reaction products. Etching processes may be diffusion-limited if the rate is limited by arrival of reactants (or removal of reaction products) at the solid–liquid interface, or may be activation-limited if the rate is determined by the chemical reactivity of the species involved. Both diffusion-limited and activation-limited etching processes are widely used in the semiconductor industry.

In diffusion-limited processes, the rate of etching can be influenced by agitation and by the viscosity of the liquid etchant system. In activation-limited processes, doping type and level, crystallographic orientation of the substrate and temperature are critical variables.

In etching of semiconductor materials, photons can play an important role in influencing the rate of etching. Photons can generate hole–electron pairs in the semiconductor material; the available holes can then increase the etch rate of *n*-type semiconductor materials.

Kinematic phenomena

Three aspects of kinematic phenomena have been discussed in connection with wet-chemical etching of solid surfaces: effects of motion of the liquid phase relative to the solid surface, the effects of crystallographic orientation on motion of the liquid–solid interface as single-crystal materials are etched, and the lateral motion of atomic or molecular steps along the surface when a single-crystal material is etched.

The effect of motion of the etchant, due to agitation, gas evolution or other means of increasing solution flow along the surface being etched, depends on the type of reaction and the rate-limiting step. Agitation increases the rate of diffusion-limited processes, but can decrease the rate of an exothermic activation-limited process. Very high solution velocity is attained in jet etching systems and in chemical–mechanical polishing systems.

When crystalline materials are etched, there is a tendency for certain crystallographic planes to etch more rapidly than other crystallographic planes. In activation-limited etchants, etch rates of single-crystal materials can vary by several orders of magnitude depending on the direction of etching. Orientation effects can be amplified by adsorption of impurities on surfaces, since adsorption may be specific to certain orientations. It has also been shown that work-damaged surfaces of single crystals etch faster than undamaged portions of the same crystal. The mechanisms for formation of etch pits when crystals are etched may be considered in terms of kinematics (95).

Semiconductor wafers generally have surfaces which are nominally parallel to a major set of crystallographic planes. The plane of the surface of such wafers normally intersects the crystallographic planes, such as (100), at a small angle, and thus the surface consists of a series of atomic (or molecular) steps across the surface. These atomic steps, and associated kinks, are sites of enhanced reactivity. Typical silicon wafers have the surface plane within $\pm 1°$ of a major crystallographic plane, (100) or (111). Some (111) wafers are prepared off-orientation by 2.5° (±0.5°) to improve epitaxial layer quality. The kinematic theory of etching (18, 25, 95) considers crystal units to be removed from the edge of each step at the same rate. Thus the surface retains its slope, but the steps move across the surface and the overall etch rate normal to the surface is determined by the velocity of motion of the steps.

Adsorption, desorption and complexing reactions

In wet-chemical etching of a solid, the reactant (or reactants) must approach the surface, become adsorbed, and react chemically; the product (or products) of the reaction must be desorbed and move away from the surface. The overall rate of the etching process is controlled by one or more steps of this sequence. Adsorption may be either chemical or physical. Generally, the effect of adsorbed impurities on a surface is to reduce the etch rate.

In some cases, constituents of the etchant solution form complexes with the material being etched, thus increasing the solubility of the product of the etching reaction and increasing the etch rate of surface-limited reactions. Examples of complexing agents include fluoride, which forms the fluosilicate ion with silicon (SiF_6^{2-}); iodide, which forms the auroiodide ion with gold (AuI_2^-); and ethylenediamine tetra-acetic acid salts, which form EDTA complexes with a number of metals.

In systems containing two metals (or a metal plus a semiconductor), use of a complexing agent can change the relative locations of the two metals in the electromotive series, and thus increase the relative etch rate of the metal that is more strongly complexed.

Purity of etchants and diluents

The harmful effects of chemical impurities, such as heavy metals, on the yield and performance of semiconductor devices have been recognized since the very beginning of the microelectronics era in the 1950s. Processing chemicals have been the major source of chemical contaminants, and great efforts have been made over the years to improve the purity of chemicals. Twenty years ago analytical-grade chemicals that contained impurities in the parts-per-million range were considered reasonably pure. Today one requires and uses chemicals that contain dissolved contaminants in the parts-per-billion range, and in the case of processing water, parts-per-trillion.

Semiconductor-grade or electronic-grade chemicals, which have higher purity than, for example, analytical reagent-grade chemicals, are used at present for most integrated circuit fabrication. MOS-grade low-sodium chemicals represent the next highest purity level and are frequently used in especially critical operations. Ultrahigh-purity VLSI grade wet chemicals are now available, but at a premium cost, and are being evaluated for their cost-effectiveness in terms of device yield improvement (175). The Semiconductor

Equipment and Materials Institute, Inc. (SEMI) has published standards (108) on high-purity chemicals.

In the semiconductor industry, both proprietary etchants and etch formulations based on mixtures of electronic-grade chemicals are used. Device manufacturers generally prefer to know the exact composition of their etchants, and to be able to obtain the chemical ingredients from a number of sources. On the other hand, proprietary etchants may consist of compositions, sometimes patented, which have advantages for some applications.

Glass, the conventional and time-honoured container material, is no longer adequate for superclean liquid reagents because impurities from the glass are continuously being leached. PTFE (Teflon) followed by certain grades of polypropylene/polyethylene are claimed to be superior to glass as container materials (175). Piping used to distribute ultrapure processing water should be made of polyvinylidene fluoride (PVDF) or perfluoro-alkoxy (PFA) rather than polyvinyl choride (PVC) whose additives tend to contaminate the water (176).

So far we have discussed aspects of dissolved impurities which, in general, are now under good control. Of much greater concern are particle contaminants present in all processing chemicals, including the high-purity deionized water used for etchant preparation and rinsing. With the advent of VLSI technology, which requires linewidths of $2\,\mu m$ and is approaching $1\,\mu m$, particles are of major concern because they can mask etching, oxidation, metallization, and ion implantation. Conductive particles between interconnect lines can cause electrical shorts. Particles of sizes of 10–25% of the minimum linewidth are considered potentially failure-producing and should be reduced to a front surface density of less than $0.01/cm^2$ for particles of $1\,\mu m$ size (177). Particles in this submicron range are especially difficult to detect, monitor, remove, and prevent from depositing on device surfaces. Semiconductor-grade liquid chemicals contain between 10^2 and 10^5 particles greater than $1\,\mu m$ per millilitre (178), and much larger numbers of smaller-sized particles. Once deposited on a surface, particles are more difficult to remove as their size decreases because of rapidly increasing adhesion force. Consequently, the removal of particles from processing chemicals and deionized water by ultrafiltration through chemically-resistant filters has become an important technology (179). Manufacturers of electronic chemicals now supply ultralow-particle/high-purity etchants, solvents, photoresists, and resist stripping agents that are claimed to have 10^3 to 10^4 times fewer micron-size particles than conventional semiconductor-grade chemicals (175). Finally, point-of-use filtration of all liquid chemicals, as well as deionized water (180) and all process gases (181) through 0.1 to $0.2\,\mu m$ chemically-resistant final filters is absolutely essential if particles generated from containers, piping, pumps, and during transfer and handling are to be removed (179, 182). Very substantial yield improvements of VLSI devices have been achieved by implementation of these purification techniques (178).

Deposition of contaminants

The deposition of impurities on the surfaces of electronic materials is a major problem in microelectronic technology, and its prevention and control are imperative for attaining high yields and reliable performance of integrated circuits and other microelectronic devices. Contamination can occur before, during, and after wet-chemical etching operations, and originates from many sources. Contaminants are either dissolved chemical impurities or particle materials. Chemically adsorbed ionic species have reacted with the surface and are tightly bound, and hence can be removed only by chemical agents that react and solubilize them. Atomic or elemental contaminants, mainly heavy metals on semiconductor surfaces, also require chemical agents for their removal. Certain chemisorbed species, such as alkali ions and metal deposits, are the most dangerous to solid-state devices. Extensive studies have been reported on the types and concentrations of contaminants and reagent components deposited on semiconductor and oxide surfaces from etchant solutions under various processing conditions (34, 183–188). For future ULSI circuits, control of metallic contaminants to required levels of less than $10^{10}\,cm^{-2}$ in the device surface region have been estimated, and concentrations of specific metallics in the silicon bulk should not exceed 0.01 ppba (177).

Particles are critically important for two reasons: (i) they may block or otherwise interfere mechanically with the device structure during or after processing, and (ii) they may constitute a source of chemical contaminants, depending on their size and composition.

Types of etching processes

Isotropic and anisotropic etching. Vitreous, amorphous, and polycrystalline materials of uniform composition etch isotropically or non-preferentially at equal rate in all directions since they lack long-range order. Single-crystal materials can etch isotropically or anisotropically, depending on the etchant composition and the reaction kinetics. Isotropic etchants are selected if the crystalline material is to be polished to a smooth, structureless surface. Diffusion-limited etchants and those which etch at a high rate and increase the temperature locally assist this process because they tend to etch uniformly in all directions. Anisotropic or preferential etchants are reaction-rate limited and generally etch at low rates that depend on the exposed crystallographic plane of the crystal. They are selected if the object is to delineate crystal planes by chemical shaping and milling to create grooves or sloped faces, or to separate integrated circuit chips from wafers. Another use of anisotropic etchants is to detect and delineate crystal defects for analytical purposes.

Selective etching. In a technical context selectivity refers to the differences in etch rate between materials of different composition or structure, such as density. Selectivity is one of the most important factors in applied etching because the material to be etched is almost always part of a structure that consists of several other materials that must not be attacked to an appreciable extent.

Complete selectivity is usually most desirable, as for instance in pattern etching where neither the photoresist masking layer nor the substrate beneath the film to be etched are to be attacked. Partial selectivity is required for some applications. For example, controlled partial etching selectivity of dielectric layer composites is important in taper etching where a desired edge contour can be attained on the basis of etch-rate ratios of the component layers. In this case, a faster etching dielectric 'taper-control' layer is formed over the dielectric to be beveled (73).

Examples of dependence of selectivity on material composition are the etch rate of silicon in hot KOH–H_2O–isopropanol, which is strongly decreasing with increasing boron concentration in the range of 10^{18} to 10^{19} B cm^{-3} (82), and the buffered-HF etch rate of boron-containing glasses, which decreases with increasing boron content and is useful in analytical applications (73).

Electrochemical and electrolytic etching. All wet-chemical etching reactions of semiconductors are fundamentally electrochemical in their mechanism and typically involve oxidation/reduction followed by dissolution of the oxidation products. Oxidation of semiconductor atoms occurs at the microscopic anodic sites on the surface, while the oxidant is reduced at microscopic cathodic sites. In a more limited context, electrochemical etching is termed electrolytic etching or electro-etching if it involves the application of a controlled external electrode potential where the semiconductor or metal substrate is made the anode of an electrolytic cell. The rate of etching depends mainly on the applied electrode potential, the current density, the electrolyte solution, and the temperature of the system.

Silicon can be selectively electro-etched at rates that depend on the electron donor and acceptor concentrations. The process is therefore especially well suited for the preparation of thin single-crystal films of silicon from epitaxial substrates. *P*-type and heavily doped *n*-type silicon can be dissolved anodically at low potentials in dilute HF solutions, while *n*-type silicon is not attacked, so that etching stops at the interface. Selective electrolytic thinning of n^+-type silicon can also be achieved in alkaline electrolyte solutions where the etch rate is strongly dependent on the applied electrode potential. Electropolishing of silicon in suitable electrolyte solutions can produce highly polished surfaces. The same basic processes can be applied to compound semiconductors and metals. In fact, electro-etching and electropolishing of metals are by far the most important electrolytic etching processes used industrially (82).

A process which has recently been explored for applications to solid-state device technology is photoelectrochemical wet etching, or photo-etching (107,

114, 119, 189). Here a reverse-biased semiconductor electrode in a corrosive, oxidizing, and acidic electrolyte solution is optically excited by illumination with a laser light source that produces photons above the bandgap of the semiconductor. Etching proceeds by oxidation of the surface and dissolution of the oxide formed.

Analytical etching. Wet-chemical etching for analytical applications requires highly selective etchants. The composition of a material can be readily determined if the etch rate of a material can be correlated with the material component that is to be analysed. Defects in films of amorphous materials can be detected by exposure of the structure to an etchant that selectively etches the film's substrate, thereby creating a readily detectable decoration effect. Crystallographic etchants are important for delineating defects in single-crystal materials. Demarcation of a material with abrupt changes in composition, such as a semiconductor *p–n* junction, can be achieved by staining with an etchant that differentially plates out metal on part of the substrate structure.

Other types of etching processes. Cleaning and surface decontamination (other than simple physical dissolution) are special types of etching processes. An interesting process that is being explored for applications in solid-state technology is wet-chemical photo-etching by use of laser illumination (107, 114, 119, 189) and is usually used in combination with electrochemical etching. Deep-ultraviolet-induced electrodeless wet-chemical rapid etching using laser light has been used to form via holes in GaAs and semi-insulating substrates without masks in aqueous etchants at room temperature (190).

Technology of etching

Immersion etching

Immersion of the object to be etched in the etchant liquid is the simplest technique for wet-chemical etching. Silicon wafers are loaded in carriers of plastic or fused silica and submerged as a unit in the etchant. To avoid excessive depletion of the reactant, a relatively large ratio of etchant to material being etched should be employed so that the etch rate remains reasonably constant and controllable. Heating usually increases the etch rate by increasing the reactivity of the system, but requires good control if a constant etching rate is to be maintained. Agitation is necessary to constantly transport fresh etchant to and spent etchant from the wafer surface, to prevent local overheating in exothermic reactions, to release gas bubbles that may form as an etch product from the wafer surface, and to maintain an adequately constant etchant composition so as to ensure uniform etching of the entire wafer batch. Agitation can be implemented by mechanical stirring, by moving the wafer holder, or by bubbling an inert gas through the liquid. Ultrasonics can also be used for this purpose and for improving the uniformity of etching.

Addition of an etchant-resistant surface-active agent can often improve the uniformity of etching by enhancing the wetting properties and by preventing gas bubbles from clinging to the substrate surface and thereby decreasing local etch rates. Surface-active agents are likely to affect the etch rate, usually increasing it. Fluorinated surfactants are especially resistant to strong acids and alkalis used in many etchants, and are also extremely effective even at concentrations as low as 0.001–0.1%. The presence or absence of illumination during etching of patterned semiconductor device wafers can strongly affect the etch rate in localized areas of differing potentials due to the effects of photon injection. Adjacent areas of strong *p*- and *n*-type doping are especially prone to this type of reaction.

Commercial machines for immersion etching usually comprise a corrosion-resistant etch tank, a rinse station, and associated electronic controls for automated processing (66, 106).

Spray etching

Spray etching has many advantages over the immersion technique. The etch rate is increased because fresh etchant is constantly supplied to the substrate surface while the reaction products are immediately removed. The result is superior process control and improved uniformity. The etch rate can be regulated by the spray pressure and the droplet size. If a sufficiently high pressure is applied to force the impinging spray on to the substrate surface, a limited degree of directionality in etching a pattern-masked surface can be attained, resulting in steeper feature walls than those obtained by immersion etching of amorphous substrates. The economics of the process are improved because smaller quantities of etch liquid are needed (106).

Spray etch machines for the electronics industry have been considerably refined in recent years. Some systems are designed for single-wafer processing with *in-situ* end-point monitoring so that adjustments in the etch time can be made individually for each wafer. Large production processing machines for in-line processing of wafers have been in use for a number of years. Most of these machines have been designed for spray etching aluminium alloys used for integrated circuit metallization interconnects or for spray etching to develop photoresist patterns.

Chemomechanical etching

The final step in the preparation of semiconductor wafers that have been processed to meet the critical tolerances of flatness, taper, parallelism, and bow by grinding, lapping or chemical etching operations is the attainment of pit-free, haze-free, and flat surfaces of polished appearance by chemical–

mechanical polishing (136–146). Specially-designed polymeric pad materials are used with polishing slurries in machines for single-sided or, more recently, double-sided polishing of one or many wafers at a time. A great variety of machines is commercially available (142, 143, 145).

Chemomechanical polishing of silicon wafers is achieved under optimal pressure by use of a slurry of colloidal silica of small particle size in an alkaline water solution of optimally adjusted pH and temperature that results in simultaneous mild abrasion coupled with slight chemical etching. The exact mechanism of silicon etching in these reactions is not clearly understood, but there is evidence that Si–H groups are generated on the surface of polished silicon (144).

Electrolytic etching

Electroetching is done in an electrolyte solution by application of an external EMF potential to an electrically conductive substrate and a suitable counter-electrode, usually the cathode of the electrolytic cell. The rate and selectivity of etching can be controlled by the potential and/or the current density. Electrolytic etching is a powerful technique for selective removal of layers on a semiconductor and for fabricating membranes of single-crystal semiconductors. Electrolytic etching played an important role in early semiconductor technology but is now rarely used in integrated circuit fabrication, mainly because of the technical difficulty of attaching insulated electrodes to semiconductor device wafers in large-scale batch operations. Literature references were listed and detailed operating conditions described in our previous review (82).

Etching for pattern delineation

In contrast to etching the entire substrate, etching for pattern transfer requires a suitable masking material with which the substrate is coated and processed into a mask pattern by photolithography. The fabrication of integrated circuits depends on controlled deposition and selective removal of various films of semiconductors, insulators, and conductors. Microlithographic techniques are employed to create the desired patterns at each level, of which ten or more may be required in a typical integrated circuit fabrication sequence that comprises about 100 steps.

The basic microlithography process, the resist materials, the photomasks, the exposure techniques (including electron-beam and x-ray sources), and the equipment have been refined over the years to a high degree of sophistication (33, 36, 53, 99, 105, 112, 191–197) (see Chapters 9 and 10).

Etching for pattern delineation in high-resolution VLSI fabrication is one area where wet-chemical etching has been largely replaced by dry-etching techniques due to several shortcomings. The undercutting effect and resulting line thinning with sloped edges inherent in isotropic wet etching severely limits control of feature size. In fact, the line disappears when the linewidth is twice the thickness of the film being etched. The situation is further aggravated by the lack of perfect adhesion of photoresist films, combined with the capillary action of the etchant and mechanical stresses at the resist–film interface leading to undercutting. The adhesion problem can be eliminated by using a thin mask of photolithographically delineated metal or dielectric material, which can more effectively withstand the corrosive etching process. However, the isotropic etching effect of the substrate material cannot be eliminated. Clearly, dry etching for generating precise micron and submicron patterns is generally superior to wet etching. But there are exceptions even here: chromium-coated glass plates with chromium films of 60–90 nm in thickness for the fabrication of photomasks are wet-etched with better results (fewer defects) than is possible by dry etching (198).

Contour etching

Contour etching is used in some applications where the device topography needs to be shaped to attain gradual slopes or tapers to improve the step coverage of a subsequently deposited layer. Several techniques can be used to achieve tapering (47, 51, 73). In the usual isotropic wet-chemical pattern etching of an amorphous material that is patterned with a well-adherent etch-resistant mask, the resulting wall profile is determined primarily by the etching mechanism. If the rate-limiting step is activation-controlled, the resulting profile will curve cylindrically towards the bottom but will exhibit a near-vertical slope at the top edge because the etching species will dissolve the material equally in all directions from any point along the bottom edge of the mask (199). If etching continues after delineation, the wall profile steepens rapidly, increasing from the initial slope, averaging 45°, to 75° at double the break-through time. The expected slope angle can be calculated on a fixed centre of curvature axis on the assumption of an ideal model (200).

A different approach to contour taper etching is based on etch-rate differences of the component layers of a structure. A faster-etching 'taper-control' layer is formed over the material to be beveled (47). A well-adherent etch mask must be deposited over the taper-control layer, such as a photoresist polymer coating with hexamethyl-disilazane adhesion promoter. In the case of thermal SiO_2 tapering, the taper-control layer can be a CVD SiO_2 film (201), a spun-on oxide layer (202), a doped oxide (203), or a plasma-exposed oxide (204). Alternatively, ion-implantation-damaged SiO_2 (205), PSG (206) or Si_3N_4 (207) that exhibit enhanced etch rates can be used as the fast-etching control layer.

Tapered aluminium patterns can be formed by using a control layer of CVD-SiO_2 (47) or PSG (208, 209) or anodic Al_2O_3 and an aluminium etchant containing HF or NH_4F (210). Controlled photoresist interface undercutting (selective lateral attack at the photopolymer/dielectric interface) results in a decreased slope angle of the layer undergoing delineation, but this type of process has generally proved to be difficult to control in production.

Lift-off techniques

Lift-off processes (211–219) have been used to achieve high-resolution metal electrode patterns on GaAs microwave devices (166, 168), and in other applications in which narrow lines are required. In this type of process, lateral dissolution of an underlying patterned film is used to lift off portions of a deposited metal layer. While there are many variations in the lift-off processing sequence, the technique is based on a patterned sacrificial layer, typically an organic resist or deposited dielectric, in which the defined edge is re-entrant with an overhanging lip (Figure 11.1).

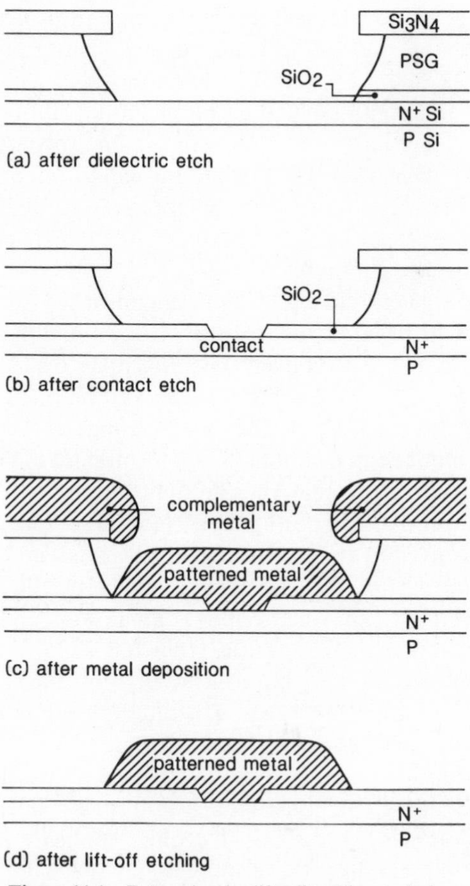

Figure 11.1 Patterning by lift-off etching technique

Surface decontamination and wafer-cleaning processes

Semiconductor device wafers before and after the various processing steps are invariably contaminated with impurities that must be removed to achieve high quality and good yield of the integrated circuit devices for which they are intended. The importance of very clean substrate surfaces in the fabrication of solid-state microelectronic devices in general cannot be overemphasized. Effective techniques for (i) avoiding deposition on, and (ii) eliminating impurities from critical surfaces are essential, since about 50% of yield losses in integrated circuit fabrication can be ascribed to contamination.

Contaminants take many forms and originate from a host of sources. However, metals, metal compounds, and plastic polymeric materials from process equipment are now the major source of contamination at the present state of sophistication in semiconductor wafer processing.

Inorganic particles and debris from organic coatings that cling to the surface of wafers are generally removed by strictly physical cleaning techniques that include mechanical brush scrubbing, high-pressure fluid scrubbing, blowing with a compressed gas jet, and immersion in a liquid using sonic techniques (103, 122, 220–222). Hydrocarbon films, such as residues left from waxes, many oils, and contaminated gases (including laboratory air) can be dissolved by immersing in or refluxing with organic solvents. Oxidation by ultraviolet radiation which generates ozone and causes bond scission can also be effective (220, 221, 223). Ionic contaminants that are lightly adsorbed or trapped in depressions can be desorbed to a large extent by rinsing in hot deionized water for a few minutes.

The removal of strongly adherent films, chemisorbed contaminants, and heavy metals that are plated out on the semiconductor surface requires chemical surface treatments to convert the impurity into a soluble compound which can be dissolved by the etching medium and retained in soluble form. The most effective and proven procedure for cleaning bare or oxidized silicon wafers consists of immersion in a hot (75–85° C) mixture of H_2O_2–NH_4OH–H_2O (1:1:5). Films of organic materials and photoresist residues are thereby oxidized to soluble products, exposing the surface. Trace metals such as nickel and copper are dissolved by the powerful oxidizing agent and kept in solution through complex formation. The second step in this procedure is immersion in hot H_2O_2–HCl–H_2O (1:1:6) which removes alkali ions and any remaining trace metals and other residual impurities (186–188, 222, 224). The procedure can also be carried out by spraying in a wafer centrifuge (225, 226) or by immersion in a megasonic tank that generates ultrahigh-frequency sonic waves which remove particles as well. In megasonic cleaning, lower temperatures and higher dilutions are adequate to perform an excellent job and, in addition to wafer cleaning, can be used for cleaning glass plates and other nonmetallic parts. Commercial systems for this application are available (227, 228); applications details of the techniques have been described in several papers (188, 222, 224, 226, 227).

Another widely used wet-chemical cleaning procedure for silicon, especially if heavy metal contaminants are present, is treatment in a mixture of HNO_3–HF–$HClO_4$–CH_3CO_2H (100:14:10:10), which removes a small amount of the silicon surface (229). Oxidized silicon wafers can be effectively cleaned by etching off a thin surface layer of about 5 nm with HF–H_2O (1:50) for 45–60s (230). Surface conditions for silicon molecular-beam epitaxial growth have been improved by a sequential treatment of degreasing, HF etching, and regrowth of a new protective oxide film (231).

Similar considerations hold for the cleaning of compound semiconductor wafers. Organic contaminants on starting wafers can be effectively removed by rinses in strong organic solvents, such as hot trichloroethylene followed by acetone and then by methanol for 5–10 minutes each (103). Removal of a surface layer by chemical etching is generally most effective for eliminating inorganic impurities. For example, GaAs can be treated briefly with HCl–H_2O_2–H_2O (1:10:80) at 70° C, which etches at the rather high rate of about $67\,nm\,s^{-1}$ (232). Chemical etching and cleaning procedures for silicon, germanium, and some III–V compound semiconductors have been reported for preparing very clean and smooth surfaces, mainly by use of bromine–methanol chemomechanical polishing (see also Chapter 3) (89).

Rinsing and drying treatments following etching and cleaning steps are very critical and must be carried out with great care to avoid recontamination. Wafers must never be allowed to dry between wet chemical steps. Deionized high-resistivity (e.g. 18 megohm at 23° C) water filtered through submicron filters is most commonly used with rinsing equipment designed for counter-current product flow (222, 224, 227).

The final step in wafer cleaning is drying. One approach is exposure of the wet wafers to high-purity isopropanol vapour in a reflux system to displace the water and allow the warm wafers to dry in a clean and controlled environment. Spin-dryers specifically designed for wafer processing are commonly used in production but are not ideal because moving parts generate particles. A preferred techniques is drying in a high-velocity forced stream of hot air that has been filtered through ultrahigh-efficiency point-of-use membrane gas filters (227).

Common systems for wet-etching electronic materials

Insulators and dielectrics

General aspects. Insulating and dielectric materials are essential to silicon and gallium arsenide integrated circuit technology, and to semiconductor microelectronics in general. They serve a large variety of purposes including insulation, isolation, passivation, planarization, diffusion, and topographic contouring. They may be used as impurity masks or barriers, impurity getters, capacitor elements, field and gate materials in MOS devices, dopant sources, and metallization protect coatings, primary passivation layers over *p–n* junctions, secondary passivation layers, and as substrate material such as sapphire for SOS integrated circuits. Except for substrates, insulators and dielectrics for integrated circuits are always used as films ranging from a few nanometres up to a few micrometres in thickness, but most of the layers are less than one micrometre thick.

The materials are silicon dioxide (SiO_2); binary silicate glasses including doped oxides, such as phosphosilicate glasses (PSG), borosilicate glasses (BSG), and arsenosilicate glasses (AsSG); ternary silicate glasses, such as borophosphosilicate glasses (BPSG) and germanophosphosilicate glasses (GPSG); silicon nitrides, comprising Si_3N_4 and plasma-deposited SiN_xH_y; and silicon oxynitrides, SiN_xO_y and plasma-deposited $SiN_xO_yH_z$. In addition, aluminium oxide (Al_2O_3), tantalum pentoxide (Ta_2O_5), and spin-on layers of organic-derived glasses and of polyimides are used in some applications. Silicon power devices often utilize relatively thick (multimicron) coatings of various multicomponent silicate glasses for passivation. Photovoltaic devices, such as silicon solar cells, may use titanium dioxide (TiO_2) or other high-refractive-index metal oxides as anti-reflection layers.

The vast majority of insulator and dielectric films are prepared by chemical vapour deposition (CVD), including plasma-enhanced or plasma-assisted chemical vapour deposition (PECVD, PACVD). Physical vapour deposition methods, such as magnetron sputtering, are employed less frequently. Radiolytic deposition processes, namely photo-CVD (PHCVD) and laser-CVD (LCVD), are still under development and are applied to experimental devices. Thermal growth is extensively used for forming SiO_2 films on silicon. Thermal nitridation of SiO_2 films is used to modify SiO_2 layers for some applications. Anodization by plasma reactions is useful for oxide growth on compound semiconductors. As already mentioned, mechanical deposition techniques, mainly by spin-on, are used for forming organo-glass mixtures and organic polymers, mostly polyimides. All of these films must be etched in the course of device processing, primarily for the purpose of patterning. The wet etchants for inorganic insulators and dielectrics used in integrated circuit fabrication are based on HF- or fluoride-containing solutions and H_3PO_4, with few exceptions.

Reviews and bibliographies pertaining to the etching of dielectric and insulator materials are included in references 24, 33, 36, 39, 42, 44, 47, 49, 51, 52, 55, 58, 72, 73, 78, 81, 82, 92, 96, 97, 98, 103, 109 and 125. The compositions of wet etchants frequently used for these materials are listed in Table 11.2.

Silicon oxides. Thermally grown SiO_2 layers are essential for passivation of *p–n* junctions in bipolar silicon integrated circuits and as gate and field dielectrics in MOS integrated circuits. Films of SiO_2 deposited by chemical and physical vapour deposition methods are used extensively, including films

Table 11.2 Etchants for dielectrics and insulators used in silicon microelectronic devices

Material	Form	HF–H_2O	BHF	H_3PO_4 (hot)	H_2SO_4 (hot)	NaOH (30%, hot)	Other liquid etchants (usually hot)
SiO_2	Amorphous or crystalline	H	H	L	0	L	
SiO	Amorphous	0	0	0	0		$HF–HNO_3$
							$NH_4F–KOH$, NH_4OH
Al_2O_3	Amorphous	H	H	M	0	0	$H_3PO_4–CrO_3–H_2O$
	Crystalline	0	0	M	0	0	V_2O_5 melt
	Sapphire	0	0	0	0	0	$H_3PO_4–H_2SO_4$; melts of $K_2S_2O_7$, $PbO–PbF_2$, and V_2O_5
Ta_2O_5	Amorphous	M	L				$KOH–H_2O_2–H_2O$
TiO_2	Amorphous	H	M	L	L		
	Crystalline	L	0	L	L		
AsSG	Amorphous	H	M				
BSG	Amorphous	H	L	L	0	L	
BPSG	Amorphous	H	H	L	0	L	
PSG	Amorphous	H	H	L	0	L	
Si_3N_4	Amorphous	L	L	M	0	0	$H_3PO_4–H_2SO_4$
SiN_xH_y	Amorphous	H	M	H	0	0	$H_3PO_4–H_2SO_4$
$SiO_xN_yH_z$	Amorphous	H	M	L	0	0	

Etch rates: H, high; M, medium; L, low; 0, nil.

formed by PECVD where low temperatures (280–380° C) are required. The quantitative (but not the fundamental) etch properties of SiO_2 films formed by various techniques vary greatly due to differences in structure and physical properties and variations in the exact stoichiometric composition. Wet-chemical etching of silicon dioxide (silica) is almost exclusively done with aqueous hydrofluoric acid solution (HF), without and with the addition of ammonium fluoride (NH_4F). The latter etchant is known as 'buffered HF,' or BHF. The exact chemical reaction mechanism and kinetics are very complex and are still not fully understood, even though several detailed studies have been conducted (44, 72, 97, 103, 199, 200, 223). Factors in the reaction include ionic strength, solution pH, etchant composition, and temperature. These factors determine the types and concentrations of available reactant species, such as HF_2^-, HF, F^-, H^+, and various fluoride polymers. It appears that the free acid is the major etching species in aqueous HF solutions, whereas the ionized fluoride associate HF_2^- is the major etchant species in buffered HF solutions (72, 233).

The mole fractions of fluoride as HF and HF_2^- may be calculated from known dissociation constants as a function of pH. For example, in a BHF solution containing 15 *M* total F^- ions, at pH 7 all of the fluoride exists as uncombined fluoride ion. With decreasing pH the HF_2^- concentration increases gradually and reaches a maximum at about pH 3.5, then declines with further pH decreases toward pH 0. For HF species, the concentration is insignificant at pH 7 and begins to increase gradually from pH 4 to pH 3, then increases rapidly as the pH is lowered toward pH 0. A commonly used BHF composition of 7:1 NH_4F (40%)–HF (49%) has a pH of about 4.5 and appears to contain virtually only HF_2^- and F^-, with very little HF. The SiO_2 etching rate for HF_2^- is four to five times as fast as that for HF. Raman spectroscopy of BHF solutions containing dissolved SiO_2 shows a very large number of absorption maxima, indicating a multiplicity of reaction products. Measurements of activation energies suggest that higher polymeric species of the type $H_xF_{x+1}^-$ are active in the etching process involving more concentrated solutions (72, 233).

Pattern delineation of SiO_2 films by wet etching must be carried out with BHF, usually 6:1 or 7:1 NH_4F (40%)–HF (49%), to minimize attack of the photoresist masking layer and the resist–dielectric interface, both of which would be damaged by strongly acidic HF solutions. Addition of NH_4F increases the pH to 3–5, maintains the concentration of fluoride ions, stabilizes the etching rate, and produces highly reactive HF_2^- ions as described above. It has been shown that NH_4F acts as a buffering agent only until the solubility limit of the reaction product $(NH_4)_2SiF_6$ is attained and then acts as a precipitating or complexing agent for the SiF_6^{2-} ion in H_2SiF_6 (234). When pattern-etching SiO_2 films over aluminium metallization, attack of the metal can be minimized by adding a dihydroxyalcohol (235), glycerol (236) or acetic acid (237) to the BHF etchant.

A typical composition consisting of NH_4F–HF–CH_3CO_2H (7.6:1:1) etches APCVD SiO_2 films deposited at 400° C from nitrogen-diluted SiH_4–O_2 (O_2:SiH_4 = 12.5) at a rate of 5.5 nm s^{-1}, practically independent of CVD conditions (deposition temperature, O_2:SiH_4 ratio, SiH_4 flow rate, total gas flow rate) (238). Good pattern definition of bias-sputtered SiO_2 for inter-aluminium insulation has been obtained with an etchant of ethylene diamine–NH_4F–HF (25:20:4), which at 24° C etches at a rate of 1.1 nm s^{-1} (239). An etchant for SiO_2 atop silicon or metallic silicides is comprised of 1–4% HF (52% by weight) in glycerol or other polyhydric alcohol plus a wetting agent. This etchant composition, which is substantially free of unbound water, is particularly suitable for removing SiO_2 from a structure where silicon is exposed and thus BHF would erode n or n^+ conductivity regions (240). The etch depth and the amount of lateral undercutting of SiO_2 (and Si_3N_4) films can be monitored during etching by use of a grating test pattern. A laser beam is aimed at the grating and the reflected first-order diffraction intensities are recorded and evaluated (241).

Polysiloxane polymer solutions, such as GR-650 by Owens-Illinois, are sometimes used as a spin-on planarizing material for integrated circuits. After the steps of spin-coating, thermal drying, and curing treatments, the polymerized polysiloxane resin is not attacked in BHF, but can be completely converted to a BHF-etchable SiO_2-like material similar to plasma-deposited SiO_2 by pyrolysis or by treatment in an oxygen plasma at 1 Torr without external heating. Plasma conversion also improves the adhesion of the spin-on glass to substrates (242). A typical 'conventional' spin-on organosilica coating, cured thermally at less than 200° C, has an etch rate in HF-H_2O (1:30) that is at least 200 times faster than that of thermally grown SiO_2 (243).

Several physical and chemical properties of SiO_2 films affect the etch rate to an appreciable extent. Low density, porosity, residual stress, structural defect density, hydrogen content (as in plasma-deposited films), and defects generated by electron-beam irradiation or ion bombardment all increase the etch rate. Heat treatments of SiO_2 films deposited at low temperatures decrease the etch rate due to densification of the structure, stress release, and possible healing of microstructural defects. Non-stoichiometric SiO_2 films with excess silicon lower the etch rates in HF and BHF solutions. The etch rates of SiO_2 films also vary depending on the various growth or deposition conditions. Films grown on silicon in oxygen at high temperatures (1000–1200° C) have among the lowest etch rates (close to fused silica) (73, 81, 82).

Normalized etch rates for various types of amorphous SiO_2 films have been published (244) as well as the effects of HF concentration and temperature (41, 233, 245–250) and of agitation (199, 247, 251) on etching of SiO_2. The etch rate of crystalline quartz plates in BHF and NH_4F solutions has been studied as a function of etchant composition and temperature (252). The etch rate of high-pressure thermal SiO_2 (500 atm O_2, 800° C) in BHF (9:1) has been found to be significantly lower (due to higher density) than that of atmospheric

1000° C-oxides (253). Additional etching mechanism studies for SiO_2 with a variety of HF-and fluoride-etchants (254, 255) and with deuterated acidic fluoride solutions (256) have been reported.

A comprehensive list of references up to 1978 on wet-chemical etching studies of variously prepared SiO_2 films has been presented in a previous review of etching (82). Additional more recent representative references are as follows: for films grown thermally (41, 72, 81, 233, 246, 249, 250, 253); formed by chemical vapour deposition at atmospheric pressure (APCVD) (81, 98, 238, 251–261) and at low pressure (LPCVD) (257, 258, 262); formed by plasma-enhanced chemical vapour deposition (PECVD) (239, 257, 261, 263–268) and by photochemical vapour deposition (PHCVD) (269–273); deposited by physical methods such as RF sputtering and electron-gun evaporation (38, 81, 239), by ionized cluster beam deposition (274); and by spin-on techniques of organic glass-forming solutions (242, 243); and finally of quartz plates (252).

In addition to HF-containing etchants, SiO_2 is slightly soluble in hot H_3PO_4 and in hot caustic solutions. For example, thermally grown SiO_2 is etched by 5 *M* KOH at 85° C at about 5.0 nm min^{-1} (275), and by 0.1 *M* NaOH at room temperature at 0.02 nm min^{-1} (276) or less. The etch rate in 10 wt% NaOH at 55° C is 0.5 nm min^{-1}, and at 90° C, 50 nm min^{-1} (277). Several unusual etchants for SiO_2 have been described (254).

The etch rate of oxygen-deficient SiO_2 in HF solutions decreases, and SiO requires the addition of HNO_3 to attain etchability. Alternatively, hot solutions of concentrated NH_4F mixed with NH_4OH or alkali hydroxides can be used for etching SiO films (33). Oxygen-doped polysilicon, SIPOS, a semi-insulating material used in silicon devices, can be etched in HF–H_2O_2–NH_4F (1:6:10) or, preferably with HF–H_2O–HNO_3 (1:20:50) (278).

Metal oxides. Various metal oxides of interest in electronic device technology include films of Al_2O_3, Ta_2O_5, TiO_2, ZrO_2, Nb_2O_5, HfO_2, GeO_2, V_2O_5, SnO_2 (actually a non-stoichiometric semiconductor), oxides grown (mostly anodically) from compound semiconductor substrates, and epitaxial $Gd_3Ga_5O_{12}$ (used in bubble domain devices). Bulk oxide materials include sapphire (α-Al_2O_3), spinel ($MgAl_2O_4$), and beryllia (BeO) used as insulating substrates. Most of these materials can be wet-chemically etched in strong etchants containing either HF, H_3PO_4, H_2SO_4, HCl, or caustics. For details and references our previous survey should be consulted (82).

The only metal oxides that are used to a limited extent in the production of silicon integrated circuits are Al_2O_3 and Ta_2O_5. Aluminium oxide films have been used as gate insulators in radiation-hard MOS devices. Wafers of sapphire (single-crystal α-Al_2O_3) are used as substrate for hetero-epitaxial silicon layers in the fabrication of SOS (silicon-on-sapphire) integrated circuits. Films of tantalum oxide (Ta_2O_5) have been used traditionally as excellent high-dielectric constant dielectric for forming thin-film capacitors.

Etchants for high-temperature deposited Al_2O_3 films usually contain H_3PO_4 and are used at elevated temperature. Films deposited at low temperature by various processes are less dense and, as a consequence, can frequently be etched in solutions containing HF, NH_4F, H_3PO_4, or mineral acids. Films of Al_2O_3 for which etch-rate data are available include those prepared by CVD at low temperatures (400–500° C) from organometallic reactants (244, 279–281), by plasma oxidation (282), by anodic growth (283), by anodizing-conversion of a polycrystalline film to an amorphous modification (284), by evaporation techniques (285), by sputtering (286), or from aluminium metal in boiling water (287). All of these low-temperature films can be etched in HF, BHF, hot H_3PO_4, and also etchants containing H_3PO_4. Anodically-grown Al_2O_3 on aluminium for multilevel integrated circuits can be selectively pattern-etched in a H_3PO_4–CrO_3–H_2O mixture (288). Densification at 700–800° C tends to transform the films to crystalline modifications with much lower rates of wet-chemical etching (279, 280). Al_2O_3 films deposited by hydrolysis of $AlCl_3$ at 900–1000° C, being very dense and polycrystalline, are nearly unetchable in HF and require boiling H_3PO_4 as the etchant (244, 289–290). Patterning can be accomplished with CVD SiO_2 as etch mask (291).

Sapphire, spinel, and beryllia are slowly etchable in boiling H_3PO_4–H_2SO_4 mixtures (292, 293). Removal of surface irregularities from crystalline Al_2O_3 can be achieved with melts of V_2O_5 above 800° C (294). Melts of $K_2S_2O_7$, PbO–PbF_2, and V_2O_5 have also been used for etching sapphire (295). Films of Ta_2O_5 from pyrolytic CVD at 500° C can be etched in dilute HF (296). Amorphous Ta_2O_5 grown by anodic oxidation is etchable in BHF (297, 298), whereas high-temperature crystalline films are not (299). Ta_2O_5 films can be pattern-etched with a gold mask in NaOH(30%)–H_2O_2 or KOH(30%)–H_2O_2 (9:1) at 90° C at a rate of 100–200 nm min^{-1} (300, 301). Electron irradiation of Ta_2O_5 and Al_2O_3 films decreases their etch rates (302), in contrast to SiO_2 (205), PSG (206), or Si_3N_4 (207).

Silicate glasses. Of the numerous silicate glasses that have been explored for applications in microelectronics technology, only two have attained major importance in the fabrication of integrated circuits: phosphosilicate glasses (PSG) and borophosphosilicate glasses (BPSG). Both materials are denser than SiO_2 and have lower intrinsic stress than deposited SiO_2 films and a favourably higher coefficient of thermal expansion. They are excellent dielectrics with the unique and extremely important capability of gettering and immobilizing sodium ions, thereby enhancing the device stability and reliability. PSG films have been used for many years as a secondary passivation protective coating over aluminium-metallized integrated circuits and as an interconductor insulator in multilevel VLSI devices. An additional important application for both glasses is the smoothing of the device topography by a viscous thermal fusion flow that provides gradually sloping steps, thereby ensuring continuity of the subsequently deposited metal interconnects. The advantage of BPSG over PSG is that fusion can be achieved at temperatures as low as 750–800° C rather than the 1050° C or higher necessary for typical PSG layers. Processing at reduced temperature is imperative for VLSI and VHSIC devices where lateral diffusion redistribution of dopants must be minimized to avoid detrimental effects on the shallow junctions of these complex circuits.

PSG and BPSG films can be prepared by a variety of processes, but are most frequently synthesized by CVD processes based on APCVD, LPCVD, and PECVD. Thermally formed PSG layers result by reacting vapours of $POCl_3$ (303) or P_2O_5 (304) with SiO_2, typically at 1000° C. Wet-chemical etching of PSG and BPSG films is invariably based on aqueous HF and HBF, but the etching properties are markedly different from those of SiO_2. Thermal densification decreases the etch rates of films deposited at low temperature, as is the case for other oxide and silicate glasses.

The etch rates of PSG films in all HF–H_2O mixtures and P-etch increase approximately logarithmically with weight percent phosphorus (73, 237). In typical BHF etchants the rate of etching is affected to a much lesser extent by the phosphorus concentration than with unbuffered solutions, but still varies logarithmically with the phosphorus content (73, 103, 237). The exact etch rates are influenced by the density and stress of the films, which are in turn the result of the CVD parameters (such as temperature and rate of deposition, and the chemical process used for CVD) and the thermal post-CVD treatments. If PSG films prepared by APCVD at 445° C from SiH_4–PH_3–O_2 in N_2 are heated at 800° C for 15 minutes in N_2 the etch rate in P-etch decreases substantially due to densification; this effect becomes smaller as one proceeds from 0% P (SiO_2) to higher phosphorus concentrations. At 14 wt% P there is no further change, which may be explained by self-densification (73, 237). The etch rates of as-deposited PSG films containing 0–9 mol% P_2O_5 prepared by APCVD at 350° C from SiH_4, PH_3, and O_2 in Ar exhibit a maximum when etched in 1:1 diluted BHF (HF–NH_4F–H_2O, 1:10:11) (305).

Several etch rate studies for PSG films have been reported: for films deposited by oxidation of the hydrides by APCVD at 300–550° C (98, 261, 306–310) and at 700–900° C (311); by LPCVD (262, 312); by PECVD based on hydride oxidation at 300–400° C (261, 313); by CVD of organometallics (314); by several LPCVD reactions (315); by photo-enhanced CVD (273, 316); and by thermal transformation of SiO_2 with $POCl_3$ or P_2O_5 at 1000–1100° C (248, 303, 304, 317, 318). The effect of various types of densification treatments of CVD PSG films on the etch rate have been reported in a systematic study (319). The phosphorus in PSG (as well as in BPSG) films deposited at low temperature by LPCVD and especially by PECVD may exist as P_2O_5, P_2O_3, PH_3, and a fourth yet unidentified species (320).

BSG films are etchable in HF and BHF. In non-buffered HF solution the etch rate increases with increasing B_2O_3 content, whereas in BHF it decreases sharply, reaches a minimum, and then increases at concentrations greater than 18–20 mol % B_2O_3 in typical BHF etchant compositions (305, 308, 321–325). Boron oxide at moderate concentrations in BSG appears to protect the silica network from attack by BHF; at concentrations greater than 18–20 mol % B_2O_3 bonding as a borosilicate may be impaired due to excess B_2O_3, and solubility in the water component of the etchant becomes a predominant factor, resulting in an increase in the etch rate (305, 308). Microphase separation could be another reason. Additional similar explanations of the etching mechanism have been proposed but not conclusively proven (72, 321, 322). Addition of HNO_3 to HF solutions, such as in P-etch (81), or to BHF (324) tends to render the etchants more selective towards boron, making them particularly suitable for analytical applications. Other etch-rate studies with BSG include characterization for diffusion sources (325) and of boron-implanted SiO_2 (326); effects of thermal densification (98, 308); use of low-acidic fluroboric acid etchant (327), and detailed investigations of BSG films prepared by various CVD processes (73, 98, 303–309, 311).

Etch-rate data have been reported for BPSG films synthesized from inorganic and organometallic reactants by APCVD (328–331), by PECVD (332, 333), by LPCVD (315, 334, 335) and formed from solution by spin-on techniques (336). As may be expected, BPSG films exhibit wet-etching properties of both BSG and PSG. In non-buffered HF solutions the etch rate increases with increasing boron and phosphorus contents. In BHF etchants the etch rate increases with increasing phosphorus at constant boron concentration, but decreases with increasing boron at fixed phosphorus concentration, as in BSG. Thermal densification lowers the etch rates as usual. Composition testing by etch rate analysis of densified BPSG film is useful if either the B or the P content is known to be constant. Patterning by wet-etching is readily achieved with BHF (6:1) (329, 331).

Chemically vapour-deposited germanophosphosilicate glasses (337–341) have been proposed as a possible substitute for PSG or BPSG. However, the water-soluble GeO_2 component in these ternary glasses tends to be leached out on exposure to water, and the etch rates in HF etchants are high because of the large percentage of GeO_2. Etch-rate data for germanophosphosilicate glasses have been reported in two of the above-noted papers (337, 339).

Among a number of other binary and ternary silicate glasses (73, 81, 82, 98, 342) only CVD arsenosilicates are of limited use in integrated circuit technology as a diffusion source for arsenic, and possibly as an insulator and passivation layer in multilevel VLSI silicon circuits (343, 344). Etch rates for AsSG films as-deposited at 450° C and after 800° C densification have been reported for various BHF mixtures, which can also be used for pattern etching (345). The BHF etch rates of films densified at 1100° C increase approximately linearly with increasing As_2O_3 concentration up to 5 mol % (305, 346). Etch rates of APCVD arsenosilicates and etch rate ratios of $AsSG/SiO_2$ in NH_4F solution have been compared with those for CF_4 plasma etching (analogous measurements were also made with BSG and PSG films) (309).

Multicomponent silicate glasses prepared by conventional glass technology are now used in integrated circuit fabrication to a limited extent only for lid sealing of ceramic packages. Devitrifying solder glasses are used for this application, which does not involve etching of any sort. Frits of high-purity electronic glasses of low melting temperatures are used as passivation material for silicon power devices where relatively thick layers of an insulator are required to protect the high-voltage *p–n* junctions. The etching properties of multicomponent silicate glasses have been extensively reviewed in several papers (24, 38, 42, 44, 58, 81, 82, 97) and will not be further discussed here.

Silicon nitrides and silicon oxynitrides. Thin films of silicon nitride, Si_3N_4, are important for semiconductor devices because of their effectiveness as an alkali diffusion barrier. Thermal (as opposed to plasma) Si_3N_4 films are usually prepared by thermally activated LPCVD from SiH_2Cl_2 or SiH_4 plus NH_3 at about 800° C; they are etchable at room temperature in HF or BHF, in H_3PO_4 at 140–200° C, in HF–HNO_3 (3:10) at 70° C, and in molten NaOH at 450° C (33, 39). The etch rates are strongly affected by oxygen linkages that may be incorporated in the film structure and increase the rates in HF and BHF with increasing oxygen content, while decreasing them in H_3PO_4. The rate law in acidic HF solutions has been identified and found to be linear in both [HF] and $[HF_2^-]$, but is independent of $[F^-]$ (347). The reaction equations for the two etchants can be summarized as follows:

$$Si_3N_4 + 18HF \rightarrow H_2SiF_6 + 2(NH_4)_2SiF_6$$

$$3Si_3N_4 + 27H_2O + 4H_3PO_4 \rightarrow 4(NH_4)_3PO_4 + 9H_2SiO_3$$

The rate of etching depends on the exact stoichiometry of the films, the substrate temperature during deposition, and the density of the films.

Patterning of thermal Si_3N_4 films by wet chemical etching is usually carried out by reflux boiling in H_3PO_4 at 180° C using CVD SiO_2 as an etch mask. The etch rates under these conditions are approximately 10 nm min^{-1} for CVD Si_3N_4, but only 0–2.5 nm min^{-1} for the CVD SiO_2 etch mask and 0.3 nm min^{-1} for any exposed single-crystal silicon; etch rate plots for all three materials as a function of temperature and H_3PO_4 concentration have been published (348). Boiling H_3PO_4 must be treated with special care because of its complex chemistry (348, 349). At elevated temperatures, HF of optimal concentration can etch composite layers of Si_3N_4 and SiO_2 at an equal rate, which is required for applications where the etched composite structure must have pattern walls of a uniform taper angle (246, 250). An HF solution of 0.25 wt% at 90° C etches thermally grown SiO_2 and typical CVD Si_3N_4 films at an equal rate of 7.0 nm min^{-1}, whereas an HF solution of 0.20 wt% at 90° C etches SiO_2 at 4.5 nm min^{-1} and Si_3N_4 at 6.0 nm min^{-1}, resulting in an etch rate ratio that is generally preferable for film taper etching (347, 350). The activation energies for Si_3N_4 and SiO_2 etching suggest (347) that equal etch rates can also be achieved by use of low-pH etchants that maximize the concentration of HF and minimize that of HF_2^- species. Mixtures of H_3PO_4 and fluoboric acid at controlled temperature have also been reported for this purpose (351). A mixture of H_2SO_4 and H_3PO_4 widely used for wet etching Si_3N_4 oxidizes the silicon surface and consequently prevents attack by the H_3PO_4; in addition, this etchant prevents selective plating on n^+ silicon regions and etches Si_3N_4 at the same rate as H_3PO_4 (352).

In contrast to thermal Si_3N_4, films of 'silicon nitride' deposited by plasma-enhanced CVD at low (300–400° C) temperature (from SiH_4 plus NH_3 and/or N_2) have much higher etch rates than stoichiometric high-temperature CVD Si_3N_4. The rates depend strongly upon the film composition, which can be expressed as $Si_xN_yH_z$ with major amounts of hydrogen (353). The material should really be regarded as a polysilazane with unique properties rather than as a variety of silicon nitride (354). Silicon nitride films deposited by sputtering may also have lower etch rates than high-temperature CVD-Si_3N_4, if they are less dense, contain large quantities of entrapped gases, or are non-stoichiometric. The hydrogen originates from the SiH_4 and the NH_3 and has a pronounced effect on the etch rates and the chemical, physical, electrical, mechanical and stress properties. The hydrogen content can vary from about 13 to 39 atomic percent, depending on deposition conditions, and generally decreases with increasing substrate temperature of deposition; LPCVD Si_3N_4 films, for comparison, contain about 4–5 at % of hydrogen (355). The etch rates of plasma silicon nitrides in 13:1 BHF at 25° C range from 1–500 nm min^{-1} for 13–39 at % hydrogen, respectively, for plasma-deposited nitride films prepared from SiH_4 and NH_3 at low temperature (355). In addition to the hydrogen content, the etch rates depend on the silicon to nitrogen ratio and the density of the films. The etching mechanism can be ascribed to the bond breaking by HF_2^- ions in BHF, resulting in reaction products such as H_2SiF_6 and $(NH_4)_2SiF_6$ (72, 356). Fast-etching regions of plasma–silicon nitride films have been reported over wet-chemically etched steps of aluminium, a phenomenon that is mainly due to localized stress (357). The fast-etching region in this case is prevalent at the upper edge of a sloped aluminium step. On more vertical steps resulting from dry etching, the fast-etching region of a dielectric overcoat is at the heel or bottom corner of the structure (358). Etch rates of boron- and phosphorus-doped plasma silicon nitride films in HF, BHF, and H_3PO_4 have also been reported (359).

Extensive reviews of the literature on etching of thermal silicon nitrides, plasma silicon nitrides, and silicon oxynitrides are available (39, 73, 82, 103, 244, 353, 360). In addition to the references already cited, more recent results of etching studies have been reported on thermal Si_3N_4 concerning etching in NH_4OH–H_2O_2–H_2O solutions during cleaning (188, 222, 361), and for etching of thermally nitrided silicon (362) and SiO_2 (362, 363). The etch rate of silicon oxynitrides, which are a combination of oxide and nitride of the general formula $Si_xO_yN_z$, depends on the stoichiometry, the deposition conditions, and the film density. For example, HF etch rates for thermal CVD $Si_xN_yO_z$ films (prepared from SiH_4, NH_3, and NO) range from 30–500 nm min^{-1}, as compared to 3000–5000 nm min^{-1} for SiO_2 and 13–15 nm min^{-1} for Si_3N_4; in H_3PO_4 at 180° C the etch rates are 1–10 nm min^{-1} for $Si_xN_yO_z$, 0.8–1.0 nm min^{-1} for SiO_2, and 6–10 nm min^{-1} for Si_3N_4 (364). Additional etch rate results of this type have been published (365).

Numerous papers have been published on wet-chemical etching studies of plasma silicon nitride. In addition to the combined etching reviews on nitrides noted above, reviews on etching specifically of plasma silicon nitride and oxynitride films are available (265, 268, 366–381). Etch rates of silicon nitride prepared by low-temperature photochemical vapour deposition have also been reported (382–384).

Etch rate studies of plasma silicon nitride films in HF and BHF as a function of various deposition parameters (such as substrate temperature, system pressure, RF power, gas flow rates and ratios, N_2 and O_2 additions) have been examined by many investigators. The following selected references are particularly informative or have been published recently: 98, 355–357, 359, 372–381.

Semiconductors

General considerations. Wet etching of semiconductors differs from wet etching of either dielectrics or metals in several important respects; these differences are attributable to two conductivity types, varying resistivity levels, and the presence of exposed *p–n* junctions. Electrochemical reactions at semiconductor electrodes, in contrast to those at metal electrodes, may involve valence-band electrons as well as conduction-band electrons. Moreover, most semiconductors are single-crystalline, and thus crystallographic orientation can significantly influence etch rates. The etching properties of semiconductors may also be influenced to a considerable extent by the presence of light during immersion into the etchant.

Etching of semiconductors requires that covalent bonds be broken in order to permit the dissolution of surface atoms. Accordingly, *p*-type semiconductors are more easily etched than *n*-type semiconductors. The holes necessary for etching of *n*-type semiconductors can arise from a chemical reaction which extracts electrons at an electrolyte–semiconductor interface, from illumination which generates hole–electron pairs, or from injection from an adjacent forward-biased *p*-type region. The photon energy required to generate hole–electron pairs depends on the optical energy gap of the semiconductor. Energy gaps (385) range from 0.18 eV for InSb to 1.1 eV for Si, 1.35 eV for GaAs, and 2.4 eV for CdS.

Etching of single-crystal semiconductors may be either isotropic or anisotropic. Isotropic etches are used to remove work damage and to produce a polished surface. A number of publications have described polishing etchants for silicon (4, 5, 15, 31, 56, 84, 103, 386–393) and for other semiconductors (8, 15, 21, 56, 59, 65, 82, 84, 103, 394–404). Anisotropic etchants are used to produce special device structures (84, 103, 115, 170, 405–416), and to assess the crystallographic quality of semiconductor materials. In the case of compound semiconductors, the crystal face representing one compound can be considerably more reactive. In gallium arsenide, for example, the (111)As face etches more rapidly than the (111)Ga face in several types of etchants.

Wet-chemical etching of silicon and of other semiconductor materials may involve either activation-rate-limited or diffusion-rate-limited processes. Chemical polishing to achieve smooth surfaces is performed by use of relatively concentrated solutions which are diffusion-limited systems; anisotropic etching, including etching to reveal crystallographic dislocations, is performed using solutions which are activation-limited.

Semiconductor silicon. Silicon is by far the most widely used semiconductor material. Most silicon devices are fabricated using single-crystal silicon with the surface plane of the silicon parallel to a major crystallographic plane, for example (100) or (111). Single-crystal silicon with (100) surface orientation is widely used in fabrication of both MOS and bipolar integrated circuits. Some MOS and bipolar circuits also use (111) orientation. Polycrystalline silicon and amorphous silicon are also used in a number of semiconductor devices (417–427), and thus the etching properties of each of these types of silicon, with various donor and acceptor levels, are of interest. Wet etchants for silicon typically include an oxidizing agent, such as nitric acid, plus a complexing agent, such as hydrofluoric acid to ensure that the etching products are soluble. In alkaline etchants, the oxidizing agent can be water, resulting in hydrogen evolution, or can be dissolved oxygen from the atmosphere. The product in either case is a soluble silicate.

Single-crystal silicon. Silicon is most commonly etched with either nitric acid/hydrofluoric acid etchants (17, 56, 82, 84, 103, 116, 386–393), or by alkaline etchants (428–441). Both types of systems have been extensively studied. The reaction for etching of silicon in a nitric acid/hydrofluoric acid etchant mixture may be written as

$$3Si + 4HNO_3 + 18HF \rightarrow 3H_2SiF_6 + 4NO + 8H_2O$$

Etching of silicon in an alkaline system may be represented by the equation

$$Si + 2KOH + H_2O \rightarrow K_2SiO_3 + 2H_2$$

Table 11.3 lists commonly used polishing etchants for silicon. An example of a

Table 11.3 Etchants for silicon and GaAs

Etchant	Semi-conductor	Composition	Comments	Reference
CP-4A	Si	3:5:3 $HF–HNO_3–CH_3COOH$	Polishing, 80 μm min^{-1}	103
Planar etch	Si	2:15:5 $HF–HNO_3–CH_3COOH$	Polishing, 5 μm min^{-1}	103
White etch	Si	1:3 $HF–HNO_3$	Polishing	17
Polysilicon etch	Poly-Si	3:50:20 $HF–HNO_3–H_2O$	0.8 μm min^{-1}	103
Alcohol-KOH	Si	50g KOH, 200g *n*-propanol, 800g H_2O	Anisotropic	431
Br_2-methanol	GaAs	1:9 $Br–CH_3OH$	Polishing	394
Peroxide-sulphuric	GaAs	5:1:1 $H_2SO_4–H_2O_2–H_2O$	Polishing	401

chemical polish etchant (82, 84, 103) for silicon is CP-4A, consisting of 3:5:3 $HF–HNO_3–CH_3COOH$. The etch rate of silicon (103) is about 80 μm min^{-1}.

The kinetics of etching of Si in HF, HNO_3 and H_2O have been studied (386) as a function of composition at 25° C. In the region of high HNO_3 compositions, etch rates are a function only of the hydrofluoric acid concentration; in the high-HF region, nitric acid concentration determines the etch rate. In the high-HNO_3 region and with small amounts of diluent, the reaction is diffusion-limited, with an activation energy (388) of 17 kJ mol^{-1}. In the high-HF region, the activation energy is about 46 kJ mol^{-1}.

The role of acetic acid as a diluent has been investigated in detail (387). A comparison of the temperature dependence of the etch rate of (111)-oriented *n*-type silicon in $HNO_3–HF–CH_3COOH$ with the temperature dependence of the viscosities of the etching solutions showed that the etching processes were diffusion-controlled reactions (389).

In some cases, and particularly with boron-doped wafers, a stain tends to form during wet-chemical etching. A typical etchant for chemical lapping of silicon wafers, 4:1:2 $HNO_3–HF–CH_3COOH$, was studied (391), and it was concluded that the stain was due to a suboxide of silicon. The stain is removed with 2:1 HF–0.038*M* potassium permanganate solution (391). In some device fabrication processes, a high ratio of silicon etch rate to SiO_2 etch rate is desired; this is achieved by use of an HNO_3-rich $HNO_3–HF$ etchant composition, or by the use of alkaline etchants.

Flat semiconductor wafer surfaces are produced by a chemical-mechanical process (139–146) which uses a slurry of solid in liquid, with a porous, resilient pad to achieve a high slurry velocity across the wafer surface. Silicon wafers are polished using a suspension of silica gel particles in an aqueous etchant (139, 142, 143, 145, 146) adjusted to a pH of approximately 11. The chemical–

Table 11.4 Bond density and surface energy of silicon *v.* orientation (148)

Silicon surface crystal-lographic orientation	Bond density ($\times 10^{14}$ cm^{-2})	Surface energy ($\times 10^{14}$ eV cm^{-2})
(100)	6.8	13.3
(110)	9.4	9.4
(111)	11.8	7.6

mechanical polishing of silicon is generally accomplished in two steps, stock removal and final polish (145). The final polish step, performed at a lower temperature and with a different slurry composition, removes haze from the wafer surfaces.

The etch rate of single-crystal semiconductors in certain etchants can vary considerably with orientation. In alkaline etchants, there are large differences in the etch rates of silicon of the major crystallographic orientations. These anisotropic etchants are used to prepare a variety of structures, including dielectric-isolated integrated circuits (103, 170). They have also been used to micromachine silicon to create specialized structures (408–410, 412, 414–416). Wet chemical etchants have been developed which etch silicon 50 times as fast in the [100] direction than in the [111] direction. For pattern etching, the etch mask is aligned parallel to or perpendicular to the [110] flat of the silicon wafer.

Silicon oxidation processes, which may be considered as a special type of etching process, have been extensively characterized (442–449) in terms of kinetics, effects of orientation, and effects of temperature.

The bond density and surface energy of the major silicon orientations are shown in Table 11.4. These differences qualitatively correlate with observed differences in etch rates of surface-limited reactions. Silicon wafers which are heavily doped with phosphorus have oxidation rates that are enhanced compared to lightly-doped silicon (443).

Polycrystalline and amorphous silicon. Polysilicon and amorphous silicon are frequently used in commercial semiconductor devices (153, 156, 417–427). Typically, the polysilicon used for interconnects and gates of MOS integrated circuits is deposited by low-pressure chemical vapour deposition (418) at a temperature of about 620° C. Subsequent processing steps involve heat treatments at temperatures from 900–1100° C; at these temperatures the grain size of the polycrystalline silicon grows (418) to about 0.1 μm. While the crystallites in polysilicon for integrated circuits tend to have a preferred orientation, etchants used to delineate polysilicon generally are chemical polishes which dissolve various orientations at approximately equal rates.

Silicon chemical etchants are frequently used as post-etchants after patterning of aluminium–silicon alloys in integrated circuits. For example,

Al–1% Si subjected to elevated temperatures tends to precipitate silicon at grain boundaries. Patterning of the Al–Si alloy in etchants such as phosphoric–nitric–acetic acid mixtures results in complete dissolution of the aluminium, but leaves silicon precipitates ('freckles') between the patterned Al lines. These residual silicon crystallites are removed by either a chemical etch (HNO_3–HF mixture) or by a plasma-etching step.

Generally, the polycrystalline silicon in MOS integrated circuits is n^+ (phosphorus-doped), but in some cases p^+ (boron-doped) polycrystalline silicon is used (in some CMOS integrated circuits). Wet chemical etching of polycrystalline silicon is generally used for integrated circuits with linewidths of 5 μm and greater, while reactive-ion dry etching is typically used for integrated circuits with linewidths of 3 μm or smaller.

Amorphous silicon is used in low-cost photovoltaic devices (solar cells) and in developmental thin-film silicon transistors for display applications (419–421, 423, 424, 426, 427). There is also a trend toward fabrication of silicon-gate MOS integrated circuits by processes in which the gate conductor layer is deposited in amorphous form (418) by LPCVD at 560° C; subsequent heat treatments in the integrated circuit fabrication process convert the patterned amorphous silicon films into polycrystalline silicon.

Compound semiconductors. Wet-chemical etching of compound semiconductors has been described in detail (21, 34, 54, 82). Available publications describe polishing etchants (56, 65, 82, 84, 103, 140, 394–404), anisotropic etchants (450–453), selective etchants (450, 451) and etchants to reveal dislocations. Compositions for etching gallium arsenide (65, 84, 103, 169, 395, 398–404), gallium phosphide (84, 396, 397), gallium arsenide–phosphide and a large number of other compound semiconductors have been reported (54, 82, 452, 453). Wet etchants for compound semiconductors must be able to dissolve each of the constituent elements. Etchants typically contain an oxidizing agent.

Etchants commonly used for gallium arsenide (34, 54, 65, 84, 103, 394, 395, 398, 399, 401, 450) include bromine–methanol, H_2SO_4–H_2O_2–H_2O and H_3PO_4–H_2O_2–H_2O. Several polishing etchant compositions for GaAs are shown in Table 11.3. A typical polishing etchant (401) for GaAs is 8:1:1 H_2SO_4–H_2O_2–H_2O. Etch rate is 1.5μm min^{-1} for all orientations except ⟨111⟩Ga, for which it is slower. In preferential etchants, the etch rates of major crystallographic planes of GaAs, in decreasing order, are ⟨110⟩, ⟨111⟩B, ⟨100⟩, ⟨111⟩A. Bromine–methanol solutions containing colloidal silica are used for chemical–mechanical polishing of gallium arsenide wafers (140, 145).

Information is also available on wet-chemical etchants for a variety of other compound semiconductors (21, 34, 35, 54, 452, 453), including GaSb, InP, InSb, InAs, CdS, CdSe, CdTe, HgCdTe, ZnS, Bi_2Te_3, PbS, PbSe and many other compound semiconductors.

Conductors

Conductor materials for solid-state microcircuit applications are heavily doped polycrystalline silicon, pure metals, metal alloys, and refractory metal silicides. The superconducting material niobium stannide (Nb_3Sn), which is used for some electronic devices, can be electrochemically polished in an H_2SO_4–HNO_3–HF mixture (454). Tin oxide, SnO_2, a transparent electric conductor used for photovoltaic devices, can be pattern-etched with HCl and zinc powder (455, 456) or electrochemically etched in dilute HCl (457).

Metals. Wet etching of metals is carried out with etchants that contain a sufficiently strong oxidizing agent to oxidize the metal to the ionic state. Concentrated or dilute mineral acids and acid mixtures are the most widely used chemical etchants for metals. Solutions of strong alkalis, or of oxidizing agents such as Fe^{3+}, Ce^{4+}, I_3^-, O_2^{2-}, $[Fe(CN)_6]^{3-}$, $[S_2O_8]^{2-}$, or $[ClO_4]^-$ are also frequently used in etchant and polishing formulations (74).

Electrochemical etching procedures are available for virtually every conductor (7, 458). As in semiconductor electro-etching, the material to be etched is made the anode of an electrochemical cell. The etch rate and surface finish are determined by the cell voltage, current density, cathode material, etching medium, and other factors such as temperature, complexing agents, surfactants, etc. The potential-scan method for etching metals (74) involves the determination of the potential range for oxidation using electrochemical methods, followed by substituting a chemical reagent of suitable oxidizing potential for the applied voltage.

Galvanic effects in the wet-chemical etching of metal films in the presence of a second layer of different electrochemical free energy must not be overlooked. When a metal in contact with another more noble overlaying metal is etched, galvanic effects can cause a substantial change in the etch rate unless the correct etchant is used (75, 459). The result of uncontrolled etching may be undesirable undercutting. Etchant compositions to avoid these effects are very similar to electrolytic polishing etchants for metals. On the other hand, bevelled edges in metal films can be formed by utilizing galvanic effects; in this application a thin film of a less noble metal is deposited on the metal to be bevelled with an etchant of appropriate composition (459, 460).

Numerous general and specific references for metal etching are available from the literature and from metallographic etching handbooks (7, 12, 33, 36, 63, 74, 75, 82, 83, 85, 91, 92, 100, 101, 103, 110, 117, 121).

A comprehensive survey of specific etchants and procedures for metals of importance in the electronics industry has been presented in a previous chapter on chemical etching (82). Our present account will therefore be limited to the three major metals used in the fabrication of silicon integrated circuits: aluminium, aluminium alloys, and tungsten.

Wet-chemical pattern delineation etching of pure aluminium and aluminium alloys containing small percentages of silicon, copper, and/or titanium (primarily to suppress electromigration and to reduce crystallization and hillock formation of Al) is mainly based on various combinations of H_3PO_4, HNO_3, and acetic acid that are frequently used at elevated temperatures (33, 47, 460–462). Gas evolution, which occurs in these acid media, can be avoided by the use of certain alkaline etchants (74), or by the addition of a surface-active agent to the etching solution (463). Anisotropic wet etching of aluminium by use of an evacuated etching system to achieve aluminium lines with 1.0–1.2 μm linewidth has been reported (465). Laser-controlled chemical etching of aluminium for high spatial resolution is possible by localized heating with a focused Ar^+ laser beam in a solution of H_3PO_4, HNO_3, and potassium dichromate (464). Twelve etchant compositions were tabulated in the previous survey on aluminium etching (82). Numerous etchants for aluminium alloys have been reported (83).

Tungsten, like the other Group VIB metals, chromium and molybdenum, can be patterned in alkaline $[Fe(CN)_6]^{3-}$ solutions (36, 466–468). Tungsten can also be readily pattern-etched using electrochemical methods (466) or chemical oxidizing agents (468). The anodic behaviour of tungsten (as well as molybdenum and chromium in methanol solution has also been investigated (469). Other etchants for tungsten and certain tungsten alloys include H_2O_2–H_2O, H_2O_2–H_2O–NH_4OH; for electrolytic etching NaOH–H_2O or H_2SO_4–HF (40)%–CH_3OH (95%) mixtures have been employed (83). Delineation etching of tungsten films using photoresist masking layers can be accomplished with 0.25 *M* KH_2PO_4–0.24 *M* KOH–0.1 *M* $K_3Fe(CN)_6$ solution (116).

Metal silicides. Refractory metal silicides have attained considerable importance in VLSI technology as a contact and interconnect material to replace doped polycrystalline silicon (470–476). Selective etchants for $TiSi_2$, $MoSi_2$, and $NbSi_2$ are NH_4OH–H_2O_2; for $CoSi_2$, 3HCl–H_2O_2; for PtSi, 3HCl–HNO_3; for Pd_2Si, KI:I_2; and for $NiSi_2$, HNO_3(476). $NiSi_2$, $TiSi_2$, WSi_2, $MoSi_2$, $CoSi_2$, and $TaSi_2$ are known to also etch in HF. Pattern delineation is usually accomplished by dry-etching techniques. However, selective etching is frequently used to remove unreacted metal from silicides.

Organic materials

Wet etching of organic materials is a very important part of semiconductor device technology. The most significant applications are in development of positive resists, and in stripping of resists after patterning is completed. Wet etching is also used to pattern polyimide films (171), and for deflashing of plastic-encapsulated devices.

Microlithography is a key process in the fabrication of integrated circuits (see Chapters 9 and 10). While electron-beam lithography and x-ray lithography are being extensively investigated for use on wafers, optical photolithography is the main technique used for high-volume production of integrated circuits, including devices with feature sizes smaller than 2 μm. Electron-beam lithography is extensively used in preparation of high-resolution photomasks. In any of these lithographic processes, developer solutions are used, after selective exposure of resists, to remove selected areas of resist to produce resist patterns (477–487). Both negative and positive etch resists are widely used, but positive types are predominant for high-resolution applications.

Wet etching in aqueous alkaline solutions is used to develop positive resists. Positive photoresist photoproduct is soluble in alkaline aqueous solutions because of the acidic nature of the phenolic groups produced by the action of light on the base-insoluble sensitizer (inhibitor). Metal-ion-free developers are used for developing positive resists on devices which are susceptible to metal ion contamination. A number of studies have been made on the factors which determine rates for developing resists (482–487). Positive photoresist developers and other alkaline etchants are used to pattern partially-cured polyimide layers (488, 489).

After the resist has been used as a mask to selectively etch an underlying layer, the resist must be removed in an etching step which does not attack the materials in unmasked regions, and does not dissolve the underlying material. The patterned resist is polymerized, with cross-linking having occurred as a result of the hard bake, uv flood step, ion implantation or other processing step after exposure. Photoresist films are removed by wet etching, oxygen plasma etching ('ashing'), or by a combination of plasma etching followed by wet etching. Plasma etching tends to leave metal oxide residues on the wafer surface, and thus it is desirable to follow the plasma oxidation step with an acid-etch step to ensure removal of alkali metal ions (principally sodium) and other contaminants which would have an adverse effect on devices.

Hot peroxide–sulphuric acid compositions have been widely used to remove photoresist materials; a typical etchant composition is 1:2 H_2O_2–H_2SO_4 at 120° C. A number of proprietary resist strippers are also widely used in the industry; their chemical composition is specified by the manufacturers only in very general terms. Phenol has been used in combination with organic solvents in some stripper compositions. Proprietary strippers are frequently used at 90–100° C.

Most of the integrated circuits fabricated for commercial purposes are molded in novolac epoxy compositions. Deflashing of plastic-encapsulated devices is accomplished by mechanical abrasion and/or use of solvents such as *N*-methyl-2-pyrrolidone (490).

Etching for materials characterization and defect detection

Wet etches are very valuable for characterizing composition and quality of deposited dielectrics, for revealing localized structural defects in thick dielectric films, and for analysing factors which influence yield and/or reliability of semiconductor devices. They are used in the analysis of the devices fabricated by suppliers and/or competitors (reverse engineering) to assess designs, materials, and fabrication processes which may have cost, performance, or other implications.

Selectivity–key to analytical etching of dielectrics

While selectivity is important to many etching processes in microelectronic technology, it is an essential feature required in analytical etching to determine composition and other properties of dielectric films. Selectivity is achieved by proper choice and variation of etchant composition within the constraints of the systems so that maximal etch-rate sensitivity results for the particular material to be etched in the presence of some other material for which etching should be minimal. For example, the selective etch rate of silicate glasses is a sensitive measure of their chemical composition (38, 73, 81) and can be conveniently utilized for process control. It is sufficiently accurate and precise for most practical requirements if used with calibration curves established for films on polished silicon wafers. The etch rate is also sensitive, to a lesser extent, to structural film properties (73, 81, 491) whose effects can be reduced or eliminated by a high-temperature annealing treatment (800–1000° C). Annealing can therefore render the etch rate a unique function of the chemical film composition (73, 297, 391). Etch rates of films as-deposited and after densification can thus furnish valuable information on both chemical and physical film properties. The etch rate is determined either integrally, until the entire film is dissolved, or differentially, to remove a portion. In integral etching, the hydrophobic property of the semiconductor substrate in HF can usually be used as endpoint (25). Etch-time measurements must be performed under isothermal conditions to obtain accurate results. Variations of a few degrees Centigrade can cause considerable differences in the etch rate, which varies for different materials or film-compositions due to the specific activation energies (73).

Etch-rate plots of phosphosilicate glass compositions for non-selective BHF and selective non-buffered HF etchants with high sensitivity for the phosphorus content demonstrate that the logarithm of the etch rate varies linearly with the weight percent of phosphorus (73, 103, 237, 303). Because the etch rate of dielectric films is also affected by the film density and stress, which are functions of the deposition conditions, films should be annealed at typically 800° C (15 minutes, N_2) prior to etch rate measurements if the data are to be utilized for composition testing based on calibration plots.

Etch rate *v.* composition data have been established also for borosilicate glasses (305, 308, 311, 321–325), borophosphosilicates (328, 329, 332–335, 492), arsenosilicates (345, 346), and germanosilicates (337). Etch rate plots of glasses as a function of the reactant gas composition in CVD work are especially useful in process control applications (73, 237, 316, 321, 322, 323, 329). The etch rate of plasma-enhanced silicon nitride (SiN_xH_y) films in BHF has been correlated with the hydrogen concentration and the film stoichiometry (268, 256, 355).

The selective etch-rate technique has been widely used in analytical applications where composite dielectric films must be resolved. It is possible to identify composition and thickness of component layers in bilayers and in multilayer structures from plots of composite film thickness *v.* time of etching, using graphical resolution techniques (73, 237, 303, 318, 325, 493, 494).

Detection of defects in dielectric layers

Localized structural defects consisting of pinholes, hairline cracks, and partial microholes in dielectric films on semiconductors and metallized substrates, such as integrated circuits, can often be revealed by selective wet-chemical etching. The etchant must penetrate through the defect opening and selectively attack the substrate material on which the dielectric film was grown or deposited. A sufficient period of time must be allowed for the etchant to remove enough substrate material so that the substrate demarcation becomes visible under a microscope as an enlarged area surrounding the defect (49, 52, 495, 496).

Partial discontinuities, such as voids and fast-etching latent imperfections in SiO_2 films, can be detected by briefly etching the dielectric film prior to selective substrate etching. Successive oxide and substrate etching has been used to reveal partial pinholes of decreasing depth (495–497).

Several wet-etching techniques have been developed for detecting pinholes in SiO_2 layers on silicon substrates. Pyrocatechol–ethylenediamine water mixtures (498) are restricted to elemental silicon substrates and etch slowly. Catechol–hydrazine in aqueous or anhydrous media is applicable to both elemental and lightly doped silicon substrates (499). The most widely used method for most dielectric films on silicon is based on single-component alkali solutions (500), such as a 10 wt% aqueous solution of NaOH below 70° C or, preferably, a 10 wt% solution of sodium metasilicate (Na_2SiO_3) at 55° C (30 minutes) with improved etch rate selectivity for Si/SiO_2 (277). All these alkaline etchants are anisotropic for single-crystalline silicon and form readily detectable orientation-dependent etch pits in the oxide defect areas that expose the silicon substrate.

Cracks in PSG films (but not pinholes in undoped SiO_2 films) on silicon substrate wafers can be made distinctly visible under a microscope by etching for about 30s in H_3PO_4–HF–H_2O (1:3:60). The cracks become darker than surrounding areas; in a substantial number of cases a white region is formed immediately adjacent to the dark crack, further enhancing the detectability (501).

Electrochemical anodization in a weak electrolyte solution has been used for locating defects in SiO_2 protective coating over nickel–chromium resistors with aluminium leads (502). Ceric sulphate solution can also be used for selectively etching nickel-chromium under defective SiO_2 layers. Detection of discontinuities in dielectric layers over aluminium metallization can be accomplished readily by immersion of the sample in aluminium etchants (502–504). Etching in HF-containing solutions to remove thin layers of passivation glass and subsequent exposure to aluminium etchants reveals partial pinholes and areas of insufficiently thick metal step coverage (495, 496).

The metal demarcation etching technique is applicable for testing other types of metallization and insulator systems as long as the metal etchant does not attack the insulator coating or the substrate. We have successfully analysed localized structural defects in layers of SiO_2, silicate glasses, Al_2O_3 and Si_3N_4 over numerous metal substrates or delineated metal films. Patterns of tungsten and molybdenum metallization were etched with an aqueous solution of potassium ferricyanide and KOH; platinum with aqua regia; gold with an aqueous solution of potassium iodide and iodine; chromium in a mixture of saturated ceric sulphate solution and HNO_3, copper with ferric chloride solution, and nickel with diluted HNO_3 (495).

The defect detection limit for dielectric films can be substantially increased by selective and isotropic etching of the substrate beneath a film with defect openings, and exposure of the rinsed and dried sample to a solution of a suitable fluorescing reagent solution which fills the voids. The sample can then be examined, with enhanced sensitivity, under a fluorescence microscope (505–507).

The detection of defects in plasma-deposited silicon nitride films over vertical steps in the substrate by wafer etching has been discussed earlier (357–359). The fast-etching regions in coatings over steps indicate localized differences in the film structure that can be due to a degradation in the insulating properties and lead to device failure or reliability problems (357–359, 378, 508). Plasma deposition in a nitrogen-free gas system appears to eliminate anomalous etching phenomena, at least for specific conditions of plasma CVD (508).

Finally, several dislocation etchants have been described for characterizing crystallographic defects in plates of sapphire (single-crystal α-aluminium oxide). These etchants include H_3PO_4, KOH, and potassium bisulphate ($KHSO_4$) at temperatures up to 600° C (509).

Characterization of structure and defects in semiconductors

Wet-chemical etching has historically provided a very valuable means for assessing crystallographic quality of single-crystal semiconductors in terms of lattice defects which intersect the surface (5, 17, 18, 21, 22, 25, 28, 34, 35, 54, 65, 84, 136). Etching continues to be widely used to characterize a variety of semiconductor materials and devices (103, 116, 137, 154). Etching is used to delineate dislocations, stacking faults, and other crystallographic defects in semiconductor wafers, and to determine the effects of various processing steps on crystallographic quality. Wet-chemical etching is a very effective technique for delineating junctions (54), either by staining, by chemical plating of a metal such as copper, or by selective etching followed by microscopic observation of the resulting topology. Anisotropic etching provides a convenient means for determining crystallographic orientation relative to the surface of a wafer or device (54).

Single-crystal silicon. A number of specialized dislocation etchants have been developed to permit characterization of the defect density and presence of surface work damage in silicon crystals (54, 103, 392, 510–520). The earlier etchants were formulated for (111) orientation silicon (510, 512), which was widely used for initial bipolar and MOS devices. Other etchants have been developed to assess the quality of (100) orientation silicon wafers (513–516). Table 11.5 lists a number of wet chemical etchants which are used to assess crystallographic quality of silicon samples, or are used to delineate junctions in silicon (103, 521–523).

Defect etchants are also effective in studying the quality of polycrystalline silicon (524). Junction delineation (34, 35, 54, 103, 521–523, 525–527) is used to reveal the location of *p–n* junctions or high-impurity/low-impurity boundaries in semiconductor samples, either on the surface, in angle-lapped samples, or in cross-section. Considerable skill is involved in attaining high-quality results. A large amount of information has been published on the use of wet-

Table 11.5 Etchants used for silicon materials and device studies

Etchant	Composition	Uses	References
Dash etch	1:3:10 $HF–HNO_3–CH_3COOH$	Crystallographic defects	510
Sirtl etch	1g CrO_3, 2mL H_2O, 1mL HF	Crystallographic defects	512
Secco etch	2:1 HF–0.15 M $K_2Cr_2O_7$	Delineation of grain boundaries	513
Wright etch	2:1:1:2:2 $HF–HNO_3$–5M $CrO_3–CH_3COOH$–0.14M $Cu(NO_3)_2$	Crystallographic defects	514, 515
Schimmel etch	2:1 HF–1M CrO_3	Crystallographic defects	516
Nitric/hydrofluoric etch	199:1 HNO_3–HF	Junction delineation	103
Copper sulphate staining solution	8g/L $CuSO_4 \cdot 5H_2O$, 10 mL HF	Junction delineation	521
Periodic/hydrofluoric etch	5 g H_5IO_6, 50 mL H_2O, 2 mL HF, 5 mg KI	Junction delineation	522, 523

chemical etchants to characterize various types of silicon materials and devices (528–539).

Single-crystal compound semiconductors. Wet etchants have been developed for revealing defects in a wide variety of compound semiconductors (21, 22, 25, 28, 34, 35, 54, 65, 103, 540–542), including GaAs, GaP, InSb, CdTe, and many other materials. Wet etchants have also been used to indicate crystal polarity (54, 103), to determine surface orientation (54, 543), to reveal *p–n* junctions or hetero-epitaxial layer interfaces (103), and to evaluate the depth of surface work damage (544).

Delineation of structural defects in metals

Wet-chemical etchants have been used to determine precipitation effects in metal alloy films such as Al-Si (545) or Al–Cu–Si (546), to delineate grain boundaries, to determine deposited metal film step coverage in structures with steep underlying steps, and to remove metals to examine underlying structures such as contact regions (545, 547–554). Coverage of interconnect metal over steps in the topography of integrated circuits is especially critical to device performance (47) and is best examined by chemical etch-back techniques combined with scanning electron microscopic examination at various stages of etching (555).

Wet-chemical selective etching and electrochemical etching techniques are applied extensively in metallographic analysis where structural defects in the microstructure and compositional inhomogeneities of polished metal specimens and cross-section samples are to be made visible for microscopic examination. The literature cited for metallographic etching should be consulted for specific information on this topic (1, 60, 62, 83, 91, 100, 101, 110, 117).

Preparation of samples for analysis

Selective etching is widely used in assessing device-yield limitations (556), in developing an understanding of reasons for failure of completed devices during accelerated reliability tests or in field usage (557–562), in analysing device structures to determine performance limitations, and in assessment of suppliers' or competitors' devices. Wet-etch techniques are extensively used to selectively remove portions of samples (563–567) in preparation for optical microscopy, scanning electron microscopy, or transmission electron microscopy. Sectioned samples are frequently prepared for these purposes, which may involve both mechanical grinding and lapping, chemophysical etch-polishing, or chemical etching.

Problem areas and future prospects

Wet-chemical etching will continue to be widely used in the fabrication of integrated circuits. In some patterning applications, it will be replaced by dry-etching processes (plasma etching or reactive-ion etching), particularly for high-resolution devices with feature sizes of 3 μm or smaller. The choice of wet *v.* dry etching depends upon many factors. Dry etching can provide a number of advantages (112, 127, 164, 568, 569), such as freedom from undercutting, vertical walls of etched regions, and no liquid disposal problems. By contrast, wet-etching techniques provide total freedom from charge build-up or from radiation damage effects, better etch-rate ratios of the materials to be etched relative to underlying materials, and freedom from polymer formation effects (570).

Some of the most recent applications of wet etching are in conjunction with dry etching, to achieve the advantages of dry etching while avoiding some of the limitations. For example, reactive ion etching of openings in thin, thermally grown oxide films on silicon has been shown to damage the silicon and to adversely affect the dielectric integrity of a subsequently grown oxide. The use of reactive ion etching to remove most of the SO_2 layer, followed by a short wet chemical etch to remove the balance of the SiO_2, can provide steep walls without silicon damage. Similarly, removal of most of an aluminium metal layer by reactive ion etching, followed by wet-chemical etching, can provide high-resolution metal patterns without exposing substrates to vacuum–ultraviolet radiation effects. Reactive ion etching of conductor lines over steep topology steps tends to leave conductive stringers between lines; application of a brief wet-chemical etch after reactive-ion etching results in complete removal of stringers with relatively little lateral etching of edges of metal lines (since the stringer is near but not at the step, and thus can be laterally etched from both sides during wet etching). Another area of increasing use for wet chemical etching could be in implementation of self-aligned silicide gates, interconnects and contacts of high-density integrated circuits.

Wet etching also has many analytical applications, and will continue to be widely used to characterize semiconductors, dielectrics and device structures.

Conclusions

Wet-chemical etching processes for inorganic and organic materials of importance in the manufacturing of integrated circuits have been systematically reviewed, classified, and referenced. No attempt has been made to provide a comprehensive or complete coverage of the subject; rather, typical and representative references have been cited, emphasizing recent work to complement the previous chapter on etching published in 1978 (82).

A survey of wet-etching applications in the semiconductor industry has been presented, followed by a discussion of the chemistry underlying etching and the technology to implement the processes in the plant and laboratory. The most common systems for wet etching of electronic materials have been discussed, including insulators, dielectrics, semiconductors, conductors, and organic materials. Finally, a section has been devoted to the etching for materials characterization and defect detection, where selectivity plays a particularly important role.

The wide range of material presented in this chapter should amply demonstrate the crucial importance of wet-chemical etching in the electronics industry. Although detailed scientific understanding of many etching processes is certainly less than complete, the practical applications and technology have been refined to a high degree over the past decades. Wet etching will continue to be a key process in semiconductor technology and effectively complement competing processes in micro-pattern generation based on dry etch techniques.

Acknowledgement

The authors gratefully acknowledge the assistance of K. Fabian and L. Sanders in preparation of the manuscript, and L. K. White in reviewing the manuscript.

References

A. *Books and review articles* (*in chronological sequence, 1949–86*)

1. G. L. Kehl, in *The Principles of Metallographic Laboratory Practice*, McGraw-Hill, New York (1949) 60–79.
2. W. J. McG. Tegart, *The Electrolytic and Chemical Polishing of Metals in Research and Industry*, Pergamon, London (1956).
3. J. N. Shive, in *Transistor Technology*, Vol. I, eds. H. E. Bridgers, J. H. Scaff, and J. N. Shive, Van Nostrand, Princeton (1958) 350–361.
4. J. W. Faust, Jr., in *Methods of Experimental Physics*, Vol. 6, Part A, eds. K. Lark-Horovitz and V. A. Johnson, Academic Press, New York (1959) 147–175.
5. N. B. Hannay (ed.), *Semiconductors*, ACS Monogr. No. 140, Reinhold, New York (1959).
6. J. F. Dewald, in ref. 5, 727–752.
7. W. J. McG. Tegart, *Electrolytic and Chemical Polishing of Metals*, revised edn., Pergamon, New York (1959).
8. H. C. Gatos, J. W. Faust, Jr., and W. J. Lafleur (eds.), *The Surface Chemistry of Metals and Semiconductors*, John Wiley, New York (1960).
9. J. W. Faust, Jr., in ref. 8, 151–173.
10. H. C. Gatos, in ref. 8, 381–406.
11. N. Hackerman, in ref. 8, 313–325.
12. C. V. King, in ref. 8, 357–380.
13. P. Lacombe, in ref. 8, 244–284.
14. D. R. Turner, in ref. 8, 285–310.
15. P. J. Holmes (ed.), *The Electrochemistry of Semiconductors*, Academic Press, London (1962).
16. J. I. Carasso and M. M. Faktor, in ref. 15, 205–255.
17. P. J. Holmes, in ref. 15, 320–377.
18. B. A. Irving, in ref. 15, 256–289.
19. J. I. Pankove, in ref. 15, 290–328.

20. D. R. Turner, in ref. 15, 155–204.
21. J. W. Faust, Jr., in *Compound Semiconductors*, Vol. I, eds. R. K. Willardson and H. L. Goering, Reinhold, New York (1962) 445–468.
22. W. G. Johnston, in *Progress in Ceramic Science*, Volume 2, ed. J. E. Burke, Pergamon, The MacMillan Co., New York (1962) 1–75.
23. E. A. Efimov and I. G. Erusalimchik, *Electrochemistry of Germanium and Silicon*, translation, ed. A. Peiperl, Sigma Press, Washington (1963).
24. L. Holland, in *The Properties of Glass Surfaces*, John Wiley, New York (1964) 290–347.
25. H. C. Gatos and M. C. Lavine, in *Progress in Semiconductors*, Vol. 9, eds. A. F. Gibson and R. E. Burgess, Temple Press Books, London (1965) 1–45.
26. Anon., *Integrated Silicon Device Technology*; X. Chemical/Metallurgical Properties of Silicon, ASD-TDR-63-316, Vol. X; AD 625, 985; Research Triangle Institute (Nov. 1965).
27. P. J. Boddy, *J. Electroanal. Chem.* **10**(3) (1965) 199–244.
28. M. Aven and J. S. Prener (eds.), in *The Physics and Chemistry of II–VI Compounds*, North-Holland, Amsterdam (1967) 141, 155, 772, 773.
29. A. F. Bogenschütz, *Etching Semiconductors* (in German); *Atzpraxis für Halbleiter*, Carl Hanser Verlag, München (1967).
30. V. A. Myamlin and Y. V. Pleskov, *Electrochemistry of Semiconductors*, Plenum, New York (1967).
31. S. K. Ghandhi, in *The Theory and Practice of Microelectronics*, John Wiley, New York (1968) 154–161.
32. H. Gerischer, in *Physical Chemistry: an Advanced Treatise*, Vol. 9, *Electrochemistry*, Part A, ed. W. Jost, Academic Press, New York (1970) 463–542.
33. R. Glang and L. V. Gregor, in *Handbook of Thin Film Technology*, eds. L. I. Maissel and R. Glang, McGraw-Hill, New York (1970) 7–1 to 7–66.
34. P. F. Kane and G. B. Larrabee, *Characterization of Semiconductor Materials*, McGraw-Hill, New York (1970).
35. W. R. Runyan and S. B. Watelski, in *Handbook of Materials and Processes for Electronics*, ed. C. A. Harper, McGraw-Hill, New York (1970) 7–1 to 7–113.
36. R. J. Ryan, E. B. Davidson, and H. O. Hook, in *Handbook of Materials and Processes for Electronics*, ed. C. A. Harper, McGraw-Hill, New York (1970) 14–1 to 14–133.
37. P. V. Shchigolev, *Electrolytic and Chemical Polishing of Metals*, Freund Publ. House, Holon, Israel (1970).
38. W. A. Pliskin and S. J. Zanin, in *Handbook of Thin-Film Technology*, eds. L. I. Maissel and R. Glang, McGraw-Hill, New York (1970) 11–1 to 11–54.
39. J. T. Milek, in *Silicon Nitride for Microelectronic Applications*, Part 1, Preparation and Properties; *Handbook of Electronic Materials*, Vol. 3, IFI/Plenum, New York (1971) 40–53.
40. D. A. Vermilyea, in *Ann. Rev. of Materials Science*, Vol. 1, ed. R. A. Huggins, Annual Reviews Inc., Palo Alto (1971) 373–398.
41. H. F. Wolf, in *Semiconductors*, Wiley-Interscience, New York (1971) 130–135.
42. F. M. Ernsberger, in *Ann. Rev. of Materials Science*, Vol. 2, ed. R. A. Huggins, Annual Reviews Inc., Palo Alto, (1972) 529–572.
43. Anon., E3, E340, and E407, in *ASTM Standards, Part 31*, American Society for Testing and Materials, Philadelphia (1972).
44. R. H. Doremus, in *Glass Science*, John Wiley, New York (1973) 229–252.
45. Y. V. Pleskov, in *Progress in Surface and Membrane Science*, Vol. 7, eds. J. F. Danielli, M. D. Rosenberg, and D. A. Cadenhead, Academic Press, New York (1973) 57–93.
46. A. K. Vijh, *Electrochemistry of Metals and Semiconductors*, Marcel Dekker, New York (1973).
47. W. Kern, J. L. Vossen, and G. L. Schnable, *11th Ann. Proc. Rel. Phys.* (1973) 214–223.
48. A. H. Agajanian, *Solid State Technol.* **16** (Dec. 1973) 73–74, 104–105.
49. W. Kern, RCA Rev. **34** (Dec. 1973) 655–690.
50. T. C. Harman and I. Melngailis, in *Applied Solid State Science*, Vol. 4, ed. R. Wolfe, Academic Press, New York (1974) 1–94.
51. J. L. Vossen, G. L. Schnable, and W. Kern, *J. Vac. Sci. Technol.* **11** (1974) 60–70.
52. W. Kern, *Solid State Technol.* **17** (Mar. 1974) 35–42, and **17** (Apr. 1974) 78–84, 89.
53. W. S. DeForest, *Photoresist Materials and Processes*, McGraw-Hill, New York (1975).
54. W. R. Runyan, chapters 1, 2, 7 and 9 in *Semiconductor Measurements and Instrumentation*, McGraw-Hill Book Co., New York (1975).

55. H. C. Nathanson and J. Guldberg, in *Physics of Thin Films*, Vol. 8, eds. G. Hass, M. H. Francombe, and R. W. Hoffman, Academic Press, New York (1975) 251–333.
56. B. Tuck, *J. Mater. Sci.* **10** (1975) 321–339.
57. A. H. Agajanian, *Solid State Technol.* **18** (Apr. 1975) 61–65.
58. G. L. Schnable, W. Kern, and R. B. Comizzoli, *J. Electrochem. Soc.* **122** (1975) 1092–1103.
59. A. H. Agajanian, *Semiconductor Devices, A Bibliography of Fabrication Technology, Properties, and Applications*, IFI/Plenum, New York (1976).
60. M. Beckert and H. Klemm, *Handbook of Metallographic Etching Processes*, 3rd edn, Deut. Verlag Grundstoffind., Leipzig (1976).
61. W. T. Harris, *Chemical Milling: The Technology of Cutting Materials by Etching*, Oxford Series on Advanced Manufacturing, Clarendon Press, Oxford (1976).
62. G. Petzow, J. Back, and T. Mager, *Materialkundlich-Techn. Reihe* **1** (1976) 1–120.
63. M. J. Pryor and R. W. Staehle, in *Treatise on Solid State Chemistry*, Vol. 4, ed. N. B. Hannay, Plenum, New York (1976) 457–620.
64. L. L. Shreir (ed.) *Corrosion*, 2nd edn., Newnes-Butterworth, London (1976).
65. D. J. Stirland and B. W. Straughan, *Thin Solid Films* **31** (1976) 139–170.
66. H. W. Markstein, *Electron Packag. and Prod.* **16** (Feb. 1976) 24–32.
67. V. F. Dorfman, *Microélectronika* (*Soviet Microelectronics*) **5**(2) (1976) 99–124; (Tr. Plenum, New York).
68. Anon., *Circuits Manufact.* **16** (Apr. 1976) 42–46; A. Zafiropoulo, *ibid.*, 42, 44, 46.
69. J. W. Faust, *Electrochem. Soc. Ext. Abstr.* **76–1** (1976) 140–141.
70. H. G. Hughes and M. J. Rand (eds.), *Symp. on Etching for Pattern Definition*, Electrochemical Society, Princeton (1976).
71. J. W. Faust, Jr., in ref. 70, 198–199.
72. J. S. Judge, in ref. 70, 19–36.
73. W. Kern, in ref. 70, 1–18d.
74. D. MacArthur, in ref. 70, 76–90.
75. L. Romankiw, in ref. 70, 161–193.
76. R. Tijburg, *Physics in Technol.* **7** (1976) 202–207.
77. G. L. Schnable and P. F. Schmidt, *J. Electrochem. Soc.* **123** (1976) 310C–315C.
78. A. H. Agajanian, *Solid State Technol.* **20** (Jan. 1977) 36–48.
79. C. M. Wilmsen and S. Szpak, *Thin Solid Films*, **46** (1977) 17–45.
80. H. Sieter, *Electrochem. Soc. Ext. Abstr.* **77–1** (1977) 455–457.
81. W. A. Pliskin, *J. Vac. Sci. Technol.* **14** (1977) 1064–1081.
82. W. Kern and C. A. Deckert, in *Thin Film Processes*, eds. J. L. Vossen and W. Kern, Academic Press, New York (1978) 401–496.
83. G. Petzow, *Metallographic Etching*, American Society for Metals, Metals Park, OH (1978).
84. W. Kern, *RCA Rev.* **39** (1978) 278–308.
85. E. F. Duffek and E. Armstrong, in *Printed Circuits Handbook*, 2nd edn., ed. C. F. Coombs, Jr., McGraw-Hill, New York (1979) 8–1 to 8–45.
86. J. J. Kelly and G. J. Koel, *Philips Tech. Rev.* **38** (1979) 149–157.
87. G. L. Schnable, R. B. Comizzoli, W. Kern and L. K. White, *RCA Rev.* **40** (1979) 416–445.
88. S. R. Morrison, *Electrochemistry at Semiconductor and Oxidized Metal Electrodes*, Plenum, New York (1980).
89. D. E. Aspnes and A. A. Studna, *Appl. Phys. Lett.* **39** (1981) 316–318.
90. Anon., in *Metals Handbook*, 9th edn., Vol. 5: *Surface Cleaning, Finishing, and Coating*; American Society for Metals, Metals Park, OH (1982) 1–103, 303–309, 569–677.
91. Anon., in *Metals Handbook*, 9th edn., Vol. 9: *Metallography and Microstructures*; American Society for Metals, Metals Park, OH (1982) 21–162, 163–597.
92. M. J. Collie (ed.), *Etching Compositions and Processes*, Chemical Technology Review No. 210, Noyes Data Corporation, Park Ridge, NJ (1982).
93. D. J. Elliot, *Integrated Circuit Fabrication Technology*, McGraw-Hill, New York (1982).
94. J. Grabmaier (ed.), *Silicon Chemical Etching*, Vol. 8 of *Crystals: Growth, Properties, and Applications*, ed. H. C. Freyhardt, Springer Verlag, Berlin (1982).
95. R. B. Heimann, in ref. 94, 173–224.
96. J. S. Judge, in *Proc. Tutorial Symp. on Semiconductor Technology*, eds. D. A. Doane, D. B. Fraser, and D. W. Hess, Proceedings Volume 82–5, The Electrochemical Society, Inc., Pennington, NJ (1982) 173–175.

97. A. Paul, in *Chemistry of Glasses*, Chapman and Hall, London (1982) 108–147.
98. W. Kern, *Semiconductor Internatl.* **5** (Mar. 1982) 89–103.
99. D. M. Allen, *Electron. Prodn.* **11** (Sept. 1982) 113, 115, 117.
100. Anon., in *ASM Metals Reference Book*, 2nd edn., American Society for Metals, Metals Park, OH (1983) 133–163.
101. E. A. Brandes (ed.), in *Smithells Metals Reference Book*, 6th edn., Butterworths, London (1983) 10–1 to 10–72.
102. D. J. Ehrlich and J. Y. Tsao, in *VLSI Electronics: Microstructure Science*, Vol. 7, ed. N. G. Einspruch, Academic Press, Orlando, FL (1983) 129–164.
103. S. K. Ghandhi, in *VLSI Fabrication Principles*, John Wiley, New York (1983) 477–531.
104. Y. Y. Gurevich and Y. V. Pleskov, in *Semiconductors and Semimetals*, Vol. 19, *Deep Levels, GaAs, Alloys, Photochemistry*, eds. R. K. Willardson and A. C. Beer, Academic Press, New York (1983) 255–310.
105. L. F. Thompson, C. G. Willson and M. J. Bowden, *Introduction to Microlithography: Theory, Materials, and Processing*, ACS Symp. Ser. 219, American Chemical Society, Washington, (1983).
106. P. S. Burggraaf, *Semiconductor Internatl.* **6** (1983) 48–53.
107. R. Tenne and G. Hodes, *Surface Sci.* **135** (Feb. 1983) 453–478.
108. Anon, *Book of SEMI Standards 1984*, Vol. 1, Chemical Division, Semiconductor Equipment and Materials Institute, Inc., Mt. View, CA (1984).
109. Anon., Index Listing, in *1984 Annual Book of ASTM Standards*, American Society for Testing and Materials, Philadelphia (1984) 196–197.
110. Anon., ASTM Designation E340–68 (reapproved 1980); and pp. 387–406, ASTM Designation E407–70 (reapproved 1982), in *1984 Annual Book of ASTM Standards*, Vol. 03.03: Metallography; Nondestructive Testing; Section 3., American Soc. for Testing and Materials, Philadelphia (1984) 330–346.
111. D. J. Elliot, *Integrated Circuit Mask Technology*, McGraw-Hill, New York (1984).
112. J. B. Kruger, M. M. O'Toole and P. Rissman, in *VLSI Electronics: Microstructure Science*, Vol. 8, ed. N. G. Einspruch, Academic Press, New York (1984) 91–136.
113. D. V. Podlesnik, H. H. Gilgen, and R. M. Osgood, in *Laser Controlled Chemical Processing of Surfaces Symposium*, eds. A. W. Johnson, D. J. Ehrlich, and H. R. Schlossberg, North-Holland, Amsterdam (1984) 161–165.
114. T. J. Chuang, Ingo Hussla, and W. Sesselmann, in *Springer Ser. Chem. Phys., 39*, eds. V. I. Goldanskii, R. Gomer, F. P. Schafer, and J. P. Toennies, Springer Verlag, Berlin (1984) 300–314.
115. V. G. I. Deshmukh, T. I. Cox, and J. D. Benjamin, *Phys. Technol.* **15** (1984) 301–307, 314.
116. W. E. Beadle, J. C. C. Tsai and R. D. Plummer (eds.), *Quick Reference Manual for Silicon Integrated Circuit Technology*, John Wiley, New York (1985).
117. H. E. Boyer and T. L. Gnall (eds.), in *Metals Handbook*, Desk Edition, 9th edn., Vol. 9, *Metallography and Microstructures*, American Society for Metals, Metals Park, OH (1985) 35–1 to 35–60.
118. C. W. Wilmsen, in *Physics and Chemistry of III–V Compound Semiconductor Interfaces*, ed. C. W. Wilmsen, Plenum, New York (1985) 403–420.
119. Y. Rytz-Froidevaux, R. P. Salathe, and H. H. Gilgen, *Appl. Phys.* **A37**(3) (1985) 121–138.
120. J. O'M. Bockris, *J. Electrochem. Soc.* **132** (1985) 2648–2655.
121. Anon., *Metal Finishing*, Guidebook and Directory '86, 54th edn., Metals and Plastics Publications, Inc., Hackensack, NJ (1986).
122. F. J. Fuchs, in Ref. 121, 152–160.
123. N. Hall and J. B. Mohler, in ref. 121, 406–416.
124. W. P. Innes, in ref. 121, 118–140; J. F. Jumer, in ref. 121, 397–405.
125. W. Kern, *RCA Rev.* **47** (1984) 186–202.
126. L. A. Casper (ed.), *Microelectronics Processing: Inorganic Materials Characterization*, ACS Symp. Ser. 295, American Chemical Society, Washington (1986).

B. Additional references

127. S. J. Fonash, *Solid State Technol.* **28** (Jan. 1985) 150–158.

128. S. J. Fonash, *Solid State Technol.* **28** (Apr. 1985) 201–205.
129. S. W. Pang, *Solid State Technol.*, **27** (Apr. 1984) 249–256.
130. J. Dieleman and F. H. M. Sanders, *Solid State Technol.* **27** (Apr. 1984) 191–196.
131. T. Watanabe and Y. Yoshida, *Solid State Technol.* **27** (Apr. 1984) 263–266.
132. M. Valente and G. Queirolo, *J. Electrochem. Soc.* **131** (1984) 1132–1135.
133. G. S. Oehrlein, R. M. Tromp, J. C. Tsang, Y. H. Lee and E. J. Petrillo, *J. Electrochem. Soc.* **132** (1985) 1441–1447.
134. J. E. Spencer, *Solid State Technol.* **27** (Apr. 1984) 203–207.
135. R. Lachenbruch, T. Wicker and J. Peavey, *Semiconductor Internatl.* **8** (May 1985) 164–168.
136. T. M. Buck and R. L. Meek, in *Silicon Device Processing*, ed. C. P. Marsden, NBS Spec. Publ. 337, National Bureau of Standards, Washington (1970) 419–430.
137. J. E. Lawrence and H. R. Huff, in *VLSI Electronics: Microstructure Science*, Vol. 5, ed. N. G. Einspruch, Academic Press, New York (1982) 51–102.
138. J. A. Moreland, in *Silicon Materials*, Vol. 12 of *VLSI Electronics: Microstructure Science*, eds. N. G. Einspruch and H. Huff, Academic Press, Orlando (1985) 63–87.
139. L. H. Blake and E. Mendel, *Solid State Technol.* **13** (Jan. 1970) 42–46.
140. E. W. Jensen, *Solid State Technol.* **16** (Aug. 1973) 49–52.
141. R. B. Herring, *Solid State Technol.* **19** (May 1976) 37–43.
142. P. S. Burggraaf, *Semiconductor Internatl.* **2** (Oct. 1979) 58–65.
143. Anon., *Semiconductor Internatl.* **7** (13) (1984) (1985 Master Buying Guide) 20–24; **8** (13) (1985) (1986 Master Buying Guide) 20–23.
144. A. Schnegg, I. Lampert and H. Jacob, *Electrochem. Soc. Ext. Abstr.* **85–1** (1985) 394.
145. C. Murray, *Semiconductor Internatl.* **8** (Jul. 1985) 94–103.
146. A. Schnegg, M. Grunder and H. Jacob, *Electrochem. Soc. Ext. Abstr.* **86–1** (1986) 197.
147. H. F. Wolf, *Silicon Semiconductor Data*, Pergamon, Oxford (1969).
148. H. F. Wolf, *Semiconductors*, Wiley-Interscience, New York (1971).
149. A. B. Glaser and G. E. Subak-Sharpe, *Integrated Circuit Engineering*, Addison-Wesley, Reading, MA (1977).
150. A. G. Milnes, *Semiconductor Devices and Integrated Electronics*, Van Nostrand Reinhold, New York (1980).
151. E. H. Nicollian and J. R. Brews, *MOS* (*Metal Oxide Semiconductor*) *Physics and Technology*, John Wiley, New York (1982).
152. H. E. Oldham and S. L. Partridge, *GEC J. of Research* **1** (1983) 2–14.
153. C. M. Botchek, *VLSI-Basic MOS Engineering*, Vol. 1, Pacific Technical Group Inc., Saratoga, CA (1983).
154. S. Ghandhi, *VLSI Fabrication Principles*, John Wiley, New York (1983).
155. P. L. Shah and R. H. Havemann, in *VLSI Electronics: Microstructure Science, Vol.* 7, ed. N. G. Einspruch, Academic Press, Orlando, FL (1983) 39–127.
156. S. M. Sze (ed.), *VLSI Technology*, McGraw-Hill, Princeton (1983).
157. R. M. Warner, Jr. and B. L. Grung, *Transistors—Fundamentals for the Integrated-Circuit Engineer*, John Wiley, New York (1983).
158. A. Cardon and L. J. L. Fransen, *Dynamic Semiconductor RAM Structures—A Patent Oriented Survey*, Pergamon, Oxford, (1984).
159. D. G. Ong, *Modern MOS Technology: Processes, Devices and Design*, McGraw-Hill, New York (1984).
160. E. Noguchi, K. Toyoda, T. Fukushima, F. Baba, M. Higuchi and K. Aoyama, *Fujitsu Sci. Tech. J.* **21** (1985) 337–369.
161. W. M. Bullis and S. Broydo (eds.), *VLSI Science and Technology/1985*, The Electrochemical Society, Pennington, NJ (1985).
162. N. G. Einspruch (ed.), *VLSI Handbook*, Academic Press, Orlando, FL (1985).
163. P. Stroeve (ed.), *Integrated Circuits: Chemical and Physical Processing*, ACS Symp. Ser. **290**, American Chemical Society, Washington (1985).
164. S. M. Sze, *Semiconductor Devices—Physics and Technology*, John Wiley, New York (1985).
165. J. V. DiLorenzo and D. D. Khandelwal, *GaAs FET Principles and Technology*, Artech House Inc., Dedham, MA (1982).
166. T. Andrade, *Solid State Technol.*, **28** (Feb. 1985) 199–205.
167. A. G. Rode and J. G. Roper, *Solid State Technol.* **28** (Feb. 1985) 209–215.
168. N. G. Einspruch and W. R. Wisseman (eds.), *GaAs Microelectronics*, Vol. 11, in *VLSI*

Electronics: Microstructure Science, ed. N. G. Einspruch, Academic Press, Orlando, FL (1985).

169. EMIS, *Properties of Gallium Arsenide*, EMIS Datareviews Series, No. 2, Electronic Materials Information Service, INSPEC Division, Institution of Electrical Engineers, Hitchin, Herts, UK (1985).
170. K. E. Bean and W. R. Runyan, *J. Electrochem. Soc.* **124** (1977) 5C–12C.
171. K. L. Mittal (ed.), in *Polyimides*, Vol. 2, Plenum, New York (1984) 713–954.
172. C. E. Jowett, *Materials and Processes in Electronics*, Hutchinson, London (1982).
173. C. A. Steidel, in *VLSI Technology*, ed. S. M. Sze, McGraw-Hill, New York (1983) 551–598.
174. W. H. Schroen and J. M. Pankratz, in *VLSI Electronics: Microstructure Science*, Vol. 9, ed. N. G. Einspruch, Academic Press, Orlando, FL (1985) 185–206.
175. R. Iscoff, *Semiconductor Internatl.* **6** (Apr. 1983) 88–92.
176. R. Iscoff, *Semiconductor Internatl.* **9** (Feb. 1986) 74–82.
177. H. R. Huff and F. Shimura, *Solid State Technol.* **28** (Mar. 1985) 103–118.
178. D. W. Johnson, *Semiconductor Internatl.* **8** (Mar. 1985) 168–172.
179. D. L. Tolliver, *Semiconductor Internatl.*, **7** (Jul. 1984) 99–103.
180. C. Misky and J. Gilliland, *Microcontamination* **3** (May 1985) 71–88.
181. M. A. Accourazzo, K. Laubow and B. Y. H. Liu, *Solid State Technol.* **27** (Mar. 1984) 141–146.
182. P. Burggraaf, *Semiconductor Internatl.* **8** (Mar. 1985) 86–89.
183. W. Kern, *RCA Rev.* **31** (1970) 207–233.
184. W. Kern, *RCA Rev.* **31** (1970) 234–264.
185. W. Kern, *RCA Rev.* **32** (1971) 64–87.
186. W. Kern, *Solid State Technol.* **15** (Jan. 1972) 34–38, (Feb. 1972) 39–45.
187. W. Kern and D. Puotinen, *RCA Rev.* **31** (1970) 187–206.
188. W. Kern, *RCA Engineer* **28–4** (1983) 99–105.
189. R. Tenne, V. Marcu and Y. Prior, *Appl. Phys.* **A37** (1985) 205–209.
190. D. V. Podlesnik, H. H. Gilgen and R. M. Osgood, *Jr., Appl. Phys. Lett.* **45** (1984) 563–565.
191. C. A. Deckert and D. L. Ross, *J. Electrochem. Soc.* **127** (1980) 45C–56C.
192. L. K. White and D. Meyerhofer, *RCA Rev.* **44** (1983) 110–134.
193. S. Mackie and S. P. Beaumont, *Solid State Technol.* **28** (Aug. 1985) 117–122.
194. H. L. Stover (ed.), *Optical Microlithography IV*, Proc. SPIE, Vol. 538, International Society for Optical Engineering, Bellingham, WA (1985).
195. L. F. Thompson (ed.), *Advances in Resist Technology and Processing II*, SPIE Vol. 539, International Society for Optical Engineering, Bellingham, WA (1985) and preceding volumes on photolithography.
196. S. K. Ghandhi, in ref. 154, 533–566.
197. M. J. Bowden and J. Frackoviak, in *Silicon Processing*, ASTM Spec. Publ. 804, American Society for Testing Materials, Philadelphia (1983) 125–143.
198. V. K. Sharma and M. J. Wheeler, Microelectronic Engineering **3** (1985) 313–320.
199. J. Lawrence, *Electrochem. Soc. Extended Abstr.* **72–2** (1972) 466–468.
200. R. G. Brandes and R. H. Dudley, *J. Electrochem. Soc.* **120** (1973) 140–142.
201. L. K. White, *J. Electrochem. Soc.* **127** (1980) 2687–2693.
202. Y. I. Choi, C. K. Kim and Y. S. Kwon, *IEE Proc.* **133** Pt. 1 (1986) 13–17.
203. L. H. Hall and D. L. Crosthwait, *Thin Solid Films* **9** (1972) 447–455.
204. H. Ono and H. Tango, *J. Electrochem. Soc.* **126** (1979) 504–506.
205. R. A. Moline, R. R. Buckley, S. E. Haszko and A. U. MacRae, *IEEE Trans. Electron Devices* **ED-20** (1973) 840.
206. J. C. North, T. E. McGahan, D. W. Rice and A. C. Adams, *IEEE Trans. Electron Devices* **ED-25** (1978) 809–812.
207. P. D. Parry and S. P. Bristol, *J. Vac. Sci. Technol.* **15** (1978) 664–667.
208. M. M. Schlacter, R. S. Keen, Jr. and G. L. Schnable, *IEEE J. Solid-State Circuits* **SC-6** (1971) 327–334.
209. G. L. Schnable and R. S. Keen, Jr., in *Advances in Electronics and Electron Physics*, Vol. 30, ed. L. Marton, Academic Press, New York (1971) 79–138.
210. T. Agatsuma, A. Kikuchi, K. Nakada and A. Tomozawa, *J. Electrochem. Soc.* **122** (1975) 825–829.
211. J. M. Frary and P. Seese, *Semiconductor Internatl.* **4** (Dec. 1981) 72–88.

212. L. Land, L. Mercer and S. Miller, *Proc. Interface '80 Microelectronics Seminar*, Eastman Kodak Co. Publ. No. G-130, Rochester, NY (1981).
213. G. C. Collins and C. W. Halsted, *IBM J. R&D.* **26** (1982) 596–604.
214. A. A. Milgram, *J. Vac. Sci. Technol. B1* (1983) 490–493.
215. K. Ehara, T. Morimoto, S. Muramoto and S. Matsuo, *J. Electrochem. Soc.* **131** (1984) 419–424.
216. T. Yachi and T. Serikawa, *J. Electrochem. Soc.* **132** (1985) 2775–2778.
217. H. J. Geelen, M. Deschler and A. M. Krings, *Microelectronic Engg.* **3** (1985) 499–505.
218. Y. Mimura, *J. Vac. Sci. Technol. B* **4** (1986) 15–21.
219. K. Ishibashi and S. Furakawa, *IEEE Trans. Electron Devices* **ED-33** (1986) 322–326.
220. D. M. Mattox, in *Deposition Technologies for Films & Coatings: Developments and Applications*, ed. F. Bunshah, Noyes Publications, Park Ridge, NJ (1982) 63–82.
221. D. M. Mattox, *Thin Solid Films* **124** (1985) 3–10.
222. W. Kern, *Microelectronics Tech. Symp. on Process Chemical Contamination Control Technology*, Millipore Corporation, May 20, 1985, Santa Clara, CA.
223. J. R. Vig, *J. Vac. Sci. Technol. A* **3** (1985) 1027–1034.
224. W. Kern, *Semiconductor Internatl.* **7** (Apr. 1984) 94–99.
225. D. Burkman, *Semiconductor Internatl.* **4** (Jul. 1981) 103–116.
226. P. S. Burggraaf, *Semiconductor Internatl.* **4** (Jul. 1981) 91–95, 97–100.
227. S. Shwartzman, A. Mayer and W. Kern, *RCA Rev.* **46** (1985) 81–105.
228. Anon., *Semiconductor Internatl.* **7** (13) (1985) (1985 Master Buying Guide) 78–81; **8** (13) (1985) (1986 Master Buying Guide) 76–77.
229. R. E. Blaha and W. R. Fahrner, *J. Electrochem. Soc.* **123** (1976) 515–518.
230. M. Polinsky and S. Graf, *IEEE Trans. Electron Devices* **ED-20** (1973) 239–244.
231. Y. H. Xie, K. L. Wang and Y. C. Kao, *J. Vac. Sci. Technol. A* **3** (1985) 1035–1039.
232. D. L. Partin, A. G. Milnes and L. F. Vassamillet, *J. Electrochem. Soc.* **126** (1979) 1581–1583.
233. J. S. Judge, *J. Electrochem. Soc.* **118** (1971) 1772–1775.
234. S. A. Harrell and J. R. Peoples, Jr., *Electrochem. Soc. Extended Abstr.* **112** (1965) 247.
235. D. S. Herman, M. A. Schuster and H. G. Oehler, *Electrochem. Soc. Extended Abstr.* **71–1** (1971) 167–169.
236. J. J. Gajda, *12th Ann. Proc. Reliab. Physics*, Institute of Electrical and Electronics Engineers, New York (1974) 30–37.
237. W. Kern, *RCA Rev.* **37** (1976) 78–106.
238. C. Pavelescu, C. Cobianu and E. Segal, *J. Mater. Sci. Lett.* **3** (1984) 643–646.
239. N. S. Alvi, K. E. Stone and J. R. Weaver, *Electrochem. Soc. Extended Abstr.* **81–2** (1981) 596–598; and N. S. Alvi, private communication.
240. J. J. Gajda, US Patent 4, 230, 523, Oct. 28, 1980.
241. H. P. Kleinknecht and H. Meier, *J. Electrochem. Soc.* **125** (1978) 798–803.
242. A. D. Butherus, T. W. Hou, C. J. Mogab and H. Schonhorn, *J. Vac. Sci. Technol. B* **3** (1985) 1352–1356.
243. N. Endo and S. Matsui, *Jap. J. Appl. Phys.* **22** (1983) L109–L111.
244. T. L. Chu, *J. Vac. Sci Technol.* **6** (1969) 25–33.
245. W. Kern, unpubl.
246. R. Herring and J. B. Price, *Electrochem. Soc. Extended Abstr.* **73–2** (1973) 410–412.
247. J. Dey, M. Lundgren and S. Harrel, in *Kodak Photoresist Seminar Proceedings*, Vol. II, Eastman Kodak Co., Rochester, NY (1968) (P-192-B) 4–11.
248. W. A. Pliskin and R. P. Esch, in ref. 70, 37–46.
249. C. C. Mai and J. C. Looney, *SCP & Solid State Technol.* **9** (Jan. 1966) 19–24.
250. V. Harrap, *Semiconductor Silicon 1973*, eds. H. R. Huff and R. R. Burgess, The Electrochemical Society, Princeton (1973) 354–362.
251. W. Kern, *RCA Rev.* **29** (1968) 557–565.
252. C. R. Tellier, *J. Mater. Sci.* **17** (1982) 1348–1354.
253. E. A. Irene, D. W. Dong and R. J. Zeto, *J. Electrochem. Soc.* **127** (1980) 396–399.
254. R. K. Iler, in *The Chemistry of Silica: Solubility, Polymerization, Colloid and Surface Properties*, John Wiley, New York (1979) 3–115.
255. E. F. Duffek and D. Pilling, *Electrochem. Soc. Extended Abstr.* **111** (1965) 244–246.
256. L. K. White, *Thin Solid Films* **79** (1981) L73–L76.
257. A. C. Adams, in *VLSI Technology*, ed. S. M. Sze, McGraw-Hill, New York (1983) 93–129.
258. K. Watanabe, T. Tangaki and S. Wakayama, *J. Electrochem. Soc.* **128** (1981) 2630–2635.

259. C. Pavelescu, C. Cobianu, L. Condriuc and E. Segal, *Thin Solid Films* **114** (1984) 291–294.
260. C. Cobianu and C. Pavelescu, *J. Mater. Sci. Lett.* **3** (1984) 979–982.
261. F. Gualandris, G. U. Pignatel and S. Rojas, *J. Vac. Sci. Technol. B* **3** (1985) 1604–1608.
262. A. C. Adams and C. D. Capio, *J. Electrochem. Soc.* **126** (1979) 1042–1046.
263. M. Vandenberg, *Electrochem. Soc. Extended Abstr.* **79–1** (1979) 262–264.
264. S. M. Ojha, in *Physics of Thin Films*, Vol. 12, Academic Press, New York (1982) 237–296.
265. A. C. Adams, *Solid State Technol.* **26** (Apr. 1983) 135–139.
266. B. Gorowitz, T. B. Gorczya and R. J. Saia, *Solid State Technol.* **28** (Aug. 1985) 197–203.
267. P. Pan, L. A. Nesbit, R. W. Douse and R. T. Gleason, *J. Electrochem. Soc.* **132** (1985) 2012–2019.
268. A. C. Adams, *Electrochem. Soc. Extended Abstr.* **85–2** (1985) 377.
269. R. F. Sarkozy, *1981 Symposium on VLSI Technology, Digest of Technical Papers*, IEEE Cat. No. 81 CH 1711–1, 68–69.
270. P. K. Boyer, C. A. Moore, R. Solanki, W. K. Ritchie and G. J. Collins, *Ext. Abstr. 15th Conf. Solid State Devices Mater. 1983*, (1983) 109–115.
271. Y. Tarui, J. Hidaka and K. Aota, *Jap. J. Appl. Phys.* **23** (1984) L827–L829.
272. J. Y. Chen, R. C. Henderson, J. T. Hall and J. W. Peters, *J. Electrochem. Soc.* **131** (1984) 2146–2151.
273. R. Solanki, C. A. Moore and G. J. Collins, *Solid State Technol.* **28** (June 1985) 220–227.
274. Y. Minowa and K. Yamanishi, *J. Vac. Sci. Technol. B* **1** (1983) 1148–1151; T. Kubota, *Jap. J. Appl. Phys.* **11** (1972) 1413–1420.
275. L. E. Katz and W. C. Erdman, *J. Electrochem. Soc.* **123** (1976) 1249–1251.
276. S. C. H. Lin and I. J. Pugac-Muraszkiewicz, *J. Appl. Phys.* **43** (1972) 119–125.
277. I. J. Pugacz-Muraszkiewicz and B. R. Hammond, *J. Vac. Sci. Technol.* **14** (1977) 49–53.
278. R. V. Prasad, F. A. Selim and D. D. L. Chung, *Mater. Lett.* **4** (1986) 71–76.
279. J. A. Aboaf, *J. Electrochem. Soc.* **114** (1967) 948–952.
280. M. T. Duffy and W. Kern, *RCA Rev.* **31** (1970) 754–770.
281. M. Mutoh, Y. Mizokami, H. Matsui, S. Hagiwara and M. Ino, *J. Electrochem. Soc.* **122** (1975) 987–992.
282. H. Katto and Y. Koga, *J. Electrochem. Soc.* **118** (1971) 1619–1623.
283. A. J. Learn, *J. Appl. Phys.* **44** (1973) 1251–1258.
284. K. Iida, *J. Electrochem. Soc.* **124** (1977) 614–623 (See also ref. 70, 56–62).
285. E. Ferrieu and B. Pruniaux, *J. Electrochem. Soc.* **116** (1969) 1008–1013.
286. R. S. Nowicki, *J. Vac. Sci. Technol.* **14** (1977) 127–133.
287. H. Harada, S. Satoh and M. Yoshida, *IEEE Trans. Reliability* **R-25** (1976) 290–295.
288. G. C. Schwartz and V. Platter, *J. Electrochem. Soc.* **122** (1975) 1508–1516.
289. E. L. MacKenna, *Proc. 1971 Semicond./IC Proc. and Prod. Conf*, Ind. and Sci. Conf. Management, Chicago, (1971) 71–83.
290. P. J. Tsang, R. M. Anderson and S. Cvikevich, *J. Electrochem. Soc.* **123** (1976) 57–63.
291. K. M. Schlesier, J. M. Shaw and C. W. Benyon, Jr., *RCA Rev.* **37** (1976) 358–388.
292. A. Reisman, M. Berkenblit, J. Cuomo and S. A. Chan, *J. Electrochem. Soc.* **118** (1971) 1653–1657.
293. M. F. Ehman, *J. Electrochem. Soc.* **121** (1974) 1240–1243.
294. M. Safdar, G. H. Frischat and H. Salge, *J. Amer. Ceram. Soc.* **57** (1974) 106–107.
295. B. Siesmayer, R. Heimann, W. Franke and L. Lacmann, *J. Cryst. Growth* **28** (1975) 157–161.
296. E. Kaplan, M. Balog and D. Frohman-Bentchkowsky, *J. Electrochem. Soc.* **123** (1976) 1570–1573.
297. J. P. S. Pringle, *J. Electrochem. Soc.* **119** (1972) 482–491.
298. P. W. Wyatt, *J. Electrochem. Soc.* **122** (1975) 1660–1666.
299. W. H. Knausenberger and R. N. Tauber, *J. Electrochem. Soc.* **120** (1973) 927–931.
300. J. Grossman and D. S. Herman, *J. Electrochem. Soc.* **116** (1969) 674.
301. H. M. Day, A Christou, W. H. Weisenberger and J. K. Hervonen, *J. Electrochem. Sec.* **122** (1975) 769–772.
302. B. H. Hill, *J. Electrochem. Soc.* **115** (1969) 668–669.
303. J. M. Eldridge and P. Balk, *Trans. Metallurg. Soc. AIME* **242** (1968) 539–545.
304. E. H. Snow and B. E. Deal, *J. Electrochem. Soc.* **113** (1966) 263–269.
305. A. S. Tenney and M. Ghezzo, *J. Electrochem. Soc.* **120** (1973) 1091–1095.
306. L. Hall, *J. Electrochem.* Soc. **118** (1971) 1506–1507.
307. T. Tokuyama, T. Miyazaki and M. Horiuchi, in *Thin Film Dielectrics*, ed. F. Vratny, The

Electrochemical Society, Inc., New York (1969) 297–326.
308. W. Kern and R. C. Heim, *J. Electrochem. Soc.* **117** (1970) 562–568, 568–573.
309. K. Jinno, H. Kinoshita and Y. Matsumoto, *J. Electrochem. Soc.* **124** (1977) 1258–1262.
310. K. Chow and L. G. Garrison, *J. Electrochem. Soc.* **124** (1977) 1133–1136.
311. A. C. Adams, C. D. Capio, S. E. Haszko, G. I. Parisi, E. I. Povilonis and McD. Robinson, *J. Electrochem. Soc.* **126** (1979) 313–319.
312. R. A. Levy, S. M. Vincent and T. E. McGahan, *J. Electrohem. Soc.* **132** (1984) 1742–1749.
313. A. Takamatsu, M. Shibata, H. Sakai, and T. Yoshimi, *J. Electrochem. Soc.* **131** (1984) 1865–1870.
314. K. Sugawara, T. Yoshimi and H. Sakai, *Chemical Vapor Deposition*, Fifth Internatl. Conf. 1975, eds. J. M. Blocher, Jr. and H. E. Hinterman, The Electrochemical Soc., Inc., Princeton (1975) 407–412.
315. G. Smolinsky and T. P. H. F. Wendling, *J. Electrochem. Soc.* **132** (1985) 950–954.
316. H. Itoh, M. Hatanaka, Y. Akasaka and H. Nakata, Ext. Abstr., 16th (1984 Internatl.) Conf. on Solid State Devices and Matl., Kobe, 437–440.
317. P. F. Schmidt, W. van Gelder and J. Drobek, *J. Electrochem. Soc.* **115** (1968) 79–84.
318. P. Balk and J. M. Eldridge, *Proc. IEEE* **57** (1969) 1558–1563.
319. W. Kern, *RCA Rev.* **37** (1976) 55–77.
320. J. Houskova, K.-K. N. Ho and M. K. Balazs, *Semiconductor Internatl.* **8** (1985) 236–241.
321. A. H. El-Hoshy, *J. Electrochem. Soc.* **117** (1970) 1583–1584.
322. D. M. Brown and R. P. Kennicott, *J. Electrochem. Soc.* **118** (1971) 293–300.
323. F. N. Schwettmann, R. J. Dexter and D. F. Cole, *Electrochem. Soc. Extended Abstr.* **72–2** (1972) 623–624.
324. L. Rankel Plauger, *J. Electrochem. Soc.* **120** (1973) 1428–1430.
325. K. O. Schwenker, *J. Electrochem. Soc.* **118** (1971) 313–317.
326. F. N. Schwettman, R. J. Dexter and D. F. Cole, *J. Electrochem. Soc.* **120** (1973) 1566–1570.
327. S. S. Chang, US Patent 3, 784, 424; Jan. 8, 1974.
328. C. Ramiller and L. Yau, SEMICON/WEST 82, May 25, 1982, San Mateo, CA; Technical Program Proceedings, 29–37.
329. W. Kern and G. L. Schnable, *RCA Rev.* **43** (1982) 423–457.
330. W. Kern, W. Kurylo and C. J. Tino, *RCA Rev.* **46** (1985) 117–152.
331. W. Kern and R. K. Smeltzer, *Solid State Technol.* **28** (June 1985) 171–179.
332. I. Avigal, *Solid State Technol.* **26** (Oct. 1983) 217–224; see also *Errata*, **27** (1984) 123, 139.
333. J. E. Tong, K. Schertenleib and R. A. Carpio, *Solid State Technol.* **27** (Jan. 1984) 161–170.
334. F. S. Becker and D. Pawlick, *Electrochem. Soc. Extended Abstr.* **85–2**, (1985) 380–383; and *Proc. Symp. on Reduced Temperature Processing for VLSI*, eds. R. Reif and G. R. Shrinivasan, the Electrochemical Society Inc., Pennington, NJ (1986) 148–159.
335. A. J. Learn and W. Baerg, *Thin Solid Films* **130** (1985) 103–111.
336. S. L. Chang, K. Y. Tsao, M. A. Meneshian and H. A. Waggener, in VLSI Science and Technology/1985, Proc. Third Internatl. Symp., Proc. Vol. 85–5, eds. W. M. Bullis and S. Broydo, the Electrochemical Society, Inc., Pennington, NJ (1985) 231–236.
337. T. Abe, K. Sato, M. Konaka and A. Miyazaka, *Suppl. J. Jap. Soc. Appl. Phys.* **39** (1970) 88–93.
338. A. Igbal, W. I. Lehrer and J. M. Pierce, *Electrochem. Soc. Extended Abstr.* **83–2** (1983) 359–360.
339. G. Burton, P. Tunstasood, F. Chien, R. Kovacs and M. Vora, *IEDM Technical Digest*, IEDM **84** (1984) 582–585.
340. K. Nassau, R. A. Levy and D. L. Chadwick, *J. Electrochem. Soc.* **132** (1985) 409–415.
341. R. A. Levy and K. Nassau, in *Proc. Symp. on Reduced Temperature Processing for VLSI*, eds. R. Reif and G. R. Shrinivasan, The Electrochemical Society, Inc., Pennington, NJ (1986) 132–147.
342. D. M. Brown, M. Garfinkel, M. Ghezzo, E. A. Taft, A. Tenney and J. Wong, *J. Cryst. Growth* **17** (1972) 276–287.
343. G. W. B. Ashwell and S. J. Wright, *Semiconductor Internatl.* **8** (Jan. 1985) 132–134; *Proc. First Internatl. IEEE VLSI Multilayer Interconnection Conf.*, New Orleans, LA, June 21–22, 1984, Institute of Electrical and Electronics Engineer, New York (1985) 122–129.
344. G. W. B. Ashwell and S. J. Wright, *1985 Proc. Second Internatl. IEEE VLSI Multilayer Interconnection Conf.*, June 25–26, 1985, Santa Clara, CA, Institute of Electrical and Electronics Engineers, New York (1985) 285–291.
345. J. Wong, *J. Electrochem. Soc.* **119** (1972) 1071–1080.

346. M. Ghezzo and D. M. Brown, *J. Electrochem. Soc.* **120** (1973) 110–116.
347. C. A. Deckert, *J. Electrochem. Soc.* **125** (1978) 320–323.
348. W. van Gelder and V. E. Hauser, *J. Electrochem. Soc.* **144** (1967) 869–872.
349. D. C. Miller, *J. Electrochem. Soc.* **120** (1973) 1771–1774.
350. C. A. Deckert and G. L. Schnable, US Patent 4, 269, 654; May 26, 1981.
351. A. S. Squillance, A. E. Martin and J. J. Rudmann, US Patent 3, 811, 974, July 19, 1971.
352. V. D. Wohlheiter, *J. Electrochem. Soc.* **122** (1975) 1736–1739.
353. W. Kern and R. S. Rosler, *J. Vac. Sci. Technol.* **14** (1977) 1082–1099.
354. W. A. Lanford and M. J. Rand, *Electrochem. Soc. Extended Abstr.* **77–2** (1977) 421–422.
355. R. Chow, W. A. Lanford, W. Ke-Ming, and R. S. Rosler, *J. Appl. Phys.* **53** (1982) 5630–5633.
356. W. A. P. Claassen, W. G. J. N. Valkenburg, W. M. v. d. Wijgert and M. F. C. Willemsen, *Thin Solid Films* **129** (1985) 239–247.
357. A. Hiraiwa, K. Mukai, S. Harada, T. Yoshimi and S. Itoh, *Jap. J. Appl. Phys.* **18** (1979) 191–192.
358. W. Kern, unpubl. report.
359. Y. K. Fang, C. F. Huang, C. Y. Chang and R. H. Lee, *J. Electrochem. Soc.* **132** (1985) 1222–1225.
360. C.-E. Morosanu, *Thin Solid Films* **65** (1980) 171–208.
361. M. Watanabe, M. Harazono, Y. Hiratsuka and T. Edamura, in *Silicon Nitride Thin Insulating Films*, Symp. Proc. Vol. **83–8**, eds. V. J. Kapoor and H. J. Stein, The Electrochemical Society, Inc. Pennington, NJ (1983) 488–496.
362. J. A. Nemetz and R. E. Tessler, *Solid State Technol.* **26** (Feb. 1983) 79–85; **26** (Sept. 1983) 209–215.
363. M. M. Moslehi, C. J. Han, K. C. Saraswat, C. R. Helms and S. Shatas, *J. Electrochem. Soc.* **132** (1985) 2189–2197.
364. D. M. Brown, P. V. Gray, F. K. Heumann, H. R. Philipp and E. A. Taft, *J. Electrochem. Soc.* **115** (1968) 311–317.
365. M. J. Rand and J. F. Roberts, *J. Electrochem. Soc.* **120** (1973) 446–453.
366. J. R. Hollahan and R. S. Rosler, in *Thin Film Processes*, eds. J. L. Vossen and W. Kern, Academic Press, New York (1978) 335–360.
367. A. R. Reinberg, *Ann. Rev. Mater. Sci.* **9** (1979) 341–372.
368. T. D. Bonifield, in *Deposition Technologies for Films and Coatings*, eds. R. F. Bunshah *et al.*, Noyes Publications, Park Ridge, NJ (1982) 365–384.
369. H. Y. Kumagai, in *Chemical Vapor Deposition 1984*, Proc. Ninth Internatl Conf., Proc. vol. **84–6**, eds. McD. Robinson, C. H. J. van den Brekel, G. W. Cullen and J. M. Blocher, Jr., The Electrochemical Society Inc., Pennington, NJ (1984) 189–204.
370. Y. Catherine, in *Plasma Processing*, Proc. Fifth Symp., proc. vol. **85–1**, eds. G. S. Mathad, G. C. Schwartz, and G. Smolinsky, The Electrochemical Society, Inc., Pennington, NJ (1985) 317–344.
371. P. N. Kember, S. C. Lidell and P. Blackborrow, *Semiconductor Internatl.* **8** (1985) 158–161.
372. E. A. Taft, *J. Electrochem. Soc.* **118** (1971) 1341–1346.
373. A. K. Sinha, H. J. Levinstein, T. E. Smith, G. Quintana and S. E. Haszko, *J. Electrochem. Soc.* **125** (1978) 601–608.
374. H. J. Stein, V. A. Wells and R. E. Hampy, *J. Electrochem. Soc.* **126** (1979) 1750–1754.
375. F. Sequeda and R. E. Richardson, Jr., *J. Vac. Sci. Technol.* **18** (1981) 362–363.
376. W. A. P. Claassen, W. G. J. N. Valkenburg, F. H. P. M. Habraken and Y. Tamminga, in *Silicon Nitride Thin Insulating Films*, Symp. Proc. Vol. 83–8; eds. V. J. Kapoor and H. J. Stein, The Electrochemical Society, Inc., Pennington, NJ (1983) 430–441.
377. A. D. Weiss, *Semiconductor Internatl.* **6** (July 1983) 88–94.
378. C. H. Ling, C. Y. Kwok and K. Prasad, *Phys. Stat. Sol.* (a) **84** (1984) K1–K4.
379. C. Blaauw, *J. Electrochem. Soc.* **131** (1984) 1114–1118.
380. D. M. Chen and A. W. Swanson, *Electrochem. Soc. Extended Abstr.* **84–1** (1984) 157–158.
381. P. W. Bohn and R. C. Manz, *J. Electrochem. Soc.* **132** (1985) 1981–1984.
382. J. W. Peters, F. L. Gebhart and T. C. Hall, *ISHM 1979 Proc.*, Los Angeles (1979) 128–136.
383. J. W. Peters, F. L. Gebhart and T. C. Hall, *Solid State Technol.* **23** (Sept. 1980) 121–126.
384. Y. Numasawa, K. Yamazaki and K. Hamano, *Jap., J. Appl. Phys.* **22** (1983) L792–L794.
385. T. S. Moss, *Phys. Stat. Sol.* (*b*) **131** (1985) 415–427.
386. H. Robbins and B. Schwartz, *J. Electrochem. Soc.* **106** (1959) 505–508.

387. H. Robbins and B. Schwartz, *J. Electrochem. Soc.* **107** (1960) 108–111.
388. B. Schwartz and H. Robbins, *J. Electrochem. Soc.* **108** (1961) 365–372.
389. A. F. Bogenschütz, W. Krusemark, K.-H. Löcherer and W. Mussinger, *J. Electrochem. Soc.* **114** (1967) 970–973.
390. B. Schwartz and H. Robbins, *J. Electrochem. Soc.* **123** (1976) 1903–1909.
391. D. G. Schimmel and M. J. Elkind, *J. Electrochem. Soc.* **125** (1978) 152–155.
392. D. G. Schimmel, in ref. 116, 5–1 to 5–15.
393. L. D. Dyer, *Electrochem. Soc. Extended Abstr.* **86–1** (1986) 329–330.
394. C. S. Fuller and H. W. Allison, *J. Electrochem. Soc.* **109** (1962) 880.
395. J. C. Dyment and G. A. Rozgonyi, *J. Electrochem. Soc.* **118** (1971) 1346–1350.
396. L. R. Plauger, *J. Electrochem. Soc.* **121** (1974) 455–457.
397. A. Milch, *J. Electrochem. Soc.* **123** (1976) 1256–1258.
398. Y. Mori and N. Watanabe, *J. Electrochem. Soc.* **125** (1978) 1510–1514.
399. S. Adachi and K. Oe, *J. Electrochem. Soc.* **131** (1984) 126–130.
400. D. E. Aspnes and A. A. Studna, *Appl. Phys. Lett.* **46** (1985) 1071–1073.
401. J. Massies and J. P. Contour, *J. Appl. Phys.* **58** (1985) 806–810.
402 J.-A. Chen, S.-C. Lee and T.-I. Ho, *J. Electrochem. Soc.* **132** (1985) 3016–3019.
403. J. van de Ven, J. E. A. M. van den Meerakker and J. J. Kelly, *J. Electrochem. Soc.* **132** (1985) 3020–3033.
404. J. van de Ven, J. E. A. M. van den Meerakker and J. J. Kelly, *J. Electrochem. Soc.* **133** (1986) 799–806.
405. K. E. Bean and P. S. Gleim, *Proc. IEEE* **57** (1969) 1469–1476.
406. A. I. Stoller, *RCA Rev.* **31** (1970) 271–275.
407. H. I. Smith, *Proc. IEEE* **62** (1974) 1361–1387.
408. H. C. Nathanson *et al.*, Special Issue on Three-Dimensional Semiconductor Device Structures, *IEEE Trans. Electron Devices* **ED–25** (1978) 1177–1270.
409. E. Bassous, *IEEE Trans. Electron Devices* **ED–25** (1978) 1178–1185.
410. K. E. Petersen, *Proc. IEEE* **70** (1982) 420–457.
411. D. V. Podlesnik, H. H. Gilgen, R. M. Osgood, Jr. and A. Sanchez, *Appl. Phys. Lett.* **43** (1983) 1083–1085.
412. B. Petit, J. Pelletier and R. Molins, *J. Electrochem. Soc.* **132** (1985) 982–984.
413. S. Hava and R. G. Hunsperger, *IEEE Trans. Electron Devices* **ED–32** (1985) 993–996.
414. G. Kaminsky, *J. Vac. Sci. Technol. B* **3** (1985) 1015–1024.
415. L. Csepregi, *Microelectronic Engg.* **3** (1985) 221–234.
416. H. H. Busta and R. D. Cuellar, *Electrochem. Soc. Extended Abstr.* **86–1** (1986) 403–404.
417. L. L. Kazmerski (ed.), *Polycrystalline and Amorphous Thin Films and Devices*, Academic Press, New York (1980).
418. G. Harbeke, L. Krausbauer, E. F. Steigmeier, A. E. Widmer, H. F. Kappert and G. Neugebauer, *J. Electrochem. Soc.* **131** (1984) 675–683.
419. T. P. Brody, *IEEE Trans. Electron. Devices* **ED–31** (1984) 1614–1628.
420. M. J. Thompson, *J. Vac. Sci. Technol. B* **2** (1984) 827–834.
421. Y. Hamakawa (ed.) *JARECT*, Vol. 16, OHM, Tokyo and North-Holland, Amsterdam (1984).
422. S. C. Jain and S. K. Mehta, *Solar Cells* **15** (1985) 285–305.
423. K. W. Mitchell and K. J. Touryan, in *Annual Review of Energy*, Vol. 10, Annual Reviews Inc., Palo Alto, CA, 1985.
424. M. H. Brodsky, *Amorphous Semiconductors*, (*Topics in Applied Physics*, Vol. 36) Springer Verlag, New York (1985).
425. G. Harbeke (ed.), *Polycrystalline Semiconductors*, Springer Verlag, Berlin (1985).
426. H. F. Bare and G. W. Neudeck, *J. Vac. Sci. Technol. A* **4** (1986) 239–241.
427. K. Takahashi and M. Konagal, *Amorphous Silicon Solar Cells*, John Wiley, Somerset, NJ (1986).
428. J. B. Price, in *Semiconductor Silicon 1973*, eds. H. R. Huff and R. R. Burgess, The Electrochemical Society, Princeton, NJ (1973) 339–353.
429. D. F. Weirauch, *J. Appl. Phys.* **46** (1975) 1478–1483.
430. M. J. Declercq, L. Gerzberg and J. D. Meindl, *J. Electrochem. Soc.* **122** (1975) 545–552.
431. K. E. Bean, *IEEE Trans. Electron. Devices* **ED–25** (1978) 1185–1193.
432. I. Barycka, H. Teterycz and Z. Znamirowsky, *J. Electrochem. Soc.* **126** (1979) 345–346.
433. A. Reisman, M. Berkenblit, S. A. Chan, F. B. Kaufman and D. C. Green, *J. Electrochem. Soc.* **126** (1979) 1406–1415.

434. D. L. Kendall, *Ann. Rev. Mater. Sci.* **9** (1979) 373–403.
435. M. M. Abu-Zeid, *J. Electrochem. Soc.* **131** (1984) 2138–2142.
436. O. J. Glembocki, R. E. Stahlbush and M. Tomkiewicz, *J. Electrochem. Soc.* **132** (1985) 145–151.
437. E. D. Palik, V. M. Bermudez and O. J. Glembocki, *J. Electrochem. Soc.* **132** (1985) 871–884.
438. M. M. Abu-Zeid, D. L. Kendall, G. R. DeGuel and R. Galaezzi, *Electrochem. Soc. Extended Abstr.* **85–1** (1985) 400–401.
439. D. L. Kendall, G. R. DeGuel and S.-M. Park, *J. Electrochem Soc.* **132**, Abstract 739 RNP (June 1985) 221C.
440. S. Polchlopek, R. Kwor, P. L. F. Hemment, K. J. Resson and D. Trumble, *Electrochem. Soc. Extended Abstr.* **85–2** (1985) 473.
441. M. M. Abu-Zeid, *Electrochem. Soc. Extended Abstr.* **86–1** (1986) 325–326.
442. B. E. Deal and A. S. Grove, *J. Appl. Phys.* **36** (1965) 3770–3778.
443. C. P. Ho, J. D. Plummer, J. D. Meindl and B. E. Deal, *J. Electrochem. Soc.* **125** (1978) 665–671.
444. J. D. Plummer, in *Semiconductor Silicon 1981*, eds H. R. Huff and R. J. Kriegler, the Electrochemical Society, Pennington, NJ (1981) 445–454.
445. R. B. Fair, in *Silicon Integrated Circuits*, Part B, ed. D. Kahng (Supplement 2B of *Applied Solid State Science*, ed. R. Wolfe), Academic Press, New York (1981) 1–108.
446. L. E. Katz, in *VLSI Technology*, ed. S. M. Sze, McGraw-Hill, New York (1983) 131–167.
447. E. A. Taft, *J. Electrochem. Soc.* **132** (1985) 2486–2489.
448. H. Z. Massoud, J. D. Plummer and E. A. Irene, *J. Electrochem. Soc.* **132** (1985) 2685–2700.
449. S. A. Schwarz and M. J. Schulz, in *VLSI Electronics: Microstructure Science*, Vol. 10, eds. N. G. Einspruch and R. S. Bauer, Academic Press, Orlando, FL (1985) 29–77.
450. S. Iida and K. Ito, *J. Electrochem. Soc.* **118** (1971) 768–771.
451. L. Hollan, J. C. Tranchart and R. Memming, *J. Electrochem. Soc.* **126** (1979) 855–859.
452. S. Adachi and H. Kawaguchi, *J. Electrochem. Soc.* **128** (1981) 1342–1349.
453. L. Edwards-Shea, *GEC J. of Research* **3** (1) (1985) 55–57.
454. E. A. James, RCA Laboratories, Private communications.
455. R. Muto and S. Furuuchi, *Rep. Res. Lab., Asahi Glass Co. Ltd.* **23** (1973) 27–44.
456. J. Kane, H. P. Schweizer and W. Kern, *J. Electrochem. Soc.* **123** (1976) 270–277.
457. B. J. Baliga and S. K. Ghandhi, *J. Electrochem. Soc.* **124** (1977) 1059–1060.
458. R. Weiner, *Metalloberfläche* **27** (1973) 441–447.
459. J. J. Kelly and G. J. Koel, *Philips Tech. Rev.* **38** (6) (1978/79) 149–157.
460 J. J. Kelly and C. H. de Minjer, *J. Electrochem. Soc.* **122** (1975) 931–936.
461. T. Agatsuma, A. Kikuchi, K. Nakada and A. Tomozawa, *J. Electrochem. Soc.* **122** (1975) 825–829.
462. W. Kern and R. B. Comizzoli, *J. Vac. Sci. Technol.* **14** (1977) 32–39.
463. C. W. Halsted and M. W. Haller, *Electrochem. Soc. Extended Abstr.* **72–2** (1976) 748–750.
464. T. Hara, T. Hirayama, H. Ando and M. Furukawa, *J. Electrochem. Soc.* **132** (1985) 2973–2975.
465. J. Y. Tsao and D. J. Ehrlich, *Appl. Phys. Lett.* **43** (1983) 146–148.
466. W. Kern and J. M. Shaw, *J. Electrochem. Soc.* **118** (1971) 1699–1704.
467. Pamphlet P-91, Eastman Kodak Co., Rochester, NY (1966).
468. T. A. Shankoff and E. A. Chandross, *J. Electrochem. Soc.* **122** (1975) 294–298.
469. G. S. Kelsey, *J. Electrochem. Soc.* **124** (1977) 927–933.
470. S. P. Murarka, *Silicides for VLSI Applications*, Academic Press, New York (1983).
471. Y. Yamamoto, H. Miyanaga, T. Amazawa and T. Sakai, *IEEE Trans. Electron Devices* **ED-32** (1985) 1231–1239.
472. S. P. Murarka, *Solid State Technol.* (Sept. 1985) 181–185.
473. D. R. McLachlan and J. B. Avins, *Semiconductor Internatl.* **7** (Oct. 1984) 129–138.
474. P. Burggraaf, *Semiconductor Internatl.* **8** (May 1985) 293–298.
475. R. A. M. Wolters and A. J. M. Nellissen, *Solid State Technol.* **29** (Feb. 1986) 131–136.
476. C. Y. Ting and S. S. Lyer, *1985 Proc. Second Internatl. IEEE VLSI Multilevel Interconnection Conf.*, June 25–26, 1986, Santa Clara, CA. Institute of Electrical and Electronics Engineer, New York (1985) 307–318.
477. F. H. Dill, in ref. 70, 194–197.
478. D. Burkman and A. Johnson, *Solid State Technol.* **26** (May 1983) 125–129.
479. C. G. Willson, in ref. 105, 87–159.
480. L. F. Thompson and M. J. Bowden, in ref. 105, 161–214.

481. M. J. Bowden and J. Frackoviak, in *Silicon Processing*, ASTM STP 804, ed. D. C. Gupta, American Society for Testing Materials, Philadelphia (1983) 125–143.
482. A. C. Ouano, in *Polymers in Electronics*, ACS Symp. Ser. 242, ed. T. Davidson, American Chemical Society, Washington, (1984) 79–90.
483. D. S. Soong, in *Advances in Resist Technology and Processing II*, SPIE Proc. **539**, ed. L. F. Thompson, International Society for Optical Engineering, Bellingham, WA (1985) 2–5.
484. W. D. Hinsberg, G. G. Willson and K. K. Kanazawa, in ref. 483, 6–13.
485. F. Rodriquez, R. J. Groele and P. D. Krasicky, in ref. 483, 14–20.
486. J. S. Petersen, A. Kozlowski, K. Eastwood, M. K. Iwan and M. Stan, in ref. 483, 284–290.
487. F. Rodriquez, P. D. Krasicky and R. J. Groele, *Solid State Technol.* **28** (May 1985) 125–131.
488. D. J. Belton, P. van Pelt and A. E. Morgan, in ref. 481, 273–289.
489. A. Endo and T. Yada, *Jap. J. Appl. Phys.* **23** (1984) 384–385.
490. P. H. Singer, *Semiconductor Internatl.* **8** (June 1985) 66–72.
491. W. A. Pliskin and H. S. Lehman, *J. Electrochem. Soc.* **112** (1965) 1013–1019.
492. F. S. Becker, D. Pawlik, H. Schäfer and G. Staudigl, *J. Vac. Sci. Technol. B.* **4** (1986) 732–744.
493. J. Kraitchman and J. Oroshnik, *J. Electrochem Soc.* **114** (1967) 405–406.
494. W. A. Pliskin, *J. Electrochem. Soc.* **114** (1967) 620–623.
495. W. Kern and R. B. Comizzoli, NBS Spec. Publ. **400–31** US Govt. Printing Office, Washington (1977).
496. W. Kern and R. B. Comizzoli, *J. Vac. Sci. Technol.* **14** (1977) 32–39.
497. A. D. Lopez, *J. Electrochem. Soc.* **113** (1966) 89–90.
498. R. M. Finne and D. L. Klein, *J. Electrochem. Soc.* **114** (1967) 965–970.
499. J. M. Crishal and A. L. Harrington, personal communication.
500. I. J. Pugacz-Muraszkiewicz, *IBM J. Res. & Dev.* **16** (1972) 523–529.
501. S. G. Garg, J. H. Nevin, R. A. Bailey and S. A. Sefick, *Microelectronics and Reliab.* **22** (1982) 611–613.
502. W. F. Keenan and W. R. Runyan, *Microelectronics and Reliab.* **12** (1973) 125–138.
503. D. W. Flatley and J. T. Wallmark, *J. Electrochem. Soc.* **114** (1967) 275C.
504. V. Y. Doo and V. M. L. Sun, *Metallurg. Trans.* **1** (1970) 741–745.
505. W. Kern, R. B. Comizzoli and G. L. Schnable, *Industr. Res. Devl.* **24** (1982) 151–160.
506. W. Kern, *Industr. Res. Devel.* **24** (June 1982) 131–134.
507. W. Kern, R. B. Comizzoli and G. L. Schnable, *RCA Rev.* **43** (1982) 310–338.
508. M. Maeda and H. Nakamura, *Thin Solid Films* **112** (1984) 279–288.
509. R. G. Vardiman, *J. Electrochem. Soc.* **118** (1971) 1804–1809.
510. W. C. Dash, *J. Appl. Phys.* **27** (1956) 1193–1195.
511. P. J. Holmes, *Proc. IEE* **106B**, Suppl. 15 (1959) 861.
512. E. Sirtl and A. Adler, *Z. Metallkde.* **52** (1961) 529–531.
513. F. Secco d'Aragona, *J. Electrochem. Soc.* **119** (1972) 948–951.
514. M. Wright Jenkins in ref. 70, 63–75.
515. M. Wright Jenkins, *J. Electrochem. Soc.* **124** (1977) 757–762.
516. D. G. Schimmel, *J. Electrochem. Soc.* **126** (1979) 479–483.
517. V. D. Archer, *J. Electrochem. Soc.* **129** (1982) 2074–2076.
518. R. B. Heimann, *J. Mater. Sci.* **19** (1984) 1314–1320.
519. K. H. Yang, *J. Electrochem. Soc.* **131** (1984) 1140–1145.
520. K. H. Yang, in *Semiconductor Processing*, ASTM STP 850, ed. D. C. Gupta, American Society for Testing and Materials, Philadelphia, (1984) 309–319.
521. C. P. Wu, E. C. Douglas, C. W. Mueller and R. Williams, *J. Electrochem. Soc.* **126** (1979) 1982–1988.
522. J. Patel and A. D. Trigg, *GEC J. Research* **2** (1984) 240–246.
523. Anon., *Semiconductor Internatl.* **8** (June 1985) 20.
524. B. L. Sopori, *J. Electrochem. Soc.* **131** (1984) 667–672.
525. P. A. Iles and P. J. Coppen, *Br. J. Appl. Phys.* **11** (1960) 177–184.
526. SEMI, *Book of SEMI Standards 1984*, Vol. 3, Materials Division, Semiconductor Equipment and Materials Institute, Mt. View, CA (1984).
527. M. C. Roberts, *Electrochem. Soc. Extended Abstr.* **85–2** (1985) 677–678.
528. D. C. Miller and G. A. Rozgonyi, in *Handbook on Semiconductors*, Ser. ed. T. S. Moss, Vol. 3, ed. S. P. Keller, North-Holland, Amsterdam (1980) 217–246.
529. E. Spenke and W. Heywang, *Phys. Stat. Sol. (a)* **64** (1981) 11–44.
530. K. V. Ravi, *Imperfections and Impurities in Semiconductor Silicon*, John Wiley, New York (1981).

531. V. D. Archer, *J. Electrochem. Soc.* **129** (1982) 2074–2076.
532. P. L. Fejes, H. M. Liaw and F. S. d'Aragona, *IEEE Trans. Comp. Hybr. Mfg. Technol.* **CHMT-6** (1983) 314–322.
533. J. H. Matlock, *Solid State Technol.* **26** (Nov. 1983) 111–116.
534. N. G. Einspruch and G. B. Larrabee (vol. eds.), *Materials and Process Characterization*, Vol. 6, *VLSI Electronics: Microstructure Science*, ed. N. G. Einspruch, Academic Press, Orlando, FL (1983).
535. J. H. Matlock, in *Silicon Processing*, ASTM STP 804, ed. D. C. Gupta, American Society for Testing and Materials, Philadelphia (1983) 332–361.
536. H. R. Huff and F. Shimura, *Solid State Technol.* **28** (Mar. 1985) 103–118.
537. M. K. El-Ghor and G. Rozgonyi, *Electrochem. Soc. Extended Absz*, **85–1** (1985) 398–399.
538 N. M. Johnson, S. G. Bishop and G. D. Watkins (eds.), *Microscopic Identification of Electronic Defects in Semiconductors*, MRS Vol. 46, Materials Research Society, Pittsburgh, (1985).
539. H. R. Huff, T. Abe and B. Kolbesen (eds.), *Semiconductor Silicon 1986*, The Electrochemical Society, Inc., Pennington, NJ (1986).
540. J. Nishizawa, Y. Oyama, H. Tadano, K. Inokuchi and Y. Okuno, *J. Crystal Growth* **47** (1979) 434–436.
541. W. R. Wagner, L. I. Greene and L. I. Koszi, *J. Electrochem. Soc.* **128** (1981) 1091–1904.
542. W. F. Tseng, H. Lessoff and R. Gorman, *J. Electrochem. Soc.* **132** (1985) 3067–3068.
543. G. P. Yablonskii, *Phys. Stat. Sol. (a)* **92** (1985) 431–441.
544. D. F. Weirauch, *J. Electrochem. Soc.* **132** (1985) 250–254.
545. G. L. Schnable and R. S. Keen, *Proc. IEEE* **57** (1969) 1570–1580.
546. T. Hara, N. Ohtsuka, T. Takeda and T. Yoshimi, *J. Electrochem. Soc.* **133** (1986) 1489–1491.
547. E. Philofsky and E. L. Hall, *IEEE Trans. Parts Hybr. Pkg.* **PHP-11** (1975) 281–290.
548. A. J. Learn, *J. Electrochem. Soc.* **123** (1976) 894–906.
549. L. A. Berthoud, *Thin Solid Films* **43** (1977) 319–329.
550. P. B. Ghate, *Thin Solid Films* **83** (1981) 195–205.
551. S. Vaidya, *J. Electronic Mtls.* **10** (1981) 337–346.
552. D. Pramanik and A. N. Saxena, *Solid State Technol.* **26** (Jan. 1983) 127–133; **26** (Mar. 1983) 131–138.
553. T. E. Wade *et al.*, *Proc. IEEE VLSI Multilevel Interconnection Conference* (IEEE Catalog 86 CH2337–4), June 1986, Inst. of Electrical and Electronics Engineers, New York (1986).
554. S. S. Cohen and G. S. Gildenblat, *Metal-Semiconductor Contacts and Devices*, Vol. 13 of *VLSI Electronics: Microstructure Science*, ed. N. G. Einspruch, Academic Press, Orlando, FL (1986).
555. R. W. Belcher and G. P. Hart, *20th Ann. Proc. Reliab. Phys. 1982* (1982) 202–206.
556. J. F. Casey, J. W. Meredith and G. M. Oleszek, *J. Electrochem. Soc.* **129** (1982) 354–357.
557. M. J. Howes and D. V. Morgan (eds.), *Reliability and Degradation: Semiconductor Devices and Circuits*, John Wiley, Chichester (1981).
558. J. A. Wurzback, F. J. Grunthaner, and J. Maserjian, *J. Vac. Sci. Technol.* **20** (1982) 962–967.
559. N. D. Stojadinovic, Microelectron. Reliab. **23** (1983) 609–707.
560. B. P. Richards and P. K. Footner, *Microelectronics J.* **15** (1) (1984) 5–25.
561. *Proceedings ISTFA 1985* (International Symposium for Testing and Failure Analysis—1985), The International Society for Testing and Failure Analysis, Torrance, CA (1985) and previous issues.
562. *24th Ann. Proc. Reliability Physics*, IEEE Cat. No. 86CH2256–6, Inst. of Electrical and Electronics Engineers New York (1986) and previous issues.
563. C. J. Varker and L. H. Chang, *Solid State Technol.* **26** (Apr. 1983) 143–146.
564. R. B. Marcus and T. T. Sheng, *Transmission Electron Microscopy of Silicon VLSI Circuits and Structures*, Wiley-Interscience, New York (1983).
565. M. C. Roberts, G. R. Booker, S. M. Davidson and K. J. Yallup, Inst. Phys. Conf. Ser. No. 67, Sect. 10 (1983) 467–472.
566. C. J. Werkhoven, C. W. T. Bulle-Lieuwma, B. J. H. Leunissen and M. P. A. Viegers, *J. Electrochem. Soc.* **131** (1984) 1388–1391.
567. L. Rivaud and G. Hawkins, *Rev. Sci. Inst.* **56** (1985) 563–566.
568. R. A. Morgan, *Plasma Etching in Semiconductor Fabrication*, Elsevier, Amsterdam (1985).
569. T. Sugano (ed.), *Applications of Plasma Processes to VLSI Technology*, translated by H.-G. Kim, John Wiley, New York (1986).
570. C. Murray, *Semiconductor International*, 9 (May 1986) 80–85.

12 Polyimides

Y. K. LEE and M. FRYD

Introduction

The movement in the semiconductor industry towards very-large-scale integration (VLSI) leads to multilayer construction and feature sizes of 1 μm or less. This development requires that the interlevel dielectric layer be able to planarize the underlying topography sufficiently to provide good step coverage for the next layer of metal.

Vapour-deposited silicon nitride and oxide, which are currently used in almost all microelectronic devices as the dielectric or passivating layers, coat the substrate in a conformal fashion, faithfully replicating the underlying metal topography upon which they are coated. Because of their lack of flexibility, they are prone to stress cracking at step edges, a problem which is aggravated by the higher aspect ratio of submicron size features. Finally, their relatively high dielectric constant (*c*. 5) requires thicker coatings which limit the speed of the device.

One solution to the above problem is the use of polyimides which have the following advantages:

(i) Good thermal, mechanical and electrical properties
(ii) Commercial availability of high-purity materials of various grades
(iii) Ability to be processed by standard photolithographic and various dry etching schemes
(iv) Good adhesion to silicon dioxide substrate (after priming), aluminium alloys and themselves.

Synthesis

Aromatic polyimides are usually synthesized by allowing an aromatic dianhydride and an aromatic diamine to react in an aprotic solvent with formation of the appropriate soluble polyamic acids. The reaction occurs at room temperature and is mildly exothermic, external cooling usually being necessary to minimize imide formation. The polyamic acid precursors are readily converted to polyimides by dehydration, using heat or reagents such as acetic anhydride–pyridine.

Polyimides can also be prepared from aromatic diisocyanates and aromatic dianhydrides. This reaction requires heat and a trace of water. When a high-

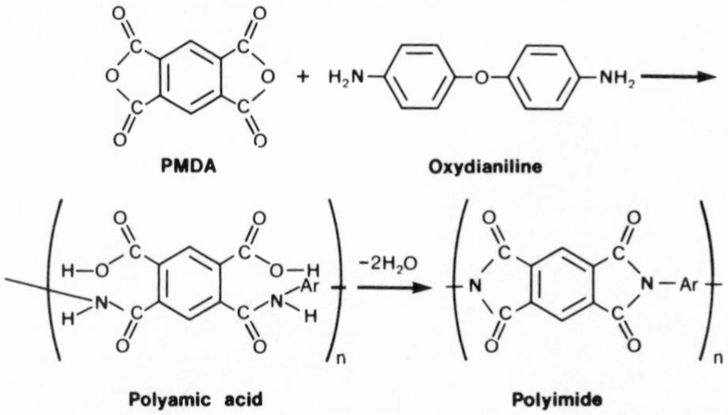

Figure 12.1 Ar = oxydianiline in the two lower structures

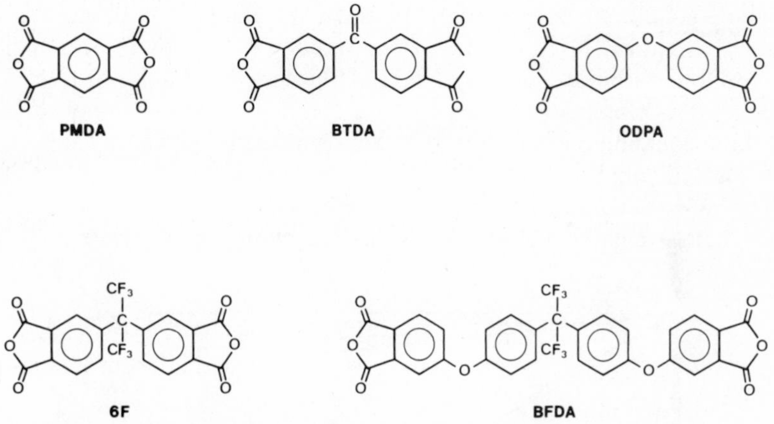

Figure 12.2 Aromatic dianhydrides used in the synthesis of polyimides

solids solution is required, polyimides can also be formed by heating an aromatic diamine with diesters of aromatic tetraacids. By far the most commercially important polyimide is that based on the condensation product of pyromellitic dianhydride, PMDA, and oxydianiline, ODA, (Figure 12.1) which is usually sold as the polyamic acid solution in *N*-methyl pyrrolidone (NMP).

A number of other polyimides, based on various combinations of the dianhydrides and diamines shown in Figures 12.2 and 12.3 below are also used in the microelectronic industry. Each one provides a different balance of properties, e.g., greater solubility, lower glass transition temperature, different electrical properties.

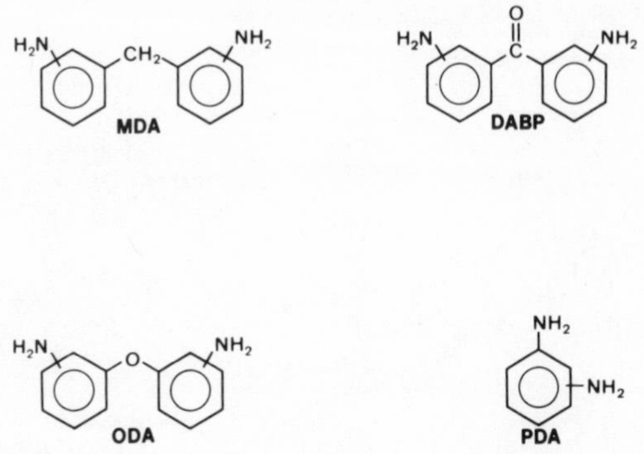

MDA DABP

ODA PDA

Figure 12.3 Aromatic diamines used in the synthesis of polyimides

Structure–property relationships

Thermal properties

Aromatic polyimides have outstanding thermal stability in both inert gases and air, as shown in Figure 12.4. This stability is due to the resonance stabilization of the aromatic system and the lack of readily abstractable hydrogens. A comparison of the thermal stability of all-aromatic polyimides and those containing even short aliphatic segments is shown in Table 12.1. This high thermal stability is critical because of the high processing temperatures encountered during sintering (430–450° C for 30 minutes in inert atmosphere) which is required for formation of reliable contacts and the

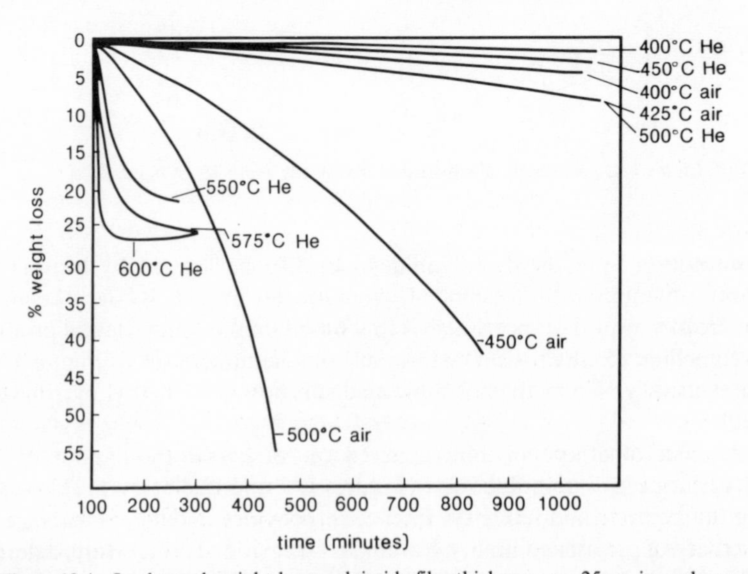

Figure 12.4 Isothermal weight loss: polyimide film thickness was 25 μm in each case

Table 12.1 Isothermal weight loss of pyromellitimides (1)

Diamine	% wt loss at 325° C			
	100 h	200 h	300 h	400 h
1, 3-diaminobenzene	3.3	4.3	5.0	5.6
4, 4′-diaminodiphenyl	2.2	3.6	5.1	6.5
4, 4′-diaminodiphenylether	3.3	4.0	5.2	6.6
4, 4′-diaminodiphenyl sulphide	4.8	5.8	6.8	7.9
4, 4′-diaminodiphenylmethane	9.4	12.9	14.7	16.8
2, 2′-bis (4-aminopheny) propane	16.1	26.2	31.0	36.0

removal of radiation damage in the inorganic oxides. Any material that is to be used as an interlevel dielectric must be able to survive the sintering step.

Mechanical properties

Like most polymeric materials the tensile strength and elongation of polyimides has been shown by Wallach (2, 3) to be a function of molecular weight. Optimum properties occur at a relatively low degree of polymerization (DP = 100) which usually suggests strong intermolecular interactions. In the case of polyimides, this interaction is thought to be due to a charge transfer complex between the dianhydride and diamine segments of neighbouring chains (4). Thus, commercially available polyimides such as those based on PMDA/ODA have elongations of 20–100% and tensile strengths of 14–25 000 psi and moduli of 400–500 000 psi.

For semiconductor applications which usually do not require such high values, W. Volksen (5) has shown that polyamic acid solutions need only exceed MW 10 000 and should preferably be close to 18 000.

Electrical properties

Like most organic polymers, polyimides have a low dielectric constant (3–3.6 at 100 and 1000 Hz) compared to inorganic materials. They have a dielectric strength of 3–4 000 volts/mil (1 mil ≡ 25 μm), a volume resistivity of 10^{18} ohm cm at 25 °C and 10^{14} ohm cm at 200 °C. The dissipation factor (1 000 Hz) is 0.003 at 25° C when the polyimide has been cured at 300° C for 1 hour. Unlike most organic materials, the dissipation factor of polyimides is relatively independent of frequency and temperature up to 200° C. This is due to the symmetry of the molecule which leads to cancelling of the dipole moments and to the high glass transition temperature (T_g) of the polyimides. For microelectronic applications, there is a large literature addressing the electrical properties of polyimides (6–14) with an equally large diversity among the reported results on the mechanism of conductivity. However, the conductivity of polyimides is so low at normal device operating temperatures that they can be considered good insulators.

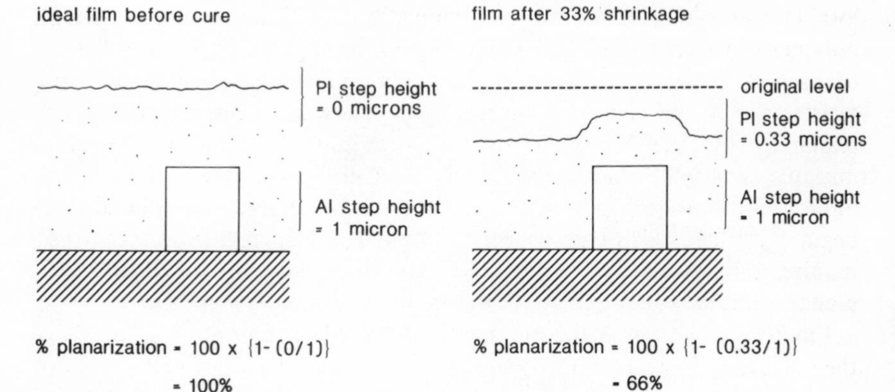

Figure 12.5 Percentage planarization. As a result of a 33% film shrinkage, the film goes from 100% planarization to 66%, representing a 33% loss of planarization

$$\% \text{ planarization} = 100\% \times \left(1 - \frac{\text{PI step height}}{\text{Al step height}}\right)$$

Planarity

As previously mentioned, a major advantage of using polyimides, rather than inorganics, as interlevel insulators in integrated circuits is their ability to planarize underlying topographical features. The degree of planarization is defined, as suggested by Rothman (15), as the percentage of reduction in step height achieved by the polyimide coating. Figure 12.5 illustrates the concept.

Ideal planarization would yield a perfectly flat surface regardless of the complexity of the underlying topography, but the ideal is, as usual, never achieved in practice. However, a fairly high degree of planarity can be achieved by manipulating three key sets of variables: percent solids, type of polyimide and number of layers.

In general, the higher the percent solids the greater the degree of planarity (Table 12.2). This is because high solids result in less shrinkage due to solvent

Table 12.2 Effect of solids on planarity

Polyimide	% of solids in solution used to coat 2 μm film	Degree of planarization over 12.7 μm line
A	13	0.17
G	15	0.32
B	18	0.30
C	32	0.72
E	36	0.78
D	47	0.78
F	60	0.82

loss. However, high solids can be obtained only with low-molecular-weight polymer which results in films with very poor mechanical properties. The use of multiple layers can result in very high planarity while maintaining satisfactory film properties. Senturia (16) has shown that one can obtain 86% planarity with only three coats and 91% with four. The problem here is that multiple coats may lead to excessively thick insulating layers. This can be minimized by using a polyimide which gives reasonably good planarity to begin with. Thus a polyimide which gives 20% planarity in one coat will require eight layers to achieve 80% planarity, while one which has 50% planarity initially requires only three coats to reach 88%.

Finally, planarity is influenced by the thermoplasticity or ability to flow of the polyimide. In general polyimides with lower glass transition temperatures will tend to exhibit a higher degree of planarity because they can continue to flow after all the solvent has evaporated and the coating has stopped shrinking.

Applications of polyimides

Currently, high-purity polyimide coatings are being used or are under development in the fabrication of semiconductor devices in the following areas: as protective overcoats; as interlayer dielectrics for multilevel devices; as alpha particle barriers; as ion-implant masks; as junction coatings; and miscellaneous applications, as described below:

Protective overcoat. The presence of pinhole defects in a passivation layer for an integrated circuit may be reduced or eliminated by the use of a polyimide topcoat over the commonly used passivation coatings of phosphosilicate glass (PSG) or silicon nitride. The chance of two defects from two separate coats occurring at the same position is highly unlikely. Therefore, the use of polyimide as a second protective coating leads to improved yield and enhanced reliability.

Interlayer dielectric. Vapour-deposited SiO_2 can be used as an interlevel dielectric in multilevel structures on monolithic integrated circuits (17–21). However, with this construction, reliability and yield become problems due to the topography of metal edges and via holes or oxide windows for interlayer connections. Polyamic acid is a viscous liquid which will flow into the cavities and produce a relatively flat surface for the next level metallization. Since multilevel metallization may be the key to the construction of very-large-scale-integration (VLSI) devices, this application probably provides the greatest value-in-use among all polyimide applications. Multilevel metallization technologies using polyimide as the between-metal insulator have been reported where aluminium is the primary metallization. A 64 K memory chip known as SAMOS was introduced by IBM in late 1978. Recent reviews by

Table 12.3 Accelerated soft error rate versus barrier thickness

Polyimide thickness (μm)	Device hours	Failures/h^{-1}
63	135	0.563
88	406	0
103	406	0
108	406	0

Wilson and Jenson (17–18), and additional recent publications (19–22), document the use of polyimides in these areas.

Alpha particle barrier. Alpha radiation emitted by trace amounts of naturally-occurring thorium and uranium isotopes in packaging materials is a source of non-destructive soft error problems in charge-coupled devices and dynamic memories. As the device geometries shrink and critical charge levels diminish, this problem becomes much more acute. Thus, the chance for a soft error to occur increases greatly in going from 64 K to 256 K to the one megabyte RAM. G. Samuelson (23) has shown that a coating of polyimide with a thickness of no less than 80 μm over the memory chip will practically eliminate this problem, as shown in Table 12.3. PI coatings can be applied by dispensing via the syringe technique or by using a cured film with a PI adhesive.

Ion-implant mask. The current approach in making ion-implant masks involves the use of photoresists or metal masks. However, the use of conventional photoresist restricts the energy and ion beam density to low levels because high temperature will cause the resist to deform and flow. The use of metal masks is a complex and expensive process involving the use of expensive equipment. The heat resistance and etchability of polyimide either by dry or wet processes makes it feasible for this application. T. Herndon (24) reported that with reactive ion etching the polyimide film gave vias with vertical walls, a highly desirable characteristic for ion implantation.

Junction coating. Polyimide is used as a junction coating, usually on discrete devices to provide mechanical protection and to strengthen wire bonding. It is dispensed with the syringe technique over the whole device including the bonding pad area.

Polyimide coatings are also used as binders for silver in die-attach applications (25), as films for lift-off metallization (26), and as dielectric coatings for digital gallium arsenide integrated circuits (27).

Fabrication methods

The most complicated process involved in the use of a polyimide occurs when it is used as an interlayer dielectric. Die attaching, junction coating and alpha-particle protection involve only the dispensing of a solution or a paste by means of a syringe, while passivation requires only some of the steps involved in the interlayer dielectric application.

A typical series of processing steps for interlayer dielectric applications is depicted in the following scheme and described in more detail below.

Processing steps

Priming the wafer → applying the PI coating → B-staging the PI film → applying the photoresist → aligning and exposing → etching and developing → neutralizing and rinsing → stripping the photoresist → final cure → surface cleaning → metallization.

Priming the wafer. If a polyamic acid solution is spun on to a wafer, the cured coating can be readily stripped. Excellent adhesion, however, can be obtained if the wafer surface is first primed with an aluminium alcoholate, a colloidal alumina, or an organo-silane. γ-Aminopropyltriethoxysilane is the most commonly used adhesion promoter because of its ease of application, commercial availability and effectiveness at very low concentrations.

Applying PI coating. The polyamic acid solution is dispensed onto the wafer, followed by spinning to give a smooth uniform coating.

B-staging the PI film. The coating must be prebaked to remove solvent, and in the process the polyamic acid coating is partially imidized. The degree of this bake determines the conditions for etching and stripping.

Applying photoresist. A solution of photoresist is spun and dried on top of the PI coating.

Aligning and exposing. The wafer is aligned with a mask and exposed to uv radiation.

Etching and developing. The etching of PI coating and developing of the photoresist can be accomplished in one step with a dilute aqueous alkaline solution when a positive photoresist is used. Typical etchants are dilute solutions of NaOH, KOH, tetra-alkyl ammonium hydroxide, etc.

Neutralization and rinse. It is imperative that alkali-metal ions are totally removed from the system. This can be accomplished by an acetic acid solution wash followed by rinsing with deionized water.

Stripping of photoresist. The photoresist can be stripped with solvents such as acetone, butyl acetate, or higher acetates.

Final cure. The polyimide coating is baked at 300° C or higher for 1 to 2 hours in order to complete the imidization to obtain the full film properties.

Surface cleaning. The surface of cured polyimide film is cleaned by oxygen plasma, or other chemical and physical means to improve adhesion for the next level of metal.

Metallization. Aluminium or Al/Cu alloy is vapour-deposited. Chu and Tang (28) have shown that adhesion between metals and polyimides can be predicted from the metal's ability to react with the polyimide.

If a negative photoresist is used, the steps are the same except that the resist must be developed prior to the etching step. After rinsing, the partially imidized film is then etched with a dilute aqueous alkali solution. The final difference is that the PI film must be baked to 200° C before stripping because the negative PI strippers tend to attack unimidized PI film.

Polyimides may also be developed by dry etching using glow-discharge plasmas or reactive-ion etching (see Chapters 15, 16).

New developments in polyimides

Several new products which are just coming into the field will enhance even further the usefulness of polyimides in microelectronic applications.

Photosensitive polyimides which can act as their own permanent photoresist, and polyamic acids with improved adhesion which do not require primers, reduce the number of processing steps. Preimidized soluble polyimides eliminate the need for high curing temperatures. Polyimides with a low coefficient of thermal expansion (TCE) reduce stress cracking of wafers produced by mismatches in TCE. Finally, high-planarity polyimides provide the ideal route to the multilevel structures and fine geometries required for VLSI. These unique new properties are balanced by a loss of some others. In general, it is imperative that the practitioner in this field recognizes that each polyimide has its own unique balance of properties. It is necessary to choose the specific polyimide which best fits the particular processing and product property requirements.

References

1. G. M. Bower and L. W. Frost, *J. Polymer Sci. A* **1** (1963) 3135.
2. M. L. Wallach, *J. Polymer Sci. A2* **6** (1968) 953–960.
3. M. L. Wallach, *J. Polymer Sci. A2* **5** (1967) 653.
4. M. Fryd, in *Polyimides*, ed. K. L. Mittal, Plenum, New York (1984) 377–385.
5. W. Volksen and P. M. Cotts, Extended Abstract of Polyimide Meeting (November 10–12, 1982), Mid-Hudson Section of SPE, 154.
6. G. A. Brown, Reliability Implications of Polyimide Multilevel Insulators, *IEEE 1981 Reliability Phys. Symp.*, 282–286.
7. J. H. Nevin and G. L. Summe, in *Microelectronics and Reliability* **21** (1981) 699–705.
8. H. K. Chang, W. M. Shen and J. Yu, *IEEE 1982 Conf. on Electr. Insul. and Dielectric Phenomena*, 108–113.
9. G. Sawa, K. Iide, S. Nakamusa and M. Ieda, *IEEE Trans. Electr. Insul.*, **EI-15** (1980) 112–119.
10. G. Sawa, S. Nakamusa, K. Iida and M. Ieda, *Jap. J. Appl. Phys.* **19** (1980) 453–458.
11. B. L. Sharma and P. K. C. Pillai, *Polymer* **23** (1982) 17–20.
12. E. Sacher, *IEEE 1976 Conf. on Elect. Insul. and Dielectric Phenomenon*, 33–37.
13. F. W. Smith, Conduction in polyimide between 20 and 300°, Unpubl. Thesis, MIT, 1985.
14. E. Sacher, *IEEE Trans. Electr. Insul.* **EI-14** (1979) 85–93.
15. L. B. Rothman, *J. Electrochem. Soc.* **12** (1980) 2216.
16. D. R. Day, D. Ridley, J. Mario, and S. D. Senturia, Polyimide planarization in integrated circuits, in *Polyimides*, ed. K. L. Mittal, Plenum, New York (1982) 767–781.
17. A. M. Wilson, *Thin Solid Films* **83** (1981) 145–163.
18. R. J. Jenson, J. P. Cummings and H. Vora, *IEEE 1984 Electronic Components Conference*, 73–81.

19. E. Gonauser, B. Unger, R. Raushert, A. Glast and K.-R. Schoen, *IEEE J. of Solid-State Circuits* **SC-19** (1984) 299–305.
20. K. Moriya, T. Ohsaki and K. Katsura, *IEEE 1984 Electronic Components Conference*, 82–87.
21. R. M. Goffken, Multi-level metallurgy for master image structured logic, *IEEE 1983 IEDM Technological Digest*, 542–545.
22. M. Gold, *Electronic News*, 16 Nov. 1978.
23. G. Samuelson, *Organic Coatings and Plastic Chemistry* **43** (1980) 446.
24. T. Herndon, private communication.
25. F. K. Moghadam, *Solid State Technol.* **27** (1984) 149–157.
26. Y. Homma, A. Yajima and S. Harada, *IEEE Trans. Electron. Dev.* **ED-29** (1982) 512–517.
27. R. L. Van Tuyl, V. Kumar, D. C. D'Avanzo, T. W. Taylor, V. E. Peterson, D. P. Hornbuckle, R. A. Fisher and D. B. Estreich, *IEEE Trans.* **MIT-30** (1982) 935–942.
28. N. J. Chou and C. H. Tang, *J. Vac. Sci. Technol.* **A2** (2) (1984) 751.

13 Molecular electronics

R. W. MUNN

Introduction

Preceding chapters have already shown that chemistry is essential to much of current semiconductor technology, a fact perhaps too little realized outside the industry. The present chapter aims to show the central role that chemistry will have in exciting new developments foreseen under the name of *molecular electronics*.

There is more than one definition of molecular electronics. A broad definition is the systematic use of molecular materials to produce new or improved electronic devices. On this definition, the contents of Chapters 5–7 and 9, 10 and 12 all constitute part of molecular electronics. A narrower definition is the systematic *incorporation* of molecular materials into electronic devices as active elements. On this definition, only Chapter 12 on polyimides is classed as molecular electronics. For the purposes of the present chapter, therefore, the narrower definition will be adopted. However, the term molecular electronics will be understood to encompass opto-electronics and related aspects of information technology, where significant benefits are expected to accrue in the fairly near future.

The impetus to study possible uses of molecular materials in electronics comes from two sources. One is simply the observation that there is an enormous number of little-studied molecular materials, which should, if only on statistical grounds, offer a potentially fruitful extension of the range of materials available to semiconductor technology. The other is that those molecular materials already studied have been found to have a wide range of properties comparable to those of conventional solid-state materials, plus some specific advantages.

Among the specific advantages foreseen from the use of molecular materials are the following.

High packing density. With molecular dimensions of the order of 10^{-9} m, the density of separate elements in a molecular material is very high. If the size of individual components in an integrated circuit is to continue to fall, one possible way of achieving this may be to use individual molecules as electronic components.

Controllability. By chemical modification of molecular structure, the molecular shape and electronic properties can be controlled. The shape determines how the molecules pack in the material, and this packing in turn determines how the properties of the material arise from those of the molecules. It is also possible to intervene directly to control this packing process by depositing a series of layers, which need not all be the same.

Reproducibility. Similarly, because the structure of the material is implicit in that of the constituent molecules, its dimensions (e.g. the thickness of a monolayer) are fixed and reproducible. The same is obviously true of molecular crystals at one extreme and biologically replicated materials at the other.

Selectivity. Molecular materials may be physically selective for molecular size or shape, or chemically selective in their reaction (or lack of one) with other molecules or with light. This selectivity is naturally both controllable and reproducible.

These advantages are of course not to be gained without some concomitant constraints, among which are the following.

Synthesis. Even if a molecular material has been designed to make use of one or more of the above advantages, it still has to be synthesized. To date, the design of molecules to achieve a specific function has proved more a barrier to progress than their synthesis, as might perhaps have been expected considering the wealth of experience available in targeted synthesis, for example in pharmaceuticals. Nevertheless, continued progress is likely to require substantial investment in personnel and facilities for synthesis.

Stability. Many of the advantages of molecular materials accrue because the molecules largely retain their separate identities. However, this is so because the molecules interact only weakly, and these weak interactions tend to make molecular materials rather fragile mechanically, with low melting points. There can also be problems arising from the reactivity of the constituent molecules with one another, with oxygen, with water, or with light.

Added to these specific constraints are the usual ones of cost, toxicity, and so on.

Even within the narrow definition of molecular electronics adopted for this chapter, there is a great deal of reported information which has not yet been satisfactorily reproduced between laboratories, leading to some confusion and over-optimistic expectations. It is preferable, therefore, to be rather general and descriptive at this stage of the subject. Three classes of materials will be singled out for consideration: polymers, Langmuir–Blodgett films, and optical materials.

Polymers

Polymers are already in widespread use for their mechanical properties, and are familiar in clothing, household and consumer goods, automobile parts, etc. The polymer backbone provides strength and flexibility, of a nature controlled by the chemical constitution of the molecules. In addition to this advantage of controllability, polymers have the advantages over metals and ceramics that they can be processed into fibres and films from the melt or solution with relatively low energy input.

These mechanical properties also find application in electronics. The important case of *polymer resists* has already been described in detail (Chapter 9): here the polymer itself is designed to be suitably reactive, but its ability to form a coherent impermeable layer of uniform thickness is also essential.

The mechanical properties of polymers can also be used to confer tractability on some electroactive molecule by incorporating it as a pendant group. An example is provided by poly(*N*-vinylcarbazole) or PVK (**1**). It has long been known that aromatic and heteroaromatic crystals are photoconducting, with possible applications in photocopying. However, the crystals do not lend themselves to the preparation of the necessary large-area thin films, so that for example carbazole itself cannot be used in a practical photocopier. As a side-group in PVK, though, carbazole still gives adequate photoconducting properties such as carrier yield, lifetime and mobility, while the polymer backbone allows thin films to be prepared by standard techniques (1).

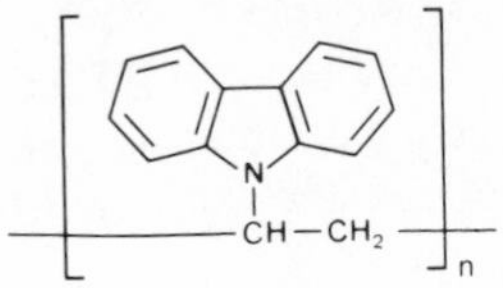

$$\left[\mathrm{CH(N\text{-}carbazolyl)}-\mathrm{CH_2} \right]_n$$

(1)

This strategy of polymerization may be used to confer stability as well as processability on electroactive molecules. One class of molecules widely used in this connection are the *polydiacetylenes* or PDAs (**2**). With suitable end-groups R, single-crystal diacetylenes R—C≡C—C≡C—R are found to undergo polymerization to single-crystal PDAs when heated or exposed to light or ionizing radiation (2). These conjugated polymers are of interest in their own right for applications in electronics, especially in opto-electronics as noted below. However, the small lattice disruption on polymerization means that ordered arrays of electroactive molecules containing diacetylene functional groups can often be polymerized to increase their mechanical strength and thermal stability.

$$\left[\mathrm{CR{-}C{\equiv}C{-}CR} \right]_n$$

(2)

Polymerization can thus be seen as an important enabling technique which allows electroactive molecules to be developed into applicable materials for electronics. However, a more central role in molecular electronics is likely to be played by polymers with conjugated backbones as electroactive molecules in their own right, and in particular by *conducting polymers* (3). These may have a variety of applications as electrodes and conductors in molecular assemblies, and also as interfaces between molecular materials and conventional electronic materials such as metals and semiconductors; the role of polyimides has already been discussed.

Probably the most intensely-studied conducting polymer is *polyacetylene*, $(CH)_x$ (**3**) (4). Thin films of this polymer can be prepared by admitting acetylene to a glass reactor coated with a film of Ziegler–Natta catalyst. The resulting low-density fibrous product is highly crystalline, but insoluble and unstable in air. When handling problems are overcome, it is found that polyacetylene can attain high conductivities on exposure to oxidizing agents such as halogens or AsF_5 or reducing agents such as sodium metal, a process referred to as 'doping'.

$$-\!\!\left[CH{=}CH\right]_n\!\!-$$

(3)

One direction of research has been to try to improve the tractability of polyacetylene. In particular, this has been achieved by devising the soluble precursor polymer (**4**) (and related polymers) which can be purified and characterized in solution and then cast into a film, which can be drawn to orient it. On heating, the polymer film loses bis(trifluoromethyl)benzene to afford polyacetylene (5). The product is no longer fibrous and so has a higher density than that prepared directly from acetylene. This slows down the diffusion of dopants into the bulk, but high conductivities can still be obtained.

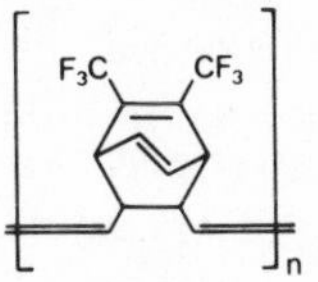

F_3C CF_3

(4)

Other research has rung the changes on polyacetylene using conventional techniques of polymer chemistry. Examples are the preparation of random copolymers of acetylene with methylacetylene; of block copolymers of polyacetylene with a soluble carrier polymer such as polystyrene; and of graft copolymers of polyacetylene on to a polymer such as isoprene. Composites of polyacetylene within a matrix of polyethylene or an elastomer have also been prepared to improve its mechanical properties (3).

A major feature of research into conducting polymers has been the development of electrochemical methods of doping and synthesis. In polyacetylene, the highly conducting state can be prepared by electrochemical oxidation or reduction, and this process is found to be both reversible and controllable. This leads to the preparation of a polymer battery system, which with appropriate counter-ions need contain no metals. The energy density of such systems can be very high, aided by the large surface area of the fibrous form of polyacetylene and the low electrode density. It has also been found that polyacetylene can act as catalytic electrode in a fuel cell (6). However, potential applications in these areas and others such as photovoltaic devices are hampered by the reactivity of the basic polymer.

Much effort has gone into the study of other conjugated polymer systems, particularly polymers of aromatic and heteroaromatic molecules. These include poly(*p*-phenylene), poly(*p*-phenylene sulphide), polypyrrole, polythiophene, polyaniline, and so on. Many of these can be doped to achieve conductivities within an order of magnitude of that of polyacetylene. They may have advantages of ease of preparation, tractability, or chemical stability.

As an example, *polypyrrole* (**5**) can be prepared by electrochemical polymerization. This process, available for many polymers, is especially convenient for conducting polymers: being insoluble, they deposit on the electrode, but being conducting, they allow ready passage of charge through to the electrode. Extensive work has been undertaken on the effect of the preparation conditions and the choice of counter-ion on the properties of the product. Polypyrrole is noteworthy for having been used to demonstrate behaviour analogous to that of a transistor (7). Three gold microelectrodes were deposited 1–2 μm apart on a silica substrate (using conventional techniques), a layer of polypyrrole was deposited on top, and the whole assembly was immersed in electrolyte. The current passing between the outermost electrodes increased greatly once the potential applied to the central gate electrode exceeded the threshold for oxidation of the polypyrrole, thus providing amplification of the gate voltage. The analogy with transistor action should not, however, be pressed too far, differences including the presence of an electrolyte and a response time of the order of seconds.

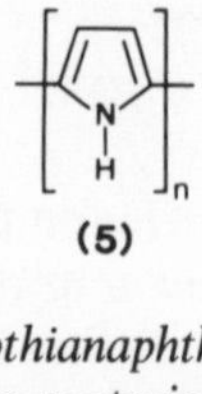

N
H
n

(5)

Finally among polymers, *polyisothianaphthene* or PITN (**6**) illustrates the use of chemical reasoning to design a material with controlled properties. The simpler parent polymer polythiophene (**7**) absorbs strongly in the visible, like polypyrrole. Fusing the aromatic ring onto the thiophene tends to favour the resonance form shown with double bonds between the repeat units, because this preserves the aromaticity. Compared with polythiophene (where the form with single bonds is favoured), this results in a greater degree of conjugation along the polymer backbone and a much lower excitation energy. Indeed, the effect is so large as to render PITN transparent in the visible in the highly conducting state, allowing possible new applications to be envisaged (8).

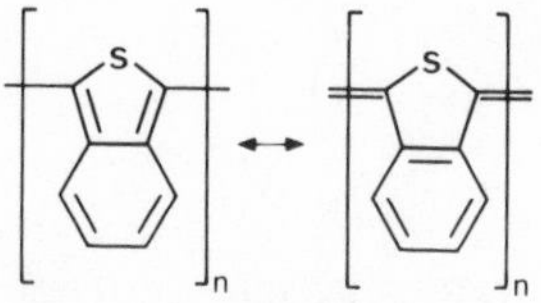

(6) 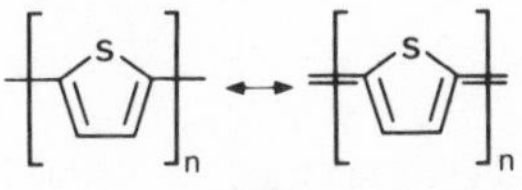 (7)

Langmuir–Blodgett films

Amphiphilic molecules contain one group which is hydrophilic and one which is hydrophobic. In aqueous solution, such molecules may adsorb at the air/water interface with the hydrophobic groups pointing out of the water. Compression of the adsorbed surface layer then eventually yields a closely packed monolayer, as shown for example by a sharp increase in the required surface pressure. This film can be transferred to a substrate dipped into the trough containing the solution, and if the surface pressure is maintained the resulting *Langmuir–Blodgett film* (LB film for short) consists of a highly ordered and perfect but non-crystalline monolayer (9, 10).

Langmuir–Blodgett films have a number of features which make them promising materials for electronics. Layers can not only be free of holes, but also have a thickness which is determined by the dimensions of a molecule, and so is accurately reproducible. Many layers may be superimposed by repeated dipping to give a wide range of thicknesses, and this process is now so automated that troughs can be left to run unattended once the required number of layers has been entered. The dipping process can also be used to provide direct control over the structure of the film by depositing alternating layers of different species (11). This yields non-centrosymmetric structures, whereas the usual tendency of a single species is to alternate in orientation between layers so that hydrophilic groups are adjacent in one pair of layers and hydrophobic groups in the next pair, giving a centrosymmetric structure.

Added to the advantages of physical controllability is the prospect of controlling chemical activity by means of appropriate structural features. The classical LB film-forming molecules were the alkanoic acids with long straight chains (the fatty acids), for example *stearic acid*, $C_{17}H_{35}CO_2H$. Improved film properties may result from using not the acid itself but the corresponding salt with a divalent ion such as Ca^{2+} or Cd^{2+}, and if a magnetic ion such as Mn^{2+} is used the film also becomes magnetic. While a hydrophobic head group and a hydrophilic tail group were initially retained as essential, it was found that considerable variation was allowed in between. An early example was the incorporation of the anthracene ring system between an *n*-butyl head group and a propanoic acid tail group at the 9- and 10-positions (**8**); the anthracene confers photoconductive and electroluminescent properties on the films (12).

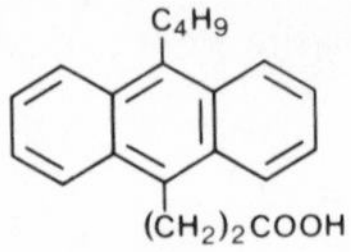

C_4H_9

$(CH_2)_2COOH$

(8)

Considerable effort has been devoted to the production of polymerizable LB films. Two goals make this desirable: one is simply to stabilize the LB film once it is formed, and the other is to polymerize it selectively for use as a resist or in integrated optics. The difficulty is to polymerize the film without destroying its other properties of perfection, uniformity, and so on. It has been found that incorporating the unsaturated group at the end of either the head or tail group is effective in this regard, for example in $C_{17}H_{35}CO_2CH{=}CH_2$ (vinyl stearate), which is polymerized by gamma rays. Another example is omega-tricosenoic acid, $CH_2{=}CHC_{20}H_{40}CO_2H$, which is polymerized by an electron beam and proves to be an attractive resist material (13), the more so since it gives good-quality LB films when dipped at a much faster rate than other film-forming species. Polymerizing the head or tail group presumably minimizes the disruption of the interaction between adjacent molecules in the film. Another way of achieving the same effect is to make use of the small disruption in the diacetylenes already referred to, and in this case the diacetylene function can be incorporated in the middle of the hydrocarbon chain (10).

The long hydrocarbon chain features in all the molecules so far mentioned as forming LB films, although it is not always saturated and sometimes contains an interruption such as the anthracene ring in molecule (**8**). However, it does appear to make the resulting LB film rather unstable. This is perhaps not surprising, given the waxy nature of the corresponding pure hydrocarbon crystals. Moreover, having to attach such a chain to some molecule containing the electronic function of real interest is time-consuming and expensive, and may also destroy or at least dilute the desired function. There are therefore attractions in trying to form LB films from a wider selection of molecules. One example of this is *quinquethienyl*, which is just the $n = 5$ oligomer of polythiophene (**7**). Another example is *phthalocyanine* (**9**) and its derivatives, which have a number of attractive features: the films resemble the crystals in having high thermal and mechanical stability, a variety of substituted forms can be prepared relatively easily, and the isolated phthalocyanine ring can accommodate various central metal atoms which modify its electronic states. The phthalocyanine LB films show that elongated molecules are not in fact necessary for film formation; what is essential may rather be molecules which have a suitable shape to pack closely and a suitable electron density to attract sufficiently strongly.

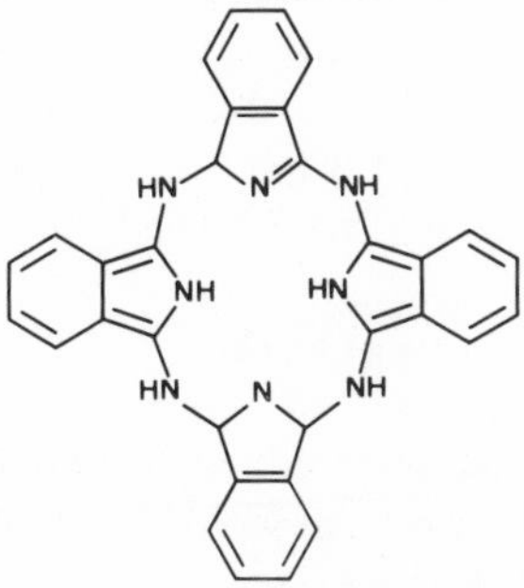

(9)

Applications of LB films in electronics are foreseen in a number of areas, some of which have already been alluded to. In general, LB films are insulating unless specifically designed otherwise. Coupled with their perfection and controllable thickness down to the nm scale, this makes them useful insulating layers, particularly on materials other than silicon which lack a suitable insulating oxide or similar easily-produced compound. The mild conditions of deposition, notably the low temperature, also minimize disturbance of the substrate. The use of LB films in *metal-insulator-semiconductor (MIS) devices* exemplifies such applications (9). Compared with the corresponding simple Schottky barrier device without an insulating layer, a solar cell based on cadmium telluride separated from a gold electrode by a controlled number of monolayers of the anthracene-substituted species (**8**) has an enhanced efficiency for 1–3 monolayers, the enhancement being some 50% for two monolayers. Similarly, in an *MIS electroluminescent diode* based on gallium phosphide, an LB film coating enhances hole injection and increases the luminescence intensity several-fold, the optimum effect again being obtained with about three layers. The existence of an optimum in these cases is associated with the rapidly decreasing efficiency of tunnelling through thicker layers. The extent to which performance may be optimized depends on a judicious choice of film material and deposition conditions.

LB films have also been investigated in *field-effect transistors.* The film is deposited between source and drain regions, and a gate electrode is evaporated on top. As the layer can be thin and pinhole-free, a small voltage applied to the gate electrode readily modifies the charge distribution and greatly effects the current–voltage characteristics between source and drain. Various semiconductors where conventional insulating layers may be unsatisfactory or unavailable have been treated in this way, and even for silicon there may be advantages in using an LB film to enhance the electrical strength of the usual silicon oxide layer. The versatility of LB films also opens up other possibilities in which the layer is not merely passive but responds to external influences which are then detected by their effect on the FET characteristics. This is the essence of a *sensor* device like those already known for ions and other molecules, but here the LB film might sense thermal, optical or mechanical stimuli as well as chemical ones. It may also be that LB films will prove useful as membrane materials to incorporate active molecules for biosensors, or might ultimately incorporate the biological sensing function themselves.

Other potential applications of LB films span a wide range. Magnetic thin films prepared by incorporation of suitable ions (as outlined above) offer possibilities in magnetic memory devices. Films incorporating luminescent groups may yield new displays. Conductivity is highly anisotropic in the films and may sometimes be quite high, so that some films could be used as two-dimensional conductors. These could well find applications in controlling electron flow in assemblies of molecular elements.

Optical materials

The control of optical properties of molecular materials is already the stock-in-trade of the dyestuffs industry. Optical properties are also of increasing importance in the electronics industry for displays, for data storage and for information transmission. It is therefore natural that molecular materials should begin to assume growing significance in electronics for their optical properties as well as their electronic properties. Indeed, there are indications that in electro-optical and nonlinear optical applications, molecular materials may be superior to conventional inorganic materials in most respects.

Photochromics change colour on illumination. This provides the basis of a mechanism for information storage by producing a pattern of coloured spots using a light beam. High densities of information can be stored compared with photographic methods, because the resolution is limited only by the physical optics of the write beam and not by the grain size of the film, which is essentially the molecular dimension. Photochromic information storage also has the advantage of being instant, without requiring a separate developing and fixing stage, and thus offers the possibility of direct reading of the information after writing it and so correcting it at once if necessary.

Realizing these advantages in practice requires optimization of many factors. The colour change which stores the information must occur at a convenient wavelength, preferably one provided by an inexpensive laser, but must not occur spontaneously, that is to say thermally. Once changed, the colour must not revert to its original (usually transparent) state spontaneously, i.e. thermally, but should ideally be reversible at another convenient wavelength, for error correction and updating and to provide a re-usable storage medium. Furthermore, it must be possible to read the coloured form and the information it conveys at a third convenient wavelength where neither the forward nor the reverse colour change occur. Hence the absorption spectra of the colourless and coloured forms must not overlap.

These requirements have been successfully met by a systematic development programme leading to the synthesis of the *fulgides* (**10**), which undergo a photochromic ring closure as shown (13). According to the Woodward–Hoffman rules, electrocyclic reactions like this follow different stereochemical pathways, depending on whether they are induced photochemically or thermally. In the fulgides, the methyl groups adjacent to the bond to be formed in the ring closure are so disposed as to make the thermal reaction highly hindered, without affecting the photochemical reaction. The same considerations apply to the reverse reactions, so that both the original and coloured forms are thermally stable.

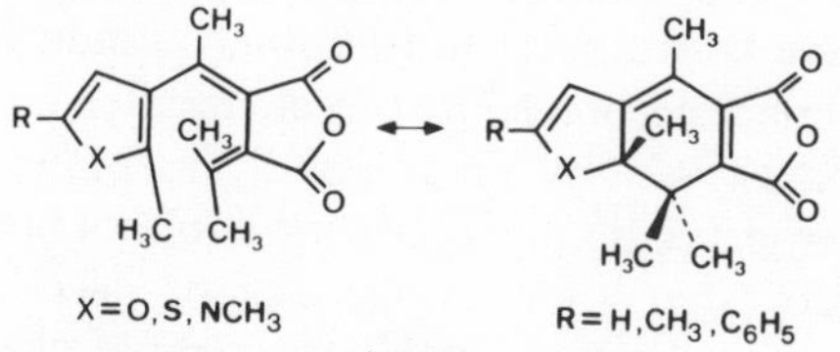

$X=O, S, NCH_3$ $R=H, CH_3, C_6H_5$

(10)

Electrochromics change colour on application of a voltage sufficient to effect some suitable oxidation or reduction. They therefore offer possibilities for new large-area directly-addressed displays. Important characteristics to be optimized then include the colour obtained and its intensity, the size of the voltage required, and the response time. Electrochromic displays can be made using inorganic ions, but another interesting possibility is the use of polypyrrole. When polymerized anodically, polypyrrole is obtained in the conducting form, which is very dark and highly absorbing. Cathodic reduction yields the insulating form, which is green and transparent. As the oxidation and reduction are reversible, photochromic switching is possible.

Optical information storage is not restricted to photochromic materials. One simple way of storing information using a laser is to physically burn holes in a polymer film; this is obviously irreversible and so provides write-once storage. A fundamental limitation on optical methods is the diffraction limit beyond which physical optics no longer applies, which represents a lower bound to the size of the region in which information can be stored for a given wavelength. This restriction can be overcome by storing information at several different wavelengths within the same physical location, using *persistent spectral hole-burning*.

Persistent spectral hole-burning entails burning a hole not in the sample but in its optical spectrum. The sample consists of molecules which are photoactive in some suitable way (as described below), dispersed in an amorphous medium. The medium contains a distribution of different sites, each of which gives a slightly different energy to the photoactive molecule it contains. As a result, the optical spectrum of the sample is composed of a distribution of individual spectra each slightly displaced from one another, giving a large inhomogeneous linewidth. Light from a laser of much narrower linewidth is then absorbed only by those few molecules in sites which give them just the right energy to be in resonance with the laser. The molecules are chosen to undergo some photochemical or photophysical change to a product which does not absorb at the laser frequency; this may occur simply because the sites at which the initial absorption takes place are not all equivalent and so broaden the absorption of the product. For a· sufficiently intense laser beam, the photoprocess removes a sizeable fraction of the absorbing molecules from resonance with the laser beam. If the photoproduct is stable over long periods, then a subsequent measurement of the spectrum of the sample will reveal a 'hole' or dip in absorption where the molecules originally in resonance with the laser frequency have been converted to non-absorbing products.

With a tunable laser, this procedure can be repeated for various frequencies each separated from the next by more than the laser linewidth. The result is a spectrum with a distinct pattern of holes burnt (or not burnt) at a sequence of frequencies, so that binary information can be encoded. All this can be accomplished within the area of a single laser spot down to one μm or so in diameter. Thus by using the frequency domain as well as the spatial one, the density of information storage is increased, by a factor which can be as large as 1000.

Engineering and materials requirements for frequency-domain optical information storage have been reviewed (14); among those where chemical input is required are the choice of the system to undergo hole-burning, at a suitable frequency, in a suitably short time, with the formation of a stable photoproduct. Systems in which hole-burning has been investigated include phthalocyanine (**9**) and its derivatives, where the photoeffect is the tautomeric change which transfers the protons from one pair of central nitrogen atoms to the other; colour centres in inorganic crystals, where the effect is believed to involve an electron tunnelling away and being trapped; and a variety of systems showing photoprocesses such as transfer of charge or protons, or change of geometry in the molecule or site.

Optical properties are increasingly being studied not only for information storage and display purposes, but also for information transmission, where optical fibres offer high information densities and low sensitivity to electromagnetic interference. Here molecular materials are under investigation as means of manipulating light beams, one goal being the construction of integrated optical devices akin to electronic integrated circuits.

The optical properties are useful for different applications. The first is simply the *refractive index*, differences in which serve to guide light beams, as by internal reflection in an optical fibre. The refractive index describes the linear response of a medium; the other useful properties are nonlinear. The *second-order susceptibility* (first nonlinear susceptibility) $\chi^{(2)}$ governs phenomena such as second-harmonic generation or frequency doubling, with input frequency ω and output frequency 2ω; optical mixing of different input frequencies ω_1 and ω_2 to produce output at their sum or difference; parametric oscillation and amplification; and electro-optic modulation using one static input frequency $\omega = 0$ to control output at the input frequency. The *third-order susceptibility* (second nonlinear susceptibility) $\chi^{(3)}$ governs a still wider range of phenomena, including third-harmonic generation, four-wave mixing, and optical bistability, which can be used for optical gates and amplifiers.

Organic solids offer significant advantages in the magnitude and controllability of these properties (15). A further important advantage is that, contrary to what might have been supposed, organic materials show high resistance to laser damage, a property essential to the utilization of nonlinearity. This high resistance is probably associated with the large number of molecular degrees of freedom available for the dissipation of energy.

The refractive index of an organic material can be altered in various ways. Among those of potential importance is the deposition of LB films with different refractive indices. Another is polymerization, where the polymer and monomer refractive indices may differ significantly, for example in the polydiacetylenes (which might be incorporated in an LB film structure). This process has the further advantage that complex patterning of the polymer is possible by photolithographic or electron-beam techniques suitable for preparing integrated optical devices.

Utilization of $\chi^{(2)}$ requires three factors to be optimized: the molecular response governed by the first hyperpolarizability β, the material response governed by the arrangement of the molecules, and the device response governed by the physical form of the material (16). Both theoretical and experimental work show that large-scale charge transfer helps to enhance β, and here systematic development seems possible. The first requirement for a non-zero $\chi^{(2)}$ (or β) is a non-centrosymmetric structure. For example, *p*-nitroaniline (**11**) has a high β value, but crystallizes in a centrosymmetric structure, whereas *m*-nitroaniline (**12**) has a lower β value but crystallizes in a non-centrosymmetric structure. By adding a methyl group to *p*-nitroaniline in the *meta* position to produce MNA (**13**), it proves possible to retain the high β value of the parent compound but induce a non-centrosymmetric structure. The resulting crystal is clearly superior to the standard material lithium niobate in its second-order properties, with a parametric figure of merit some 40–50 times higher. Another way of inducing a non-centrosymmetric structure is by depositing different layers alternately in an LB film, as already noted.

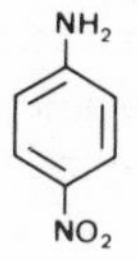

NH_2 / NO_2 (**11**) NH_2 / NO_2 (**12**) NH_2 / CH_3 / NO_2 (**13**)

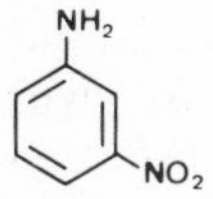

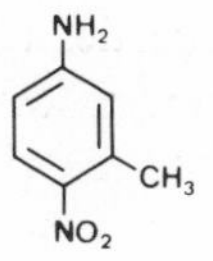

All materials have a non-zero $\chi^{(3)}$, but molecular materials offer the possibility of enhancing it through extensive conjugation. In this respect, the polydiacetylenes are the front runners, showing values which can be a factor of 10 or so higher than the standard material gallium arsenide. Once again, this could be achieved within an LB film assembly.

The definition of suitable physical forms for the materials is initially a task for device physicists and engineers, but chemical aspects remain. One is again the deposition of LB films. On the other hand, for some purposes it may be necessary to produce highly perfect single crystals to avoid light scattering and hot spots leading to laser damage. Surface device configurations have advantages which can be exploited if suitably perfect surfaces can be prepared, while other useful configurations include crystal growth in grooves or capillaries. At present, however, the main effort is still concentrated on optimization of materials properties (16).

Conclusions

This chapter has considered possible new roles for chemistry in the semiconductor industry of the future. The areas selected have been far too broad to give more than a flavour of each. However, they do illustrate how ingenious synthetic capabilities can find new sources of inspiration. Equally important in the long run will be the optimization of material properties for a given molecule. As always in materials science, this will require painstaking study of conditions of preparation, deposition, and so on.

An encouraging example where chemistry has played a new role in the semiconductor industry is the liquid crystal display. Some of the areas described here may blossom in this way—perhaps all of them will eventually—but at present many of them lack the extensive background research available when the display applications of liquid crystals began to be developed. Much can nevertheless be learned from the example of liquid crystals, in particular the need for chemists to collaborate with physicists and engineers both in development and in fundamental research (17).

References

1. J. Mort, *Science* **208** (1980) 819.
2. R. H. Baughman, *J. Polym. Sci. Polym. Phys. Edn* **12** (1974) 1511.
3. J. R. Reynolds, *J. Mol. Electronics* **3** (1986) 1.
4. J. C. W. Chien, *Polyacetylene—Physics, Chemistry and Material Science*, Academic Press, New York (1984).
5. J. H. Edwards and W. J. Feast, *Polymer* **21** (1980) 595.
6. R. J. Mammone and A. G. MacDiarmid, *J. Chem. Soc. Faraday Trans. I* **81** (1985) 105.
7. H. S. White, G. P. Kittlesen and M. S. Wrighton, *J. Am. Chem. Soc.* **106** (1984) 5375.
8. M. Kobayashi, N. Colaneri, M. Boysel, F. Wudl and A. J. Heeger, *J. Chem. Phys.* **82** (1985) 5717.
9. G. G. Roberts, *Contemp. Phys* **25** (1984) 109.

10. M. Sugi, *J. Mol. Electronics* **1** (1985) 3.
11. M. F. Daniel and G. W. Smith, *Mol. Cryst. Liq. Cryst. Lett.* **102** (1984) 193; I. R. Girling, P. V. Kolinsky, N. A. Cade, J. D. Earls and I. R. Peterson, *Optics Comm.* **55** (1985) 289.
12. G. G. Roberts, T. M. McGinnity, W. A. Barlow and P. S. Vincett, *Solid State Comm.* **32** (1979) 683; *Thin Solid Films* **68** (1980) 223.
13. P. J. Darcy, H. G. Heller, P. J. Strydom and J. Whittall, *J. Chem. Soc. Perkin I* (1981) 202.
14. W. E. Moerner, *J. Mol. Electronics* **1** (1985) 55.
15. A. F. Garito, K. D. Singer and C. C. Teng, ACS Symp. Ser. **233** (1983) 1.
16. J. Zyss, *J. Mol. Electronics* **1** (1985) 25.
17. G. W. Gray, *Proc. Roy. Soc. Lond. A* **402** (1985) 1.

14 Computer modelling of semiconductor interfaces

J. Q. BROUGHTON and P. S. BAGUS

I. Introduction

This chapter seeks to review what we have learned about the capabilities and the future directions of two historically separate methodologies for calculating interface structure and dynamics; those of 'ab-initio' quantum calculations and of Monte Carlo (MC) and Molecular Dynamics (MD) simulations. Our emphasis will be on semiconductor surfaces. Our bias will be towards the quantum chemist's approach to the problem, namely the utilization of cluster calculations as opposed to the band structure calculations of the physicists. The 'ab-initio' methods rely upon solving Schrödinger's equation within the Born–Oppenheimer approximation, often at the level of the one-electron (orbital) approximation but increasingly at the configuration interaction (CI) or multiconfigurational level. The simulation methods involve marching through the configuration space of the system *assuming pre-defined interatomic potentials* either by using forces generated from total system energies and solving Newton's laws of motion (MD) (1) or by throwing up new configurations in a Markov chain with probabilities given by Boltzmann factors involving total energy differences (MC) (2). Traditionally, quantum chemists have been concerned with zero-temperature properties while statistical mechanicians have been concerned with many-body phenomena at finite temperatures. Ideally, of course, we would wish to perform an 'ab-initio' total energy calculation for each atomic configuration thrown up by our MC or MD calculations. As we shall see, the first tentative steps are being made towards this goal.

In many ways there is a reversal of the ease with which covalent systems may be handled by quantum mechanical (QM) and statistical mechanical (SM) methods. In the former case, since the form of bonding at low temperatures is well defined, the boundaries of the clusters may be terminated to an excellent approximation by hydrogen atoms (to saturate dangling bonds) or 'frozen' sp^3 hybrids (3). Boundary conditions for metal clusters are less simple. However, in SM calculations, the simple metals and inert gases may be handled efficiently and accurately using pair potentials, the former being described by pseudopotentials derived by perturbing about the free electron gas (4). It is the covalent systems which provide the trouble now because a simple analytic form for the variation of interatomic potential with coordination is non-trivial to find. We wish a potential here capable of describing interactions between semiconductor atoms where hybridization varies from *sp* and sp^2 (necessary at crystal–vapour (CV) surfaces) to sp^3 in the crystal and amorphous phases to metallic in the melt. Many guesses have recently been made at the form of this interaction. It is only very recently, therefore, that simulations of silicon surfaces have begun to appear in print. The important question of impurity trapping in semiconductors has not been tackled using continuous coordinate simulation although insights into the mechanism are available from lattice-gas simulations.

In the next section, we give a preamble to the simulation aspects of this review. Section III is devoted to a simple description of the MC and MD methods. Section IV describes the potentials developed to describe covalent bonding. We go into some detail here since this is where most of the action has been of late. Section V describes what we have learned of crystal growth using such potentials starting firstly with what we can infer from simple interface models and continuing to full-blown calculations involving many-body potentials. Section VI discusses the determination of wave functions to describe the electronic structure of semiconductor surfaces at an atomic level. The focus is on the use of clusters to represent the interface and Molecular Orbital, MO, theory to obtain the wave function computationally. The unique merits of this approach are discussed and it is briefly contrasted with more traditional band-structure modelling of the interface. The major part of the section is devoted to illustrating these merits with a review of the results of MO cluster model studies of the interaction, adsorption and penetration, of halogen and hydrogen atoms on a Si surface. Finally, Section VII gives our thoughts on how further progress may be made and where we think the field is heading.

II. Simulation preamble

Sophisticated continuum mechanical simulations of semiconductor growth from the melt exist (5, 6), dealing with conditions close to equilibrium, specific to experimental processes, and concerned with solving heat-flow equations. However, few calculations of atomic-level growth processes have been performed. Certainly there have been few serious attempts to study impurity incorporation or even the growth of two-component semiconductor systems using continuous coordinate potentials, although MC lattice gas models have been of use here. Also, although numerous simulations of scattering of atoms, ions, and molecules off dynamic inert gas/simple model substrates exist in the literature, we are aware of only one piece of work in this area for semiconductor surfaces (7). The reason is very simple; the interatomic potentials describing the covalent bond are more complex than simple two-body systems and very little is known about their form.

III. Simulation procedures

Given a potential which is a continuous function of distance between atoms in the system, it is possible to advance through phase-space in (commonly) two ways: either by following the 'true' dynamics of the system or by ensuring that the right configurational averages are produced. We assume that the motion of the atoms obeys classical laws. The total potential energy (U) of the system may be written

$$U = \sum_{i} \phi_1(i) + \sum_{i<j} \phi_2(ij) + \sum_{i<j<k} \phi_3(ijk) + \cdots \quad (14.1)$$

where subscript n implies n-body and the sums are over all particles i or j or k in the system. We have truncated the series at the three-body term, although rigorously all orders should be included in the sum for any given system. From (14.1) may be obtained the force (f_i) on each atom $-\nabla_i U$. Newton's laws of motion may then be integrated using a finite difference scheme in which, given the known present position, velocity and acceleration (force) it is possible to predict new positions and velocities a small time increment later. A popular simple yet stable algorithm is due to Verlet (18):

$$\bar{r}_i(t + \Delta t) = -\bar{r}_i(t - \Delta t) + 2\bar{r}_i(t) + [(\Delta t)^2/m]\bar{f}_i(\bar{r}_i(t)) \quad (14.2)$$

where m is the mass of the atom, $\bar{r}$ is the position vector and Δt is a small time increment. Typically, Δt is 0.01 of an Einstein period. Velocities ($\bar{v}$) may be obtained by the central difference formula:

$$\bar{v}_i(t) = [\bar{r}_i(t + \Delta t) - \bar{r}_i(t - \Delta t)]/(2\Delta t). \quad (14.3)$$

Thus, given a starting configuration and some velocity distribution, it is possible to follow the detailed trajectories of every particle in the system as a function of time. In its simplest form, this molecular dynamics technique represents a microcanonical ensemble (constant number of particles, N, constant volume, V, and constant energy, E), but recent methods have been derived for altering trajectories slightly in order to maintain constant N, pressure (P) and temperature (T) (9). MD allows the detailed analysis of both time-dependent (e.g., diffusion coefficients, phonon frequencies) and time-independent properties (e.g., pressure, energy, density, positional distribution functions).

MC techniques, on the other hand, are usually concerned with obtaining time-independent properties, the simplest case being the canonical average (constant N, V, T). The ensemble average value of a property X which is a function of all the atomic coordinates $\{\bar{r}\}$ of the system is given by (10):

$$\langle X \rangle = \frac{\int X(\{\bar{r}\}) \exp(-U(\{\bar{r}\})/k_B T)\, d\{\bar{r}\}}{\int \exp(-U(\{\bar{r}\})/k_B T)\, d\{\bar{r}\}} \quad (14.4)$$

where integrals are over all configuration space and k_B is Boltzmann's constant. An efficient way of obtaining such averages is through the use of the Metropolis algorithm (2). (Clearly we cannot sample *all* phase space since these are multidimensional integrals.) Particles are selected at random and moved an arbitrary distance; if the energy of the move decreases then the move is accepted, otherwise the move is accepted with a probability given by $\exp(-\Delta U/k_B T)$ where $\Delta U = U_{\text{final}} - U_{\text{initial}}$. After the system relaxes from its initial configuration, the properties of interest are obtained as a simple average over each configuration (whether accepted or not) thrown up by the Metropolis procedure. We usually aim for an acceptance probability of approximately 50%.

Typically, in simulations of bulk systems, properties in either MD or MC are obtained from averages over 1–10 000 moves per atom at a given point in the phase diagram. Usually studies are performed as a function of some intensive property like T, P or ρ; thus many runs are necessary. As a rule of thumb, we require that the CPU time requirement to move every particle in the system one step is ~ 1 second or less. At present, systems with up to 10 000 atoms are computationally accessible. The total evolved time in a typical MD run is 10^{-10} seconds. Thus it is important to focus on properties which relax on this time scale. Present computer power enables the handling of long-range pair-potentials and short-range three-body (up to next-nearest neighbour) potentials for the above system sizes.

When considering the dynamic process of crystal growth, these limits of computer time are stretched even further. That is, we may examine only growth rates which are significantly fast enough to see on MD or MC time scales.

MC techniques can be modified so that we look at the stochastic dynamics of the system. We do this by considering the grand canonical ensemble (constant V, T and chemical potential μ) where the partition function is defined by (10)

$$\Xi = \sum_m \exp[-(U(m) - N(m)\mu)/k_B T] \tag{14.5}$$

where m labels the various possible states of the system. The number of particles in the system is allowed to fluctuate. Particles may be moved, created or destroyed by the Metropolis scheme such that the probability P of finding the system in a given state n is given by

$$P(n) = \exp[-(U(n) - N(n)\mu)/k_B T]/\Xi. \tag{14.6}$$

The way in which this may be done is non-unique. Such is the case at equilibrium. In deviating from equilibrium, it is necessary to make physically reasonable assumptions about how the rates of adsorption and desorption vary with supersaturation of the fluid phase, but with the proviso that as equilibrium is approached, the correct grand canonical partition function is again achieved.

One way of doing this which has proven useful in the study of roughening is via the so-called solid-on-solid (SOS) kinetic Ising model (11). Particles are allowed to exist only on lattice sites (Ising models are examples of lattice-gas systems) and usually a nearest-neighbour Hamiltonian is assumed. Thus the interparticle potential is defined only at discrete distances. Overhangs in the interface profile are not allowed. The fluid phase is assumed to be featureless, i.e. a mean-field fluid. We make the physically reasonable assumption that the deposition frequency is (12, 13)

$$k^{+} = \nu \exp(\mu / k_{\mathrm{B}} T) \tag{14.7}$$

where ν is a frequency factor and μ is the chemical potential of the fluid. Therefore, k^{+} is site-independent. Evaporation, on the other hand, is site-specific and occurs with a frequency

$$k_{n}^{-} = \nu \exp(-n\phi / k_{\mathrm{B}} T) \tag{14.8}$$

where n is the number of bonds to the other atoms of the crystal and ϕ is the energy required to break one bond (relative to the mean fluid bond energy). Equations (14.7) and (14.8) are consistent with the correct SOS equilibrium partition function (14) and can be modified to handle two-component systems. The great advantage of lattice-gas simulations is that, by virtue of the simple Hamiltonian, large systems may be handled. Note, however, that growth rates are known not in an absolute sense but relative to k^{+} (in a dense fluid this is not well defined).

Lastly, we note that edge effects are minimized in all these interface simulations by assuming periodic boundary conditions in the plane of the interface.

IV. Interatomic potentials

Part of the reluctance towards developing good general-purpose potentials for semiconductor systems stems from the computational requirements of using accurate potentials. Simple estimates of computer time would predict that the time spent in a pairwise calculation would scale as N^2 and that that spent in a three-body calculation would increase as N^3. In fact, particle 'book-keeping' techniques (8, 15) reduce the scaling to linear in N. But the constant of proportionality is generally much higher in high-n rather than low-n n-body systems, the reason being that even with short-ranged n-body potentials the number of terms that have to be evaluated for each move increases with n.

Historically, perhaps the favourite potential for simulations has been the Lennard–Jones (LJ):

$$\phi(r_{ij}) = 4\varepsilon\left[\left(\frac{\sigma}{r_{ij}}\right)^{12} - \left(\frac{\sigma}{r_{ij}}\right)^{6}\right] \tag{14.9}$$

where σ and ε are parameters chosen to fit experimental observables. The behaviour of systems containing inverse power, hard-sphere and Morse potentials has also been widely studied. We mention the LJ here because it has been employed in crystal-growth simulations with ramifications for Si growth.

Turning to potentials for covalent systems, an early successful one is due to Keating (16). He chose to use a bond-stretching potential (two-body) and a bond-bending term (three-body) each of which couples only nearest neighbours:

$$U_{\text{tot}} = \frac{1}{2}\sum_i \left[\frac{\alpha}{4a^2} \sum_{j=1}^{4} (r_{ij}^2 - 3a^2)^2 + \frac{\beta}{2a^2} \sum_{j,k<i}^{4} (\bar{r}_{ik} \cdot \bar{r}_{ij} + a^2)^2 \right] \quad (14.10)$$

where a is a lattice constant and α and β are chosen to fit the elastic constants of Si, Ge and diamond. Calculations of phonon dispersion curves with this model show good agreement with experiment except for the transverse acoustic (TA) branches which lie at too-high frequencies. On the other hand, valence force field potentials (17) (similar to those commonly applied to the theory of molecular vibrations) and shell-models (systems which comprise nuclei coupled to shells of electrons of zero mass) such as that due to Weber (18), are able to get even the TA branch correct. All of these potentials, however, suffer from the limitation of lack of generality; that is, they are useful for describing systems in which the deviation from local tetrahedrality is not too great.

Recently, attempts have been made to correct for this. Smith (19) used a large-amplitude generalization of Keating's potential in a study of tunnelling states in amorphous Si and Ge. This idea was continued by Stillinger and Weber (SW) (20) who sought a potential which would fit not only the lattice parameter and elastic constants of crystalline Si but also the liquid radial distribution function $g(r)$ and the bulk melting temperature T_m. They assumed

$$U_{\text{tot}} = \varepsilon\left[A \sum_{i<j} \left(B\left(\frac{\sigma}{r_{ij}}\right)^4 - 1 \right) \exp\left(\frac{\sigma}{r_{ij} - a}\right) + \sum_{i<j<k} \phi_3(\bar{r}_i/\sigma, \bar{r}_j/\sigma, \bar{r}_k/\sigma) \right] \quad (14.11)$$

where

$$\phi_3 = h(r_{ij}, r_{ik}, \phi_{jik}) + h(r_{ji}, r_{jk}, \phi_{ijk}) + h(r_{ki}, r_{kj}, \phi_{ikj})$$

and

$$h(r_{ij}, r_{ik}, \phi_{jik}) = \lambda \exp[\gamma(r_{ij} - a)^{-1}] \exp[\gamma(r_{ik} - a)^{-1}] \times [\cos\phi_{jik} + \tfrac{1}{3}]^2.$$

Here, A, B, ε, σ, λ, γ and a are constants. If $r \geqslant a$, the two- and three-body functions vanish. ϕ_{jik} is the angle between j and k subtended at vertex i. The three-body term is identically zero at the tetrahedral angle and is positive otherwise.

The Pearson-Takai–Halicioglu–Tiller potential (21) which has been used for the modelling of Si and SiC surfaces consists of a two-body Mie (LJ-like) potential and a three-body Axilrod–Teller term (22). It was parameterized against experimental bond lengths and energies of both bulk and cluster Si and SiC systems. Considering that the Axilrod–Teller potential is usually used for describing triple-dipole terms in van der Waals' systems, the potential does surprisingly well.

Biswas and Hamann (BH) (23) have a potential which also relies upon a two- and three-body separation of the total energy. It fits extremely accurately a database of 'ab-initio' calculations of the energy of bulk and surface Si structures. The potential is distinct from that of SW in that it does not explicitly 'force' a tetrahedral angle at low *T*. The three-body part reduces to a product of two-body sums, but even so the amount of computer time required to generate moves in configuration space is still large compared with the SW potential, mainly because of the larger cut-off radius of the BH potential.

Lastly, we turn to Tersoff's model (24) for Si. It does not rely on an expansion into *n*-body terms; rather, it has the form of a Morse pair potential but with a bond strength parameter depending upon local environment:

$$U = \sum_{i<j} u_{ij}$$

$$u_{ij} = f_c(r_{ij})[A\exp(-\lambda_1 r_{ij}) - B_{ij}\exp(-\lambda_2 r_{ij})]. \qquad (14.12)$$

Here, u_{ij} is the interaction between atoms *i* and *j*. A, B, λ_1 and λ_2 are all positive with $\lambda_1 > \lambda_2$. f_c is a cut-off function. B_{ij} depends upon local environment:

$$B_{ij} = B_0\exp(-z_{ij}/b)$$

$$z_{ij} = \sum_{k\neq i,j} \exp[-m\lambda_2(r_{ik} - r_{ij})]/[c + \exp(-d\cos\phi_{ijk})] \qquad (14.13)$$

B_0, b, m, c and d are constants. z_{ij} is a weighted measure of the number of bonds competing with bond *ij*. The parameter *m* determines how closer neighbours are favoured over more distant ones in forming bonds. The range of the potential is short, which gives it computational efficiency. Properties of bulk and dimer silicon are used to parameterize its six unknowns. This intuitively appealing potential has been shown to stabilize the Si(111)(7 × 7) surface preferentially in the Takayanagi *et al.* structure (25), but it does have one (it is hoped) minor flaw. It gives a slightly lower energy for a simple hexagonal system with large c/a ratio than for diamond-cubic Si.

Concomitantly with all this work, groups have been establishing databases against which future potentials might be parameterized. Raghavachari and Logovinski (RL) have used Hartree–Fock calculations with electron-correlation corrections to find absolute minimum energy configurations for Si clusters with $n = 2$–10. Tomanek and Schluter (TS) (27) used local density functional (LDF) theory to find similar minima for clusters up to $n = 14$. The two different calculational approaches show good agreement in the structures found. The database of energy *v.* lattice parameter for Si in many hypothetical bulk structures used to parameterize the BH potential is due to Yin and Cohen (28, 29). Despite the elegance of this potential, there is a problem in transferability in the potential parameters when describing geometries of Si clusters, that is, when comparison is made with the ab-initio results of RL and

TS. We should say, however, that this statement is almost certainly true of all potentials developed to date and that the BH potential is the only one which has been put to such a severe test. The moral would seem to be that although a potential may be chosen to fit bulk properties arbitrarily well, this does not necessarily imply that low-coordination Si systems will be fit equally well.

More recently still, attempts have been made to calculate in an 'ab-initio' sense the total energy of each configuration thrown up in a MD calculation. The method is known as simulated annealing and is due to Car and Parrinello (30). The technique unifies into one classical–mechanical scheme both the atomic and electronic degrees of freedom of the system. The atomic degrees of freedom are the nuclear coordinates and the electronic are the coefficients which multiply the basis functions in the Kohn–Sham orbitals of the system. A Lagrangian (L) is defined in which the potential energy of the system is defined using LDF theory and *both* atomic and electronic degrees of freedom are given a classical kinetic energy. Atoms move on a Born–Oppenheimer surface:

$$L = \frac{1}{2}\mu \sum_i \int_\Omega d\bar{r} |\dot{\psi}_i|^2 + \frac{1}{2}\sum_I M_I \dot{R}_I^2 - E[\{\psi_i\}, \{\bar{R}_I\}] \tag{14.14}$$

and the ψ_i are subject to the constraints:

$$\int_\Omega d\bar{r} \psi_i^*(\bar{r}, t) \psi_j(\bar{r}, t) = \delta_{ij}. \tag{14.15}$$

The sums are over occupied-orbital index and atom index, respectively. R denotes nuclear coordinates and M physical atomic masses while μ is an arbitrary parameter with the dimensions of mass. The terms ψ are orthonormal one-electron eigenfunctions and E is given by the LDF approximation to the electronic Schrödinger equation. The kinetic energy of the electronic degrees of freedom is therefore fictitious, but it transpires that that set of coefficients which minimizes the potential energy of the system for given atomic configuration may be thought of as that in which the fictitious temperature has been annealed to zero. Equation (14.14) may be used to derive laws of motion for both atomic and electronic degrees of freedom. The dynamics of atomic motion over the Born–Oppenheimer surface are followed by keeping the fictitious electronic temperature at or near zero.

This technique involves the diagonalization at each time step of a matrix whose size is expressed as (number of occupied orbitals) rather than (number of basis functions) (2). When a plane wave basis is used, this results in considerable savings of computer time. Nevertheless, at the time of writing, handling systems of size greater than 64 particles for more than a few thousand time steps is impractical. Computer time scales as the cube of the number of atoms in the problem and requires fast-Fourier transform techniques. Thus system sizes in keeping with those typically used in crystal growth modelling are not yet attainable. However, it is undoubtedly true that methods like this will provide the way forward to the modelling of real systems.

V. Simulation results

In the ensuing results section, all quantities will be given in reduced units; that is in units of ε, σ, k and m for the LJ and SW systems. Given in the order argon (LJ), silicon (SW), the unit of distance is 0.3405 (0.20951) nm and the unit of energy is $1{\cdot}653 \times 10^{-23}$ (3.4723×10^{-19})J per atom. The masses of argon and of the stable ^{28}Si isotope are 6.628 and 4.6457×10^{-26} kg per atom, respectively. The unit of temperature is $1.197 \times 10^{2}(2.515 \times 10^{4})$K, and the time step in the MD simulations is 1.08×10^{-14} (3.83×10^{-16}) seconds.

Growth from the melt

Most materials for which crystallization rate data are available exhibit thermally activated growth. This classical behaviour is described by the Wilson–Frenkel equation (32) for growth rate (v):

$$v = (Da/\lambda^2) f_0 [1 - \exp(-\Delta\mu/k_B T)] \tag{14.16}$$

where D is the diffusion coefficient in the liquid and λ is the mean free path. Atoms in an adjacent liquid layer of thickness a are assumed to impinge upon the surface at a rate proportional to D/λ^2. f_0 is a steric factor (< 1) to take account of some collisions being more effective than others for growth. Equation (14.16) is sometimes written in terms of the viscosity (η) of the liquid rather than the diffusion coefficient, the two being approximately linearly related.

Since the diffusion coefficient can reasonably be represented by the Arrhenius equation

$$D = D_0 \exp(-Q/k_B T) \tag{14.17}$$

equation (14.16) has the form of being linear in the undercooling near T_m and of dropping off exponentially fast when the high viscosity in the liquid takes over at temperatures typically near 0.5 T_m.

The Wilson–Frenkel equation accurately predicts the temperature dependence of the growth of GeO_2 over a 400° range in which the viscosity changes by five orders of magnitude (33). Similar behaviour over more limited temperature ranges has been observed for systems with high-viscosity melts such as polymers and two-component network glass-formers (34).

For low-viscosity melt materials, growth rates are high even at low undercoolings and equation (14.16) has never been verified. For example, the growth rate of dendrites in supercooled nickel has been measured as a function of the initial bath temperature (35), which is not the same as the interface temperature. Rates up to 40 ms^{-1} at an undercooling of 175 K were measured; the actual interface cooling is probably a factor of 10 smaller than this. Interfaces travelling at these velocities produce cavitation behind the shock-wave front which further complicates the verification of equation (14.16).

Turnbull and co-workers (36, 37) on the basis of the difficulty of quenching pure metals into the glassy state and on the basis of Ruhl and Hilsch's finding (38) that metal films deposited at liquid helium temperatures undergo a transition at 20 K, have suggested that crystallization of these melts is *not* thermally activated.

It is at this point that simulation is of help. Imagine for a moment an equilibrium crystal-melt interface on a glass microscope slide directly beneath your field of view. Cooling the slide will cause the interface to propagate into the liquid. The observer may keep the interface beneath the field of view by pushing the slide in the opposite direction at exactly the interface growth velocity. The analogue of this experiment was performed by Broughton *et al.* (39) in simulations of the growth of FCC(111) and (100) crystal-melt systems. They used a LJ potential smoothly truncated to zero at a cut-off radius of 2.5σ. Liquid particles are fed from the bottom of the computational box passing through a heat bath into a dynamic zone which contains the interface. Particles pass through a second heat bath at the top of the box before being annihilated. The liquid feed velocity is chosen to keep the interface in the middle of the box. Zero pressure is maintained in the z-direction (normal to the interface) and the xy box dimensions are chosen to give zero pressure at the heat bath.

The growth rates as a function of interface temperature are given in Figure 14.1. T_m for this system is known to be in the range 0.612–0.617 from a variety of computational measurements (40) The Wilson–Frenkel growth rate is shown for comparison (the parameters such as D, λ and $\Delta\mu$ are known directly from bulk LJ calculations). For the (100) face, crystallization is

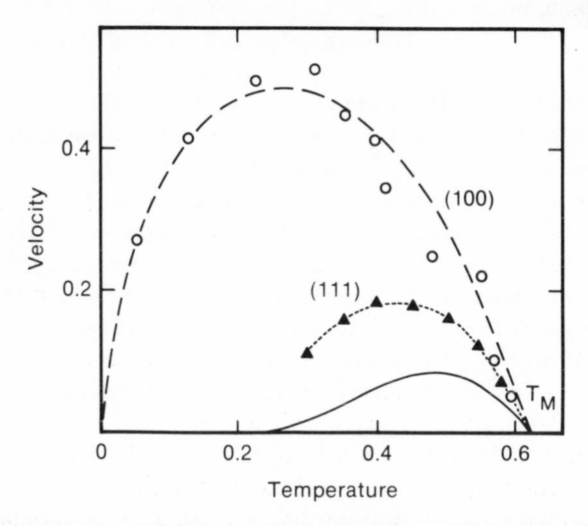

Figure 14.1 Molecular dynamics data for the velocity of the crystal-liquid interface *v.* interface temperature. The solid curve represents equation (14.16) and the long dashed curve represents equation (14.18). The short dashed curve is drawn to aid the eye

apparently not thermally activated. Even a very small activation barrier would reduce the growth rate significantly at low temperatures.

To test for the presence of a (100) energy barrier to crystallization, the two heat sinks were put at 0 K. The resultant growth rate is the point with the lowest interface temperature in Figure 14.1. As a further test, the velocities of all particles were set equal to zero and the simulation restarted. The interface velocity increased to the same rate as before. Repeated applications of the procedure also failed to stop the interface motion. This indicates that there are no energy barriers which can prevent the liquid atoms from reaching the lattice sites.

The peak in the growth rate corresponds to a linear velocity of about $80\,\mathrm{ms}^{-1}$ for argon. Near T_m, the diffusion coefficient of the liquid is $1.6 \times 10^{-5}\,\mathrm{cm}^2\,\mathrm{s}^{-1}$. At $v = 50\,\mathrm{ms}^{-1}$, the interface moves one atomic diameter in $\sigma/v = 7.6 \times 10^{-12}$ seconds. During this time, an atom in the liquid can move an average distance of $(Dt)^{1/2} \approx 0.3$ atomic diameters. At $T = 0.5T_m$, the corresponding distance is 0.06. This means that near the melting point, these atoms can move a significant distance while the interface is in their vicinity. But at lower temperatures, the liquid is nearly rigid on this time scale.

The (111) face grows less rapidly than the (100), with the maximum velocity occurring at higher temperatures. Growth here conforms more to the Wilson–Frenkel behaviour. The termination of growth near $T = 0.3$ is explained by the following argument. On FCC(111) faces, there are two almost degenerate sub-lattices; one corresponding to FCC, the other to HCP stacking. When particles crystallize from the liquid, it is (in the absence of other defects) at one or other of these sites. Thus, in-plane grain-boundaries arise which have to anneal out for growth to occur. This takes finite time and is achievable at higher temperatures where diffusion coefficients are high. At lower temperatures, this annealing process cannot occur; the in-plane boundaries are trapped and the distinction between liquid and solid is lost. Thus the absence of data below $T = 0.3$ for the (111) face signifies that the system is no longer able to support an interface. Growth now is activated, hence the greater similarity to the Wilson–Frenkel model.

The (100) face has no degenerate sub-lattice and liquid particles experience no potential energy barrier to finding perfect lattice sites on which to crystallize. In the absence of such a barrier, the rate at which atoms in the interface approach the lattice sites is determined by the average thermal velocity, $(3KT/m)^{1/2}$. They travel a distance λ, which is on average only a fraction of the interatomic spacing, a. Thus, equation (14.16) can be modified to give

$$v = (a/\lambda)(3k_\mathrm{B}T/m)^{1/2} f_0[1 - \exp(-\Delta\mu/k_\mathrm{B}T)]. \tag{14.18}$$

The dashed curve in Figure 14.1 has been plotted with $\lambda = 0.4a$, the average distance from the centre of points distributed randomly in a sphere of radius a, and f_0 was selected to give the best fit.

Equation (14.18) has profound implications. It is usually assumed that the glass-forming potential of different materials is only a matter of degree and that the value of the glass transition temperature relative to the melting point provides a good measure of this property. The present result implies that there is a class of materials with unstable glassy states since they crystallize even at very low temperatures when in contact with crystalline material. Big molecules must be reoriented to crystallize; some mixtures have to be sorted out to crystallize. Each of these requires rearrangement of the liquid structure and so there is an activation barrier to crystallization of these materials. Simple non-directionality bonded one-component systems are expected to follow non-activated growth and an equation similar to (14.18). This result, incidentally, also explains why one-component glasses have never been made.

Silicon, however, represents a system midway between these two extremes. Whereas the melt is of low viscosity and non-activated growth might be expected to apply, silicon crystal is a directionally-bonded network and has much in common with the network glass formers following Wilson–Frenkel kinetics. The growth kinetic behaviour of silicon is not known. The situation is complicated by two facts. Crystalline and liquid silicon are bonded in dissimilar ways, the former being semiconducting and the latter metallic. Also, there are two distinct low-temperature metastable disordered phases; the metallic and the amorphous, the latter being semiconducting. Briefly, the relevant facts pertaining to silicon growth are the following. In the Czochralski growth of Si from the melt, it is known that a slight undercooling exists at the (111) interface but not at any of the other faces (41). This implies that the (111) interface is just below its roughening transition and grows by a layerwise mechanism. (The (111) face, unlike the (100) or (110), has an in-plane fully connnected bond network. Consequently it costs energy to insert a step at the interface which leads to a finite roughening transition temperature T_R. Laser annealing studies (42) show the (100) to be capable of growing at up to $15\,\mathrm{m\,s^{-1}}$ before amorphization sets in. The comparable rate for the (111) is $13\,\mathrm{m\,s^{-1}}$. The (100) grows defect-free over most of this range, whereas the (111) exhibits twins and incipient lattice breakdown at velocities above $\sim 8\,\mathrm{m\,s^{-1}}$.

Part of this behaviour can be explained even in terms of the LJ growth simulations. The (100) FCC face is analogous to that of diamond cubic (DC) in that there is no degenerate sub-lattice. On the other hand, the FCC (111) offers the opportunity of either FCC or HCP stacking while that of DC offers a similar choice between wurtzite and DC. The quality of the as-grown crystal in the LJ (100) system is high, although some vacancies are trapped. In the LJ (111) system, however, not only are there vacancies but also a high concentration of stacking faults. Figure 14.2 gives the vacancy concentration as a function of interface growth temperature for both the FCC (111) and (100) faces. The solid line is due to Squire and Hoover (43) who calculated the free energy of formation of vacancies in the LJ system. We note that whereas thermodynamic arguments dictate that vacancy concentration should de-

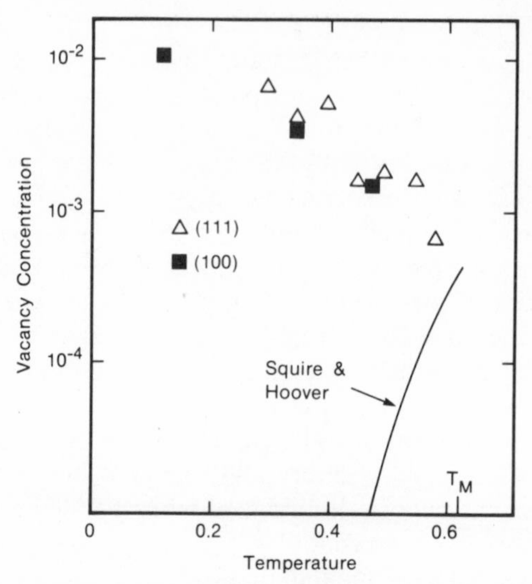

Figure 14.2 Vacancy concentration of as-grown material *v.* temperature for the (111) and (100) LJ systems. Equilibrium concentration given by continuous line

crease with decreasing temperature, the kinetic reality is the opposite of this. In both the (111) and (100) systems, the growth rates are so high and the liquid diffusion coefficient decreases so rapidly below T_m that the liquid does not have time to completely populate the crystal template before the interface sweeps past.

The concentration of HCP stacking faults in the (111) system (defined as the ratio of ABC to ABA stackings) may also be determined as a function of interface temperature. Within the scatter of the data, the concentration is almost temperature independent at a value of 0.5. On the other hand, the chemical potential difference between bulk FCC and HCP phases is also known to be temperature-independent. For these cut-off LJ systems, the HCP is actually marginally more stable than the FCC structure. Thus we expect the HCP to dominate more and more as the temperature decreases, but instead we find temperature-independent behaviour. As before, kinetic arguments explain this result.

We see, therefore, that simulations on the apparently dissimilar LJ system provide insights into what happens during Si growth. This idea of attacking complex problems by modelling simpler systems is the basis behind lattice-gas simulations of crystal growth. Gilmer has performed such simulations for the growth of three different faces of silicon, finding that growth rates follow the order (100) > (110) > (111). The simulations were performed using an effective bond energy between nearest-neighbours (ϕ_{nn}) silicon atoms of $kT_m = 0.35\phi_{nn}$ since the roughening temperature has been calculated to be $k_B T_R = 0.37\,\phi_{nn}$. The (111) growth kinetics were consistent with an undercooling of several

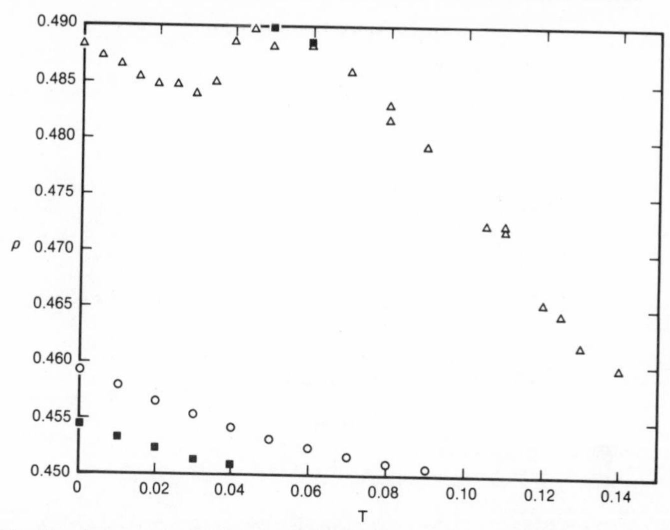

Figure 14.3 Density as a function of temperature along the $P = 0$ isobar for the SW system. Circles represent crystal, triangles represent liquid, and squares represent the amorphous phase. The crystal melts near $T = 0.10$; the amorphous near 0.045 and the liquid may be supercooled to $T = 0$ without nucleating either the crystal or amorphous phases

degrees during Czochralski growth. Unlike the behaviour of the (100) and (110) faces, the growth rate is not linear with undercooling near equilibrium.

However, these MC calculations do not give absolute rates; rather rates are given normalized by the arrival rate at the surface. Jackson (45) sought to remedy this by coupling the MC results with melt diffusion coefficient data and by *assuming* Wilson–Frenkel kinetics to give the growth rate v. temperature behaviour of the (100) and (111) faces. The (100) was predicted to have a maximum growth rate of $\sim 9\,\mathrm{ms}^{-1}$ at a temperature of $\sim 1450\,\mathrm{K}$ ($T_m = 1683\,\mathrm{K}$). The comparable values for the (111) are $3.5\,\mathrm{ms}^{-1}$ at $\sim 1250\,\mathrm{K}$. In both cases, the growth rate drops to zero near 400° C.

Broughton and Abraham (46) attempted to determine by MD simulation whether growth was activated on the Si (111) face. They used the SW potential whose $P = 0$ isobar (which closely follows crystal–vapour coexistence) is given in Figure 14.3. Note that quenching a liquid phase does not produce the amorphous phase on MD time scales. The supercooled melt phase of the SW system is a metastable extrapolation of the high temperature melt. Thus, in examining the growth behaviour of the silicon–melt interface, the complication of crystallization from an amorphous phase is avoided.

At either end of their system in the z-direction is a hard wall. Periodic boundary conditions are applied in the xy plane. The model comprises four parts (increasing in z); a crystalline reservoir region whose temperature is kept constant when necessary, using random and dissipative forces of six planes abutting one hard wall; a crystalline region of 14 planes abutting the reservoir;

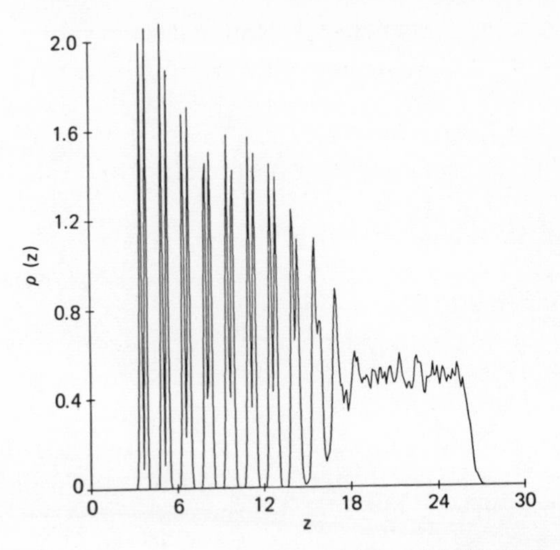

Figure 14.4 Density profile through triple-point SW (111) system

a region to be melted, comprising initially 16 crystal planes abutting the crystal, and lastly a vapour sandwiched between the liquid and the other hard wall. In the figures following, the six planes of the reservoir are not plotted and extended from $z = 0$ to ~ 3.0. The xy period of the system was adjusted to give zero pressure in the bulk crystal.

First we discuss crystal-melt coexistence. The density profile ($\rho(z)$) through the three-phase triple point-system is shown in Figure 14.4. Notice the two periods of the crystal. The shorter represents the distance between the body centre of the tetrahedron and the three-atom base. The longer is the distance

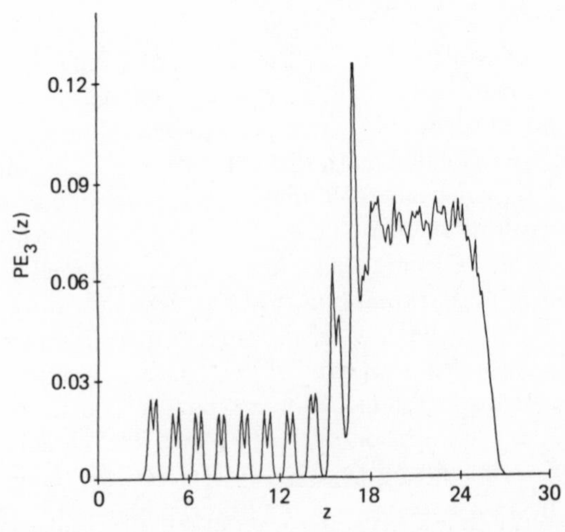

Figure 14.5 Three-body energy density profile through triple-point SW (111) system

between the body centre and the apex. Notice also how sharp is the interface. The two-body contribution to the total potential energy profile per unit volume (not shown) has almost identical features to $\rho(z)$. It also shows a sharp interface; but perhaps the most dramatic evidence for a rapid transition from liquid to crystal properties is given by the three-body potential energy profile ($PE_3(z)$) in Figure 14.5.

In the crystal, the three-body energy contribution is non-zero due to the finite mean square displacement of particles away from perfect lattice sites. In the liquid, the number of nearest-neighbour atoms is near six (clearly non-tetrahedral) and the three-body interaction is very positive. However, not only does the three-body increase in magnitude in going from crystal to liquid, but so also does the two-body, due to the increase in nearest-neighbours. The number and energy density profiles change from having periodic structure to being featureless between layers in which atoms lie in threefold hollows. The doublets are replaced by a singlet in the interface. However, the trajectory plots of Figure 14.6 taken in the xy plane as a function of z traversing the interface show the strong singlet feature to comprise highly mobile atoms. In fact these trajectory plots, apart from again showing the sharpness of the interface, would place the interface between diffusing and crystalline regions as occurring at the first non-zero minimum in the density profile ($z = 16.5$). This

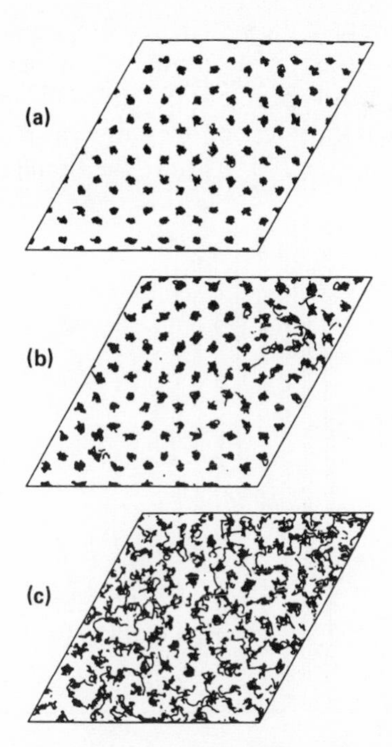

Figure 14.6 Trajectory plots in xy plane over final 10 000 Δt of equilibrium run for triple-point system. (*a*) $z = 13.5$ to 15.0; (*b*) $z = 15.0$ to 16.5; (*c*) $z = 16.5$ to 18.0

same observation has also been made for the densely-packed faces of Lennard–Jonesium (47, 48). That the first mobile layer moves in the strongly periodic field of the substrate is evinced by the sharp interface feature in the density profiles.

The $\rho(z)$ profile shows none of the persistent periodic oscillations through the interface and into the liquid that is found for the LJ system (47, 48). We conclude that the very dissimilar packing of atoms in the Si melt relative to the crystal damps out any such oscillations very rapidly. The poor packing in the Si system is particularly well demonstrated by the non-monotonic behaviour of $PE_3(z)$ through the interface. Whether this feature would be present in a yet more sophisticated potential for Si remains to be seen.

In order to determine whether the growth of Si(111) is activated, the system was cooled thus: starting with the equilibrium triple point configuration, the temperature of the reservoir was instantaneously dropped to $\sim 0.2\, T_m$. The rest of the system was allowed to time evolve naturally. In such a system the interface, if it were to follow non-activated growth kinetics, should grow at finite rate even after the entire system has achieved the reservoir temperature. Even if competition for the degenerate sub-lattice sites on the (111) face were to occur, finite growth rates of a disordered but predominantly tetrahedral phase (akin to the amorphous phase) might be expected. However, neither of these situations occurs.

The mean temperature of the system as a function of time from the start of the quench was monitored to determine when the quench was complete. Figure 14.7 gives the density profile at the end of the quench. Comparing with Figure 14.4, notice that the interface does move a small distance (of the order of a lattice spacing) over the 100 000 Δt of the simulation. That a single layer

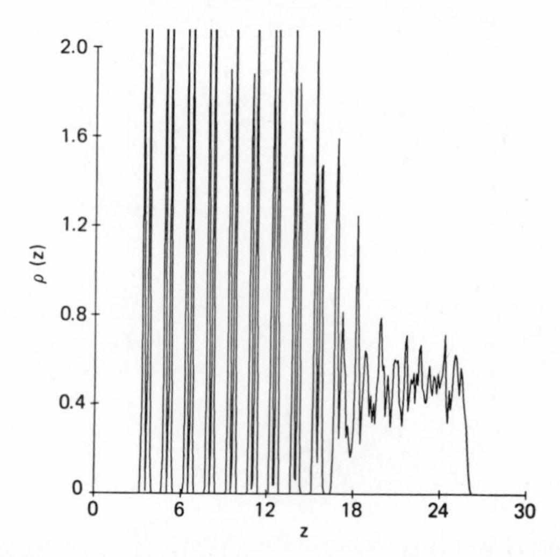

Figure 14.7 Density profile during final 5000 Δt of quench

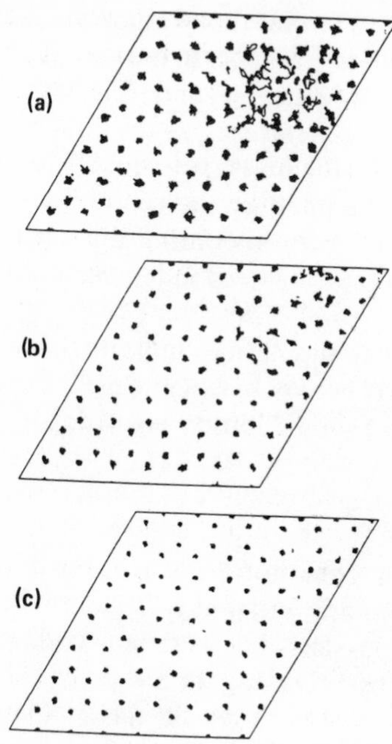

Figure 14.8 Trajectory plots in xy plane for $z = 15.0$ to 16.5 region over $10\,000\,\Delta t$ consecutive sub-intervals from start of quench. (*a*) First $10\,000\,\Delta t$; (*b*) second $10\,000\,\Delta t$; (*c*) third $10\,000\,\Delta t$

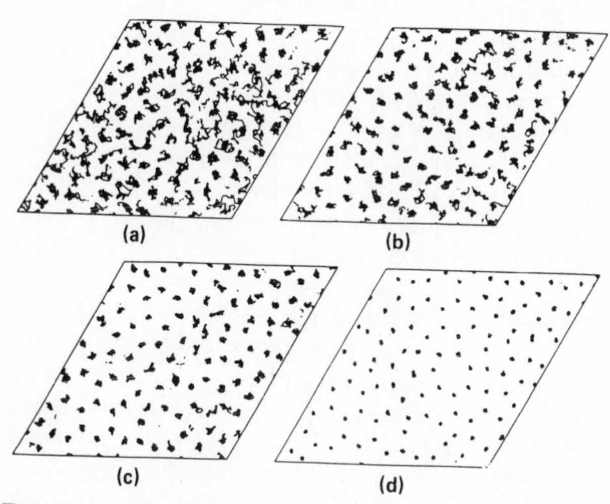

Figure 14.9 Trajectory plots in xy plane for $z = 16.5$ to 18.0 region over $10\,000\,\Delta t$ sub-intervals. (*a*), (*b*), (*c*): First, second and third $10\,000\,\Delta t$ from start of quench; (*d*) final $10\,000\,\Delta t$ of quench

crystallizes is to be expected from what we have already said of the equilibrium interface. The first mobile layer moves in the periodic field of the substrate and is obviously strongly ordered. Beyond this layer, the liquid feels little correlation with the crystal. However, as the trajectory plots illustrate for the regions $z = 15.0$ to 16.5 (Figure 14.8) and 16.5 to 18.0 (Figure 14.9), these outermost layers near the interface are not perfect. The trajectory plots are taken over 10 000 Δt sub-intervals during the quench. Notice the top-right corner of Figure 14.8*c* and the preponderance of 5-, 7- and 8-fold rings in Figure 14.9*d*.

In conclusion, then, the simulations indicate that activated growth kinetics are appropriate for Si(111) growth from its melt. We learn more of Si growth mechanisms in the next subsection on melt propagation into the crystal.

Melt propagation

This subsection pursues the notion of whether Si(111) and (100) faces, in contact with their melt, are rough near T_m and whether growth occurs in a layerwise or continuous fashion. Abraham and Broughton (49) chose to examine this problem by MD using the SW potential. Their simulation results confirm in a very direct way (after all, the simulator is Maxwell's demon) expectations which until now had to be deduced from indirect experimental observation, namely that small (111) facets are observed during the Czochralski growth of dislocation-free silicon (41). Such facets imply the surface to be undercooled and that growth occurs via the nucleation and growth of layers. Simple nearest-neighbour bond models of the (111) and (100) DC surfaces (44) predict $T_R = 0$ for the (100) and $T_R > T_m$ for the (111) (see previous subsection). The (111) interface is expected to be smooth and sharp, the (100) rough and broad.

Abraham and Broughton considered crystal-vapour films of (111) and (100) systems. These were periodically connected in the x, y and z directions with sufficient extent in the z to allow for vapour phase. After equilibration of the slabs at temperatures near T_m, the kinetic energy of the top four layers was then instantaneously raised on one side of the slab sufficient to cause approximately half the system of eventually melt. The propagation history of the melt front into the centre of these slabs was followed until equilibrium was again achieved. (These are constant-energy systems.)

After the heat pulse (time zero) it took $\sim 60\,000\ \Delta t$ for the system to reach equilibrium, that is for the total system temperature to be invariant with time and for the temperature profile through the system to be flat. Figure 14.10 gives trajectory plots in a thin x–z section for the (111) and (100) interfaces as a function of time. The (111) melt front clearly progresses a layer at a time, which we expect by microscopic reversibility from the Czochralski crystal growth experiments. Trajectory plots in the x–y plane and for other x–z sections show the front to be flat in the x–y plane. In contrast, the (100) interface grows over

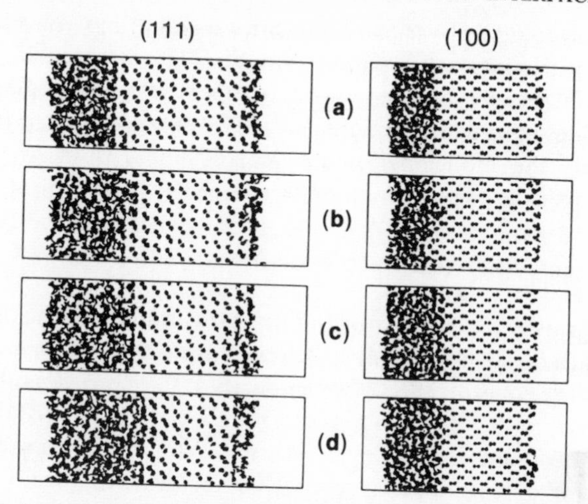

Figure 14.10 Trajectory plots of thin *xz* slice through (111) and (100) SW system covering elapsed time ranges of (*a*) (15–20) × 10^3 Δt; (*b*) (25–30) × 10^3 Δt; (*c*) (45–50) × 10^3 Δt; (*d*) (95–100) × 10^3 Δt. Times defined after initial heat pulse

several layers simultaneously. Figure 14.11 shows x–y plots over the same time sub-interval during the growth of four neighbouring layers. The region in the top left corner of the computational box shows particles vibrating on lattice sites over four layers while the lower right indicates substantial disorder in those same layers. This behaviour is symptomatic of a rough interface. Irregular hill-and-valley structures at the interface form and uniform dynamically as the system fluctuates around equilibrium. A similar trajectory analysis for the equilibrium (111) interface shows a transition from crystal to a highly

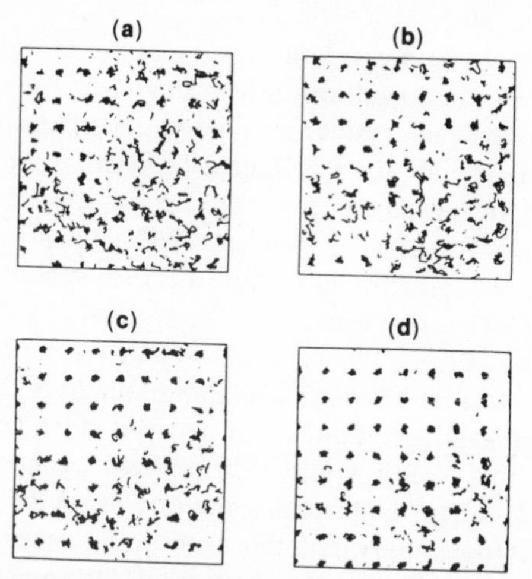

Figure 14.11 Trajectory plots in *xy* plane of adjacent layers through the (100) SW interface covering elapsed time between 45 000 Δt and 50 000 Δt after heat pulse

mobile region to occur over one layer. Such a smooth interface is in keeping with the experimental observation that Si (111) is above T_R at T_m. The temperature at which the three-phase system reaches equilibrium is ~ 1760 K, which considering the fact that the SW potential represents a first tentative step towards the modelling of covalent systems, is in very acceptable agreement with the experimental value of 1683 K.

Impurity trapping

At normal growth velocities (for systems near equilibrium) the distribution of impurities in a crystal is generally calculated using Burton, Prim and Schlicter theory (50). The concentration of the impurity at the interface is obtained using transport theory in the fluid phase and the adjacent crystal phase concentration is found from the equilibrium phase diagram. During laser annealing studies (42), however, regrowth velocities are typically five orders of magnitude faster than those operant during Czochralski growth (for example). We might expect (and do find) the distribution coefficient, which is the ratio of the concentration in the crystal to that in the melt, to deviate from that predicted by equilibrium theory. At these high velocities, there is much less time for equilibration between the crystal and the melt via the exchange of atoms at the interface. The magnitude of the deviations, however, can be very large. For example, bismuth (51) and indium (52) in silicon have equilibrium distribution coefficients K_0 of $\sim 5 \times 10^{-4}$ whereas at interface velocities of $\sim 5\,\mathrm{ms}^{-1}$, K can reach values near 0.3. These high concentrations are surprising for several reasons. Firstly, there is little disruption of the host lattice, and secondly, the energies of the impurities actually increase during crystallization. Simple rate theory would predict a *decrease* in K as the interface is undercooled.

Why does K increase by several orders of magnitude at high growth velocities? Several ideas have been advanced. The first postulates a diffuse transition region between crystal and melt (53), perhaps hundreds of lattice spacings in extent. The impurity is immobilized in a viscous medium, as the interface sweeps by, *before* its energy rises to the value typical of the bulk. Given the simulations on the crystal–melt interfaces described in the previous subsections, however, which show the transition region to be somewhere between one and four layers (depending upon the face), we can rule out the diffuse interface hypothesis.

Of greater significance perhaps are two other hypotheses. If impurities preferentially segregate to the interface between crystal and melt, the high concentration in this region would increase their probability for crystallization. Secondly, atomic size mismatch may enhance trapping. We expect the compressive stress (for impurities larger than Si, like Bi) to increase gradually as the impurity is buried deeper in the crystal. Outward relaxation relieves some of this stress when the impurity is near the surface, but such relief is not possible deeper down. The atom is trapped *before* the increased energy in the crystal is achieved.

These latter two ideas have been tested by MC lattice-gas simulation. Gilmer gives two excellent reviews of the subject (44, 54). It is his work we briefly describe below. The system is represented by a three-dimensional spin-1 Ising model; that is the lattice sites can be occupied by atoms in one of three states: crystalline a atoms, crystalline b atoms or a mean-field liquid. As in the case of the one-component solid-on-solid model described in the subsection on simulational procedures, the arrival rates of the two species are assumed to be proportional to their concentrations in the liquid phase whereas the off-rates depend upon the potential energy at the site:

$$k_a^- = k^+ \frac{z_a^l}{z_a^c} \exp\left[\frac{E_a^0 - n_a\phi_{aa} - n_b\phi_{ab}}{k_B T}\right]. \qquad (14.19)$$

Here, ϕ_{aa} and ϕ_{ab} are effective bond energies (this is a nearest-neighbour model), k^+ is the on-rate, z_a and z_b are the individual partition functions of a and b atoms at their sites excluding interactions with neighbours, superscripts c and l represent crystal and liquid, respectively; n_a and n_b are the number of nearest neighbours of type a and b and E_a^0 is the crystallization energy from the melt. The model allows for substitutional impurities only, but such is the experimental case.

Gilmer considered two situations. The results for the Bi/Si system are given in Figure 14.12. He parametrized his system thus: knowing there to be a small undercooling at the Si(111) interface during Czochralski growth determines

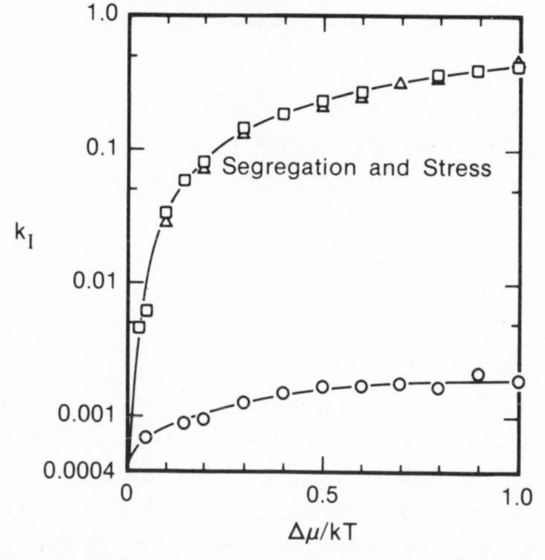

Figure 14.12 Distribution coefficient K_I *v.* driving force $\Delta\mu/kT$. Circles and squares represent (100) systems. Triangles represent the (111). See text for fuller description. (Courtesy of G. H. Gilmer.)

ϕ_{aa} quite accurately. The value of the latent heat of fusion ($L = 3.24\,kT_m$) then gives E_a^0. He then made a reasonable guess for E_b^0 which leads directly to a value for ϕ_{ab}.

In the lower part of Figure 14.12, Gilmer assumed an absence of any potential favouring the crystallization of impurities at the (100) surface (i.e., $E_a^0 = E_b^0$). K_I (the distribution coefficient at the interface) does increase with increasing undercooling even in this case, but only by an order of magnitude. This is clearly insufficient to explain the experimental observation. The enhanced trapping, nevertheless, is caused by the impurity having insufficient time to desorb from the surface before being embedded in the growing interface.

In the upper part of the figure, the combined effects of segregation and atomic stress are shown. The value of E_b^0 is thought to be less than E_a^0 which causes some preferential segregation of the impurities to the interface and enhances trapping. This effect alone provides another order of magnitude to K_I at high undercoolings (55). For the effects of surface segregation on enhancing crystal growth rates and on trapping in clustering or ordering alloys, the reader is referred to Gilmer's review (54).

However, it is the effect of atomic stress which plays the largest role (55). In addition to the crystallization energy E_b^0, a second term E_b^1 is included at the moment when the impurity is fully coordinated. Thus the impurity perceives stress only when fully incorporated. Now the limiting value of K_I is in the right range to fit the laser annealing experiments. Note that the difference between the zero values of K_I for the (111) and (100) faces is identical within the statistical certainty of the data. However, when K_I is plotted as a function of interface velocity (Figure 14.13) we see that the distribution coefficient is larger

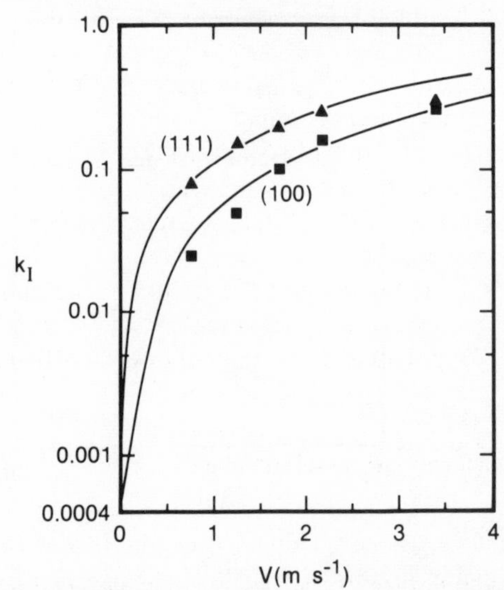

Figure 14.13 Distribution coefficient K_I *v.* interface velocity. The experimental data for bismuth (squares and triangles) are compared with simulation data. (Courtesy of G. H. Gilmer.)

for the (111) for given growth velocity and thus is in accord with experiment. A large value of $\Delta\mu$ is required on the (111) to achieve a given velocity. The unknown value of k^+ was used as a fitting parameter to optimize agreement between simulation and experiment.

Finally, note in Figure 14.13 the rapid increase in K_I for the (111) interface even at low velocities. This may help explain the observation that impurities are often found behind the (111) facet during Czochralski growth.

Other systems

Our review has not encompassed all the atomistic simulation work pertaining to crystal growth at semiconductor surfaces, although we hope that it has given the reader a fair idea of the methods used, the questions being asked and the capabilities of simulation. The interested reader is referred to the following for a wider view of semiconductor modelling.

Many attempts have been made to obtain atomic coordinates representative of amorphous silicon (a-Si). Polk and Boudreaux (56), Henderson (57), Guttman (58) and Wooten *et al.* (59) have each generated models of a-Si. Herman and Lambin (60) have developed a graph-theoretical method for getting from FCC and closely packed LJ liquid structures to DC and amorphous silicon. Andersen (61) is presently studying the glassification of germanium using MD quenches of SW-like liquids.

Vapour-phase deposition has also been modelled. Schneider *et al.* (62) have examined the epitaxial growth of LJ atoms on a close-packed (triangular lattice) substrate as a function of substrate temperature and deposition rate. A study of the vapour deposition of SW atoms on (111) and (100) DC surfaces has been performed by Dodson and Taylor (63). Proceeding to the even more complex, Singh and Madhukar (64) have proposed lattice models for GaAs growth involving not only AA, AB and BB interactions but also different possible adsorption states and their accompanying dissociation processes. Das Sarma (65) is continuing this work.

Ramirez *et al.* (66) have examined bcc–fcc epitaxy using pairwise potentials, while Grabow and Gilmer (67) have examined the role of misfit dislocations in strained epitaxial layers of simple LJ-like systems and are currently extending their work to covalent adlayers.

Lastly, Wetzel *et al.* (68) have used the Keating potential to study grain-boundaries in carbon, silicon and germanium. Also Weber (69) and Abraham (70) have used the SW potential to examine the surface dimer reconstruction on Si(100).

VI. Electronic structure and wavefunctions

General considerations

Knowledge about the nature of the electronic wavefunctions resulting from the interactions between a semiconductor surface and atoms arriving at that surface is necessary in order to understand the subsequent reactions. This is obvious from the preceding sections. In all the problems discussed above, an atom–surface interaction potential was used to determine the statistical mechanical properties. These potentials were constructed by making assumptions, albeit often rather simple ones, about the nature of the interaction. The limitations of these potentials may, and often does, lead to the results of the statistical mechanical, MC or MD, studies having mainly qualitative rather than quantitative value.

In general, rigorous quantum mechanical treatments are needed to obtain quantitative interaction potentials. These potentials may be used, by themselves, to provide a general understanding of the nature of energy barriers and wells or they may be used as input for statistical mechanical treatments. Of course, the electronic wave functions provide far more information than just interaction potentials. They directly give, for example, ionization energies that can be compared with photoemission (PES) spectra. In fact, the results of a theoretical cluster model study which gave ab-initio wave functions for F interacting with and penetrating a Si(111) surface (72) have been used to assist in the interpretation of synchrotron PES studies (73). The theory provided information about binding energy shifts which helped to establish the nature of the SiF_x reaction intermediates and the sites which they occupy in the surface.

Two general classes of approach are used for the theoretical treatment of the electronic structure aspects of the adsorbate–substrate interaction. The first involves the application of one of the many types of band-structure theory; some results obtained using these approaches are cited in references 74–78. In most of these approaches, the surface is represented by a repeated slab; in typical applications, the slab has between 3–9 atomic layers. An adsorbate layer may be added to the surface placed, for symmetry, on both sides of the slab. The periodically repeated slabs are separated by a large enough region of vacuum to ensure that interactions between slabs are negligibly small. A second class of approach involves the use of molecular orbital (MO) theory applied to finite clusters of atoms chosen to model the adsorbate–substrate system. These MO cluster models normally treat a small part of the surface and contain only one adsorbate (72, 79–82) although there have been studies with clusters containing a few (about five) adsorbates (83). The substrate may be modelled with only one atom (79, 81, 84) which, in certain cases, will describe the qualitative features of the chemical bonding between adsorbate and substrate. More usually, 10–20 substrate atoms are included and work has been done with even larger clusters (80, 83, 85–87). Various procedures have been proposed to represent the effect of the remaining atoms of the semi-infinite substrate; this is the embedding problem (3, 87). For covalently bonded semiconductors such as Si, a successful approach to embedding has been to terminate the cluster with H atoms (3, 72, 81, 88, 89). This ensures the appropriate coordination and proper hybridization of all the semiconductor atoms of the cluster.

These two general classes of approach focus on different aspects of the adsorbate–substrate interaction. The MO cluster model is appropriate mainly for the analysis of the local chemical bonding. Ab-initio wavefunctions, either self-consistent field (SCF) (81), or including electron correlation through configuration interaction (CI) (87, 88, 90), give good descriptions of the features of this interaction. These descriptions contain the key qualitative features and they may also have quantitative value.

An advantage of the cluster approach is that the wavefunctions can be analysed to elucidate the features of the chemical changes which occur when bonds are formed between the adsorbate and substrate. Techniques have been developed to permit quantitative measures of the degree of ionicity and the energetic importance of intra-unit charge transfer between substrate and adsorbate. The Constrained Space Orbital Variation (CSOV) (79, 91) makes it possible to evaluate separately the energetic importance of this charge transfer and of the polarization of the charge of one of the units, adsorbate or substrate, in response to the presence of the other unit. The use of projection operators makes it possible to determine the number of electrons which occupy the orbitals of the free, isolated adsorbate; i.e. the adsorbate ionicity (92–94). For example, F interacting, at certain sites, with a Si surface is essentially F^- or fully ionic (92). We shall return to this way of measuring ionicity later and shall demonstrate that projection operators give more definitive results for ionicity than the more usual Mulliken population analysis (95) which often makes a misleading artificial division of charge between adsorbate and substrate (96). Another major success of the MO cluster model has been to show, through the use of the CSOV analysis, for the covalent metal surface–carbonyl bond that the π back-donation is considerably more important than the σ donation (79, 96–99).

Still another advantage of the MO cluster approach is that it is possible to study the interaction as a function of the adsorbate–substrate geometry; in particular, to obtain potential surfaces or curves for the interaction. Band-structure theory has been normally used to describe the electronic structure at fixed geometries. Recently, there have been developments, for example see (76), (77), which indicate that it may be possible to compute potential surfaces for the adsorbate–substrate interaction with wavefunctions for extended surfaces obtained using a local density function for the exchange-correlation potential. Unfortunately, not many such studies have been carried out and it is premature to make an evaluation of the applicability and limitations of potential surfaces obtained with such approaches.

F, Cl, and H interaction with and penetration of silicon

In the remainder of this section, we shall review the cluster model determination of the nature of the interaction of F, Cl, and H with a Si(111) surface (72, 88, 92) at a threefold site suitable for allowing the adsorbate to penetrate below the surface layer of Si atoms. Our results show that F/Si(111) at this threefold site is essentially F.$^-$ A strong interaction develops because F^- polarizes the Si substrate; this polarization can be considered as the formation of an image charge in the surface. The electrostatic force between F^- and its image allows the F to penetrate below the Si surface without a barrier. This penetration as a first step in the etching reaction of F/Si(111) to form primarily SiF_4 as a volatile product (100, 101) is consistent with experimental evidence (73). We further show that Cl does not penetrate the surface because it is a larger ion. However, H has a barrier for surface penetration even through it is smaller than F. This is explained because the H–Si interaction is covalent and there is no electrostatic force to overcome the surface barrier as is the case for the ionic interaction of F^- with Si. Overall these results provide us with a consistent picture which explains the features which determine the character of the interaction and subsequent reaction.

We will first describe the cluster model, the adsorption site or interaction geometry and the wavefunctions which are used. We will then discuss the projection technique that is used to measure the ionicity of the interaction of F with Si. The importance of the F^- anionicity for the character of the interaction is illustrated with a CSOV analysis. Finally, we consider the nature of the interaction potential curves for F, Cl, and H along an assumed reaction path for surface penetration.

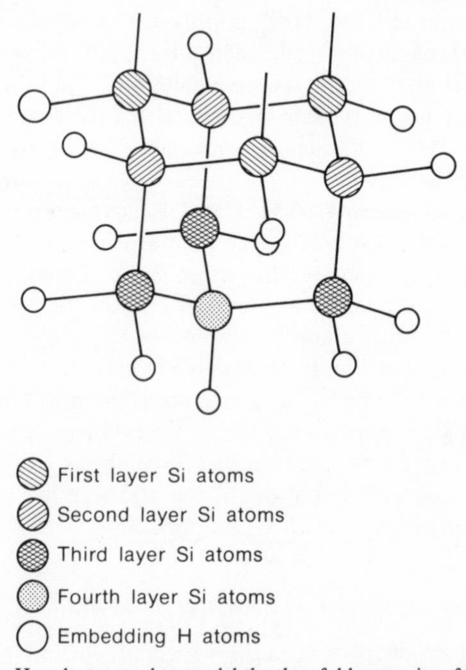

Figure 14.14 The $Si_{10}H_{13}$ cluster used to model the threefold open site of Si(111); the Si atoms form the different surface layers and the embedding H atoms are distinguished. The surface layer dangling bonds are indicated

The cluster that will be described here contains ten Si atoms; three each from the first three layers of an unreconstructed Si(111) surface and one from the fourth layer. The Si–Si distances are taken from the bulk geometry. These atoms, shown in Figure 14.14, are chosen to model one of the two different kinds of threefold sites on a Si(111) surface. In one of these, there is a second layer atom below the centre of the atoms forming the threefold site. In the other, the one which has been modelled, the first atom below the site is in the fourth layer. In Figure 14.15, these two sites are shown. The open site has been selected because it provides a reasonable path for surface penetration. The adsorbate atom is moved along an assumed reaction path which is normal to the surface and passes through the centre of the triangle of first layer atoms. On this path, the adsorbate can go quite far below the surface before it encounters, or passes close to, a Si atom. For the unreconstructed Si(111) surface, each surface layer atom is threefold coordinated and has a 'dangling bond' (81) normal to the surface where its missing bulk neighbour would be. The Si atoms of the lower layers all have the bulk T_d geometry and are fourfold coordinated. However, the ten Si atoms of the cluster do not have this number of neighbours and would not, as a consequence, be sp^3 hybridized. To model properly the true environment of the Si atoms, H atoms are added to take the place of the missing atoms and to ensure that the Si atoms of the cluster will have sp^3 hybridization (3,81, 88, 89). Including these environmental H atoms, the surface is represented by the $Si_{10}H_{13}$ cluster shown in Figure 14.14. The interaction with the adsorbate is modelled by $Si_{10}H_{13}X$ where X = F, Cl or H. Interactions with Si(111) have also been modelled with smaller clusters (72, 81, 89). As long as the Si atoms directly involved in the chemical bonding with the adsorbate were included, the essential features of the interaction were obtained. For on-top site chemisorption on Si(111), a substrate with one Si atom and three environmental H atoms, Si_1H_3, gives essentially the same results (81) as a Si_4H_9 cluster with four Si atoms, the adsorption site atom and its nearest neighbours, and nine environmental H atoms. Thus, $Si_{10}H_{13}$ is

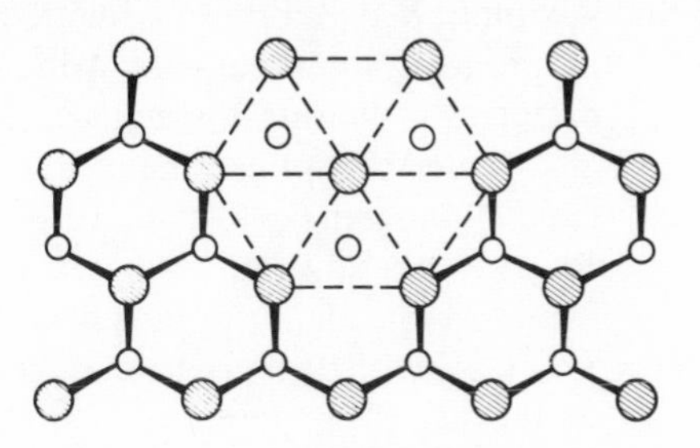

Figure14.15 Schematic representation of the unreconstructed Si(111) surface showing the two types of threefold sites. From (72)

adequately large to model the open threefold site penetration of Si(111). For $Si_{10}H_{13}$, self-consistent field (SCF) wavefunctions are obtained with the linear combination of atomic orbitals (LCAO) method using a contracted Gaussian type orbital (CGTO) basis set. A basis set is used that is sufficiently large so that no significant changes are expected with larger basis sets. For $Si_{10}H_{13}F$ and $Si_{10}H_{13}Cl$, CGTO basis set SCF wavefunctions have also been obtained. However for $Si_{10}H_{13}H$, it was necessary to go beyond the one-configuration SCF approximation, and a two-configuration multi-configuration SCF, MCSCF, wavefunction was used (72). This MCSCF dissociated correctly to neutral wave functions for $Si_{10}H_{13}$ and H when the H atom was taken to large distances. The MCSCF treatment was required because the SiH bond was covalent (72). Further details of the wavefunctions and the computational approach used are given in references 72, 81, 88, and 92.

The SCF potential curve for F normal to the surface at the threefold open site discussed above has two minima; one 1.3 Å above the surface layer atoms and one 1.4 Å below. The maximum between these two wells is ~ 1.5 eV above the minima (72, 92). This potential curve is consistent with F being significantly ionic near the surface. The electrostatic interaction between F^{-q} (where q is the effective ionicity of the electron acceptor F) and the polarized Si will assist the F to overcome the barrier to penetration of the surface which arises from a Pauli repulsion. This repulsion is related to the fact that the F and Si charge clouds interpenetrate and lead to an overlap repulsion (79). When F is in the surface layer, it is closest to the three Si atoms in this layer and the repulsion should be largest. The maximum of the potential curve is at 0.4 Å below the surface consistent with a cancellation of electrostatic attraction and Pauli, or overlap, repulsion. When a Mulliken gross population analysis is used to estimate the F ionicity, the results are somewhat surprising. While, by this measure, the F is reasonably ionic, the ionicity varies fairly strongly as a function of its distance from the surface. Near the minima above the surface, the ionicity is ~ −0.7 while near the minima below the surface, the ionicity is ~ −0.5; near the maximum between these minima, the ionicity is ~ −0.2. The Mulliken population analysis contains an arbitrary division of overlap charge, or overlap population, sharing it equally to the two atoms between which the charge is distributed (95). When the overlap populations are large, this division leads to uncertainties in the Mulliken gross charge or ionicity. For the $Si_{10}H_{13}F$ cluster (92), the overlap population between F and the neighbouring Si atoms is ~ 5 when F is near the surface. It is quite possible that the variation of the F ionicity with distance from the Si surface is not physical but is an artefact of the Mulliken population analysis. A more rigorous measure of ionicity is given by projecting the orbitals of the F atom from the $Si_{10}H_{13}F$ cluster wavefunction. The projection operator P_F is

$$P_F = \Phi_{1s}\Phi_{1s}^\dagger + \Phi_{2s}\Phi_{2s}^\dagger + \Phi_{2px}\Phi_{2px}^\dagger + \Phi_{2py}\Phi_{2py}^\dagger + \Phi_{2pz}\Phi_{2pz}^\dagger \qquad (14.20)$$

The F ionicity q is obtained from the expectation value of P_F for the cluster wavefunction, Ψ:

$$q = <\Psi|P_F|\Psi> - 9.0 \tag{14.21}$$

The value for q will depend on whether the Φ_i of equation (14.20) are optimized for free F or for F^-; however, the differences are very small (94). Since the Mulliken population analysis indicates a fairly large ionicity for F, the SCF orbitals for the F^- anion have been used to construct P_F. From this projection, the ionicity is between -0.95 and -0.97 for F at distances from the surface ranging between -1.5 and $+1.0$ Å, giving definitive evidence that F/Si(111) at this site is essentially F^- and that the bonding is ionic.

In order to understand how the bond between F and the Si surface develops, a CSOV analysis (79, 91) has been performed for the $Si_{10}H_{13}F$ cluster (92). In this analysis, the orbitals of the $Si_{10}H_{13}^+$ and F^- units are varied separately in order to determine the nature of the charge rearrangements when the units are brought together. For this purpose, the changes in three quantities are examined. The first is the interaction energy, E_{INT}:

$$E_{INT} = -E(Si_{10}H_{13}F) + E(Si_{10}H_{13}^+) + E(F^-) \tag{14.22}$$

The second is the expectation value of the position of the electrons normal to the surface:

$$\langle z \rangle = \langle \Psi|\Sigma_i z_i|\Psi \rangle \tag{14.23}$$

And the third is the expectation value of the second moment of the position of the electrons in the plane of the surface:

$$\langle x^2 \rangle = \langle y^4 \rangle = \langle \Psi|\Sigma_i x_i^2|\Psi \rangle \tag{14.24}$$

The C_{3v} symmetry of $Si_{10}H_{13}F$ ensures that $\langle x^2 \rangle = \langle y^2 \rangle$ and $\langle x \rangle = \langle y \rangle = 0$. The differences of these three quantities as the orbitals are varied provide measures of the nature and importance of the changes in the electronic structure due to the interaction of F and Si. Specific results are presented below for F 0.53 Å above the Si surface.

The starting or reference point for the CSOV analysis, CSOV step 0, is the superposed $Si_{10}H_{13}^+$ and F^- charge distributions when F is brought up to the Si surface. This is the Frozen Orbital (FO) step since no changes in the orbitals of the two units are allowed. (For computational convenience, the $Si_{10}H_{13}^+$ orbitals are Schmidt orthogonalized to those of F^-; however, this does not change the FO properties, (14.22)–(14.24).) In the next CSOV step, step 1 or V(Si), the F^- orbitals are fixed and the Si orbitals are allowed to vary (that is, self-consistently adjust themselves in the field of the 'frozen' F^- orbitals); the changes in the three properties of (14.22)–(14.24), ΔE_{INT}, $\Delta\langle z \rangle$, and $\Delta\langle x^2 \rangle$ are given in Table 14.1. From $\Delta\langle z \rangle$ and $\Delta\langle x^2 \rangle$, we see that the Si charge moves away from F^-; it moves both vertically, $\langle z \rangle$, and laterally in the plane of the surface, $\langle x^2 \rangle$. This charge motion may be viewed as the formulation of an image charge in the Si surface. The changes for an extended

Table 14.1 Changes between successive CSOV steps for the interaction energy ΔE_{INT} in eV, and for the vertical, $\Delta\langle z \rangle$, and lateral, $\Delta\langle x^2 \rangle$, positions of the electrons, in Å and Å^2 respectively, for $Si_{10}H_{13}F$; the unconstrained, or full, SCF is included as the final CSOV step. The interaction energy E_{INT} at each step is also given. Fluorine is 0.53 Å above the surface.

CSOV step	E_{INT}	ΔE_{INT}	$\Delta\langle z \rangle$	$\Delta\langle x^2 \rangle$
0. FO	+0.30	—	—	—
1. V(Si)	+3.79	+3.50	−1.41	+3.50
2. V(F)	+4.44	+0.65	−0.07	+0.11
3. Full SCF	+4.53	+0.08	+0.03	0.00

surface will be even larger than for our $Si_{10}H_{13}F$ cluster, especially for lateral charge motion, since the extended surface provides more possibilities for the Si charge to move away from F^-. However, we note that the cluster image charge leads to a 3.5 eV increase in E_{INT}! For the second CSOV step, V(F), the $Si_{10}H_{13}^+$ orbitals are fixed as given by V(Si) and the F^- orbitals are allowed to vary. There are much smaller changes in all three properties at this step. However, these changes describe the polarization of F^- in response to the polarized $Si_{10}H_{13}^+$ charge distribution. These separated, sequential, CSOV variations give results very close to those of the full, unconstrained, SCF variation; see Table 14.1. This is evidence that the CSOV analysis has included all the important features of the interaction. Although the absolute values of the properties may change for different-sized clusters or LCAO basis sets (92), the general features represent the behaviour of F/Si(111).

The projection and the CSOV analysis give a clear picture of the nature of the F-Si(111) interactions. It is essentially entirely ionic: F is F^-! The polatization of the Si surface to form an image charge is the major charge redistribution as a consequence of the interaction of $Si_{10}H_{13}^+$ with F^-.

One would like to have a potential curve for F approaching and penetrating Si with this ionic interaction. However, SCF wavefunctions have a serious limitation for the energetics of ionic bonding. They do not give correct values for electron affinities (EAs) and, to a lesser degree, for ionization potentials, (IPs); for example, the EA of F is 1.4 eV or 2.0 eV smaller than the experimental value (102). Thus, the energy of an ionic system is not given correctly relative to the separated neutral components. There is a simple procedure to correct for this error. The SCF binding or interaction energy of the ionic system is taken with respect to the SCF energies of the separated ionic units. Experimental values for the EA of the anionic unit and the IP of the cationic unit are then used to obtain an interaction energy with respect to the separated neutral units. This procedure corrects for the errors of the SCF energies for the EA and IP; it has been shown to give accurate dissociation energies for ionic diatomic molecules (103). As long as the interaction has a largely ionic behaviour, the correction is appropriate although it cannot be used in a region where the ionicity is changing. The F–Si(111) interaction at the threefold open site is

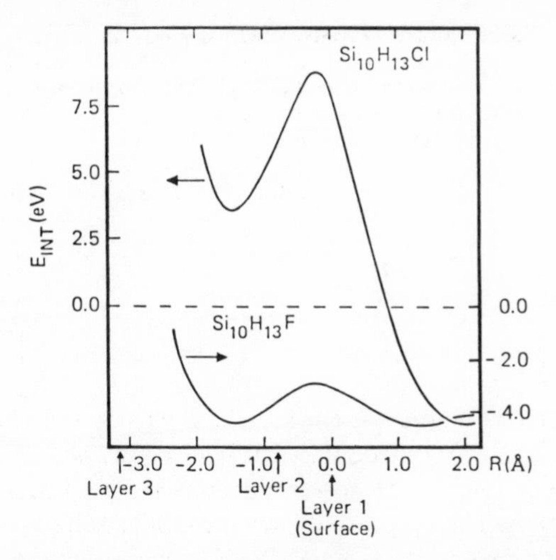

Figure 14.16 Potential curves for the $Si_{10}H_{13}X, X = F$ or Cl, clusters as a function of the distance of the halogen atom from the Si(111) surface layer of atoms. The dotted line is the energy of the separated $Si_{10}H_{13}$ and X sub-units. The positions of the second and third layers of Si(111) are marked. From (92)

ionic over a large range of F distances near, both above and below, the surface. Thus, this correction has been applied to the calculated SCF potential energy curve along the assumed reaction coordinate for surface penetration at the threefold open site on Si(111). It involves adding the difference between the experimental and SCF EAs for F and between the $Si_{10}H_{13}$ IP and the Si(111) work function to the calculated SCF interaction energies for $Si_{10}H_{13}F$ with respect to neutral F and $Si_{10}H_{13}$ (92). The corrected curve is shown in Figure 14.16. The inner and outer potential wells each have a depth of ~ 4.0 eV, and the maximum of the curve is ~ 2.5 eV below the separated neutral $Si_{10}H_{13}$ and F. If the Si atoms had been allowed to relax, the maximum separating the inner and outer minima would be even lower.

For $Si_{10}H_{13}Cl$ which models the Cl adsorption and penetration at the same open threefold site of Si(111), SCF wavefunctions have also been obtained (72). While the same orbital projection and detailed bonding analysis have not been made for the Cl–Si interaction as for F–Si (92), the Mulliken population analysis indicates that the bond has a considerable ionic character. For this reason, the computed SCF interaction energies for $Si_{10}H_{13}Cl$ have also been corrected with the experimental Cl EA and Si IP. This corrected curve is also shown in Figure 14.16. A key difference between F/Si(111) and Cl/Si(111) is that there is a large, ~ 10 eV, barrier for Cl to penetrate the Si surface. This high barrier can be easily understood as a size effect since the Cl atom is much larger than F (72). The SCF mean values of the outermost orbitals provide a simple measure of the relative sizes of F and Cl; for F, $\langle r \rangle_{2p} = 0.6$ Å while for Cl, $\langle r \rangle_{3p} = 1.0$ Å. By this measure Cl is 70% larger than F. This mean that it will

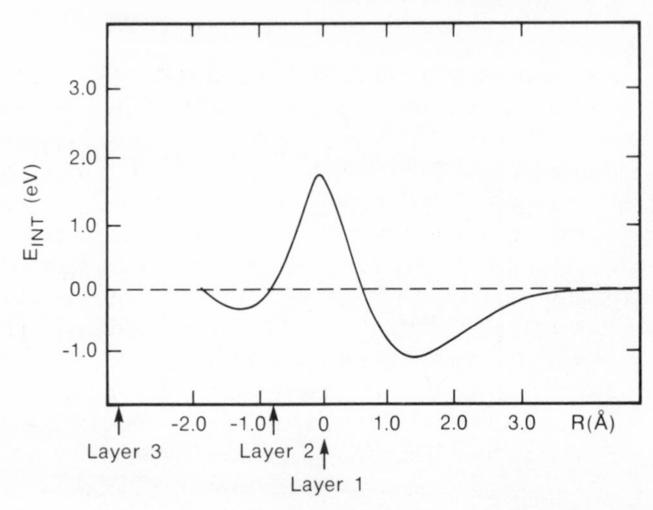

Figure 14.17 Potential curve for the $Si_{10}H_{13}$—H cluster as a function of the distance of the H atom from $Si_{13}H_{10}$; see the caption for Figure 14.16

be difficult for Cl to form the intermediates which are required as precursors to the formation of volatile $SiCl_4$. This difference in the abilities of F and Cl to penetrate the surface explains the differences in their observed reactivities. While fluorine radicals react spontaneously with Si (100, 101, 104), Cl radicals do not react sponstaneously, at any appreciable rate, under normal conditions (105).

Finally, we briefly discuss the interaction of H at this same Si(111) site as modelled by $Si_{10}H_{13}H$ (88). In this case multiconfiguration SCF(MCSCF) wavefunctions have been used. The interaction potential from (88) is shown in Figure 14.17 as a function of the distance of H from the surface. This is the directly computed potential curve and no corrections have been made, since the Mulliken population analysis indicates that the bonding is covalent (88). There is no significant transfer of charge between the Si surface and the adsorbed H atom. Above the surface, the potential curve has a minimum with a depth of ~ 1.0 eV; thus a H atom would be bound in the threefold site, although the on-top site, where the well depth is ~ 3 eV (81), is preferred energetically. There is also a shallow minimum about 1 Å below the surface Si layer.

However, the most important point to notice in Figure 14.17 is that the barrier between the outer and inner minimum, ~ 1.8 eV, is quite large even though H is a considerably smaller atom than F. The reason that the difficulty of surface penetration is so much greater for H than F is because H is covalently bound while the F bond is largely ionic. The F ionicity gives rise to electrostatic forces which overcome the Pauli repulsion due to the overlap and interpenetration of the adsorbate and substrate charge distributions. These forces do not arise for H because the bonding is covalent.

Conclusions

We have seen how cluster models may be used to predict or explain reactivities of silicon surfaces to simple adsorbates and how growth rates and mechanisms may be explored by simulation methods. Clearly, the combined approach of solving the Schrödinger equation for each configuration thrown up by the statistical mechanical algorithm is a logical way forward. (We do not mean by this, however, to negate the continued use of simpler models to explore the parameter space dependence of system behaviour.) For such an approach, one is faced with the computer time constraints that we discussed earlier. But there is also the limitation of relative accuracy. In a good total energy calculation, the relative accuracy of different points on a potential surface is normally limited to the order of a milli-Hartree. This is especially true if the chemical bonding at these points is different. The point to be *made* is that a milli-Hartree is $k_B T$ at room temperature. The point to *realize* is that for a statistical mechanical simulation to do a good job at obtaining structural properties, diffusion coefficients and phase diagrams with which to compare experiment, a relative accuracy of perhaps two or three orders of magnitude better is required.

A quantum scientist may be justifiably proud of his achievement at predicting the site at which (and with what energy) an incoming atom or molecule may prefer to stick on a surface. This is necessary for developing an understanding of the fundamental electronic structure considerations which govern the atomic interactions. However, the dynamics, growth behaviour and phase transition temperatures of such system, even if a state of the art calculation could be performed for each atomic configuration in the ensemble, require a very high accuracy for the relative energies. It is for this reason that the overwhelming majority of statistical mechanical simulations of surface phenomena have been performed on model systems. Simple inter-atomic potentials are usually chosen to demonstrate a given phenomenon for which the details of the potential are unimportant, as in the study of long-wavelength phenomena such as roughening (11), or in which the mechanism or existence is in question, as in the mechanism of 2D melting (9) or in the existence of quasi-liquid layers at high-temperature crystal surfaces (71). The more recent potentials such as that of SW represent a hybrid in that a specific system is represented but, it is hoped, the potential is constrained to be accurate in important parts of phase space by the parametrization procedure.

An approach for treating specific system behaviour could involve a marriage of the empirical potential and 'ab-initio' methods. The idea to be exploited is that parameterization forces the energy hypersurface to meet experiment at certain important points and that the 'ab-initio' structure of the method helps to ensure accurate interpolation between the points. Simplifying Coulomb and exchange integrals assists tractability. Ideas along these lines are already being pursued by R. E. Allen *et al.* (106), Davenport (107), P. B.

Allen *et al.* (108), Chadi (109), and Mele *et al.* (110). These workers have (i) attempted to reduce more complete ab-initio methods to a tight-binding framework and (ii) used tight-binding methods in total energy calculations. By way of promise we note that in the work of Tomanek and Schluter (27) an emipirical total energy tight-binding Hamiltonian gives good agreement with LDF ab-initio energies in silicon clusters. Accurate modelling of semiconductor growth at the atomic level may be possible with such approaches.

Acknowledgements

JQB would like to thank Farid Abraham and George Gilmer for helpful suggestions during the preparation of this manuscript. JQB's silicon simulation work was supported by a grant from DOE (# DE-FG02-85ER45218) and a grant from the Olin Charitable Trust Foundation of Research Corporation.

References

1. A. Rahman, *Phys. Rev.* **136A** (1964) 405.
2. N. Metropolis, A. W. Rosenbluth, M. N. Rosenbluth, A. H. Teller and E. Teller, *J. Chem. Phys.* **32** (1953) 1087.
3. B. Cartling, B. Roos and U. I. Wahlgren, *Chem. Phys. Lett.* **58** (1973) 1066.
4. W. A. Harrison, *Pseudopotentials in the Theory of Metals*, W. A. Benjamin, Reading, MA (1966).
5. J. J. Derby and R. A. Brown, submitted to *AICh Journal.*
6. M. J. Crochet, J. J. Van Schaftingen, P. J. Wouters, J. M. Marchal, S. Dupont and F. T. Geyling, submitted to *Int. J. Num. Meth. Fluids.*
7. R. R. Lucchese and J. C. Tully, *Surface Sci.* **137** (1974) 570.
8. L. Verlet, *Phys. Rev.* **159** (1967) 98.
9. F. F. Abraham, *Advances in Physics* **35**, 1.
10. T. L. Hill, *An Introduction to Statistical Mechanics*, Addison-Wesley, Reading, MA (1962).
11. J. D. Weeks and G. H. Gilmer, *Adv. Chem. Phys.* **40** (1979) 157.
12. G. H. Gilmer and P. Bennema, *J. Cryst. Growth* **13/14** (1972) 148.
13. G. H. Gilmer and P. Bennema, *J. Appl. Phys.* **43** (1972) 1347.
14. G. H. Gilmer and K. A. Jackson, in *Crystal Growth and Materials*, eds. E. Kaldis and H. J. Scheel, North-Holland, Amsterdam (1976).
15. B. Quentrec and C. Brot, *J. Comp. Phys.* **1** (1973) 430.
16. P. N. Keating, *Phys. Rev.* **145** (1966) 637.
17. R. Tubino, L. Piseri and G. Zerbi, *J. Chem. Phys.* **56** (1972) 1022.
18. W. Weber, *Phys. Rev.* **B15** (1977) 4789.
19. D. A. Smith, *Phys. Rev. Lett.* **42** (1979) 729.
20. F. H. Stillinger and T. A. Weber, *Phys. Rev.* **B31** (1985) 5262.
21. E. Pearson, T. Takai, T. Halicioglu and W. A. Tiller, *J. Cryst. Growth* **70** (1984) 33.
22. B. M. Axilrod and E. Teller, *J. Chem. Phys.* **11** (1943) 299.
23. R. Biswas and D. R. Hamman, *Phys. Rev. Lett.* **55** (1985) 2001.
24. J. Tersoff, *Phys. Rev. Lett.* **56** (1986) 632.
25. K. Takayanagi, Y. Tanishiro, M. Takahashi, and S. Takahashi, *J. Vac. Sci. Technol.* **A3** (1985) 1502.
26. K. Raghavachari and V. Logovinsky, *Phys. Rev. Lett.* **55** (1985) 2853.
27. D. Tomanek and M. A. Schluter, *Phys. Rev. Lett.* **56** (1986) 1055.
28. M. T. Yin and M. L. Cohen, *Phys. Rev.* **B26** (1982) 5668.
29. M. T. Yin and M. L. Cohen, *Phys. Rev. Lett.* **45** (1980) 1004.
30. R. Car and M. Parrinello, *Phys. Rev. Lett.* **55** (1985) 2471.
31. H. A. Wilson, *Phil. Mag.* **50** (1900) 238.
32. J. Frenkel, *Phys. Z. Sowjetunion* **1** (1932) 498.

33. P. J. Vergano and D. R. Uhlman, in *Reactivity of Solids*, eds. J. W. Mitchell, R. C. DeVries, R. W. Roberts and P. Cannon, John Wiley, New York (1969).
34. K. A. Jackson, D. R. Uhlman and J. D. Hunt, *J. Cryst. Growth* **1** (1967) 1.
35. J. L. Walker, quoted in *Principles of Solidification*, B. Chalmers (ed.), John Wiley, New York (1964) 114.
36. D. Turnbull and B. A. Bagley, *Treatise Solid State Chem.* **5** (1975) 526.
37. F. Spaepen and D. Turnbull, in *Proc. Conf. on Laser-Solid Interactions and Laser Processing*, Boston (1978), in *AIP Conf. Proc.* **50** (1979) 73.
38. R. Ruhl and P. Hilsch, *Z. Phys.* **B26** (1977) 161.
39. J. Q. Broughton, G. H. Gilmer and K. A. Jackson, *Phys. Rev. Lett.* **49** (1982) 1496.
40. J. Q. Broughton and G. H. Gilmer, *J. Chem. Phys.* **79** (1983) 5095.
41. T. F. Ciszek, *J. Cryst. Growth* **10** (1971) 263.
42. A. G. Cullis, N. G. Chew, H. C. Webber and D. J. Smith, *J. Cryst. Growth* **68** (1984) 624.
43. D. R. Squire and W. G. Hoover, *J. Chem. Phys.* **50** (1969) 701.
44. G. H. Gilmer, *Mater. Sci. and Eng.* **65** (1984) 15.
45. K. A. Jackson, in *Surface Modification and Alloying*, eds. J. M. Poate, G. Foti, D. C. Jacobson, NATO Conf. Series VI, Vol. 8, Plenum (1983).
46. J. Q. Broughton and F. F. Abraham, *J. Cryst. Growth*, in press.
47. J. Q. Broughton and F. F. Abraham, *Chem. Phys. Lett.* **71** (1980) 456.
48. J. Q. Broughton, A. Bonissent and F. F. Abraham, *J. Chem. Phys.* **74** (1981) 4029.
49. F. F. Abraham and J. Q. Broughton, *Phys. Rev. Lett.* **56** (1986) 734.
50. J. A. Burton, R. C. Prim and W. P. Slichter, *J. Chem. Phys.* **21** (1953) 1987.
51. P. Baeri, G. Foti, J. M. Poate, S. U. Campisano and A. G. Cullis, *Appl. Phys. Lett.* **38** (1981) 800.
52. J. M. Poate, in *Laser and Electron Beam Interactions with Solids, Proc. Symp. of Materials and Research Society*, eds. B. R. Appleton and G. K. Celler, North-Holland, Amsterdam (1982).
53. R. F. Wood, *Phys. Rev.* **B25** (1982) 2786.
54. G. H. Gilmer, *Science* **208** (1980) 355.
55. G. H. Gilmer, in *Laser–Solid Interactions and Transient Thermal Processing of Materials, Proc. Symp. of Materials Research Society*, eds. J. Narayan, W. L. Brown, and R. A. Lemons, North-Holland, Amsterdam (1983).
56. D. E. Polk and D. S. Boudreaux, *Phys. Rev. Lett.* **31** (1973) 92.
57. D. Henderson, *J. Non-Cryst. Solids* **16** (1974) 317.
58. L. Guttman, in *Tetrahedrally Bounded Amorphous Semiconductors*, eds. M. H. Brodsky, S. Kirkpatrick and D. Weaire, AIP Conference Proceedings, No. 20, American Institute of Physics (1974).
59. F. Wooten, K. Winer and D. Weaire, *Phys. Rev. Lett.* **54** (1985) 1392.
60. F. Herman and P. Lambin, to be published.
61. H. C. Andersen, private communication.
62. M. Schneider, A. Rahman and I. K. Schuller, *Phys. Rev. Lett.* **55** (1985) 604.
63. B. W. Dodson and P. A. Taylor in *Computer-Based Microscopic Description of the Structure and Properties of Materials*, eds. J. Q. Broughton, W. Krakow, and S. T. Pantelides, *MRS Symp. Proc.* **63** (1986).
64. J. Singh and A. Madhukar, *J. Vac. Sci. Technol.* **B1** (1983) 305.
65. S. Das Sarma, private communication.
66. R. Ramirez, A. Rahman and J. K. Schuller, *Phys. Rev.* **B30** (1984) 6208.
67. M. H. Grabow and G. H. Gilmer, private communication.
68. J. T. Wetzel, A. A. Levi and D. A. Smith, in ref. 63.
69. T. A. Weber, in ref. 63.
70. F. F. Abraham and I. P. Batra, *Surface Sci.* **163** (1985) L752.
71. J. Q. Broughton and G. H. Gilmer, *Acta. Met.* **31** (1983) 845.
72. M. Seel and P. S. Bagus, *Phys. Rev.* **B28** (1983) 2023.
73. F. R. McFeely, J. F. Moror, N. D. Schinn, G. Landgren and F. J. Himpsel, *Phys. Rev.* **B30** (1984) 764.
74. K. C. Pandey and J. C. Phillips, *Phys. Rev.* **B13** (1976) 750; K. C. Pandey, *Phys. Rev.* **B14** (1976) 1557; *IBM J. Res. and Devel.* **22** (1978) 250.
75. N. D. Lang and A. R. Williams, *Phys. Rev.* **B18** (1978) 616.

76. I. P. Batra, *J. Vac. Sci. Technol.* **B3** (1985) 750; *Phys. Rev.* **B29** (1984) 7108; I. P. Batra and S. Ciraci, *Phys. Rev.* **B29** (1984) 6419.
77. M. Weinert, E. Wimmer and A. J. Freeman, *Phys. Rev.* **B26** (1982) 4571.
78. D. Vanderbilt and S. G. Louie, *Phys. Rev.* **B29** (1984) 7099.
79. P. S. Bagus, K. Hermann and C. W. Bauschlicher, *J. Chem. Phys.* **81** (1984) 1966.
80. T. H. Upton and W. A. Goddard, *Phys. Rev. Lett.* **46** (1981) 1635.
81. K. Hermann and P. S. Bagus, *Phys. Rev.* **B20** (1979) 1603.
82. C. W. Bauschlicher, P. S. Bagus and H. F. Schaefer, *IBM J. Res. and Devel.* **22** (1978) 213.
83. C. W. Bauschlicher and P. S. Bagus, *Phys. Rev. Lett.* **54** (1985) 349.
84. K. Hermann and P. S. Bagus, *Phys. Rev.* **B16** (1977) 4195.
85. P. S. Bagus, C. W. Bauschlicher and H. F. Schaefer, *J. Chem. Phys.* **78** (1983) 1390.
86. P. S. Bagus and W. Müller, *Chem. Phys. Lett.* **115** (1985) 540.
87. J. Whitten, *Phys. Rev.* **B24** (1981) 1810; J. L. Whitten and T. A. Pakkanen, *Phys. Rev.* **B21** (1980) 4357; C. R. Fischer and J. L. Whitten, *Phys. Rev.* **B30** (1984) 6821.
88. M. Seel and P. S. Bagus, *Phys. Rev.* **B23** (1981) 5464.
89. I. P. Batra, P. S. Bagus and K. Hermann, *J. Vac. Sci. Technol.* **A2** (1984) 1075; *Phys. Rev. Lett.* **52** (1984) 384.
90. C. W. Bauschlicher, P. S. Bagus and B. N. Cox, *J. Chem. Phys.* **77** (1984) 4032.
91. P. S. Bagus, C. W. Bauschlicher, and K. Hermann, *J. Chem. Phys.* **80** (1984) 4378.
92. P. S. Bagus, in *Plasma Synthesis and Etching of Electronic Materials*, eds. R. T. H. Chang and B. Abeles, Materials Research Society Symposia Proceedings Vol. 38, Materials Research Society, Pittsburgh (1985) 179.
93. L. M. G. Pettersson and P. S. Bagus, *Phys. Rev. Lett.* **56** (1986) 500.
94. P. S. Bagus, C. J. Nelin, W. Müller, M. R. Philpott and H. Seki, *Phys. Rev. Lett.* **58** (1987) 559.
95. R. S. Milliken, *J. Chem. Phys.* **23** (1955) 1833; **23** (1955) 1841; **23** (1955) 2338; **23** (1955) 2343.
96. P. S. Bagus, C. J. Nelin and C. W. Bauschlicher, *Phys. Rev.* **B28** (1983) 5423.
97. C. W. Bauschlicher and P. S. Bagus, *J. Chem. Phys.* **81** (1984) 5889.
98. C. W. Bauschlicher, P. S. Bagus, C. J. Nelin and B. O. Roos, *J. Chem. Phys.* **85** (1986) 354.
99. P. S. Bagus, K. Hermann, W. Müller, and C. J. Nelin, *Phys. Rev. Lett.* **57** (1986) 1496.
100. Y. Y. Tu, T. J. Chuang, and H. F. Winters, *Phys. Rev.* **B23** (1981) 823.
101. H. F. Winters and F. Houle, *J. Appl. Phys.* **54** (1983) 1218.
102. H. Hotop and W. C. Lineberger, *J. Phys. Chem. Ref. Data* **4** (1975) 530.
103. P. S. Bagus, C. J. Nelin, and C. W. Bauschlicher, *J. Chem. Phys.* **79** (1983) 2975.
104. H. F. Winters and J. W. Coburn, *Appl. Phys. Lett.* **34** (1979) 70.
105. J. W. Coburn and H. F. Winters, *J. Appl. Phys.* **50** (1979) 3189.
106. M. M. Menon and R. E. Allen, *Phys. Rev.* **B33** (1986) 7099.
107. J. Davenport, private communication.
108. P. B. Allen, J. Q. Broughton and A. K. McMahon, *Phys. Rev.* **B34** (1986) 859.
109. D. J. Chadi, *Phys. Rev.* **B19** (1979) 2074.
110. E. J. Mele, D. C. Allen, O. L. Alerhand and D. P. DiVincenzo, *J. Vac. Sci. Technol.* **B3** (1985) 1068.

15 Plasma etching

D. L. FLAMM and J. A. MUCHA

Introduction

Plasma etching has largely replaced wet etching for microcircuit fabrication because of its finer resolution and adaptability to increased throughput and automation. In this chapter we survey the chemistry of plasma etching technology, show how recent theory and experimental findings can be used to select etchants, and briefly discuss the interaction of plasma conditions with this chemistry.

A simple plasma reactor consists of opposed parallel plate electrodes in a chamber that can be maintained at low pressure, typically ranging from 0.01 to 1 Torr (1.33–133 Pa) (Figure 15.1). When a high-frequency voltage is applied between the electrodes, current flows forming a *plasma* that emits a characteristic glow. Semiconductor wafers or other substrate materials on the electrode surfaces are exposed to reactive neutral and charged species from the plasma. Some of these species combine with the substrate and form volatile products which evaporate, thereby etching the substrate. For plasmas of interest to

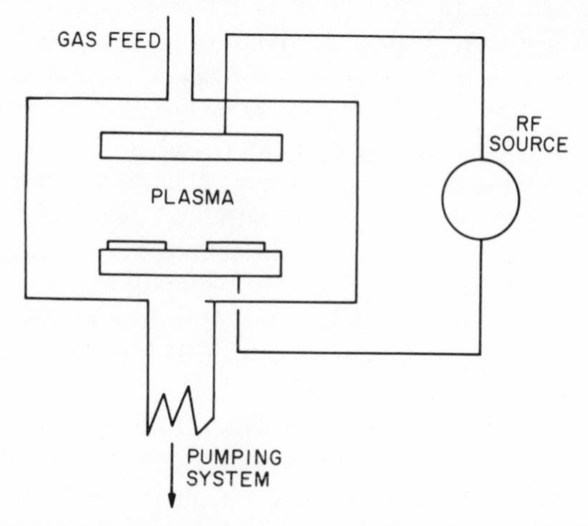

Figure 15.1 Plasma reactors consist of a chamber which can be operated at low pressure, typically between 0.01 and 1 Torr. When a high-frequency voltage is applied between two electrodes in the chamber, there is an electric discharge through which current flows, forming a plasma. Device wafers are places on one electrode and exposed to active species produced by the plasma

etching, the density of charged particles is small—only $\sim 10^9$–10^{11} cm^{-3} which at 0.1 Torr corresponds to only one charged particle per 10^4 to 10^6 neutral atoms and molecules. Positive charge is present mostly as singly ionized neutrals, and negatively charged particles are usually free electrons. In any event electrons are the main current-carriers because they are light and mobile.

Electrons in the plasma attain a high average energy from the applied field, often many electron volts (equivalent to tens of thousands of degrees Kelvin), while the neutral gas remains relatively cool. This can mostly be understood in terms of the relative mass of electrons, which is less than 0.0005 times that of other species. Since the average fraction of kinetic energy transferred in a collision is roughly twice this mass ratio, energy transfer from electrons to the surrounding neutrals is inefficient compared to heat transfer through the gas to walls. The elevated electron temperature permits electron–molecule collisions to excite high-temperature reactions that form free radicals. The coexistence of a warm gas with high-temperature species distinguishes the plasma reactor from conventional reaction systems and permits the processing of sensitive materials.

Plasma etching techniques may be characterized according to rate, anisotropy, selectivity, the degree of loading effect and texture. By convention, the term anisotropic etching refers to preferential erosion in a direction normal to the surface of a wafer. Selectivity is the ratio of etching rates between two different materials immersed in the same plasma, for example Si and SiO_2. 'Loading' is a term used to describe a measurable depletion of active etchant species from the gas phase, brought about by the consumption of this reactant in the etching process. In some plasmas, the original surface smoothness is maintained, while other plasma etchants roughen or texture an initially smooth surface. In principle, the plasma etchant feed gas would be chosen according to the type of material to be etched, for selectivity over other substrates, to minimize any loading effect, and to avoid excessive surface texture and polymer deposition. The degree of anisotropy is important when features below about 3–5 μm must be patterned.

Isotropic or anisotropic etching can be defined by reference to Figure 15.2. Usually, the desired result is a transfer of the mask pattern to the substrate, with no distortion of critical dimensions. However, in practice, the etched area extends beyond the mask openings. Invariably the mask is slowly eroded and, since the mask is usually thinnest at the boundary of an opening, the opening slowly enlarges as etching proceeds. Additionally, chemical attack (isotropic etching) of the substrate underneath the mask will enlarge a feature. When large features are desired, mask openings may be made smaller to compensate for feature enlargement (undercutting) during an isotropic etch. The feature size attainable with this technique is limited to the thickness of the film underlying the mask (typically a few μm), which is why anisotropic etching is required to make fine features.

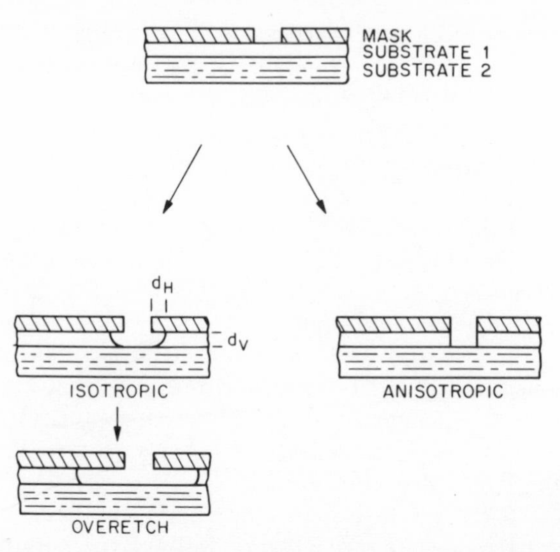

Figure 15.2 Isotropic (left) and anisotropic etching (right). Isotropic chemical etching usually has no preferred direction. This leads to isotropic circular profiles which undercut a mask. Overetching increases the undercut and radius eventually making the sidewalls nearly vertical. In anisotropic etching, the sheath field makes ions strike horizontal surfaces and ion-assisted chemistry forms profiles with vertical sidewalls that line up with the edges of the mask

Stripping (removal) of photoresist, etching large features in silicon, and removing of nitride oxidation masks are easily and economically carried out using isotropic etchants. Isotropic etching is also necessary for etching dielectric films under photoresist evaporation masks. The undercut portion of the mask 'shadows' the substrate leaving a gap between metal evaporated on to the mask and metal on the etched substrate. This facilitates 'lift-off' of the mask and overlying metal by a solvent. Isotropic etching can be used to remove the thicker vertical portion of a layer (e.g. polysilicon) at the edge of a step and for tapering features, easing the problems of deposition over steps. Undercutting is also useful to remove material under non-etchable or valuable topological features, for example fluorine atoms from a ClF_3 or NF_3 discharge can remove filaments underlying oxide in the manufacture of MOS RAMs (1).

Isotropic etchants often exhibit a loading effect that results from etchant depletion in the gas phase caused by reaction with the substrate material. If a significant portion of the reactive etchant species is consumed in etching reactions, their concentration will decrease rapidly with the area of etchable material in the reactor. The etch rate is commonly proportional to etchant concentration and for m wafers and a single etchant species, F, the loading effect is given by (2, 1):

$$\frac{R_0}{R_m} = 1 + m\Phi_F = 1 + m\frac{A_w k_{wF}}{A k_{sF}} \tag{15.1}$$

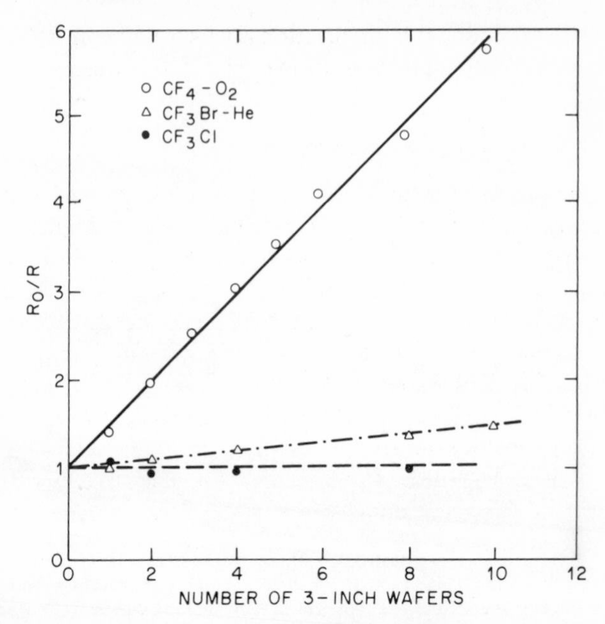

Figure 15.3 The loading effect in CF_4/O_2 (top), CF_3Br (centre) and CF_3Cl plasmas (after ref. 2). The etch rate in an empty in an reactor (R_0) divided by the etch rate with *n* units of area forms a straight line with an intercept of unity

where R_0 is the etch rate in an empty reactor (limit of zero etchable area), R_m is the etch rate with *m* wafers present, Φ_F is the ratio of F consumed by etching a wafer (rate k_{wF}) to that lost by recombination ($Ak_{sF} = Ak_{sF}^* + Vk_{VF}$), and A, A_w are areas of the reactor surface and a wafer, respectively. k_{sF}^* is the rate of etchant loss on reactor walls and k_{VF} is the volume loss rate. Φ_F is the slope of the loading effect curve (e.g. R_0/R_m *v.m*). Use of a large reactor volume (V) with high surface area (A) will minimize the sensitivity of etch rate to the area of etchable material ($\Phi_F \lesssim 1$). Equation (15.1) provides a good description of silicon etching in CF_4/O_2 and certain other discharges (see Figure 15.3). When more than one etchant species is present, for instance in ClF_3 plasmas where both Cl and F atoms participate in surface gasification reactions, a two component loading effect applies (1). Etchant loss in reaction systems with little or no loading effect is dominated by volume and heterogeneous radical recombination reactions, rather than by etching.

Loading can be avoided by employing plasmas in which the principal etchant loss process does not depend on the etching reaction ($k_{loss} \gg k_{etch}$) or in which the ion bombardment rate, rather than the etchant supply, controls the reaction (this effect is not included in equation 15.1). Thus while the concentration of fluorine atoms etching silicon in a plasma is usually limited by reaction with the substrate, giving a loading effect, chlorine atom concentrations in chlorine-containing plasmas tend to be limited to recombination rather than the etching reaction (3). Hence silicon etching in chlorine plasmas is commonly insensitive to load size and feature topology.

Basic mechanisms of plasma etching

While a wide variety of phenomena may play a role, etching mechanisms can be grouped into four categories: (i) sputtering; (ii) chemical etching (sometimes called 'isotropic' etching); (iii) *energy-driven* ion-enhanced etching; and (iv) *inhibitor* protected-sidewall ion-enhanced etching. At a basic level these categories encompass diverse elementary phenomena, many of which are not well-understood.

In sputtering, impinging particles (usually positive ions accelerated across the sheath) strike the surface with high kinetic energy. Some of the energy is transferred to surface atoms which then are ejected, leading to a net removal of material. This process is distinguished from other etching mechanisms in that the interaction is mechanical: differences in chemical bonding are important only in so far as they determine the bonding forces between surface atoms and the ballistics of dislodging them.

Chemical etching comes about when active species from the gas phase encounter a surface and react with it to form a *volatile* product. Product volatility is necessary for chemical etching since involatile products would coat the surface and protect it from further attack. In this type of etching the plasma reactor converts the feed into reactive chemical species, which are usually free radicals such as fluorine atoms (F). As we will see later, this type of etching can sometimes be done without placing the substrate in a plasma, or for that matter without operating a plasma at all. Chemical etching shares the characteristics of other chemical reactions. There is usually no directionality (as opposed to ion-induced material removal) and the etching can be specific (high selectivity) since it is governed by the relative chemical affinities between the etchant species and exposed materials. Because of this lack of directionality, chemical etching is commonly called isotropic etching (see Figure 15.2).

Plasmas differ markedly from ordinary gaseous reacting media in that a directional negative-going electric field forms along their boundaries. This sheath-field propels positive ions into boundary surfaces at normal incidence so that ion-enhanced etching is directional and perpendicular to the surface (see below).

Fast *directional* etch rates can be attained for certain plasma/substrate combinations when there are fluxes of neutral etchant species *and* ions to surfaces. The effect is synergistic—ion bombardment in the presence of reactive neutral etchant species often leads to material removal rates that greatly exceed the sum of separate chemical attack and sputtering rates. This ion-enhanced plasma etching is generally considered to proceed by one of two types of mechanism: (i) energy-driven ion-enhanced etching or (ii) inhibitor-driven ion-assisted etching.

In the energy-driven mechanism, there is usually little or no etching when the substrate surface is exposed to neutral chemical species alone in the absence of ion bombardment. Impinging ions 'damage' the substrate material by virtue of their impact energy, and thereby render the solid substrate more reactive toward incident neutral radicals. The term 'damage' is meant in a generic sense to include diverse mechanisms such as the formation of reactive dangling bonds, disruption of lattice structure and formation of dislocations, forcible injection of adsorbed reactant into a lattice by the collisional cascade, or bond-breaking in tightly-adsorbed surface intermediates. There is no evidence to support a single universal elementary mechanism for energy-driven ion-enhanced etching. Diverse elementary mechanisms are active depending on the etchant, surface material being etched and, perhaps, on the magnitudes of radical and ion fluxes to the surface (e.g. the pressure regime). The common denominator is that the ions must transfer energy and momentum to the surface to make it reactive.

In inhibitor-protected sidewall ion-enhanced etching, ion bombardment performs another function. Neutral etchant species from the plasma spontaneously gasify the substrate, and ions play a role by interacting with another component—a 'protective' inhibitor film. The role of ions in the surface-inhibitor mechanism is to 'clear' the inhibitor from horizontal surfaces that are bombarded by the flux of ions impinging in the vertical direction (Figure 15.4). The protective film is not removed from the vertical walls of masked features because these surfaces only intercept those few ions that are scattered as they cross the sheath. This protective film may originate from involatile etching products or from film-forming precursors that adsorb during the etching

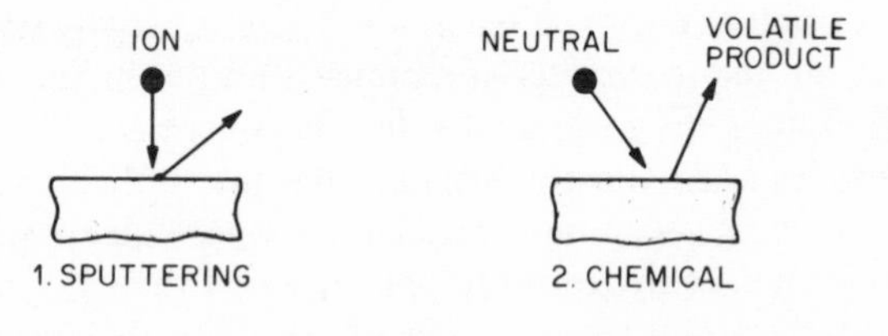

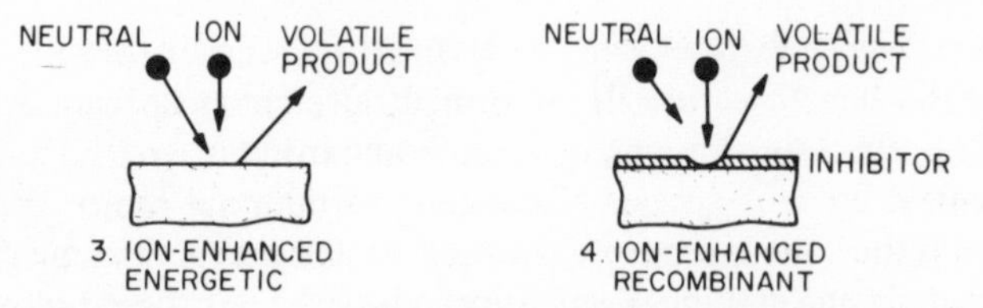

Figure 15.4 The four generic processes by which plasma etching takes place are depicted schematically from top to bottom: (1) sputtering; (2) chemical (isotropic) etching; (3) ion-enhanced energy driven etching and (4) ion-enhanced protected sidewall etching

process. Inhibitor-producing precursors can originate from a variety of sources. Some fluorocarbons (e.g. C_2F_6, CHF_3) are a source of unsaturated monomeric species in plasmas. By unsaturated species we mean the CF_2 radical (or CCl_2 radical) and derivatives (e.g. C_2F_4, C_3F_6, C_2Cl_4, etc.) formed by chain-building oligomerization. CF_2 and its oligomers tend to polymerize and form thin films on surfaces (or sometimes thick films).

While inhibitor-producing compounds are usually a deliberate additive, films can originate from more subtle sources, for example from sputtered reactor material at low frequency where sheath potentials are high or from migration of material in the resist mask. Generally, it is believed that low ion energies suffice to keep inhibitor film from forming on horizontal surfaces in practical systems of this type. By contrast, energetic ions are usually required to cause energy-driven ion-enhanced etching.

Reactor types

Plasma etching reactors have been classified according to load capacity, the position of material relative to the plasma, pressure regime, electrode geometry and generator frequency.

By load, we mainly refer to a distinction between *single wafer reactors* in which only one wafer is etched at a time as opposed to *batch reactors* in which many wafers are simultaneously etched in the same chamber. The main criteria in choosing between these alternatives are the requirements for economic throughput with high uniformity and reproducibility. Cost differences between batch and single wafer reactors intended for the same process step tend to be less than a factor of two, depending on the application being considered, and do not uniformly favour either type over the other. All other things equal, wafer to wafer reproducibility and uniformity are better when every wafer is exposed sequentially in the constant environment of a *single wafer etcher*. The drawback is that higher etch rates are required to attain the same throughput, and this generally requires a more intense plasma at higher power density. These harsher conditions usually lead to higher temperatures, less selectivity and more interface damage from ion bombardment and sputtering. Such issues and the approaches for solving them are beyond the scope of this chapter.

In most plasma etching and stripping processes used today, the semiconductor substrates are immersed in the plasma where energetic charged species are accelerated toward and impinge on substrate and reactor surfaces. While ion bombardment is an integral part of some reaction chemistries, it also can sputter-deposit material from other parts of the reactor and cause device degradation by inducing chemical or structural changes. This is aggravated when plasma and particle flux densities are high, or when ion bombardment energy exceeds the minimum level necessary for the process. An important example of structural degradation occurs in stripping resist on field effect transistors. Charge is deposited on thin insulating gate oxides because of the high electron temperature and can cause damaging electrostatic breakdown when the plasma is extinguished.

Charged species recombination is more rapid than most gas phase neutral reactions ($k_{recomb} \sim 10^{-8}$–10^{-7} cm^{-3} s^{-1} for ions with electrons around 0.1 Torr). Hence damage from the plasma environment can be avoided if the wafers are placed in a charge-free region outside the discharge. In a *downstream reactor*, neutral etchant radicals flow from the plasma to a treatment zone that is sometimes temperature-controlled. Reactors of this type have long been used for scientific studies of basic chemistry, but except for the 'barrel' etcher with a 'tunnel' (see below), commercial processing equipment based on this design is recent. Since there are no ions present to induce directional etching, applications are generally limited to the stripping of polymer resist or (infrequently) a layer of material where patterning is non-critical (since there are no ions to induce directional etching).

There is another family of etchers, closely related to these downstream etchers, which have been called '*plasma transport*' *reactors* (4, 5). The plasma source usually is operated at microwave frequencies (2450 MHz), and sometimes with an auxiliary magnetic field to increase electron confinement. These plasma transport reactors have also been referred to as microwave or *ECR* (electron-cyclotron resonance) *etchers*, but the nomenclature is unsatisfactory because there are in-plasma reactors and downstream etchers that also use microwave excitation and magnetically confined microwave transport devices that do not use resonance. These plasma transport reactors operate at low pressure, as low as 10^{-5} Torr, where charged-particle recombination is slow relative to diffusion and flow. In these devices the ions may be extracted from the plasma zone by biased grids or a magnetic field and made to impinge on the substrate with a controlled, low energy. The ions sometimes are collimated and the energy can be adjusted to promote etching, yet minimize substrate damage. However the flux of ions and neutral species that can be transported and available for etching under these conditions is often limited, and consequently the etching rate may be low. This, together with higher relative cost and complexity, has limited their use.

When plasma etching first found widespread use in the 1970s, *barrel etchers* or *volume loaded reactors* like that in Figure 15.5 were the most common type. In barrel etching wafers are simply stacked in a notched quartz support stand that is inserted into the reactor. External electrodes sustain the plasma, but there is no temperature control and wafers have no special orientation relative to the electrodes. As may be expected, these reactors are inexpensive, but uniformity is difficult to achieve and the rate of reaction can change erratically because of reactor heating from the RF power input to the plasma and exothermicity of reaction. Nevertheless, because of their large capacity and low cost, barrel reactors are still common for non-critical processes such as stripping, cleaning, and (occasionally) large feature isotropic silicon etching.

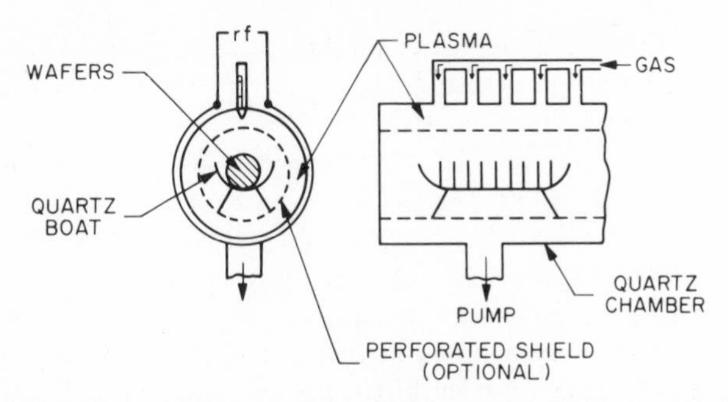

Figure 15.5 The 'volume loaded' or 'barrel' plasma etcher is commonly used for stripping resist

One notable variation on this type of reactor is the incorporation of a perforated metal insert, a '*tunnel*,' which is intended to exclude the plasma from the wafer-containing internal volume, e.g., an imperfect approximation of the downstream etcher. However the tunnel can be a source of sputtered material and damaging contamination, and there are other inherent problems, such as the lack of temperature control.

The *parallel plate*, or '*planar*', *reactor* and its variations is the most common type in use today. Originally introduced the early 1970s for chemical vapour deposition, the simple parallel plate 'Reinberg' style reactor is shown in Figure 15.6. The lower electrode supports the wafers and can be heated or cooled, while radial gas flow toward or away from the centre helps uniformity. Planar reactors are available in both single wafer and larger 'batch' sizes. Sometimes, especially in single wafer reactors, one of the electrodes contains a perforated 'shower head' gas feed which is used instead of the radial flow design.

The parallel plate reactor is often made asymmetrically, with wafers placed on the smaller electrode. The external circuit has a 'blocking' capacitor so the smaller electrode develops a negative dc 'bias' potential superimposed on the applied RF voltage waveform. At low pressure these asymmetric reactors are often called '*reactive ion etchers*,' although this term is misleading and its implications inaccurate. The term '*diode reactor*' has also been used to refer to

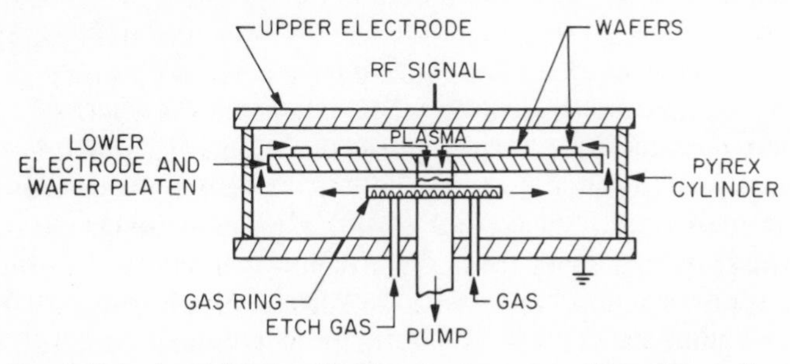

Figure 15.6 A parallel-plate 'Reinberg' style reactor with radial gas flow

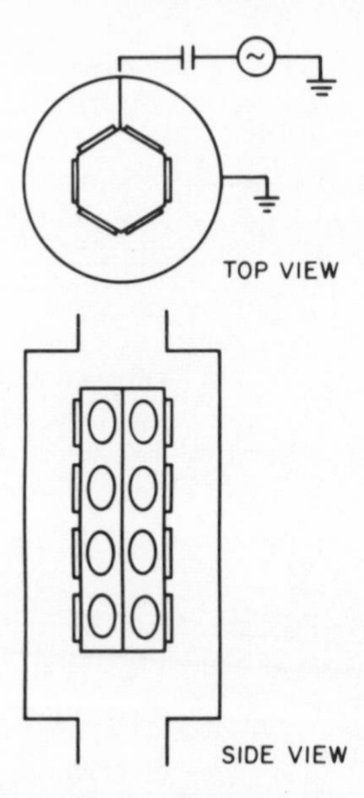

Figure 15.7 A 'hex' reactor which is developed from an asymmetric parallel plate configuration wrapped around into six 'cathode' faces. A stainless steel 'bell jar' vacuum enclosure serves as the larger 'anode' RF electrode

reactors of this type. The high-potential RF power lead is usually connected to the smaller electrode, with the larger electrode 'grounded' at common potential. This is necessary to maintain a large area ratio since there are usually grounded surfaces elsewhere, in and out of the reactor chamber, that capacitively couple to the plasma and thereby increase the effective or 'virtual' area of the grounded electrode.

The '*hexode*' *reactor* shown in Figure 15.7 is a variant of the asymmetric low-pressure parallel plate machine. Folding the electrode surfaces into a cylinder permits a larger wafer capacity for given machine dimensions, and the vertical orientation helps prevent particulate matter from settling on the wafer surface. To date, hexode reactors have been designed for batch operation at low pressure (*c.* 0.005–0.05 Torr) and high frequency (13.56 MHz).

As discussed below (p. 360), at lower pressure, high-frequency plasmas exhibit larger characteristic potentials and bias. These conditions favour ion enhanced energy-induced anisotropy. As mentioned, trade jargon often refers to asymmetric parallel plate type reactors operating in this regime as 'reactive ion' etchers, while differentiating higher pressure operation ($\gtrsim$ 0.1 Torr) by the term 'plasma etchers.' Of course, plasma etching is taking place in both regimes and ions are almost never the primary etchants (see p. 354). The term 'reactive sputter etching' has been used similarly to 'reactive ion etching.' This also is misleading since sputter etching is, by definition, the physical sputtering process, while reactive etching presumably means chemical etching.

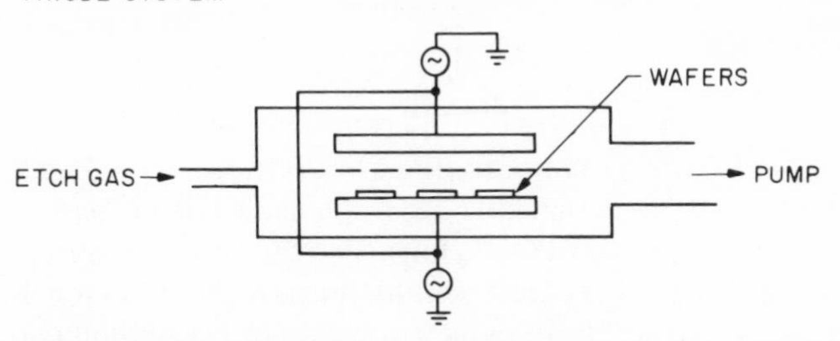

Figure 15.8 Schematic of a 'triode' reactor configuration. The upper and lower plasma zones can be driven by different excitation frequencies to suit process needs

Some reactors have employed a magnetic field to reduce the rate of electron diffusion from moderate pressure plasmas, permitting higher plasma densities at lower power and characteristic potentials. These are known as *magnetron reactors*. Finally, so-called '*triode reactors*' employ an additional electrode and power supply. The objective here is to attain high reactant generation rates in a zone that is partly shielded from the substrate (the upper zone in Figure 15.8), and control ion bombardment energies in a second zone where the substrates are located (lower region in Figure 15.8). In attempts to do this, a high excitation frequency has been used in substrate-free zone with low frequency excitation to enhance anisotropy in the substrate region.

Most of these reactor types can be used with either low (e.g. below 2 MHz) or high (2–30 MHz) RF excitation frequency. The effects of frequency in this range are surveyed below (p. 362) and covered in detail elsewhere (6). Microwave excitation, on the other hand, usually requires special equipment. Experimental large-area planar microwave plasma etchers can be made using surface wave technology (7), but there is controversy regarding their advantages, if any (8, 9). To date the chief application of microwave discharges has been as a plasma source for the downstream and plasma transport reactors, but apparently RF discharges can be used in these applications as well (8, 10, 11).

Etching species and chemistry in a plasma

As outlined, a plasma is an ionized gas with equal numbers of positive and negative charges. Usually, the negative charges are predominantly electrons, and these are the particles receiving energy from the applied RF field. Unlike neutrals and ions that are close to thermal equilibrium, electron energies are high so they can dissociate molecules into free radicals or ions and electrons. Electron–molecule reaction rate constants, k, depend strongly on the average electron energy, $\bar{\varepsilon}$, and the energy distribution function. For the process

$$e + A \rightarrow P \text{ (products)} \quad (15.2)$$

the rate of formation of P is

$$\frac{dn_P}{dt} = k(\bar{\varepsilon})n_e n_A \tag{15.3}$$

where n_e, n_A and n_P are the concentrations of electrons, A and P in the plasma. Hence discharge variables that alter $\bar{\varepsilon}$ and n_e will directly influence the rates of elementary electron–molecule reactions. The threshold energies for dissociation into charged particles are generally much higher than for dissociation into neutrals, so it is the 'tail' of the electron energy distribution that produces enough charged particles to balance diffusion and recombination losses and sustain the plasma.

A simple comparison of magnitudes shows that neutral radicals, not ions, must be the agent responsible for substrate material removal. For this reason the common phrase 'reactive ion etching' is misleading and seldom descriptive.

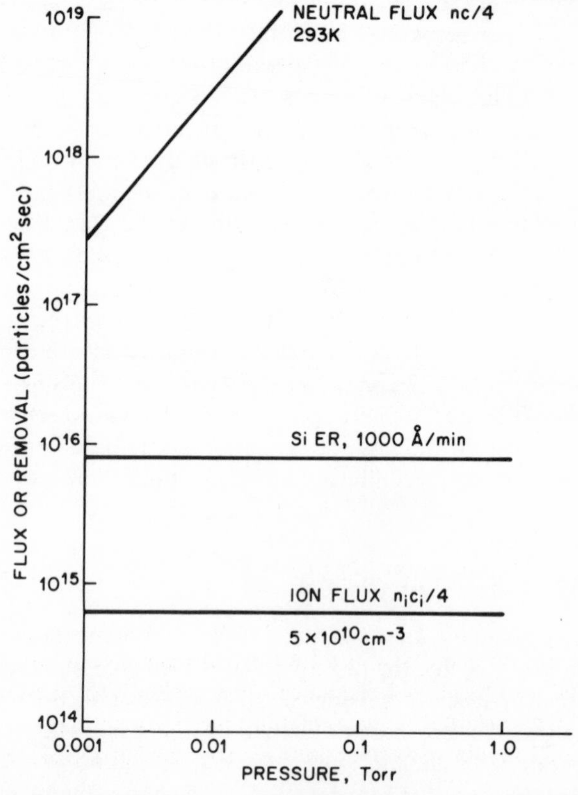

Figure 15.9 Comparison of fluxes and material removal rate for silicon etching, as a function of pressure in a typical plasma discharge. Ion flux to surfaces is usually insufficient to account for observed etch rates, even if each incident ion removed one surface atom. In practice, physical sputtering yields at moderate ion energies are less than unity

The role of ions when they participate in plasma etching is more physical than chemical—they enhance the etch rate and lend directionality, but at the etch rates commonly observed they are not the agent that combines with the substrate to form gaseous products. While a detailed calculation of plasma parameters is beyond the scope of this review, within the usual range of plasma densities, 10^{10}–$10^{11}\,\mathrm{cm}^{-3}$, the flux of ions to the wall should be no greater than $\approx 10^{15}\,\mathrm{cm}^{-2}\,\mathrm{s}^{-1}$. By contrast a moderately low etch rate, 1000 Å min^{-1}, corresponds to almost 10^{16} Si atoms removed per $\mathrm{cm}^{-2}\,\mathrm{s}^{-1}$. Therefore, even if every ion reaching a substrate surface reacted and removed a surface atom, their numbers would be insufficient to account for the observed etch rate. Indeed, 1000 Å min^{-1} is conservatively low and the assumed ion flux may be uncharacteristically intense. The flux of neutral particles, on the other hand, is more than enough to sustain this etch rate, and it is not uncommon for etchant radicals to comprise to 5–50% of the neutral population (10, 12, 13). Figure 15.9 shows these relationships graphically.

One can view the etching process as a series of steps, illustrated by the energy-driven ion assisted etching of *undoped* silicon by chlorine:

(I) Ion and electron formation (which balance loss of charged species and are part of a more complex reaction set):

$$\mathrm{e} + \left\{\begin{matrix}\mathrm{Cl} \\ \mathrm{Cl_2}\end{matrix}\right. \rightarrow \left\{\begin{matrix}\mathrm{Cl^+} \\ \mathrm{Cl_2^+}\end{matrix}\right. + 2\mathrm{e} \tag{15.4}$$

(II) Etchant formation:

$$\mathrm{e} + \mathrm{Cl_2} \rightarrow 2\mathrm{Cl} + \mathrm{e} \tag{15.5}$$

(III) Adsorption of etchant on the substrate:

$$\left\{\begin{matrix}\mathrm{Cl} \\ \mathrm{Cl_2}\end{matrix}\right. \rightarrow \mathrm{Si_{surf}} - n\mathrm{Cl} \tag{15.6}$$

(IV) Reaction to form product:

$$\mathrm{Si} - n\mathrm{Cl} \xrightarrow{\text{(ions)}} \mathrm{SiCl}_{x\mathrm{(ads)}} \tag{15.7}$$

Finally, (V) product desorption:

$$\mathrm{SiCl}_{x\mathrm{(ads)}} \rightarrow \mathrm{SiCl}_{x\mathrm{(gas)}} \tag{15.8}$$

For completeness, surface diffusion of adsorbed reactants or reactive intermediates should also be considered since some evidence for such a step has been observed (see p. 380). Chemical etching proceeds similarly, except that Step IV (equation 15.7) above would be replaced by one that does not require ion bombardment.

$$\mathrm{Si} - n\mathrm{Cl} \rightarrow \mathrm{SiCl}_{x\mathrm{(ads)}} \tag{15.9}$$

Any one of these steps can potentially be the slowest, and thereby limit the etching rate. The purely chemical reaction between undoped silicon and Cl atoms or Cl_2 is slow at ordinary temperatures. However, heavily doped silicon and heavily doped polysilicon react rapidly with Cl atoms, so that anisotropic ion-assisted etching of doped polysilicon, a conductor material in integrated circuits, must be done with an inhibitor chemistry. Note that the relative significance of ions in Step IV (equation (15.7) *v.* equation (15.9)) can vary with the chemical system. For example in the F atom etching of silicon in CF_4/O_2 plasmas, the reaction analogous to (15.9) proceeds rapidly, and the contribution of ions is minimal.

Etchant-unsaturate concepts

Halocarbons and their mixtures with oxidants such as O_2, Cl_2, NF_3, etc. are used for practically all plasma processing. Unsaturated halocarbon radicals and oligomers derived from the feed gas usually are the precursors to inhibitor sidewall protection films. When inhibitor ion-assisted etching is taking place, these films can be extremely thin, often no more than $\sim$30–150 Å at steady state (14). However, if excessive concentrations of unsaturated species are formed in the gas phase, they may leave contamination or produce gross amounts of polymer that coats all surfaces and stops etching. In halocarbon plasmas, there ordinarily is a balance between unsaturated species and the etchant/oxidant atoms such as F or O. The *etchant-unsaturate* concept provides a framework for understanding this balance in general terms and is a guide to predicting the effects of composition changes on halocarbon/oxidant plasmas.

In this reaction scheme, unsaturated fluoro- and chlorocarbon polymer precursors derived from the CF_2 or CCl_2 radical are saturated during reactions with atoms and reactive molecules. The most reactive species are preferentially removed by the saturation reactions. An ordering of this reactivity can be used to predict the dominant atomic etchants as a function of halocarbon and additive gas compositions. The general formulation is given by:

$$\text{e} + \text{halocarbon} \rightarrow \begin{matrix}\text{saturated}\\ \text{species}\end{matrix} + \begin{matrix}\text{unsaturated}\\ \text{species}\end{matrix} + \text{atoms} \tag{15.10}$$

$$\left.\begin{matrix}\text{reactive atoms}\\ \text{reactive molecules}\end{matrix}\right\} + \text{unsaturates} \rightarrow \text{saturates} \tag{15.11}$$

$$\text{atoms} + \text{surfaces} \rightarrow \left\{\begin{matrix}\text{chemisorbed layer}\\ \text{volatile products}\end{matrix}\right. \tag{15.12}$$

$$\text{unsaturates} + \text{surfaces}\left(+ \begin{matrix}\text{initiating}\\ \text{radicals}\end{matrix}\right) \rightarrow \text{films} \tag{15.13}$$

In CF_3Br, for instance, Br and F atoms are formed by electron-impact dissociation of the halocarbon and there is usually a negligible steady-state concentration of F atoms because of reaction with CF_2 radicals and unsaturated species such as C_2F_4. In heavily chlorinated plasmas (e.g. CCl_4 as the halocarbon in (15.11)), CCl_2 and C_2Cl_4 will play analogous roles. The relative reactivities of atoms and molecules in these saturation reactions are assigned a hierarchical order: $F \sim O > Cl > Br$; $F_2 > Cl_2 > Br_2$. This ordering is used to predict the predominant etchant species and reaction products as a function of feed composition. The most reactive atoms (F, O) do not coexist with unsaturates at appreciable concentrations; thus either these atoms or the unsaturates are substantially depleted. When atoms that can gasify a substrate predominate, etching takes place. Film formation is observed when unsaturates are present in excess and are adsorbed on surfaces where polymerization proceeds by reaction with saturated fluorocarbon radicals (15). Alternatively the unsaturated radicals may form sidewall films that result in anisotropic etching in the presence of ion bombardment. Polymerization of unsaturates is inhibited by surfaces that react with fluorocarbon radicals to form entirely volatile products (15, 16) (e.g. SiO_2 surfaces).

Oxidant additions to a plasma alter the balance between halogen atoms and unsaturates. More reactive oxidants will be preferentially consumed by unsaturates, tending to increase the relative concentration of less reactive halogen atoms, and while doing so they will suppress polymer formation. The effect may be illustrated by the addition of oxygen to a CF_3Br plasma. Oxygen atoms and molecular oxygen in the discharge will result in the following overall reactions:

$$\left.\begin{matrix} O \\ O_2 \end{matrix}\right\} + C_xF_{2x} \rightarrow \left\{\begin{matrix} COF_2 \\ CO \\ CO_2 \end{matrix}\right. + F, F_2 \tag{15.14}$$

and

$$\left.\begin{matrix} O \\ O_2 \end{matrix}\right\} + C_xF_{2x-y}Br_y \rightarrow \left\{\begin{matrix} COF_2 \\ COFBr \\ CO \\ CO_2 \end{matrix}\right. + F, F_2, Br, Br_2 \tag{15.15}$$

These reactions cause an increase in F, F_2, Br, and Br_2 concentrations (*cf.* refs. 17, 18, 19 and discussion below). Fluorine and oxygen combine with unsaturated species at similar rates, hence if unsaturates are present in excess, the fluorine is also consumed by saturation reactions such as:

$$\left.\begin{matrix} F \\ F_2 \end{matrix}\right\} + C_xF_{2x} \rightarrow \left\{\begin{matrix} C_xF_{2x+1} \\ C_xF_{2x+2} \end{matrix}\right. \tag{15.16}$$

Bromine-unsaturate reactions are slower and bromine removal is proportional to the unsaturated species concentration. Consequently, the consumption of unsaturates by oxygen and fluorine lowers the rate of bromine-unsaturate reactions, leading to an increase in free bromine within the discharge. Finally, if an excess of oxygen is added, unsaturated compounds are substantially depleted so fluorine and oxygen atoms will appear in the discharge along with the increased concentration of Br and Br_2. This is usually undesirable since F-atom reactions with the substrate will lead to mask undercutting, while O atoms attack photoresist as well as form surface oxides that retard etching (20). Note that oxygen itself is only one example of an oxidant that can be added. For instance NF_3 added to a CF_3Br plasma will also increase the concentration of Br atoms by liberating F and F_2 which participate in reactions analogous to equation 15, and will suppress unsaturated species (equations 14, 16).

The role of gas additives

Various additives have useful effects on plasma etching in Cl_2 and related etchants. Although the chemistry and gaseous electronics of these mixed systems are complex, an understanding of certain basic processes is helpful in anticipating, *a posteriori*, certain beneficial effects of the additions. The additives can be qualitatively grouped into the four categories summarized in Table 15.1 and discussed below.

Native oxide etchants. Often, native oxides initially cover the surface of a substrate and prevent the onset of etching. Additions of small amounts of oxide etchants—examples: C_2F_6 for silicon (21), BCl_3 for Al (22, 23, 24), or GaAs (25)—can circumvent this problem.

Radical and unsaturate scavengers. This effect is illustrated by the addition of O_2 to CF_4 plasmas (20). O_2 removes CF_x radicals, preventing their recombination with F atoms, thereby increasing the F-atom concentration. Similarly, the addition of oxidants to bromo- and chlorocarbon plasmas (e.g. CF_3Br, CF_3Cl, CF_2C_2, CCl_4—see above) will increase the concentration of chlorine and suppress the formation of polymer from unsaturated intermediates.

Table 15.1 Gas additives for plasma etching

Additive	Purpose	Example (additive-active gas:material)
Oxide etchant	Etch through material oxide to initiate etching	C_2F_6–Cl_2:SiO_2; BCl_3–Cl_2:Al_2O_3; CCl_4–Cl_2:Al_2O_3
Oxidant	Increase etchant concentration or suppress polymer	O_2–CF_4:Si; O_2–CCl_4:GaAs, InP
Rare gas, N_2	Stabilize plasma, dilute etchant, improve heat transfer	Ar–O_2:organic material; He–CF_3Br:Ti
Surface inhibitor	Induce anisotropy, improve selectivity	C_2F_6–Cl_2:Si; BCl_3–Cl_2:GaAs, Al H_2–CF_4:SiO_2

Inert gas additions. Noble gases (usually Ar or He) stabilize and homogenize plasmas that otherwise constrict, or oscillate in the reactor. The stabilization may be due to an effect on the thermal properties of the discharge gas (26) (especially with helium additions) or to a shift in the electron energy and density resulting from altered electron balance processes. Sometimes the 'inert' addition can enhance anisotropic etching by its effect on the sheath potentials and by providing non-reactive ion-bombardment that stimulates etchant-surface gasification (27).

Surface inhibitors. Several surface inhibitors (recombinants and protective film precursors) that promote anisotropic etching have been studied in detail earlier (see p. 348). Ions promote desorption of these species, so that the surface concentration is larger in regions not exposed to ion-irradiation (i.e. sidewalls of features).

Variables controlling plasma etching

Ideally we would like to know, at an atomistic level, how to control etching properties by manipulating basic chemical and plasma variables. In many basic texts and literature on electrical discharges, the plasma is examined from a fundamental point of view as a function of certain quasi-dimensionless similarity variables. These variables include the ratio of electric field to number density, E/N (sometimes written as E/P since pressure, P, is proportional to number density at standard conditions), the product of number density and a characteristic length of the reactor geometry (Nd or pd), f/N, the ratio of driving frequency (plasma RF generator) to density, and reactor shape and aspect ratios. Unfortunately, these variables are not generally useful for controlling most plasmas in the semiconductor industry. The reason for this negative conclusion is threefold. First, the similarity variables cannot be set by the engineer or scientist—relationships between the microscopic similarity parameters are determined by the plasma gas and apparatus. Moreover, there really is no well-defined E/N or n_e/N for many common processing plasmas. E and n_e may vary by a factor of ten or more with position in the plasma and can also oscillate in time with the applied electric field. Since E/N, n_e/N and related similarity parameters are neither constant nor at the disposal of an operator, the instrumental parameters or *discharge variables* are used instead. These include radio-frequency (RF) input *power*, reactor *pressure*, RF excitation *frequency*, *temperature*, *flow rate*, feed *gas composition*, reactor geometry and materials of construction. These quantities, when fixed, uniquely determine the operation of a plasma process. In the sections below, we will survey the effects of some of these variables, and refer the reader elsewhere for a more extensive treatment (6).

Those experienced with processing equipment will notice that we have not listed 'bias' as an operating parameter, although the dc bias has important effects on anisotropic etching. This is because maximum bias is determined by reactor asymmetry (6) and can only be lowered by an external circuit. We view this as part of reactor geometry, which is beyond the scope of this chapter.

Pressure effects

Pressure directly influences major phenomena that control plasma etching. Among these are: (i) the sheath potentials and the energies of ions bombarding surfaces, (ii) the electron energy, (iii) ion-to-neutral abundance ratio and fluxes to surfaces, (iv) the relative rate of higher to lower order chemical kinetics, (v) surface coverage by physisorption, and (vi) the relative rates of mass transport processes. Most of our remarks will be restricted to the pressure range generally used for plasma etching, about 1 mTorr to 5Torr.

As pressure is lowered below about 0.1 Torr, the characteristic potentials across the sheaths and the voltage applied to a discharge rise sharply, from perhaps some tens of volts up to 1000 V or more. As this happens, the mean free paths of species also increase so the larger sheath potential induces a corresponding increase in the energy of ions bombarding substrate surfaces. Sputtering does not take place until ion bombardment energy exceeds a material and ion-specific threshold (28). Sputtering rates characteristically increase rapidly with energy beyond these thresholds, although the sputtering efficiency (atoms removed per incident ion) usually remains well below unity. Similarly, it is believed that there are threshold energies and energy-dependent cross sections for damage-induced ion-enhanced etching of various materials. Since low pressure favours higher ion bombardment energies, it also favours etching by the energy-driven damage mechanisms, when such enhancement is possible. However, ion energies far above the threshold are undesirable because selectivity ratios decrease with increasing ion energy, and because there may be ion-induced electrical damage to devices.

Generally, the energy of electrons in a plasma increases monotonically with E/N, the ratio of local electrical field to particle density (or normalized pressure). The internal fields in discharges tend to increase or stay the same (but not decrease) as the applied RF potential rises. Since this potential falls with pressure, E/N decreases with pressure and the average electron energy is lower. Electron–molecule dissociation cross sections are functions of E/N, and in the range of interest it seems that electron–molecule reaction rate constants, derived by integrating the basic cross sections over the electron energy distribution function, increase with E/N and applied potential (6, 8, 29). Hence at higher pressure, specific dissociation rates tend to fall, requiring a larger electron number density to maintain the discharge. The net result of these and related basic effects is a tendency, all else equal, of electron energy and specific dissociation rates to decrease with increasing gas pressure.

Since energy transfer from electrons to neutrals and thermalization of electrons is proportional to gas pressure too, the discharge gas temperature tends to rise with pressure. At some point, usually above 10–100 Torr, the electron and gas temperatures converge and the plasma is said to be 'arc-like.' Plasma density tends to be insensitive to total pressure, being constrained to be above 10^9–$10^{10}\,cm^{-3}$ by the ambipolar limit, and below $\sim 10^{12}\,cm^{-3}$ by the glow-to-arc transition (30, 31). Most measurements to date put the density between $\sim 10^{10} - 10^{11}\,cm^{-3}$.

We pointed out that neutral species are responsible for both chemical isotropic and ion-enhanced anisotropic etching. But these processes have different kinetics. Chemical etching is frequently first order in the concentration of neutral etchant species, while ion-enhanced etching may be zero order when the flux of ions is rate limiting. Taking the chemical and ion-enhanced processes as parallel channels appears to be a good approximation (12). Thus if the plasma conditions and dissociation rates are fixed, decreasing pressure tends to decrease the concentration of neutral gas and radicals proportionately and thereby increase the significance of energetic ion-enhanced etching. This trend is also helped by the increase of sheath potentials and ion energy as pressure is lowered.

Control of polymers and inhibitor films is linked with pressure and provides a good example of pressure effects through reaction order. Simple etching reactions such as the reaction of F atoms with Si to form SiF_4 are commonly limited by a rate-limiting first-order step (see p. 369). To the extent that the etching species (F in this case) is a *constant fraction* of the gas, etching will vary with pressure ($\propto P$). Reactions that lead to inhibitor films and polymers, on the other hand, may involve homogeneous chain growth of higher order and therefore vary with higher powers of pressure. For example the dimerization of CF_2,

$$CF_2 + CF_2 \xrightarrow{(M)} C_2F_4\ (+M) \tag{15.17}$$

will have a rate law that is at least second order, $r = k_2 N_{CF_2}^2 (\propto P^2)$ and the importance of oligomer, polymer and film formation from a mixture of fixed composition containing F atoms and CF_2 will increase with pressure. As pressure is lowered, the etching rate in this simplified example will rise relative to the rate of film formation.

Increasing pressure and decreasing temperatures increases the surface concentration of physisorbed species in accordance with their adsorption isotherms, and chemical etching rates increase with surface concentration of etchant. While adsorption effects in plasma etching have not been studied, in the closely related phenomena involved in low pressure gaseous etching (LPGE) (32, 33) they can lead to an apparent 'negative activation energy.' Interesting conditions were found in which decreasing temperature led to an increase in the rate of silicon etching by XeF_2, ClF_3 and other interhalogen compounds, apparently because the surface concentration of active species increased faster than the decline in rate constant with temperature (32, 33) (see p. 374).

Finally, we note that at fixed mass flow rates, the ratio of diffusive mass transport relative to convection is constant (e.g. the Peclet number), and the rates of both transport mechanisms increase as $1/P$. Therefore the ratio of mass transport rates relative to first order surface reactions will vary as $1/P^2$ so that lower pressure tends to overcome local reactant depletion, or mass transport limitations to chemical reaction.

Excitation frequency

Most commercial plasma reactors today operate at 13.56 MHz, mainly because this is an FCC (US Federal Communications Commission) licensed industrial frequency for which RF generators are readily available and higher levels of emitted radiation are allowed. Microwave plasma equipment generally operates at 2450 MHz, another FCC industrial frequency. By the same token, the selection of lower frequencies (200–800 kHz) for processing equipment was often inspired by a manufacturer's other products or a convenient supply of induction heating power supplies; the intentional exploitation of frequency as a processing parameter for etching or deposition is a recent development.

Excitation frequency alters key discharge characteristics that have an important and direct influence on plasma chemical processing (8). We consider four separate kinds of effects, although they don't necessarily occur independently. (i) Excitation frequency can change the spatial distribution of species and electrical fields across the discharge. (ii) Frequency determines whether the energy and concentrations of species are constant in time, or whether they fluctuate during a period of the applied field. (iii) Frequency has a large effect on the minimum voltage that is required to start and operate a plasma, and on the energy with which ions bombard surfaces. And finally, (iv) frequency may change the shape of the electron energy distribution function (EEDF) and thereby control the electron–molecule reaction channels that predominate.

In general, frequency changes are associated with a defined transition in discharge characteristics when the excitation period is close to the relaxation time (τ_i) of electrical or chemical processes:

$$0.5 \lesssim \omega\tau_i \lesssim 2 \tag{15.18}$$

When $\omega\tau_i \ll 1$, the process will maintain a dynamic steady-state representative of instantaneous conditions induced by the time-varying field. On the other hand, if $\omega\tau_i \gg 1$, the process will be too slow to stay in step with the field and reaches a static steady-state in equilibrium with average conditions.

Ion transit frequency (ITF). In studies of plasma etching at $\sim$0.5 Torr (34, 35), there is a large increase in voltage and ion bombardment energy, Cl^+ ion flux to the walls, and silicon etching rate as frequency was lowered from above 10 MHz to below 1 MHz (Figure 15.10). Typically neutral reactions are too

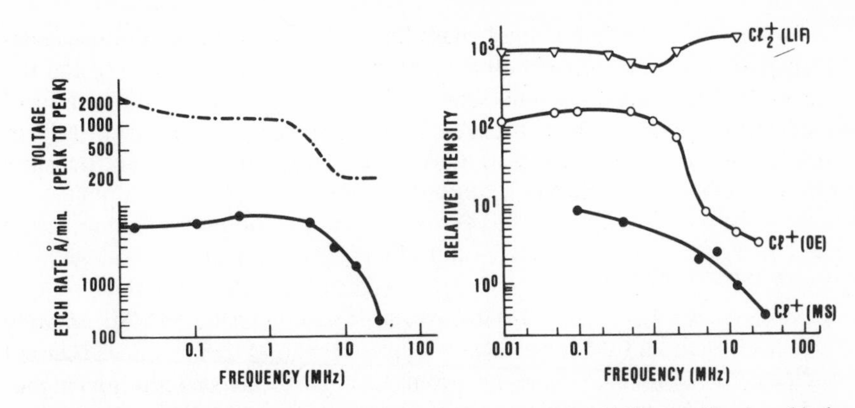

Figure 15.10 Various quantities versus excitation frequency for polysilicon etching in a chlorine plasma at 0.3 Torr and $\sim 0.4\ \mathrm{W\,cm^{-2}}$ power. Left, bottom to top: etch rate; peak-to-peak RF voltage across the reactor. Right, top to bottom: relative concentration of Cl_2 ions in the plasma measured by laser-induced fluorescence (LIF); relative intensity of optical emission from excited Cl^+ $(4d^5D_0)$ (OE); and relative flux of Cl^+ ions to the electrode measured by mass spectrometry (MS)

slow and electron motion is too fast to be associated with this transition region. These, and a variety of accompanying effects are explained by the response of ions near the plasma sheath boundary to the electric field.

For simplicity consider the case where the sheath thickness is less than the mean free path of an ion (collisionless sheath). Above the upper *ion transition frequency* (UITF), the period of the applied field is short compared to the time necessary for ions to cross the sheath and reach the electrode. We assume the potential across the sheath at the momentary positive electrode is small compared to the negative sheath potential, so ion acceleration only occurs during the negative half cycle. Hence the fastest ions with transit times many cycles long will be accelerated to

$$V_{\max}^{\mathrm{HF}} = \frac{V_0}{2\pi}\int_0^{\pi} \sin\omega t\, \mathrm{d}\omega t \approx \frac{V_0}{\pi} \tag{15.19}$$

At the UITF, ions entering the negative sheath when the momentary applied voltage is zero just cross the sheath and strike the electrode at the next zero of the applied field (π radians later), and attain the energy given by equation (15.19). As the frequency is lowered further, some ions cross in less than 1/2 cycle, thereby reaching the electrode with energies greater than V_0/π. Finally, when the average velocity of these ions permits crossing the sheath in times much less than $\pi/2\omega$ (1/4 period), some of the ions will cross and reach the electrode when the electric field, $E_0 \sin\omega t$, is at a maximum, E_0. These ions will have been accelerated to an energy equal to the peak applied potential

$$V_{\max}^{\mathrm{LF}} \cong V_0 \tag{15.20}$$

The point at which most ions rapidly traverse the sheath in a time short compared to 1/4 cycle ($\approx \pi/10\omega$) has been termed the *lower ion transit frequency* (LITF). Between these two limits, the maximum ion energy will lie within the bounds of equations (15.19) and (15.20).

While these effects on the ratio of ion energy to peak voltage account for about a three-fold increase in the energy of bombarding ions as frequency is lowered below the UITF, there is an even larger additional rise in bombardment energy caused by the increase in the discharge-sustaining voltage and sheath potential at low frequency. It has been suggested that this voltage increase is caused by a change in the discharge sustaining mechanism. At high frequency when ions move only a small distance during the RF cycle, a 'glow' discharge is maintained by volume ionization that balances recombination and diffusive loss to the walls. In this mode there is a relatively low operating voltage. By contrast, ionization at low frequency originates from the action of high energy ion impact on the negative electrode, resulting in a distribution of secondary electrons that multiply through ionizing collisions across a high potential sheath. The voltage required for this mechanism is much higher than that for the diffusion-controlled glow, so that the observed increase in potential is as expected.

Electron energy oscillation. When electrons collide with neutrals, kinetic energy is lost. Electron energy distribution functions (EEDFs) commonly have mean values far above the translational gas temperature ($\varepsilon_{gas} = \frac{3}{2}kT$) because electrons transfer a small fraction of their total energy in most collisions, while gas molecules efficiently conduct translational energy to their surroundings. If the time required for electrons to lose energy by collisions is long compared to the oscillations of the applied RF field, the mean electron energy does not respond to instantaneous changes in the field over a cycle and is instead determined by the average electric field. On the other hand when collisions are so frequent that electron energy could decay toward zero in a quarter cycle, without continuous power input, the mean electron energy will respond to the *instantaneous* field intensity and be in equilibrium with it. These two situations are depicted in Figure 15.11. The variable that determines which of these will prevail is v_u/ω, the ratio of the fraction of the average energy an electron loses per unit time to the applied frequency, ω (v_u is also called the energy loss frequency).

Thus v_u/ω should have great significance for discharge chemistry since the rates of electron–molecule reactions are most sensitive to electron energy. Electron–molecule reactions require a minimum electron energy, ε_{thresh}, to take place at all, and they reach their maximum rate at some higher characteristic energy (the cross-section for reaction resembles a bell-shaped curve as a function of electron energy). As long as the concentrations of electrons and the molecular reaction partner change slowly compared to the RF excitation period, the effective reaction rate constant will be determined from a time-

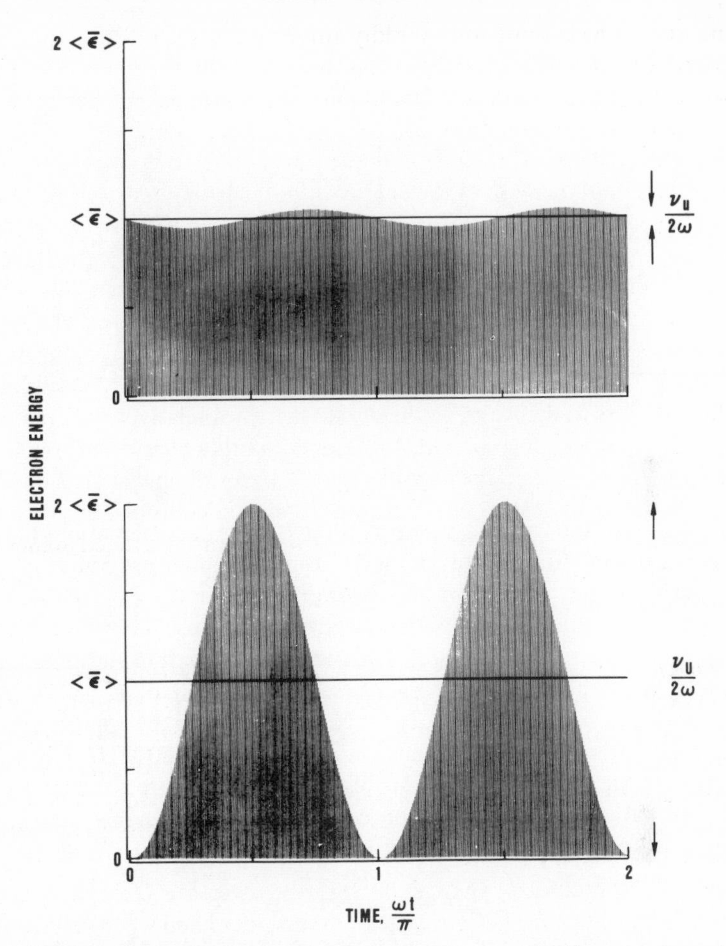

Figure 15.11 Top: electron energy as a function of ωt when $\nu_u/\omega \ll 1$ and bottom: when $\nu_u/\omega \gg 1$. The peak-to-peak ripple at top is given by $\bar{\varepsilon}\nu_u/2\omega$ (see text)

average convolution of the instantaneous electron energy excitation cross-section for reaction (36):

$$k(\langle\bar{\varepsilon}\rangle) = \frac{\omega}{2\pi}\int_0^{2\pi/\omega} K \int_0^{\infty} \varepsilon^{1/2} Q(\varepsilon) f(\varepsilon, \bar{\varepsilon})\, \mathrm{d}\varepsilon \frac{\mathrm{d}\bar{\varepsilon}}{\mathrm{d}t}\, \mathrm{d}t \tag{15.21}$$

where K is a numerical constant, $Q(\varepsilon)$ is the reaction cross-section for electrons of energy ε, and $f(\varepsilon, \bar{\varepsilon})$ is the EEDF. In general $k(\langle\bar{\varepsilon}\rangle)$ will be different depending on whether the energy remains constant over time or scans all values from zero to an energy characteristic of the peak applied voltage, E_0 (per Figure 15.11).

The ratio of the peak electron energy (sum of the dc component plus the 2nd harmonic) to the average value (dc component) is given by

$$\frac{\varepsilon_p}{\langle \bar{\varepsilon} \rangle} = 1 + \frac{\dfrac{\nu_u}{2\omega}}{\left[1 + \left[\dfrac{\nu_u}{2\omega}\right]^2\right]^{1/2}}. \tag{15.22}$$

At low frequencies or high pressure ($\nu_u \gg 2\omega$) the second right-hand term in equation (15.10) dominates and energy follows the field:

$$\bar{\varepsilon} = \varepsilon_p \sin^2 \omega t \tag{15.23}$$

while at high frequency or low pressure ($\nu_u \ll 2\omega$) the electron energy oscillates around its mean value, $\varepsilon_p/2$:

$$\varepsilon = \frac{\varepsilon_p}{2}\left\{1 + \frac{\nu_u}{2\omega}\sin 2\omega t\right\} \tag{15.24}$$

with a peak-to-peak ripple $\varepsilon_p \nu_u/4\omega$. Although some approximations have been made in this derivation, the form of the results and limiting equations (15.23) and (15.24) do not depend on the assumptions (36).

Data and calculations show that electron energy oscillation is common in RF plasma processing discharges. Time-resolved excitation and emission from Cl in a 13.2 MHz Cl_2 discharge gave (37) a ratio of $\varepsilon_p/\langle \bar{\varepsilon} \rangle \approx 38\%$, or $\nu_u \approx 63$ MHz. This is divided by 2π for comparison with the applied frequency, indicating intense energy oscillation up to about 20 MHz ($\omega\tau \approx 2$). By contrast, at the centre of a 220 KHz discharge, Cl excitation and emission are 100% modulated and the emission ceases for an appreciable time after the voltage waveform crosses zero. Other data and calculations show similar energy modulation in O_2 (38), SiH_4 (39, 40), CH_4 (40) and argon (41) discharges.

Frequency and pressure are, to a degree, interchangeable variables since they both influence the electron energy distribution in a plasma and the energy distribution of ions striking the substrate surface. Within the commonly used range of frequency and pressure (0.05–30 MHz, 0.001–1.0 Torr), either low pressure or low frequency increases the flux and energy of ions bombarding a substrate.

Power

Increasing power increases the density of radicals and ions as well as the ion energy. The concentration of charged species can saturate at high power. In every case, when power density is low (e.g. $\lesssim 0.5\ \mathrm{W\,cm^{-2}}$) raising power will increase etch rates. However at high powers, detrimental phenomena such as substrate heating and intense ion bombardment lead to gross surface damage and mask erosion.

Temperature

Temperature influences rate, selectivity, surface morphology, and the degree of anisotropy. Etch rates generally follow an Arrhenius-type dependence on substrate temperature:

$$\text{etch rate} \propto \exp(-E_a/kT) \tag{15.25}$$

The effective activation energy E_a is material-dependent; hence selectivity of one material over another will also vary exponentially with temperature. Chemical etching can also be accelerated by increasing temperature, to the point of becoming competitive with ion bombardment enhanced etching, resulting in reduced anisotropy. Substrate temperature also plays a role in determining surface morphology. For example, fluorine atoms will etch silicon with less surface roughness at low temperatures ($\lesssim 0°$ C).

Flow rate

In general, flow rates should be large enough to avoid a feed limited supply of etchant species in the reactor and to minimize polymer formation in unsaturate-rich etch mixtures. Pressure and flow interact through their effects on residence time and the comparative importance of diffusion rates, convection rates and chemical reaction rates. When feed gas is chemically transformed in the plasma, the residence time influences composition. There are two simplified cases: when the flow rate is fast compared to chemical decomposition rates for the feed (short residence time, long reaction time), the active species will be governed primarily by the feed composition. When the flow rate and etching reactions are slow compared to the homogeneous chemical reactions and mass transfer (long residence time, short reaction time), the reactant composition will be uniform and determined by elemental stoichiometry, independent of the molecular makeup of the feed (well-mixed kinetic steady-state). Rapid etching combined with long residence time should generally be avoided since gas–solid reactions, polymerization or redeposition of etched material can lead to large composition gradients and an etch rate that varies across the reactor.

Etching of different classes of material

Silicon and polysilicon

The isotropic etching of silicon (Si) and polycrystalline silicon (poly-Si) by atomic fluorine (F) is one of the oldest and most widely used processes in microcircuit fabrication. It also is the one most completely understood. Fluorine atoms etch (100) Si at a rate (Å/min) given by (10):

$$R_{F(Si)} = 2.91 \times 10^{-12} T^{1/2} n_F \exp(-0.108\,\text{eV}/kT) \tag{15.26}$$

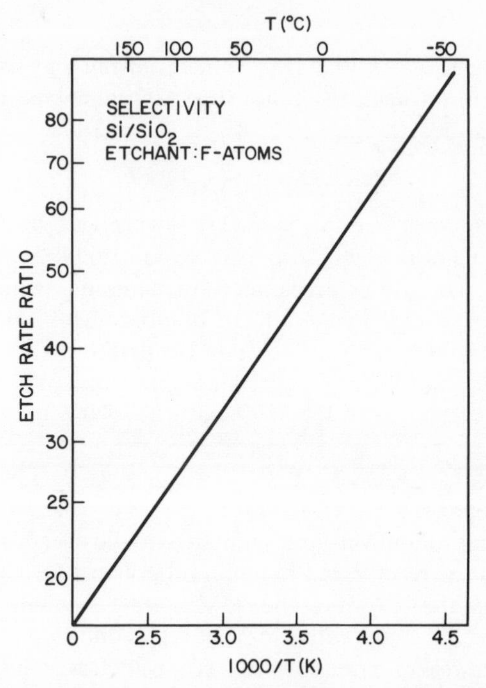

Figure 15.12 Selectivity relative to SiO_2 for etching silicon with atomic fluorine

where n_F is the F-atom concentration (cm^{-3}), T the temperature (K), and k is Boltzmann's constant. To measure this value, F atoms were produced in a RF discharge upstream of the silicon sample so the intrinsic etch rate of silicon by atomic fluorine could be determined without interference from the ion bombardment that occurs in plasma reactors. The inherent selectivity of F-atoms for etching silicon relative to SiO_2 (42) is shown as a function of temperature in Figure 15.12.

A variety of studies have led to a detailed picture of the chemistry of the F-atom/Si system. Early studies (10) of the F-atom reaction with silicon and the chemiluminescence arising from the gas phase (43) reactions

$$SiF_2 + F(F_2) \rightarrow SiF_3^* (+F) \tag{15.27}$$

$$SiF_3^* \rightarrow SiF_3 + h\nu \tag{15.28}$$

concluded that the common kinetics of the free-radical product and overall etching process require a rate-limiting step that simultaneously liberates SiF_2 from the surface and provides the precursor to higher fluoride products. Coupled with the observation of polarized Si-F infrared chemiluminescence (44), XPS data indicating an 'SiF_2-like' surface (45), and confirmation of an SiF_2 product (46), the kinetics suggested the branching reaction summarized

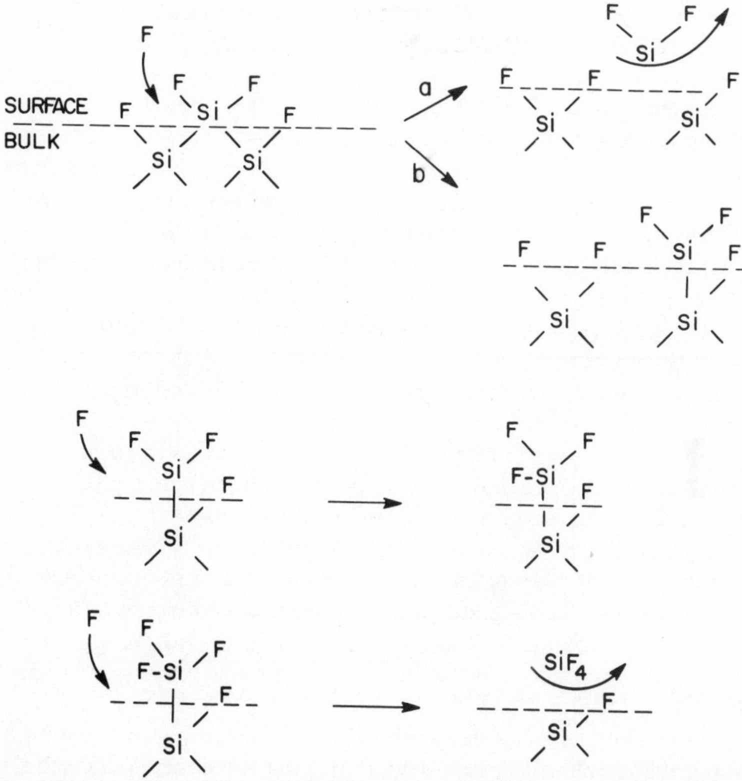

Figure 15.13 Mechanism of silicon etching by fluorine atoms. A two-channel concerted reaction results in direct formation of gaseous SiF_2 (Ia) and bound fluorosilicon radical (Ib) that is further fluorinated to higher SiF_x products. Kinetics show that (Ia) and (Ib) are branches of a single, rate-limiting reaction that may involve a common, vibrationally-excited intermediate

in Figure 15.13 to be rate-limiting. Here, the stable (SiF_2)-surface must be penetrated for etching to proceed.

Once penetration occurs, subsurface Si–Si bonds are attacked with one channel producing gaseous SiF_2 *directly* and the other leading to bound $-SiF_2-$ which is fluorinated further to produce SiF_3 and SiF_4. The latter products have been observed in mass spectrometric studies (47) using XeF_2 as a fluorinating agent. Although one might have expected these two channels to have a different temperature dependence, they do not, possibly because a vibrationally hot transition state is involved that undergoes unimolecular dissociation (channel *a*) or stabilization (channel *b*). The exothermicity of the reaction is certainly high enough for this, and the IR chemiluminescence seems to bear this out. The degree to which channel *a* or *b* occurs is statistical, depending more on the heat of reaction than substrate temperature. Thus, the relative yield is insensitive to substrate temperature and the SiF_2 product accurately tracks the etch rate, even though it may be a minor product.

This interpretation implies that the branching ratio should be sensitive to the reactant gas. Indeed this is the case. Studies of the kinetics of etching and chemiluminescence in the F_2 reaction with silicon (48) showed reasonably good tracking between etching and the SiF_2 product, but an SiF_2-consuming surface reaction was necessary to explain the results at pressures greater than 10 Torr. It is likely that the surface species was the hot intermediate that statistically dissociates; however, higher pressures enhance stabilization producing a fall-off in SiF_2 desorption. Furthermore, chemiluminescence accompanying silicon etching by XeF_2 (49) did *not* track the etch rate, suggesting lower reaction exothermicity, and improved quenching of the hot surface intermediate by physisorbed XeF_2.

More recently, a considerable body of XPS data has confirmed that a slower-etching $(SiF_x)_n$ passivation layer extends 3 to 10 monolayers into the bulk (50, 51, 52, 53). Perhaps more significantly, $-SiF-$ and $-SiF_3$ species were present in greater abundance than $-SiF_2-$ species. Further, with high doses of fluorinating agents, the layer thickens (to ~ 10 monolayers) and SiF_4 is prevalent in this 'crust'. Based on these observations, a place-exchange mechanism has been advanced (50) as the controlling factor in silicon etching by fluorinating agents. The suggestion is that F atoms are converted to anions. These anions diffuse into the bulk and move, by coulombic attraction, to the more positively charged silicon lattice sites—similar to the proposed mechanism for oxidation of silicon (54).

Although these latter studies clearly show that the mechanism of etching is more complicated than initially supposed, the fundamental concept of etching with a branching precursor, slowed by a layer that must be penetrated for gasification to proceed, is unshakeable. SiF as the dominant surface species is equally amenable to the branching mechanism discussed above, since the exothermic heat of reaction

$$[SiF]_{surf} + F \rightarrow [SiF_2]^{\dagger}_{surf} \qquad (15.29)$$

would be *directly* deposited into a hot SiF_2 surface species, rather than adjacent to it (cf. Figure 15.13). Whether the branching precursor in this layer is SiF or SiF_2 is still debatable, but that is of little practical consequence. As before, F atoms attack subsurface bonds to liberate SiF_2 directly, free trapped SiF_4, and further fluorinate other trapped SiF_x species. Of course, this attack proceeds via a branching reaction as above, so that SiF_2 is not necessarily gasified in each event.

The stability of this fluorosilicon mantle in the presence of a powerful fluorinating agent, like F atoms, is remarkable. Why the surface exhibits higher stability than the bulk, even though all reactions of F atoms to form higher SiF_x are highly exothermic, invites continued research. It is possible that anion exchange is dominant in forming the mantle and as its thickness increases coulombic forces fall off. As a result, the surface becomes more

Table 15.2 Etch rates and activation energies for silicon etching in F-source plasmas

Plasma	Etch rate at 100° C (Å/min)	E_a(eV)	Reference
$F./F_2^†$	4600	0.108	(10)
CF_4/O_2	4600	0.11	(2)
CF_4/O_2	3000	0.11	(58)
CF_4	300	0.124	(59)
SiF_4/O_2	440	0.11	(58)

†Etching downstream from RF discharge of F_2

covalent in nature and the free radical mechanism dominates at the gas–solid interface. Etching then becomes a dynamic steady state between the advancing ionic exchange front and the free radical attack which removes material. The interplay between the two governs both the thickness of the layer and the etch rate. An apparent discrepancy, however, is that molecular fluorine only forms a monolayer of SiF_2-like species during etching (53), suggesting that place exchange as a dominant factor may be overestimated. The lower reactivity of F_2 compared to F atoms, would imply that a thicker layer be formed in order for increasing covalency of the surface to allow F_2 reactions to occur.

Discharges that produce F atoms as the dominant etching species include F_2 (10), CF_4, CF_4/O_2 (20), SiF_4, SiF_4/O_2 (55), SF_6 (56, 57), SF_6/O_2 (57), NF_3 and ClF_3 (1). In all cases high selectivity over SiO_2 (10–40:1) and Si_3N_4 (5–10:1) can be achieved. The F-producing feed gases shown in Table 15.2 display similar activation energies for the etching process in line with the notion that the basic mechanism and etchant (F atoms) are the same in these chemistries. Computer modelling of CF_4 and CF_4/O_2-based plasma chemistry (60, 61) shows variations in etch rate for this system are explainable in terms of differences in F-content for different conditions. The calculations show F-atom densities in CF_4/O_2 plasmas are comparable to those measured in the downstream studies (10^{15}–$10^{16}/cm^3$), while pure CF_4 plasmas produce about a factor of ten less atomic fluorine. Finally, the strong temperature dependence of Si/SiO_2 selectivity noted above, suggests that selectivity differences among these various plasmas are attributable to temperature effects, rather than differences in active etchant.

Gases such as CF_4 and SF_6 are often preferred over pure halogen (e.g. F_2) because of their low toxicity; however, they form unsaturates such as C_xF_{2x} and fluorosulphur radicals, S_xF_y, in the discharge which can scavenge free F atoms, and in unfavourable cases, form polymeric residues. Oxygen additions have two principal effects in these plamas. First, in accord with the etchant-unsaturate model noted earlier, O_2 (and O atoms) reacts with unsaturates to promote F formation and suppress polymerization. Second, with enough O_2 present, O_2(O) reactions with silicon etched can inhibit etching, as the

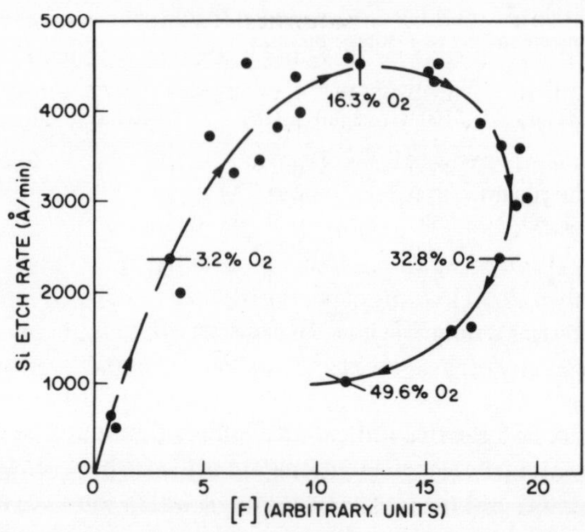

Figure 15.14 Hysteresis curve for silicon etching in a CF_4/O_2 plasma. Oxygen additions increase F-atom concentrations via reaction with fluorocarbon radicals causing the etching rate to increase. Adsorption of oxygen results in a decrease with large additions of O_2

hysteresis curve (Figure 15.14) for CF_4/O_2 etch rates shows (20). Similar results were obtained with other gases (55, 57).

Anisotropic etching of silicon in fluorine-containing plasmas is virtually impossible under most plasma conditions owing to the voracity of F atoms as an etchant. An apparent exception has been observed at low pressures with high substrate bias in feeds based on CF_4 and SF_6 and some mixed gases (the conditions sometimes referred to as 'RIE'). In this regime, the gas phase concentration of F atoms is diminished relative to the ion bombardment flux and adsorbed halocarbon species, so the mechanism and etchant species are unclear. It is interesting to note that anisotropies (ratio of undercutting to etch depth) below 0.1 (62) are achievable in this regime, albeit with a serious loss of selectivity. For example, selectivity of etching silicon relative to SiO_2, which is above 40 for isotropic F-atom etching at room temperature, drops well below 10:1 and often between 1 and 2 (63) in going to these 'RIE' conditions.

In contrast, plasmas that produce chlorine and bromine atoms are excellent for selective anisotropic Si etching. The most commonly used gases have been Cl_2, CCl_4, CF_2Cl_2, CF_3Cl, Br_2 and CF_3Br (15), and mixed feeds such as Cl_2/C_2F_6, Cl_2/CCl_4, and C_2F_6/CF_3Cl. Of these, the most effective gas mixtures appear to be Cl_2 (64), Cl_2/C_2F_6 and CF_3Cl/C_2F_6 (3, 65) mixtures with inert gas (Ar), where high etch rates (500–6300 Å/min) for undoped and doped poly-Si *and* high selectivity ($Si/SiO_2 \sim 10$–50:1) have been observed. The active etchant in these plasmas is usually Cl (or Br) atoms; however, ion bombardment plays a significant role in producing high rates and anisotropy.

In a pure chlorine plasma, ion enhancement proceeds by the energy-driven or 'damage' mechanism, while the inhibitor mechanism is more important in halocarbon feeds that generate unsaturates. Inhibitor chemistry *must* be used to etch *n*-doped polysilicon anisotropically because its chemical etch rate by Cl atoms is high (6). The importance of energy-driven anisotropic silicon etching in Cl_2 plasmas was demonstrated in one study (66) by the increase in etch rate and anisotropy with RF voltage across the reactor. In another study (34), lowering the plasma frequency from 13 MHz to 100 kHz produced a large increase in the rate of silicon etching in a Cl_2 plasma which again correlated with ion energy. Indirect evidence also supports the role of ion enhancement; for example, only a small loading effect is observed (3), while F-source plasmas (1, 2) exhibit strong loading effects. The absence of a significant loading effect and controllable anisotropy makes Cl-source plasmas most attractive for silicon etching.

A most interesting demonstration of profile control are results obtained with ClF_3/Cl_2 mixtures (1). At a fixed pressure of 0.02 Torr and 100 W of 13.6 MHz RF power, mixtures of varying composition were used to generate a continuous spectrum of profiles with varying anisotropy. Figure 15.15 shows the degree of isotropy (d_H/d_V, ratio of undercutting to vertical etching) as the ClF_3 concentration is varied. In chlorine-rich plasmas (%$ClF_3 < 10$) etching by Cl atoms is highly anisotropic ($d_H/d_V \rightarrow 0$), while in fluorine-rich plasmas d_H/d_V approaches unity because of isotropic chemical etching by F atoms. Thus, in this system, as well as others, (3, 65), arbitrary profile control is possible while maintaining high etch selectivity.

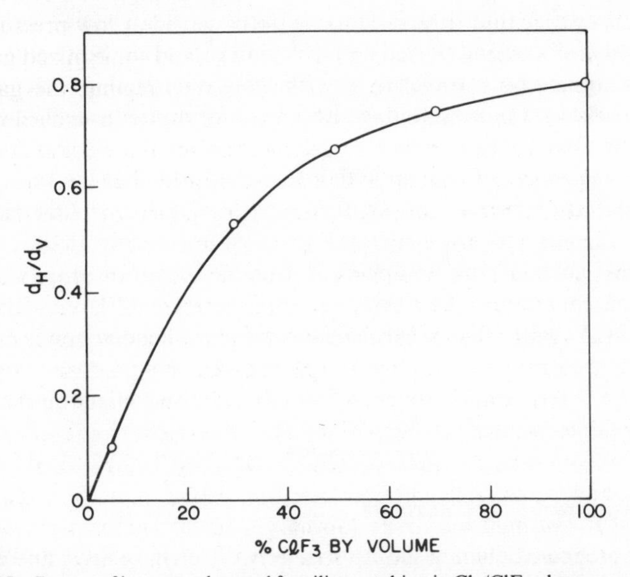

Figure 15.15 Degree of isotropy observed for silicon etching in Cl_2/ClF_3 plasma as a function of ClF_3 content of the feed. d_H/d_V is the ratio of horizontal (undercutting) to vertical etching in a parallel plate reactor at 100 W, 5 sccm total flow, and 0.02 Torr with six wafers in the reactor (after ref. 1)

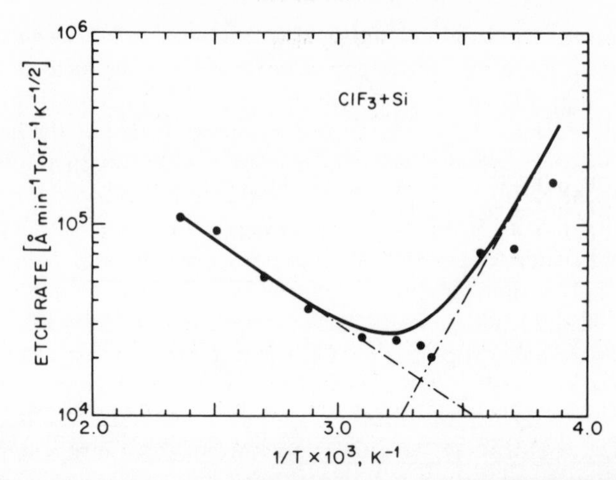

Figure 15.16 Arrhenius plot of the etch rate of silicon normalized to the flux of ClF_3 at 4.7 Torr. The smooth curve is the sum of asymptotic extrema (dashed lines). The apparent *negative* activation energy at lower temperatures suggests a physisorbed precursor to etching (after ref. 33)

In closing this discussion of silicon etching we note that interhalogen gases, such as ClF_3, BrF_3 and IF_5, achieve high selectivity over dielectrics like SiO_2 and silicon nitride (both LPCVD and plasma deposited). These gases spontaneously etch silicon *in the absence* of a plasma environment at rates approaching microns per minute (33). Although etching is isotropic, Si/SiO_2 selectivity is greater than 100:1, which is particularly attractive for damage-free removal of silicon and polysilicon over thin gate oxide. Since etching in F-containing plasmas is isotropic anyway, plasmaless interhalogen etching offers a much simpler and more selective alternative for noncritical silicon etching traditionally done in a plasma.

A surprising aspect of etching with interhalogens is that the rate becomes more rapid as the substrate temperature is reduced. That is, in contrast to the positive Arrhenius activation energies observed in virtually all other etching studies, these gases exhibit an apparent *negative* activation energy near and below room temperature. This behaviour is illustrated in Figure 15.16 for the reaction of ClF_3 with silicon. Mechanistically, this phenomenon is explained by a physisorbed precursor to chemisorption and etching. Lower temperature increases the coverage and residence time of the active etchant on the surface, enhancing the etch rate.

Silicon dioxide and silicon nitride

The earliest reported etchant gases for *selectively* etching SiO_2 and Si_3N_4 in the presence of silicon were CF_4/H_2, CF_3H, C_3F_8, and C_2F_6 (67). Selectivities as high as 15:1 (Si_3N_4/Si) and 10:1 (SiO_2/Si) were achieved as summarized in Table 15.3. These plasmas tend to be fluorine-deficient and produce CO, CO_2, COF_2, N_2, and SiF_4 as etching products (68). Fluorocarbon radicals were

Table 15.3 Etchants used for SiO_2 and Si_3N_4 films

Feed gas	Active etchant	Etch rate (Å/min)	Selectivity ($Si_3N_4/SiO_2/Si$)	Anisotropy
C_3F_8	CF_x or F + ions	6000(Si_3N_4) 2000(SiO_2)	15/5/1	—
C_2F_6	CF_x or F + ions	1200(Si_3N_4) 900(SiO_2)	8/6/1	Yes
C_2F_6/C_2H_4	CF_x or F + ions	800(SiO_2)	–/20/1	Yes
CF_3H	CF_x or F + ions	200(Si_3N_4) 300–600(SiO_2)	7/10–100/1	Yes
CF_4/H_2	CF_x or F + ions	500(SiO_2)	–/35/1	Yes

once thought to be the active etchant in these systems, and recent experiments, using photolytically generated CF_3 (69) and CF_2 radicals (70, 71) to etch oxide, support the notion that these radicals can gasify SiO_2 and Si_3N_4. However, the difficulties and slow rates observed suggest that spontaneous reactions of these species are unimportant in the overall process.

Under most conditions, oxide and nitride etching (32) takes place by energy-driven ion enhanced reaction. Accordingly these anisotropic processes show little or no loading effect (72), and are insensitive to substrate temperature (73). Electrode potential, on the other hand, has a large influence on etch rates and selectivity (74). In unsaturate-forming 'fluorine-deficient' plasmas, thin (~ 30 Å) fluorocarbon films are formed at the oxide interface and persist during etching (75, 76). Ion damage produces dangling bonds and radical groups at the fluorocarbon-substrate interface where reactions between the CF_x-film and the oxygen in the SiO_2 lattice forms volatile products like CO, CO_2, COF_2, and SiF_4. For example, the following reactions have been proposed (75):

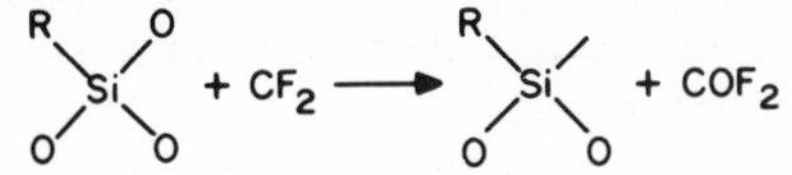

$$\mathrm{(R)(O)Si(O)(O)} + CF_2 \longrightarrow \mathrm{(R)(\cdot)Si(O)(O)} + COF_2 \qquad (15.30)$$

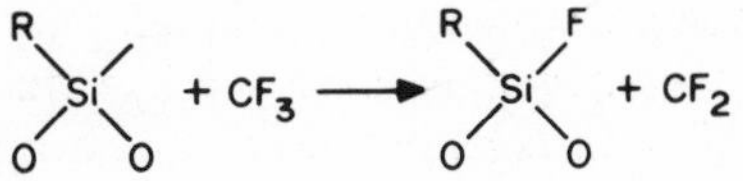

$$\mathrm{(R)(\cdot)Si(O)(O)} + CF_3 \longrightarrow \mathrm{(R)(F)Si(O)(O)} + CF_2 \qquad (15.31)$$

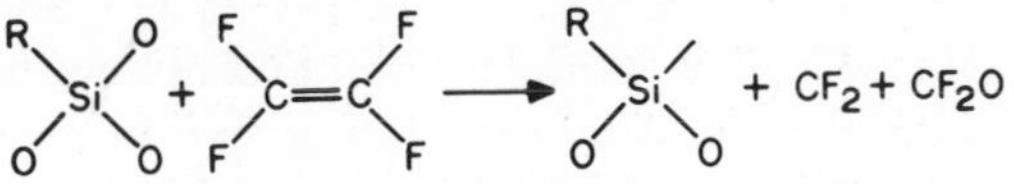

$$\mathrm{(R)(O)Si(O)(O)} + F_2C{=}CF_2 \longrightarrow \mathrm{(R)(\cdot)Si(O)(O)} + CF_2 + CF_2O \qquad (15.32)$$

$$\mathrm{(R)(O{-}SiF_3)Si(O)(O)} + CF_3 \longrightarrow \begin{cases} SiF_4 + COF_2 + \mathrm{(R)(\cdot)Si(O)(O)} \\ SiF_3 + COF_2 + \mathrm{(R)(F)Si(O)(O)} \\ SiF_4 + COF + \mathrm{(R)(F)Si(O)(O)} \end{cases} \qquad (15.33)$$

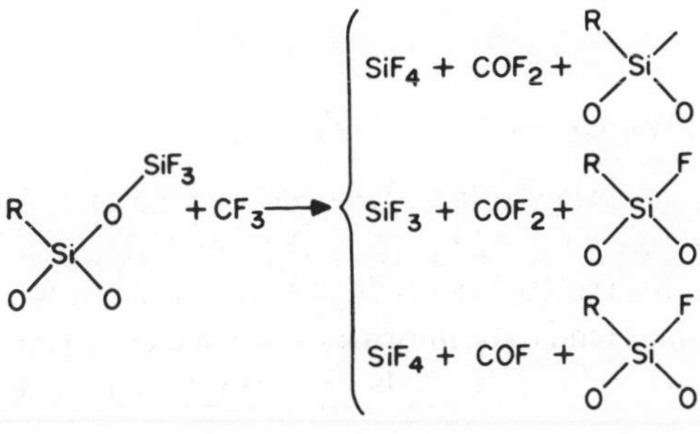

where R denotes F, O, or bound fluorocarbon. Reaction (15.32) can proceed with reactants other than C_2F_4, and C_xF_y radicals can participate in reactions similar to reaction (15.33).

Studies show that these fluorocarbon films also form on exposed Si areas during oxide etching (15, 32). There is no route to gaseous products from interactions between the silicon surface and the carbonaceous film. Thus this layer blocks attack from gas phase species and improves selectivity. Commonly, hydrogen-containing additives are combined with CF_4 to produce a F-deficient chemical environment for selective oxide and nitride etching. These include H_2, C_2H_4 and CH_4 which are efficient F scavengers. In addition, H atoms selectively convert CF_3 to CF_2 (16) which may increase the concentration of CF_2 as a film-former in the above mechanism. The amount of additive necessary remains more an art than a science because high oxide and nitride selectivity require operation close to the point at which the gas becomes strongly polymerizing.

Another way to etch silicon oxide and nitride is to use fluorine atoms. Of course, F atoms etch silicon much faster than silicon oxide or nitride, so F-source plasmas are not used to remove these dielectrics over silicon films. However they can be used to pattern or strip oxide and nitride over III–V semiconductors. When ion energy is low, there is slow chemical etching and the isotropic etch rate (Å/min) of SiO_2 by F has been determined as (10)

$$R_{F(SiO_2)} = 6.14 \times 10^{-13} T^{1/2} n_F \exp(-0.163\,eV/kT) \qquad (15.34)$$

which is in good agreement with that found in high pressure CF_4/O_2 plasmas. Under plasma conditions favouring high ion energy (low pressure or low excitation frequency) there is rapid anisotropic etching of silicon oxide by F atoms.

Adding either O_2 or H_2 to CF_4 or C_2F_6 plasmas increases the etch rate under low-frequency, high-pressure or high-pressure, high-frequency conditions. However the chemistry of these two mixtures is drastically different. As shown by the optical emission spectra in Figure 15.17, oxygen additions increase fluorine atom concentration in both the empty and Si-loaded reactors and are accompanied by a corresponding increase in SiO_2 etch rate (76). The reduced emission in the loaded reactor arises from consumption of F atoms by the presence of a large quantity of more reactive silicon. In this regime F atoms are the dominant etchant and CF_2 is consumed through reactions, such as

$$O + CF_2 \rightarrow COF + F \qquad (15.35)$$

and

$$O + CF_2 + M \rightarrow COF_2, \qquad (15.36)$$

and Si/SiO_2 selectivity is low, much as noted earlier in the discussion of silicon etching.

Hydrogen additions to a low frequency CF_4 plasma ordinarily remove free

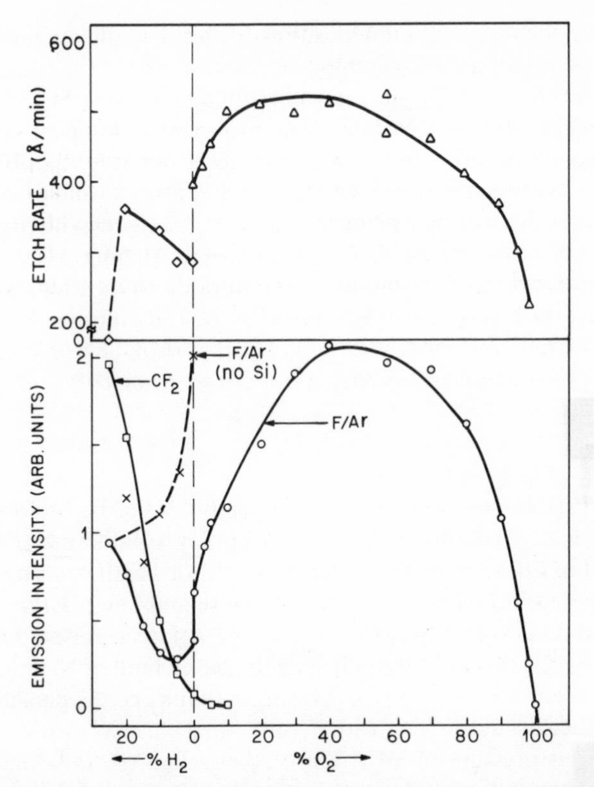

Figure 15.17 SiO_2 etch rate (top) and plasma-induced emission intensities (bottom) as a function of H_2 and O_2 additions to a CF_4 plasma. Data connected by solid lines refers to a parallel-plate reactor with lower electrode covered (i.e., loaded) with extra silicon. For comparison, data connected by the dashed line is for the same reactor without the additional silicon (after ref. 76)

fluorine by the rapid reaction

$$F + H_2 \rightarrow HF + H \tag{15.37}$$

and increase the level of the film-forming precursor CF_2 via reactions such as

$$CF_3 + H \rightarrow CF_2 + HF \tag{15.38}$$

In this case rapid etching of SiO_2 is selective over silicon and proceeds by film mechanisms discussed above. If there are substantial amounts of silicon in the reactor, formation of the protective film on silicon reduces the loading effect, so the F-atom density can actually increase (Figure 15.17) with hydrogen additions. For the conditions shown in a loaded reactor, etching of both Si and SiO_2 ceases at 25% H_2, because at this point the plasma becomes strongly polymerizing and thick films form on all surfaces in the reactor.

Silicon nitride etching characteristics vary with the deposition conditions used to make this film. The stoichiometric compound Si_3N_4 grown by high-temperature chemical vapour deposition exhibits F-atom etching rates about eight times slower than silicon and about a factor of five faster than SiO_2 in the 30–100° C range (77). Plasma-deposited silicon nitride, a SiN:H compound, tends to exhibit higher etch rates owing to incorporation of several atom-percent hydrogen and a lower degree of Si-N bonding. SiN:H with small amounts of hydrogen may exhibit etch rates as low as stoichiometric nitride (77), while with increasing hydrogen content etch rates can be comparable to, or *greater* (78) than, silicon. In CF_4/O_2 plasmas, etching is less selective for silicon nitride over SiO_2 (79) than in unsaturated oxygen-free plasmas. This suggests that the Si_3N_4 surface is oxidized as it etches, thus tending to equalize the rates. On the other hand, additions of C_2H_4 to C_2F_6 plasmas seem to affect the etching characteristics of nitride more than oxide (80).

Silicides

CF_4/O_2 mixtures have been used to etch the silicides of Ti, Ta, Mo and W (81) producing highly volatile SiF_4 and moderately volatile metal fluorides as products. The isotropic profile observed implies a significant rate of chemical etching by F atoms. However, at low pressure and low plasma frequency—conditions favouring ion bombardment of the surface—anisotropic etching of WSi_2 in CF_4/O_2 (82) and partially anisotropic etching of $MoSi_2$ in NF_3 (83) have been reported. Other work (84) shows that silicon is depleted from the silicide surface during etching with F-containing plasmas owing in part to the higher vapour pressure of SiF_4. This suggests that the rate-limiting step in silicide etching is formation and removal of the metal fluoride and that anisotropy develops from ion-enhanced desorption.

One would expect that the silicides would exhibit etching characteristics similar to silicon itself. Thus, poor selectivity is usually the case when etching a silicide in the presence of silicon or polysilicon, while selectivity over SiO_2 is in the range 5 to 20 (MSi_2/SiO_2). Much of this is because of the metal, as noted above. Molybdenum and tungsten form volatile fluorides, suggesting volatility may not be as important in these silicides as forming the hexavalent metal halide. Lower halides tend to be considerably less volatile because of interaction with neighbouring fluorides to form a network as a mode for achieving the highest degree of coordination.

Metals

Because of the high reactivity of most metals with oxygen and water vapour, plasma etching of metals often requires more attention to reactor design and process details than is necessary with other materials. Unless the metal forms a volatile oxychloride or an unstable oxide, water vapour and oxygen must be excluded from, or scavenged in the plasma reactor. In addition, the metal–oxygen bond may be extremely strong, thereby requiring ion bombardment to assist native oxide removal.

Table 15.4 Plasma etchants for metal films

Metal	Gases	Anisotropy	Selectivity
Al	CCl_4 with BCl_3	High and low ion energies	High over Al_2O_3
	Cl_2 with BCl_3, $SiCl_4$	Isotropic without surface inhibitors	High over Al_2O_3
Au	Cl_2, $C_2Cl_2F_4$, CF_3Cl	—	—
Cr	Cl_2, CCl_4 with O_2	Low ion energies	—
Mo, Nb, Ta	CF_4/O_2, SF_6/O_2	High ion energies	High over SiO_2
Ti, W	NF_3	High ion energies	High over SiO_2
Cu	Cl_2, Br_2	Likely	—

As a result of ion bombardment effects and their ability to reduce native oxides chemically, chlorocarbon or fluorocarbon gases, rather than pure halogens, are typically used to etch metal films (Table 15.4). Here, the effects are threefold. First, ion bombardment tends to damage the inhibiting oxide layer, making unoxidized metal more available to the active etchant. In addition, the chloro- and fluorocarbon radicals are probably more reactive with the damaged oxide film, providing a removal pathway for oxygen. Finally, these radicals are also effective in removing oxygen and moisture by homogeneous processes, thus minimizing oxide build-up during etching. The latter two reactions are akin to those noted earlier for oxygen additions to CF_4 plasmas. Unfortunately, halocarbon vapours are particularly susceptible to polymerization, causing residue formation that can interfere with etch processes (85).

Aluminium. Since aluminium is widely used as an interconnect layer for integrated circuits, its plasma etch characteristics have been extensively studied. As with other metals, aluminium etching is dominated by product volatility limitations. In contrast to the silicon halides that show a normal increase in boiling point with increased mass of halogen, aluminium exhibits almost the opposite trend. Most notably, the fluoride is almost refractory with exceptionally high boiling point, while the other trihalides exhibit little dependence on the mass of the halogen. The effect is primarily because of a decrease in coordination number of the metal with increasing size of the halogen (86) favouring the formation of dimers showing a 'normal' boiling-point progression with *lower* boiling points. Although Al_2F_6 is non-existent, the exceptionally high boiling point probably suggests a more extensive oligomer. Experimental evidence of such complexity has been obtained for aluminium and other metals using modulated-beam mass spectrometry to study the etch products (87). For aluminium, Al_2Cl_6 was the exclusive product near room temperature, but $AlCl_3$ is dominant at higher temperatures ($> 200°$ C). Furthermore, the change in product distribution provides strong evidence for the importance of surface diffusion and surface reactions in the overall etching mechanism.

From the above considerations, chlorine-containing gases have been principally used to etch aluminium. An induction period exists at the start of aluminium etching. This 'slow etching' period is observed because two processes must occur before aluminium etching can start (23): oxygen and water vapour must be scavenged or removed from the reactor, and the ever present thin (~ 30 Å) native aluminium oxide layer must be etched first. The former contaminants generally come from the large reservoir of moisture and oxygen adsorbed on the heavily anodized aluminium used in construction of electrodes and other reactor surfaces (88). This can be minimized by using a 'load lock' so that the chamber is not exposed to air (water vapour) between etch runs and by using an etch gas additive (BCl_3, or $SiCl_4$) that effectively scavenges water and oxygen.

Native aluminium oxide can be etched by ion bombardment of the surface to sputter the layer and by supplying chemical species capable of reacting with the oxide. Oxide-clearing radicals include CCl_x, BCl_x or $SiCl_x$. Etching of the thin native oxide layer is about 100 times slower than aluminium in a BCl_3 discharge and about 25 times slower in a CCl_4 discharge under similar conditions (22). BCl_3, however, is more effective (23) in scavenging moisture and oxygen. Results with $SiCl_4$ as an etchant gas suggest that it behaves similarly to BCl_3 (89) in scavenging ability.

Pure chlorine gas has not been successful in reproducibly etching aluminium, primarily because of its inability to etch the native oxide. After removal of the native aluminium oxide layer, molecular chlorine (Cl_2) can etch pure, clean aluminium without a plasma (90, 64). Indeed Cl_2, rather than Cl atoms, may be the primary etchant species for aluminium in a glow discharge (85).

These observations have been confirmed in other experiments. Etching of aluminium in high vacuum using a directed beam from a Cl_2-plasma discharge (91) exhibited rates that were unaffected by (i) changing the impinging ion energy, (ii) changing the degree of dissociation of Cl_2, or (iii) extinguishing the discharge. In a more quantitative study (92) where chlorine atoms were generated from Cl_2 upstream from a reaction cell and quantitated by titration with nitrosyl chloride, molecular Cl_2 was found to be four times as reactive as atomic chlorine. The observed temperature dependence suggested the rate was not limited by product desorption above 50° C. These results identify molecular chlorine as the principal etchant in aluminium etching; however, the rate-limiting step may be associated with product desorption, at least at lower temperatures. In room temperature experiments using a directed flux of Cl_2 and a Kaufman-type ion source, aluminium etching has been shown to be limited by reactant flux at low pressure (< 1 mTorr) and product desorption at pressures (> 10 m Torr) normally used in processing (93).

Copper additions to aluminium films are commonly used to increase resistance to electromigration. Unfortunately, copper chlorides and other copper halides are relatively involatile, thereby making removal during aluminium plasma etching difficult. Two methods can be used to promote copper chloride desorption: increasing the substrate temperature (consistent with the resist material being used), or increasing the ion bombardment so that significant sputtering and/or surface heating is attained.

Following etching, aluminium films often corrode on exposure to atmospheric conditions owing to the hydrolysis of chlorine or chlorine-containing residues remaining on the film sidewalls, on the substrate, or in the photoresist. Since the passivating native oxide film normally present on the aluminium surface has been removed during etching, chlorine species are left in contact with aluminium, eventually causing corrosion. Furthermore, carbon contamination and radiation damage caused by particle bombardment may enhance corrosion susceptibility (94). A water rinse or an oxygen plasma treatment after etching lowers the amount of chlorine left on the etched surfaces, but is not adequate to preclude corrosion. Low temperature thermal oxidations in dry oxygen appear effective in restoring a passivating native aluminium oxide film (94). Another method of preventing post-etch corrosion is to expose the aluminium film to a fluorocarbon plasma (95). This treatment converts the chloride residues into nonhygroscopic fluorides. Later, a nitric acid rinse can be used to remove the fluoride layer and to regrow the protective oxide.

Chlorine-based plasma etching of aluminium films causes serious degradation of photoresist materials. To some extent, these effects are a result of the etch product, $AlCl_3$. Aluminium trichloride is a Lewis acid used extensively as a Friedel-Crafts catalyst. Therefore, it is hardly surprising that this material reacts with and severely degrades photoresists (85).

Other metals. Numerous other metal films have been etched in a glow discharge. The following paragraphs present specific information on the etching of the metal films shown in Table 15.4.

Gold can be etched effectively with $C_2Cl_2F_4$ (96) or with $CClF_3$ (97), while exposure to CF_4/O_2 plasmas causes 'staining'. The observed staining is believed to be gold oxides, whose formation is enhanced by the presence of atomic fluorine (97).

Chromium is etched readily in plasmas containing chlorine and oxygen (98) because of the high volatility of the oxychloride (CrO_2Cl_2). Indeed, the high boiling point of $CrCl_2$ (1300° C) results in significantly reduced etch rates of chromium in chlorine plasmas without oxygen.

Titanium can be etched in fluorine-, chlorine-, or bromine-containing gases, because all the halides are volatile. Chlorides and bromides have been studied to a great extent since they result in high selectivity over silicon-containing films, and do not promote staining on gold (97).

Tungsten films can be etched easily in fluorine-containing plasma (99, 100). If carbon or sulphur is present in the gas atmosphere, oxygen is typically added to prevent polymer and residue formation and to increase the concentration of fluorine atoms.

Some preliminary studies of copper etching show that Cl_2 (87,101) and Br_2 (101) in the presence of ion bombardment and elevated temperatures are effective. As with aluminium, low product volatility appears to limit etching and the principal product has been identified (87) as the cuprous chloride trimer, Cu_3Cl_3.

III–V materials

Group III–V semiconductor compounds such as GaAs, GaP and InP form the basis for many new technologies, particularly optoelectronic devices. The development of plasma pattern delineation methods is an area of active research because of the parallelism with the more highly developed silicon technology. However these systems can be considerably more complicated since many applications involve binary, ternary and even quaternary alloys. Depending on the specific device and process step, etchants must be capable of selectivity between closely related substrate stoichiometries, or conversely there must be little or no etch rate differentiation between alloys. For example, in forming laser structures in $GaAs/Al_xGa_{1-x}As$ it is desirable to etch through both materials anisotropically at the same rate to form a smooth face. On the other hand, fabrication of hetero-structure transistors form these same materials requires high selectivity. These considerations demand a careful processing strategy.

As with metals, product volatility is a significant constraint in much of the strategy. The III–V compounds are particularly difficult since the Group V elements form volatile halides, whereas Group III halides, particularly the fluorides, tend to be involatile. As a result, F-source plasmas, which have been the mainstay of silicon technology, are not useful for III–V etching. Therefore, most studies have used chlorine-containing plasmas with elevated substrate temperatures to take advantage of the volatility (albeit limited) of the Group III chlorides.

The importance of the volatility concept in III–V etching relative to the reaction concepts that dominate Si etching has been shown in the etching of In P and GaAs in Cl_2 plasmas (102). Activation energies obtained by measuring etch rates as a function substrate temperature were determined to be 144 kJ/mole for InP and 44 for GaAs. These are in excellent agreement with the heats of vaporization of $InCl_3$ and $GaCl_3$, 155 kJ/mole and 50 kJ/mole, respectively, suggesting that vaporization of the Group III halide was rate-limiting. In fact, the etch rate of InP could be explained entirely on the evaporation rate of $InCl_3$. Unfortunately, this was not possible for GaAs etching, suggesting that a surface reaction

$$2GaCl_{3(ads)} \rightarrow Ga_2Cl_{6(ads)} \qquad (15.39)$$

forming the volatile dimer as a product was limiting. Confirmation of the role of surface diffusion and reaction has recently been obtained for aluminium etching which exhibits a similar enhanced volatility of the dimer over the monomer (see p. 379). Thus, the agreement between activation energy for etching and heat of vaporization was fortuitous for GaAs; however, the study does highlight the importance of product volatility in etching binary and more complex alloys.

The active etchant for III–V materials is usually atomic chlorine, and several investigations (103, 104) into appropriate sources have been conducted. An extensive study (103) has been conducted for the etching of GaAs and its oxide. Except for CCl_2F_2, fluoro- and chlorofluorocarbon plasmas were ineffective in etching GaAs, most likely because of reduced volatility caused by fluorine passivation. Chlorocarbon plasmas such as C_2Cl_4, CBr_2Cl_2 and $CHCl_3$ were also ineffective because of polymer deposition; however, CCl_4, CCl_2F_2, PCl_3 and HCl plasmas etched both GaAs and its oxide, favouring GaAs in a ratio of 400:1. Cl_2 and $COCl_2$ plasmas were found to etch GaAs exclusively with high etch rates ($5\,\mu m\,min^{-1}$). Here the results can be interpreted in terms of chemical selectivities, since gas pressure and plasma frequency were high enough to minimize ion bombardment effects.

Other studies have been conducted with Cl_2 and chloro-halocarbon gases under conditions favourable to ion-enhanced etching—very low pressure or RIBE. Anisotropic (105, 104) and directional (oblique angles of incidence of ions) etch profiles (106, 107) were observed in accord with the damage/inhibitor concepts noted earlier. More recently, $SiCl_4$ (108) has been used in the low-pressure etching of GaAs and InP to minimize polymer deposition associated with halocarbons, and reduce erosion of mask materials. In all cases, etching rates and anisotropy increased with RF power and bias voltage, and anisotropy increased with decreasing pressure.

The role of additives has not been studied to any great extent; however, oxygen additions to CCl_2F_2 plasmas (103, 106) enhanced GaAs etching and improved selectivity over native oxide, while H_2 additions (103) enhanced oxide etching. There was a strong correlation between atomic chlorine emission, added O_2 and the etching rates of GaAs, InP, and GaP in Cl_2 and CCl_4 plasmas. This invites interpretations (109) similar to those advanced in the CF_4/O_2 etching of silicon. It would not be surprising if H_2 additions to chlorocarbon plasmas in III–V etching have similar effects to those presented earlier for the selective etching of SiO_2 in fluorine-containing plasmas.

Argon additions have been shown to increase etch rates (110) under RIE conditions, while both argon and oxygen (109) tend to increase anisotropy as well. With argon, it appears that both observations are consistent with ion-enhanced clearing of lower vapour pressure Group III chloride or halocarbon polymer. Oxygen, on the other hand, is only effective in small additions, suggesting an ion-enhanced oxidation of inhibiting surface halocarbon polymers. Larger additions result in an increase in the chemical etchant, Cl, along with homogeneous oxidation of chlorocarbon radicals that are precursors to film formation.

Etching of the ternary alloy $Al_xGa_{1-x}As$ presents new problems associated with the low volatility of $AlCl_3$ and the facility of aluminium oxide formation. In achieving equi-rate RIBE (reactive ion beam from a Kaufman ion source) etching of GaAs and $Al_xGa_{1-x}As$ in Cl_2 (111), moisture levels had to be kept extremely low to prevent formation of aluminium oxide—similar to etching aluminium itself. Recently, this problem has been turned into a useful tactic in achieving high selectivity. Deliberate additions of water vapour have been found to be an effective stop-etch, permitting total selectivity in etching GaAs over the ternary aluminium alloy (112).

At present Cl_2 appears to be most suitable for etching at moderate pressures, since etch rate and anisotropy can be controlled by plasma frequency (113); however, more recent studies (102*b*, *c*) suggest that Br_2 plasmas may be superior in this aspect. One interesting feature of using Cl_2 and Br_2 is the observation of anisotropic etching under nominally isotropic plasma conditions (high pressure, high frequency, low power). Unlike the mechanisms for anisotropy discussed earlier (see p. 347), anisotropy in this instance was attributable to chemical attack along specific crystallographic planes containing a higher lattice content of the more volatile Group V element. Smooth vertical walls were observed; however, it was clear from the overhanging mask that etching was chemical.

Oxygen plasma etching of resists

Oxygen plasma etching of resists has been carried out using both isotropic 'chemical' etching and anisotropic ion-assisted processes. Purely chemical reactions with oxygen atoms are widely used for resist stripping, and isotropic chemical treatment is also used to passivate metal and silicon-bearing resists for bilevel lithography. Another application is the removal of epoxy smears from holes in fine pattern multilayer printed circuit boards (114).

At moderate pressure and high frequency, a discharge in pure oxygen produces oxygen atoms which attack organic materials: CO, CO_2 and H_2O are the eventual end products (115). The predominant degradation mechanism appears to be random chain scission (116). Stripping rates in pure oxygen plasmas are proportional to oxygen atom concentration (117, 118). Inert gas additions can help stabilize O_2 plasmas. Chemical attack of resist in oxygen plasmas is heavily influenced by the structure and substitutional groups on the polymer, and by physical variables such as temperature. Structural variables are treated in Chapter 16. The temperature dependence of attack in an oxygen plasma may also depend on the resist's glass transition temperature (T_g). For example, the apparent activation energy of poly(methyl methacrylate) (PMMA) etching in O_2 plasmas increased sharply from 18 kJ/mole below the glass transition temperature T_g (60–90° C), to ~40kJ/mole above T_g (119). Activation energies between 20–60kJ/mole have been measured for resist oxidation under chemical attack by O atoms with O_2 present), depending on the particulars of the resist and other physical conditions.

Stripping rates can be greatly enhanced by fluorine additions to the plasma. Adding even a few per cent of C_2F_6 to an O_2 plasma can bring about a large increase in the removal rate for many resists. It has been show that two mechanisms are responsible for this accelerated attack (117, 120). First fluorine atoms *abstract* hydrogen from organic polymers thereby providing more reactive unsaturated or radical sites. Second, small fluorine additions increase the concentration of atomic oxygen in the plasma. Compared to a normal saturated polymer chain, radical sites left by hydrogen abstraction are highly reactive and easily combine with oxygen. Ensuing reactions can lead to chain scission, formation of carbonyl groups and eventually, volatile oxidation products. A probable reaction sequence is

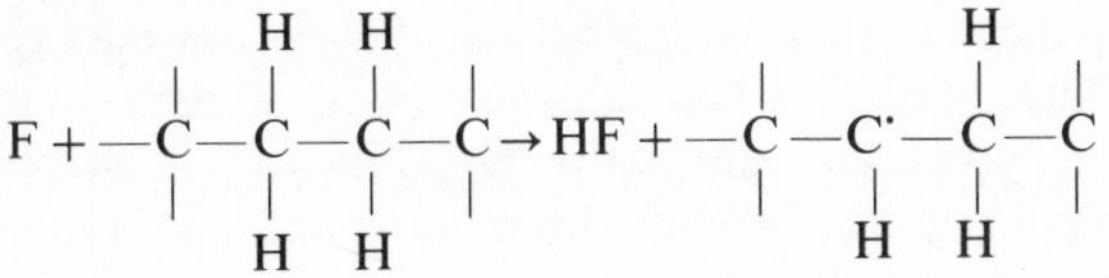

$$\begin{array}{ccccccccccccccccccc}
 & & & & \mathrm{H} & & \mathrm{H} & & & & & & & & & & \mathrm{H} & & \\
 & & \mid & & \mid & & \mid & & \mid & & & & \mid & & & & \mid & & \mid \\
\mathrm{F}+ & - & \mathrm{C} & - & \mathrm{C} & - & \mathrm{C} & - & \mathrm{C} & \rightarrow & \mathrm{HF}+ & - & \mathrm{C} & - & \mathrm{C}^{\cdot} & - & \mathrm{C} & - & \mathrm{C} \\
 & & \mid & & \mid & & \mid & & \mid & & & & \mid & & \mid & & \mid & & \mid \\
 & & & & \mathrm{H} & & \mathrm{H} & & & & & & & & \mathrm{H} & & \mathrm{H} & &
\end{array} \qquad (15.40)$$

followed by

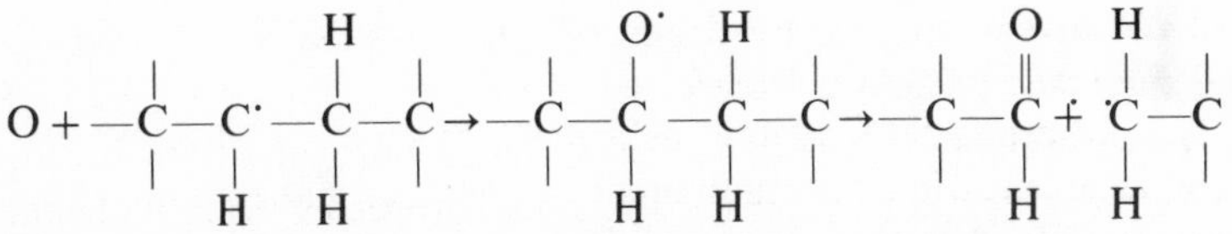

$$\begin{array}{ccccccccccccccccccccccccc}
 & & & & & & \mathrm{H} & & & & & & \mathrm{O}^{\cdot} & & \mathrm{H} & & & & & & \mathrm{O} & & \mathrm{H} & & \\
 & & \mid & & & & \mid & & \mid & & & & \mid & & \mid & & \mid & & \mid & & \| & & \mid & & \mid \\
\mathrm{O}+ & - & \mathrm{C} & - & \mathrm{C}^{\cdot} & - & \mathrm{C} & - & \mathrm{C} & \rightarrow & - & \mathrm{C} & - \mathrm{C} & - & \mathrm{C} & - & \mathrm{C} & \rightarrow - & \mathrm{C} & - & \mathrm{C}+ & & {}^{\cdot}\mathrm{C} & - & \mathrm{C} \\
 & & \mid & & \mid & & \mid & & \mid & & & \mid & \mid & & \mid & & \mid & & \mid & & \mid & & \mid & & \mid \\
 & & & & \mathrm{H} & & \mathrm{H} & & & & & & \mathrm{H} & & \mathrm{H} & & & & & & \mathrm{H} & & \mathrm{H} & &
\end{array} \qquad (15.41)$$

The exothermicity of hydrogen abstraction by fluorine can itself be enough to cleave carbon–carbon bonds (121) and, it has been suggested (122), could be a primary degradation mechanism in some plasmas where there is a high concentration of fluorine (e.g. CF_4/O_2 or SF_6/O_2 where $\gtrsim 50\%$ SF_6 in the feed).

Although unsaturated sites on the polymer chain should be reactive towards O atoms, it has been recently proposed that F atoms also assist the degradation of these sites forming radicals by addition reactions:

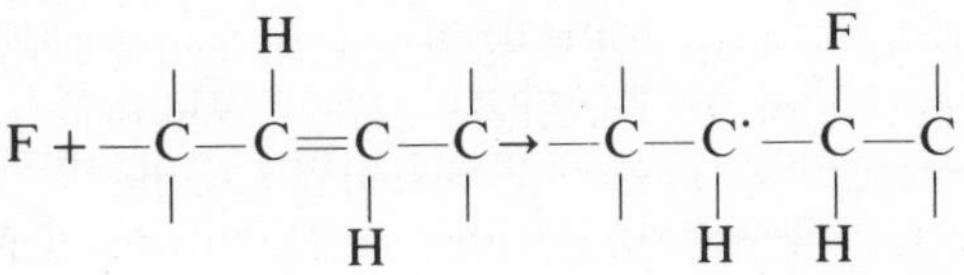

$$\begin{array}{cccccccccccccccc}
 & & & & \mathrm{H} & & & & & & & & & & \mathrm{F} & & \\
 & & \mid & & \mid & & & & \mid & & \mid & & \mid & & \mid & & \mid \\
\mathrm{F}+ & - & \mathrm{C} & - & \mathrm{C} & = & \mathrm{C} & - & \mathrm{C} & \rightarrow - & \mathrm{C} & - & \mathrm{C}^{\cdot} & - & \mathrm{C} & - & \mathrm{C} \\
 & & \mid & & & & \mid & & \mid & & \mid & & \mid & & \mid & & \mid \\
 & & & & & & \mathrm{H} & & & & & & \mathrm{H} & & \mathrm{H} & &
\end{array} \qquad (15.42)$$

which are then oxidized as before. In the etching of polyimide by SF_6/O_2, the shift from an oxygen-dominant etch mechanism to fluorine-dominant degradation leads to two maxima in the etch rate, one at low oxygen additions (at ~ 20% O_2) and one for large additions (at ~ 60% O_2) (122).

Even small fluorocarbon additions (1–8%) to an oxygen plasma can produce a dramatic increase in the concentration of atomic oxygen (123, 124) which is reflected in optical emission (17, 19). The presence of F_2 is known to lower the rate of heterogeneous Cl recombination (125). Since wall recombination of O atoms to reform O_2 is an important loss channel for O atoms, we may speculate that the increased O concentration is also because of a lower wall recombination coefficient from adsorption of F. An alternative which has been proposed is that the halocarbon additions alter the electron energy distribution function (17). Both factors may be active.

These small fluorocarbon additions also tend to lower the activation energy for attack, generally to ~ 20 kJ/mole or less. The apparent activation energy for stripping AZ340B resist is lowered from about 46 kJ/mole to 20 kJ/mole using 1% CF_4 in O_2 (126). For plasma-etching epoxy printed-circuit boards in 50% CF_4 in O_2, the Arrhenius coefficient is 18.0 ± 0.8 kJ/mole.

For purely chemical etching or stripping of resist, plasma exposure may be undesirable, since impinging ions can cause electrical damage to devices, and charging from the plasma-induced potential can cause electrostatic breakdown of thin gate oxides. Recently downstream etchers have been employed for stripping as a means to circumvent these problems.

Directional, ion assisted, resist etching is also in wide use, especially for multilevel resist processing. The 'trilevel' process is the oldest and probably best known. Conventional resist is deposited and patterned over a thin SiO_2 layer, and the resist pattern is transferred to the SiO_2 using a CHF_3-based plasma. The patterned SiO_2 is used as a mask to etch a high-aspect-ratio pattern anisotropically in a *thick* polyimide or conventional resist layer in an oxygen plasma. This layer becomes a highly resistant fine-featured mask for further plasma etching steps. Ion bombardment flux stimulates the vertical etch rate of horizontal polymer surfaces by oxygen. To achieve high aspect ratios, the process is done under plasma conditions that provide ion bombardment, with low gas phase oxygen atom concentrations—at low pressure (around a few mTorr) or at higher pressure (~ 0.1 Torr) using low frequency (e.g. ~ 200 KHz – 1 MHz) and high dilution with an inert gas (usually argon). It is has been proposed that ion bombardment assists etching by driving off H and partially oxidized carbon groups in the polymer (C—O, C=O, C—O—C) as CO (127) and that sputtering of the remaining C-rich surface may be rate-limiting. While details of the ion-assisted mechanism are uncertain, it is clear that the gas-phase concentration of O must be low in relation to the flux of ions to prevent undercutting by purely chemical reaction.

Recently, bilevel lithography schemes using silicon or metal-containing organic resists as the first layer have been studied as a means to simplify making the thick mask. The thin metal-containing resist is deposited directly on the thick polymer layer rather than on an intervening layer of SiO_2. The basic idea is for the oxygen plasma to from a protective oxide skin on the metal- (silicon)-containing mask layer that will slow or stop further erosion of the mask. Problems with developing this scheme are that passivation causes shrinkage and linewidth loss in the metal-containing resist, while the ion bombardment needed for anisotropic etching of the underlying thick polymer level erodes the mask passivation. Increasing the concentration of oxygen atoms favours growth of the passivation layer, but oxygen atoms also undercut the thick polymer level by chemical etching.

References

1. D. L. Flamm, D. N. K. Wang and D. Maydan, *J. Electrochem. Soc.* **129** (1982) 2755.
2. C. J. Mogab, *J. Electrochem. Soc.* **124** (1977) 1262.
3. C. J. Mogab and H. J. Levenstein, *J. Vac. Sci. Technol.* **17** (1980) 721.
4. T. Tsuchimoto, *J. Vac. Sci. Technol.* **15** (1978) 70, 1730.
5. T Ono, M. Oda, C. Takahashi and S. Matsuo, *J. Vac. Sci. Technol.* **B4** (1986) 696.
6. D. L. Flamm and G. K. Herb, in *Plasma Materials Interactions*, Vol. 1, eds. D. M. Manos and D. L. Flamm, Academic Press, New York (1987) in preparation.
7. J. Musil, *Vacuum* **36** (1986) 161; M. R. Wertheimer, J. E. Klemberg-Sapiana and H. P. Schreiber, *Thin Solid Films* **15** (1984) 109; J. Asmussen, *J. Vac. Sci. Technol.* **B4** (1986) 295; S. R. Goode and K. W. Baughman, *Appl. Spectrosc.* **38** (1984) 755.
8. D. L. Flamm, *J. Vac. Sci. Technol.* **A4** (1986) 729.
9. M. R. Wertheimer and M. Moisan, *J. Vac. Sci. Technol.* **A3** (1985) 2643.
10. D. L. Flamm, V. M. Donnelly and J. A. Mucha, *J. Appl. Phys.* **52** (1981) 3633.
11. G. Lucovsky, *J. Vac. Sci. Technol.* **A4** (1986) 480.
12. V. M. Donnelly, D. L. Flamm, W. C. Dautremont-Smith and D. J. Werder, *J. Appl. Phys.* **55** (1984) 242.
13. D. E. Ibbotson, D. L. Flamm, V. M. Donnelly and B. S. Duncan, *J. Vac. Sci. Technol.* **20** (1981) 489.
14. D. L. Flamm, V. M. Donnelly and D. E. Ibbotson, in *VLSI Electronics: Microstructure Science*, Vol. 8, eds. N. G. Einspruch and D. M. Brown, Academic Press, New York (1984) Chapter. 8.
15. D. L. Flamm, P. L. Cowen and J. A. Golovchenko, *J. Vac. Sci. Technol.* **17** (1980) 1341.
16. J. W. Coburn and E. Kay, *IBM J. Res. Develop* **23** (1979) 33.
17. D. L. Flamm, *Solid State Technol.* **22**(4) (1979) 109.
18. J. Heicklen, *Adv. Photochem.* **7** (1969) 57.
19. D. L. Flamm, *Plasma Chem. Plasma Proc.* **1** (1981) 37.
20. C. J. Mogab, A. C. Adams and D. L. Flamm, *J. Appl. Phys.* **49** (1979) 3796.
21. C. J. Mogab and H. J. Levenstein, *J. Vac. Sci. Technol.* **17** (1980) 1721.
22. D. W. Hess, *Solid State Technol.* **24**(4) (1981) 189; K. Tokunaga and D. W. Hess, *J. Electrochem. Soc.* **127** (1980) 928.
23. K. Tokunaga, F. C. Redeker, D. A. Danner and D. W. Hess, *J. Electrochem. Soc.* **128** (1981) 851.
24. K. Tokunaga and D. W. Hess, *J. Electrochem. Soc.* **127** (1980) 928.
25. L. A. D'Asaro, A. D. Butherus, J. V. DiLorenzo, D. E. Iglesias and S. H. Wample, in *Gallium Arsenide and Related Compounds 1980*, ed. H. W. Thim, Inst. of Physics Conf. Ser. **56**, 267–273.
26. B. E. Cherrington, in *Gaseous Electronics and Gas Lasers*, Pergamon, New York (1979) 156–158.
27. (*a*) J. W. Coburn and H. F. Winters, *J. Appl. Phys.* **50** (1979) 3189; (*b*) Y. Y. Tu, T. J. Chuang and H. F. Winters, *Phys. Rev. B.* **23** (1981) 823; (*c*) U. Gerlach-Meyer and J. W. Coburn, *Surface Sci.* **103** (1981) 177.
28. C. M. Melliar-Smith and C. J. Mogab, in *Thin Film Processes*, eds. J. L. Vossen and W. Kern, Academic Press, Orlando (1978) Chapter V-2, 521–525.
29. A. T. Bell, in *Techniques and Applications of Plasma Chemistry*, eds. J. R. Hollahan and A. T. Bell, John Wiley, New York (1974) Chapter 1.

30. A. M. Howatson, in *An Introduction to Gas Discharges*, Pergamon Press, New York (1965) Chapter 4.
31. W. P. Allis, and D. J. Rose, *Phys. Rev.* **93** (1964) 84.
32. D. L. Flamm, D. E. Ibbotson, J. A. Mucha and V. M. Donnelly, *Solid State Technol.* **24**(4) (1983) 117.
33. D. E. Ibbotson, J. A. Mucha, D. L. Flamm and J. M. Cook, *J Appl. Phys.* **56** (1984) 2939.
34. V. M. Donnelly, D. L. Flamm and R. H. Bruce, *J. Appl. Phys.* **58** (1985) 2135.
35. V. M. Donnelly, D. E. Ibbotson and D. L. Flamm, in *Ion Bombardment Modification of Surfaces: Fundamentals and Applications*, eds. O. Auciello and R. Kelley, Elsevier, New York (1984) Chapter 8.
36. D. L. Flamm, R. F. Baddour and E. R. Gilliland, *I & EC Fundamentals* **12** (1973) 276.
37. D. L. Flamm and V. M. Donnelly, *J. Appl. Phys.* **59** (1986) 1052.
38. R. M. Barnes and R. J. Winslow, *J. Phys. Chem.* **82** (1978) 1869.
39. G. de Rosny, E. R. Mosberg, Jr. J. R. Adelson, G. Devaud and R. C. Kern, *J. Appl. Phys.* **54** (1983) 2272.
40. H. Tagashira, K. Kitamori, M. Shimozuma and Y. Sakai, in *Proc. 7th Int. Symp. On Plasma Chemistry*, ed. C. J. Timmermans, Eindhoven Univ. Technol., Eindhoven, The Netherlands (1985) 1337.
41. T. Makabe, in *Proc. 7th Int. Symp. on Plasma Chemistry*, ed. C. J. Timmermans, Eindhoven Univ. Technol., Eindhoven, The Netherlands (1985) 1331.
42. D. L. Flamm, C. J. Mogab and E. R. Sklaver, *J. Appl. Phys.* **50** (1979) 6211.
43. J. A. Mucha, D. L. Flamm and V. M. Donnelly, J. Appl. Phys. **53** (1982) 4553.
44. T. J. Chuang, *Phys. Rev. Lett.* **42** (1979) 815.
45. T. J. Chuang, *J. Appl. Phys.* **51** (1980) 2614.
46. M. J. Vasile and F. A. Stevie, *J. Appl. Phys.* **53** (1982) 3799.
47. H. F. Winters and F. A. Houle, *J. Appl. Phys.* **54** (1983) 1218.
48. J. A. Mucha, V. M. Donnelly, D. L. Flamm and L. M. Webb, *J. Phys. Chem.* **85** (1981) 3529.
49. D. E. Ibbotson, D. L. Flamm, J. A. Mucha and V. M. Donnelly, *Appl. Phys. Lett.* **44** (1984) 1129.
50. H. F. Winters, J. W. Coburn and T. J. Chuang. *J. Vac. Sci. Technol.* **B1** (1983) 469.
51. (*a*) F. R. McFeely, J. F. Morar, N. D. Shinn, G. Landgren and F. J. Himpsel, *Phys. Rev. B.* **30** (1984) 764; (*b*) F. R. McFeely, J. F. Morar and F. J. Himpsel, *Surface Sci.* **165** (1986) 277.
52. (*a*) B. Roop, S. Joyce, J. C. Schultz, N. D. Shinn and J. I. Steinfeld, *Appl. Phys. Lett.* **46** (1985) 1187; (*b*) B. Roop, S. Joyce, J. C. Schultz and J. I. Steinfeld, *Surface Sci.* **173** (1986) 455.
53. C. D. Stinespring and A. Freedman, *Appl. Phys. Lett.* **48** (1986) 718.
54. F. P. Fehner and N. F. Mott, *J. Oxidation Met.* **2** (1970) 59.
55. H. Boyd and M. S. Tang, *Solid State Technol.* **25**(4) (1979) 133.
56. K. M. Eisele, *J. Electrochem. Soc.* **128** (1981) 123.
57. R. D'Agostino and D. L. Flamm, *J. Appl. Phys.* **52** (1981) 162.
58. E. P. G. T. van de Ven and P. A. Zijlstra, *Electrochem. Soc. Extended Abstracts*, May 1980, **80–1**, 253.
59. R. Horwath, C. B. Zarowin and R. Rosenberg, *Electrochem Soc. Extended Abstracts*, May 1980, **80–1**, 294.
60. D. Edelson and D. L. Flamm, *J. Appl. Phys.* **56** (1984) 1522.
61. (*a*) I. C. Plumb and K. R. Ryan, *Plasma Chem. Plasma Proc.* **6** (1986) 205; (*b*) K. R. Ryan and I. C. Plumb, *Plasma Chem. Plasma Proc.* **6** (1986) 231.
62. Y. H. Lee and M.-M. Chen, *J. Appl. Phys.* **54** (1983) 5966.
63. T. Arkado and Y. Horiike, *Jap. J. Appl. Phys.* **22** (1983) 799.
64. R. H. Bruce, *Solid State Technol.* **24**(10) (1981) 64; *J. Appl. Phys.* **52** (1981) 7064.
65. A. C. Adams and C. D. Capio, *J. Electrochem Soc.* **128** (1981) 366.
66. G. C. Schwartz, P. M. Schaible, *J. Vac. Sci. Technol.* **16** (1979) 410.
67. R. A. H. Heinecke, *Solid State Electron.* **18** (1975) 1146; **19** (1976) 1039.
68. J. W. Coburn and H. F. Winters, *J. Vac. Sci. Technol.* **16** (1979) 391.
69. N. Selamoglu, M. J. Rossi and D. M. Golden, *J. Chem. Phys.* **84** (1986) 2400.
70. G. L. Loper, M. D. Tabat, in *Proc. Int. Conf. on Lasers' 83*, ed. R. C. Powell STS, McLean, VA (1983) 31; *Proc SPIE* **459** (1984) 121.
71. J. H. Brannon, *J. Phys. Chem.* **90** (1986) 1784.
72. T. M. Mayer, *J. Electronic Materials* **9** (1980) 513.

73. S. Matsuo, *J. Vac. Sci. Technol.* **17** (1980) 587.
74. H. Toyoda, H. Komiya, and H. Itakura, *J. Electronic Materials* **9** (1980) 569.
75. D. L. Flamm and V. M. Donnelly, *Plasma Chem. Plasma Proc.* **1** (1981) 317.
76. D. L. Flamm, V. M. Donnelly and D. E. Ibbotson, *J. Vac. Sci. Technol.* **B1** (1983) 23.
77. E. P. G. T. van den Ven and P. A. Zilstra, *Proc. Electrochem. Soc* **81–1** (1981) 112.
78. R. L. Maddox and H. L. Parker, *Solid State Technol.* **21**(4) (1978) 107.
79. F. H. M. Sanders, J. Dieleman, H. J. B. Peters and J. A. M. Sanders *J. Electrochem. Soc.* **129** (1983) 2559.
80. J. A. Wenger, unpublished results (1981).
81. S. P. Murarka, *J. Vac. Sci. Technol.* **17** (1980) 775.
82. B. L. Crowder and S. Zirinsky, *IEEE Trans. Electron. Dev.* **ED-26**, (1979) 369.
83. T. P. Chow and A. J. Steckl, *IEEE IEDM Tech. Digest* **CH1616–2** (1981) 149.
84. G. S. Korman, T. P. Chow and D. H. Bower, *Solid State Technol.* **26**(1) (1983) 115.
85. D. W. Hess, *Plasma Chem. Plasma Processing* **2** (1982) 141.
86. B. E. Douglas and D. H. McDaniel, *Concepts and Models in Inorganic Chemistry*, Blaisdell, New York (1965) 238.
87. H. F. Winters, *J. Vac. Sci. Technol.* **B3** (1985) 9.
88. K. Donohoe, *7th Annual Tegal Plasma Seminar* (May 1981); *Proc. 5th Int. IUPAC Symp. Plasma Chem.*, Heriott-Watt Univ., Edinburgh (1981) 13–19.
89. G. K. Herb, R. A. Porter, P. D. Cruzan, J. Agraz-Guerena and B. R. Soller, *Electrochem. Soc. Extended Abstracts* **81–2** (October, 1981) 710.
90. R. G. Poulsen, H. Nentwich, S. Ingrey, *Proc. Int. Electron Devices Meeting*, Washington, DC (1976) 205.
91. D. L. Smith and R. H. Bruce, *J. Electrochem. Soc.* **129** (1982) 2045.
92. D. A. Danner and D. W. Hess, *J. Appl. Phys.* **59** (1986) 940.
93. N. N. Efremow, M. W. Geis, R. W. Mountain, G. A. Lincoln, J. N. Randall and N. P. Economou, *J. Vac. Sci. Technol.* **B4** (1986) 337.
94. W. Y. Lee, J. M. Eldridge and G. D. Schwartz, *J. Appl. Phys.* **52** (1981) 2994.
95. Y. T. Fok, *Electrochem. Soc. Extended Abstracts* **80–1** (May 1980) 301.
96. W. H. Legat and H. Schilling, *Electrochem. Soc. Extended Abstracts* **75–2** (October 1975) 336.
97. C. J. Mogab and T. A. Shankoff, *J. Electrochem. Soc.* **124** (1977) 1766.
98. H. Nakata, K. Nishioka and H. Abe, *J. Vac. Sci. Technol.* **17** (1980) 1351.
99. R. G. Poulsen, *J. Vac. Sci. Technol.* **14** (1977) 266.
100. J. N. Randall and J. C. Wolfe, *Appl. Phys. Lett.* **39** (1981) 742.
101. S. Park, T. N. Rhodin and L. C. Rathbun, *J. Vac. Sci. Technol* **A4** (1986) 168.
102. (*a*) V. M. Donnelly, D. L. Flamm, C. W. Tu and D. E. Ibbotson, *J. Electrochem. Soc.* **129** (1982) 2533; (*b*) D. E. Ibbotson, D. L. Flamm, V. M. Donnelly and B. S. Duncan, *J. Vac. Sci. Technol.* **20** (1982) 489; (*c*) D. E. Ibbotson, V. M. Donnelly and D. L. Flamm, *Extended Abstracts Electrochem. Soc.* **81–2** (October 1981) 650.
103. G. Smolinsky, R. P. Chang, and T. M. Mayer, *J. Vac. Sci. Technol.* **18** (1981) 12.
104. E. L. Hu and R. E. Howard, *Appl. Phys. Lett.* **37** (1980) 1022.
105. R. E. Klinger and J. E. Greene, *Appl. Phys Lett.* **38** (1981) 620.
106. L. A. Coldren and J. A. Rentschler, *J. Vac. Sci. Technol.* **19** (1981) 225.
107. M. A. Bösch, L. A. Coldren and E. Good, *Appl. Phys. Lett.* **38** (1981) 264.
108. M. B. Stern and P. F. Liao, *J. Vac. Sci. Technol.* **B1** (1983) 1053.
109 R. H. Burton and G. Smolinsky, *J. Electrochem. Soc.* **129** (1982) 1599.
110. Y. Yuba, K. Gamo, X. G. He, Y. S. Zhang and S. Namba, *Jap. J. Appl. Phys.* **22** (1983) 1211.
111. K. Asakawa and S. Sugata, *Jap. J. Appl. Phys.* **22** (1983) L653.
112. D. E. Ibbotson, D. L. Flamm and C. W. Tu, in *Proc. 7th Int. Symp. on Plasma Chemistry*, Vol. 3, ed. C.J. Timmerman, Eindhoven Inst. of Technol., Eindhoven, The Netherlands (1985) 954.
113. V. M. Donnelly, D. L. Flamm and G. J. Collins, (*a*) *Electrochem. Soc. Extended Abstracts* **81–2** (October 1982) 621; (*b*) *J. Vac. Sci. Technol.* **21** (1982) 817.
114. (*a*) D. A. Neibauer, *Electronic Packaging and Production*, (1981) 153; (*b*) B. Kegel, *Circuits Manufact.* **21** (1981) 27; (*c*) R. D. Rust, R. J. Pachter and R. J. Rhodes, *Electrochem. Soc. Extended Abstracts* **83–1**, (1983) 182–184.
115. R. F. Reichelderfer, J. M. Welty, J. F. Battey, *J. Electrochem. Soc.* **124** (1977) 1926.

116. B. J. Wu, D. W. Hess, D. S. Soong and A. T. Bell, *J. Appl. Phys.* **54** (1983) 1725.
117. J. M. Cook, B. W. Benson, *J. Electrochem. Soc.* **130** (1983) 2459.
118. J. F. Battey, *IEEE Trans. Electron. Dev.* **ED-24** (1977) 140.
119. K. Harada, *J. Appl. Polymer Sci.* **26** (1981) 1961.
120. G. N. Taylor, T. M. Wolf, *Polymer Eng. and Sci.* **20** (1980) 1086.
121. L. A. Pederson, *J. Electrochem. Soc.* **129** (1982) 205.
122. G. Turban and M. Rapeaux, *J. Electrochem. Soc.* **130** (1983) 2231.
123. J. J. Hannon and J. M. Cook, *J. Electrochem. Soc.* **131** (1984) 1164.
124. C. J. Mogab, unpublished results (1980).
125. P. Nordine, private communication (1981).
126. J. M. Cook, J. J. Hannon and B. W. Benson, in *Proc. 6th Int. Symp. on Plasma Chemistry*, ISPC-6, eds. M. I. Boulos and R. J. Muntz, Univ. Sherbrooke, Montreal (1984) 616.
127. H. Gokan, Y. Ohnishi and K. Saigo, Microelectron. Eng. **1** (1983) 251.

16 Polymers in plasmas

S. J. MOSS

Introduction

Polymers have always played a major role in the manufacture of integrated circuits as resists for patterning by means of photolithography. Other chapters in this book review the chemistry of photoresists, and x-ray and e-beam resists; almost invariably such resist systems are based on organic polymers. The resist acts as a mask for etching the underlying surface and is removed completely from the wafer surface at the end of the etching stage. The first use of plasmas, some two decades ago, in the manufacturing stages of integrated circuits was the final stripping of the resist film by exposure to an oxygen plasma.

In recent years, reactive-gas plasmas have been increasingly used to etch semiconductor materials, for a variety of reasons including the ability to etch finer lines and the advantages of a more easily automated process (see Chapter 15). Although by no means all etching is carried out with plasmas, resists are being increasingly exposed to plasmas during the manufacturing processes in what seems likely to become the dominant etching method in the future. Some finished devices also contain polymers as functional components (e.g. polyimide as dielectrics, see Chapter 12), the patterning in these materials often being carried out by reactive-ion etching.

The pattern in a resist film is usually developed by solvent extraction, taking advantage of solubility differences created in the film by exposure to ultraviolet light, x-rays or electron beams. Recently, dry-development methods have been extensively explored where the pattern in the resist is developed by exposure to an oxidizing reactive-ion etch or an oxygen plasma. The chemical properties required of such plasma-developable resists are very different from those of normal solvent-developed systems, but almost all are again based on organic polymers.

The widespread use of plasmas in integrated-circuit manufacture inevitably means that it will be increasingly important to acquire detailed knowledge and understanding of the behaviour of polymers in plasmas. This chapter surveys the published research on resists and polymers in plasmas without attempting to be fully comprehensive. The survey begins with studies of photoresist stripping by oxygen plasmas, then describes plasma-developable systems where the use of an oxygen plasma to develop the resist pattern is much more critical than simple resist stripping. This leads on to detailed studies of oxygen plasma reactions with polymers. The final section deals with reactive-ion etching of polymers. The preceding chapter presents a detailed account of fundamental aspects of plasma etching, concentrating on semiconductor materials but concluding with a brief section on resists.

Photoresist stripping

Irving first advocated the use of plasmas in integrated circuit manufacture for the removal of photoresists after an etching stage (1). Shortly thereafter, Bersin described an automated photoresist stripper based on an oxygen plasma (2).

A series of papers by Battey *et al.* (3–6) established many of the basic features of the stripping process with oxygen plasmas generated by RF at 13.56 MHz in a barrel reactor, mainly using photoresists based on poly(isoprenes). The ultimate products of the reaction were shown to be H_2O, CO_2 and CO. However, no sign of carbonyl stretching frequencies from organic materials were observed on infrared examination of the products trapped out of the reactor effluent stream. It seems likely that the primary volatile products of the stripping reaction are easily degraded completely to H_2O, CO, and CO_2 in the gaseous plasma. More direct methods of studying the constituents of plasmas are needed to detect organic materials present in the gas phase during plasma stripping.

The substrate wafers were placed on a temperature-controlled block in the reactor and the effect of changes in temperature was examined. The stripping rate showed approximately an Arrhenius dependence with activation energy of 24–30 kJ mol^{-1}.

The most significant result of the work of Battey is the clear demonstration of the central importance of oxygen atoms in determining the stripping rate (3). The oxygen atom concentration was determined at the exit point of the reactor by gas-phase titration with NO_2; under conditions of constant pressure, and power level, the O atom flow rate varied linearly with O_2 flow rate (at about 10% of O_2 rate). Pressure variations from 0.5–3.5 Torr and power levels from 50–200 W in the reactor produced a threefold change in stripping rate, but the stripping rate was directly proportional to the oxygen atom flow rate in all cases (3) (see Figure 16.1).

A recent esr study confirmed the importance of oxygen atoms in resist stripping by observing the decrease of the O atom signal when resist films placed downstream of a microwave discharge were heated (7). An activation energy of 48 kJ mol^{-1} was observed, almost double that found by Battey (3).

Degenkolb *et al.* demonstrated the value of emission spectroscopy in monitoring the stripping of photoresists (8–10), and used the method to investigate the influence of reactor variables. The process may be followed by means of the decrease in emission intensity from excited oxygen atoms (e.g. at 615.6 nm), but it is more satisfactory to observe emission from a product of the reaction. Useful wavelengths are 283.0, 297.7, 483.5, and 519.8 nm for CO*,

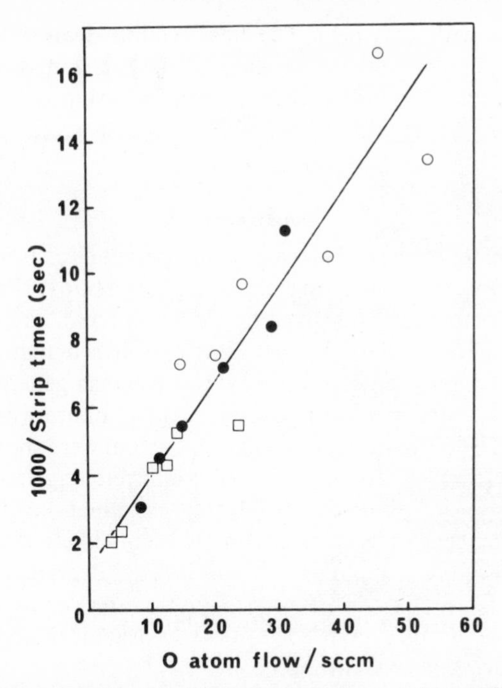

Figure 16.1 Stripping rate as a function of atomic oxygen flow rate (oxygen flow rate = 500 sccm; pressure = 0.5–3.5 Torr; temperature = 165° C). ○, 50 watts; ●, 100 watts; □, 200 watts. Redrawn from Battey (3).

283.0 and 308.9 nm for OH*, and 656.3 nm for H*. In some reactor configurations, an additional emission feature at 288–290 nm is observed during resist stripping attributable to excited CO_2^+ (11). A recent patent describes the use of optical emission for end-point control in plasma etching, including resists (12), while Roland *et al.* have reviewed methods of end-point detection (13).

Plasma-developable resists

Dry development of patterns in resists by exposure to an oxygen plasma offers two potentially valuable advantages: (i) easier access to sub-micron patterning, and (ii) a gas-phase process which would be more readily automated. An all-dry vacuum lithography process has already been described (14) in which the resist film was plasma-polymerized methyl methacrylate (a cross-linked material), patterned by electron beam, and developed by exposure to an argon–oxygen plasma.

Widespread interest in plasma development was generated by a Motorola patent (15), and the slightly later presentation of more information at a Kodak seminar (16). The basic principles of a plasma-developable resist (PDR) are

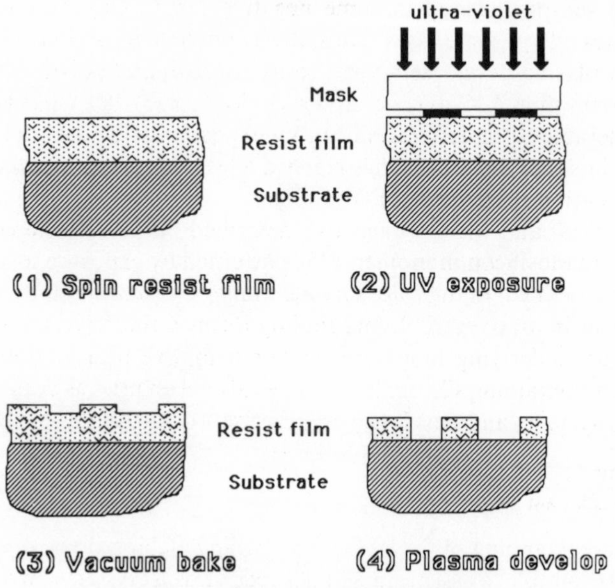

Figure 16.2 A plasma-developable resist process.

shown in Figure 16.2. A polymerizable solid monomer (A) is present in a film formed from a polymer 'binder' (B) together with other materials, e.g. photosensitizer, traces of solvent. Exposure of the resist to UV radiation causes polymerization of A, perhaps with grafting or cross-linking to B, and unpolymerized A is removed by a vacuum bake. To develop a satisfactory pattern, the polymer formed from A should have a substantially greater resistance to oxygen plasma etching than the binder, producing a negative-acting resist. The formation of an involatile polymer from A in the resist film is frequently described as 'photo-locking' following an earlier application of a similar approach to the fabrication of optical waveguide circuits (17). Other types of plasma-developable resist will be described later.

The behaviour of a PDR system based on N-vinyl carbazole (A) and poly(2,3-dichloro-1-propyl acrylate) (B) has been described in some detail by Taylor, Wolf and Goldrick (18). This system was derived from a previously reported x-ray resist (19) (dry-developed in an oxygen plasma) by adding quinone sensitizers to promote UV sensitivity. With ratios of B:A of about 4 and a few per cent of phenanthrenequinone as sensitizer, oxygen plasma development produced lines of 1 μm resolution in the resist with UV exposure times of a few seconds at 303–366 nm wavelengths. Other aromatic monomers with appropriate volatility have been studied as alternatives to N-vinyl carbazole, as well as variations on the acrylate binder.

Tsuda *et al.* have explored an alternative approach to enhancing the stability of resist materials to etching an oxygen plasma (20–24). They discovered that azide sensitizers used in some negative resists may decrease plasma etching rates (20) in some cases. Thus, it was possible to produce 0.5 μm lines by oxygen plasma development in a resist consisting of poly(methyl isopropenyl ketone) plus 4, 4′-diazide diphenylthioether (5–20%) and the pattern was transferable to underlying silicon by exposure to CF_4–O_2 plasma. The author claims that the process has reached a level suitable for practical use in microfabrication processes (24).

Taylor, Wolf and Moran have also described plasma-developable resists based on organosilicon monomers (25), patterned by exposure to x-rays. The monomer is 'locked' in the chloracrylate binder on irradiation by x-rays and development in an oxygen plasma rapidly forms a thin layer of SiO_2 which protects the underlying film from further oxidative attack. Polymers and copolymers containing silicon (26, 27) have also been used as deep-UV resists in bi-level systems and developed by reactive-ion etching in oxygen.

Plasma-oxidation of polymers

A good understanding of the factors controlling etching rates of polymers in plasmas is desirable to assist in devising superior resist systems. In the past two decades, a number of studies have been carried out directed to this end, partly inspired by the importance of plasma treatment of polymers in other fields, notably in modifying surface properties. This section surveys the main results of such studies and discusses the current state of our knowledge of the chemical mechanisms. The account will deal mainly with plasmas generated by radiofrequency discharges (usually at 13.56 MHz) under typical 'glow-discharge' conditions (i.e. at pressures around 0.1–5 Torr) in barrel or planar reactors (see Chapter 15).

All plasmas produce some bombardment of a substrate surface by more or less energetic ions. Ion bombardment may be emphasized by placing the substrate on the powered electrode, applying a dc bias to the electrode, operating at lower pressures, etc. The technique is then often referred to as *reactive-ion etching* which will be discussed further in the final section of the chapter.

Bell and Kwong examined an RF discharge in oxygen (28) and monitored conversion, oxygen atom yield (by NO_2 titration), and gas temperature (using small thermocouples). A useful model of the system including reactor geometry and flow was developed for numerical solution which gave reasonably good agreement with the results. More detailed models of oxygen plasmas have been developed recently (29, 30) for dc plasmas at around atmospheric pressures (of particular relevance to ozonizers).

Langmuir probes have frequently been used to characterize plasma properties such as electron density, plasma potentials, etc. In a recent review, Cherrington has pointed out the many complications in obtaining reliable measurements and in interpreting the results (31) when using such probes. He concludes that probe results are best used in a comparative sense, that is, in gaining insight into the effects of changes in the operating conditions on the physical properties of the plasma. However, Venugopalan responded (32) to the review by emphasizing that probe measurements provide a very valuable insight into plasma behaviour difficult to obtain in any other way, in spite of the uncertainties in absolute quantities derived from them. Langmuir probes have been used in oxygen plasmas to measure plasma potentials in parallel-plate and barrel reactors by Vossen (33) and in a reactor of a different design by Ojha (34).

Discharges through molecular oxygen produce substantial amounts of long-lived excited molecular oxygen (in $^1\Delta$ and $^1\Sigma$ states) and such states are known to play a significant role in the photo-oxidation of polymers (35). It is therefore of interest to examine the extent to which such species might be important in plasma oxidations. MacCallum and Rankin placed polymer samples downstream from a microwave discharge and examined the effects of oxygen atoms (36) (monitored by NO_2 titration) on poly(styrene), poly(vinyl chloride), poly(acrylonitrile), poly(ethylene terephthalate) and nylon 66; all the polymers reacted in the surface layers. Silver gauze was placed in the gas flow to remove oxygen atoms without significantly affecting the concentration of singlet molecules. There was no detectable change in the contact angle of water with films of poly(styrene) and poly(vinyl chloride) after several hours of exposure to singlet molecules. More recently, Cook, Hannon, and Benson (7) have also reported that excited singlet states play no significant role in the plasma oxidation of polymers.

Studies of homopolymers

An early study by Hansen *et al.* (37) examined the rates of plasma oxidation of some three dozen polymers by measuring weight loss as a function of exposure time in an oxygen plasma. They found that the most resistant polymers were highly aromatic, perfluorinated, or sulphur-vulcanized rubbers. The mechanism was regarded as similar to a standard thermal chain oxidation differing mainly in that the rapid initiation was by reaction of atomic oxygen with the polymer:

$$RH + O: \rightarrow R^{\cdot} + HO^{\cdot}$$
$$R^{\cdot} + O_2 \rightarrow ROO^{\cdot}$$
$$ROO^{\cdot} + RH \rightarrow ROOH + R^{\cdot}$$
$$ROOH \rightarrow RO^{\cdot} + HO^{\cdot}$$

The addition of relatively large amounts of a conventional thermal antioxidant (a substituted phenol present at levels up to 10%) was ineffective in reducing the etching rate. In a similar study, Lawton later confirmed that plasma oxidation rates were independent of the degree of crystallinity in samples of poly(ethyleneterephthalate) and poly(styrene) (38).

Taylor and Wolf have recently made a detailed study of plasma stripping rates of about 40 polymers in an oxygen plasma (39). The polymer samples were spin coated on 2″ Si wafers at thicknesses from 0.3–1.1 μm. An attempt was made to control the wafer temperatures by placing them in the centre of an aluminium table pre-heated to $35 \pm 1°$ C by the plasma. The temperature of the polymer film would surely be much higher than the block temperature during stripping the plasma, but temperature changes would, it was hoped, be much the same for each polymer. The samples were exposed to the plasma for two minutes and the etch rate calculated from the change in thickness. Table 16.1 shows the results obtained. Where the results overlap those of Hansen *et al.* (37) the agreement is good.

Taylor and Wolf were able to draw some valuable conclusions from their data on the relation between polymer structure and oxygen-plasma stripping rate. Firstly, they pointed out a clear parallel between stripping rate in an oxygen plasma and degradability in high-energy radiation (i.e., for polymers which are degraded on exposure to electron beams). This was interpreted as demonstrating the importance of cleavage of the polymer backbone bond; with a weak bond, where cleavage is relatively easy, the stripping rate is likely to be high. For those polymers which cross-link on exposure to e-beams, the rate-determining step in the plasma stripping process was not readily apparent. There is some evidence in the table that weakly bonded side-groups promote the stripping reaction, but the interpretation is complicated by other effects.

An unexpected finding was the pronounced ability of chlorine present in the polymer to enhance the removal rate. By some ingenious experiments using two polymer-coated wafers in the plasma, Taylor and Wolf were able to show that the effect could be attributed to the presence of chlorine in the gas-phase plasma. When a wafer coated with poly(styrene) was stripped in the presence of a wafer coated with poly(2, 3-dichloro-1-propyl acrylate), DCPA, the DCPA stripping rate was normal (i.e., 12.2) while the poly(styrene) rate was enhanced by a factor of 4.4. A similar experiment using poly(N-vinyl carbazole), PNVC, in place of poly(styrene) produced more than a threefold enhancement in the rate for PNVC. Emission spectroscopy was used to demonstrate the presence of chlorine atoms in the plasma and the effect of chlorine on the stripping rate was attributed to the increased rate of attack on the polymer backbone by chlorine atoms:

$$Cl^{\cdot} + PH \rightarrow P^{\cdot} + HCl$$

Similar experiments with two wafers where DCPA was replaced by poly(vinyl bromide) showed no significant enhancement of etch rate for poly(styrene). This finding was assumed to support the interpretation for chlorine atoms, since the corresponding abstraction reaction with bromine is endothermic. The high etch rate of poly(vinyl bromide) is presumably related to the low C–Br bond strength which allows more rapid attack on the polymer

Table 16.1 Relative rates of oxygen plasma removal of polymers (rate for poly(styrene) = 1.00) (39)

	Polymer	Rate
Aliphatic polymers		
1	Poly(methylmethacrylate)	2.37
2	Poly(ethyl acrylate)	2.78
3	Poly(isobutylene)	3.56
4	Poly(cis-butadiene)	2.78
5	Poly(cis-isoprene)	5.55
6	Poly(vinyl acetate)	5.46
7	Poly(butene-1-sulphone)	7.11
8	Poly(vinyl methyl ketone)	1.48
9	Poly(ethylenimine)	1.44
10	Poly(glycidyl methacrylate)	1.07
11	Cyclized cis-poly(isoprene)	1.24
Aromatic polymers		
12	Poly(styrene)	1.00
13	Poly(α-methyl styrene)	1.11
14	Poly(4-methyl styrene)	1.24
15	Poly(phenyl methacrylate)	1.33
16	Poly(4-vinyl pyridine)	1.70
17	Poly(acenaphthalene)	0.63
18	Poly(N-vinyl carbazole)	0.59
Halogenated polymers		
19	Poly(vinylidene fluoride)	0.83
20	Poly(4-bromo-styrene)	1.93
21	Poly(4-chloro-styrene)	2.93
22	Poly(vinyl benzyl chloride)	2.59
23	Poly(2, 3-dibromo-1-propyl acrylate)	3.04
24	Poly(2-chlorethyl acrylate)	11.67
25	Poly(1, 3-dichlor-2-propyl acrylate)	11.85
26	Poly(2, 3-dichlor-1-propyl acrylate)	12.22
27	Poly(vinyl chloride)	12.96
28	Poly(chloroprene)	13.33
29	Poly(vinyl bromide)	13.70
30	Poly(epichlorohydrin)	21.50
31	Chlorinated poly(ethylene)—42% Cl	22.22
Copolymers		
32	Poly(methyl methacrylate-co-methacrylonitrile) 94:6 mol%	2.70
33	Poly(glycidyl methacrylate-co-ethyl acrylate) 70:30 mol%	1.37
34	Poly(vinylidene chloride-co-acrylonitrile) 80:20 mol%	8.70
35	Poly(vinylidene chloride-co-acrylonitrile) 88:12 mol%	11.19
36	Poly(vinyl chloride-co-vinyl acetate) 86:14 mol%	21.50
Miscellaneous		
37	Poly(N-vinyl pyrrolidone)	0.93
38	Cresol formaldehyde resin	1.33
39	Poly(dimethylsiloxane)	0.00
40	Glass resin	0.00

backbone. However, it seems unlikely that attack by chlorine atoms could alone be responsible for the increase in etch rate. Another possible explanation might be an increase in oxygen atom concentration produced by the presence of chlorine in the gas phase. Such an effect has been demonstrated by Flamm (40) when small amounts of C_2F_6, CF_3Cl, and CF_3Br are added to an oxygen plasma, and was attributed to changes induced in the electron-energy distribution in the discharge rather than to chemical reactions (40, 41) (see also Chapter 15).

The results in Table 16.1 also clearly confirm the earlier findings of Hansen *et al.* (37) that aromatic polymers are significantly more stable to plasma oxidation. A plausible explanation for this behaviour has been difficult to find. The interpretation of Taylor and Wolf relates to their suggestion that excited oxygen atoms, $O(^1D)$, are largely responsible for the plasma oxidation. Aromatic rings are then supposed to form an excited complex with O^1D atoms which reacts with a further O^1D atom forming O_2. This explanation seems implausible in view of the known high reactivity of O^1D atoms (42), and the unlikely involvement of O^1D in the polymer stripping process.

Studies of copolymers

A recent study of the stripping rate of some homopolymers and copolymers of methyl methacrylate (MMA) with styrene by Moss, Jolly, and Tighe (43, 44) provided useful insights into structural influences. The measurements were carried out in a barrel reactor and the reactions were monitored by emission from excited CO at 297.7 nm. Figure 16.3 shows some typical emission curves.

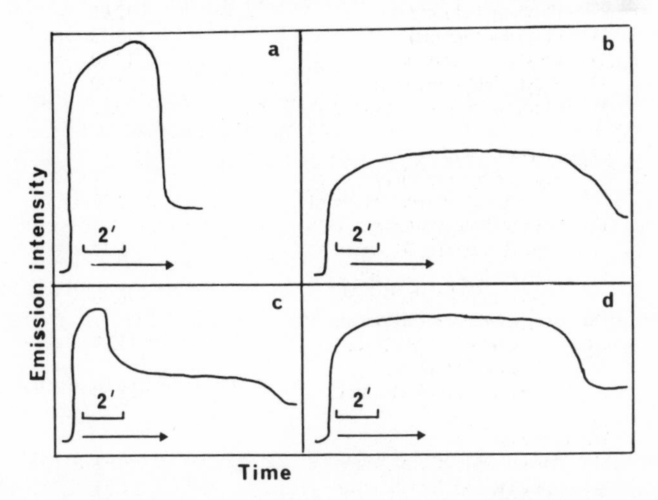

Figure 16.3 Emission profiles at 297.7 nm for polymer etching. (*a*) Poly(methyl methacrylate); (*b*) poly(styrene); (*c*) separate films on a mixed wafer; (*d*) copolymer film. Redrawn from Moss, Jolly and Tighe (43, 44).

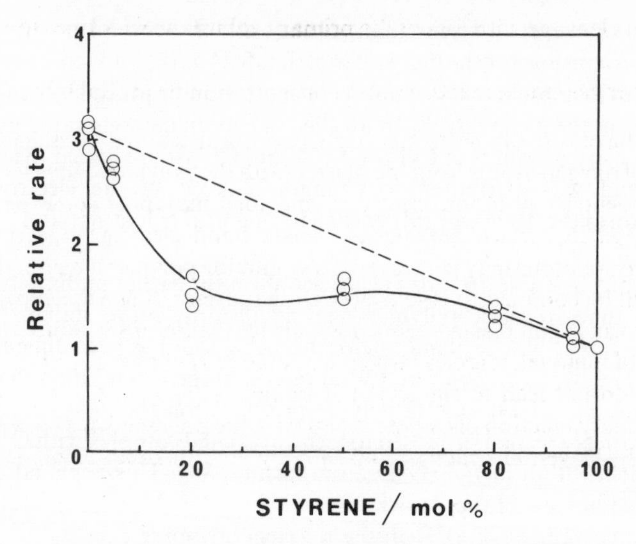

Figure 16.4 Relative stripping rates of styrene-methylmethacrylate copolymers. Redrawn from Moss, Jolly and Tighe (43, 44).

For homopolymers of styrene and MMA, there was no significant dependence of the rate on film thickness in the range studied (0.8–2.5 μm). The addition of substantial amounts of hindered phenols (e.g., Irganox 1076 up to 10%) produced no observable retardation of the stripping rates, as also noted by Hanson *et al.* (37).

Copolymers of MMA with styrene were chosen for detailed study because the two homopolymers represent extremes in plasma stripping behaviour, have similar glass transition temperatures, and the monomers are kinetically well-behaved and allow predictable control of copolymer structure in free-radical polymerizations. It is clear from the results shown in Figure 16.4 that relatively small amounts of styrene in the copolymer considerably enhance the stability to plasma oxidation. Similar, but less marked, behaviour was found with copolymers of methyl methacrylate and vinyl naphthalene (44).

The observed stripping rates for plasma oxidation of most polymers cover a relatively small range (typically within a factor of about 1–5). It is worth bearing in mind that such variations may arise through relatively small differences at several stages in the mechanism rather than in a single rate-determining step.

Four stages in the stripping process might usefully be distinguished and discussed.

(i) Surface bombardment of the polymer by species in the plasma to create free-radical sites, etc.
(ii) Reactions of free-radical sites on the polymer with O_2 leading to peroxides, alkoxy, and hydroxyl radicals

(iii) Chain cleavage with loss of the primary volatile species from the polymer surface
(iv) Further gas-phase reactions of the primary volatile products in the plasma.

As we have already seen, the most important feature of stage (i) is the reaction of oxygen atoms from the plasma with the polymer. However, surface bombardment by photons, electrons, and ions may play some part in the process, e.g., in surface heating and some bond cleavage. Electrons and, particularly, photons may penetrate deeply into the polymer film, but the effects of ions will be confined to layers close to the surface. Since the exposure of a polymer to an argon plasma under comparable conditions produces only very low rates of removal, it is clear that the interactions of photons, electrons, and ions only do not lead to significant stripping.

The primary reaction of oxygen atoms with the polymer backbone is almost certainly hydrogen abstraction, followed by an autoxidation process:

$$\mathrm{O}: + \sim\sim\sim \overset{\mathrm{H}}{\overset{|}{\mathrm{C}}} \sim\sim\sim \longrightarrow \sim\sim\sim \dot{\mathrm{C}} \sim\sim\sim + \mathrm{H\dot{O}} \qquad (16.1)$$

$$\mathrm{HO}^{\cdot} + \sim\sim\sim \overset{\mathrm{H}}{\overset{|}{\mathrm{C}}} \sim\sim\sim \longrightarrow \sim\sim\sim \dot{\mathrm{C}} \sim\sim\sim + \mathrm{H_2O} \qquad (16.2)$$

$$\sim\sim \dot{\mathrm{C}}\text{—}\mathrm{C} \sim\sim + \mathrm{O_2} \longrightarrow \sim\sim \overset{\mathrm{O—O^{\cdot}}}{\overset{|}{\mathrm{C}}}\text{—}\mathrm{C} \sim\sim \qquad (16.3)$$

$$\sim\sim \overset{\mathrm{O—O^{\cdot}}}{\overset{|}{\mathrm{C}}}\text{—}\mathrm{C} \sim\sim + \mathrm{RH} \longrightarrow \sim\sim \overset{\mathrm{OOH}}{\overset{|}{\mathrm{C}}}\text{—}\mathrm{C} \sim\sim + \mathrm{R}^{\cdot} \qquad (16.4)$$

$$\sim\sim \overset{\mathrm{OOH}}{\overset{|}{\mathrm{C}}}\text{—}\mathrm{C} \sim\sim \longrightarrow \sim\sim \overset{\mathrm{O^{\cdot}}}{\overset{|}{\mathrm{C}}}\text{—}\mathrm{C} \sim\sim + \mathrm{HO}^{\cdot} \qquad (16.5)$$

The most likely source of cleavage of the polymer chain is the alkoxy radical produced in reaction (16.5) by the reaction

$$\sim\sim \overset{\mathrm{O^{\cdot}}}{\overset{|}{\mathrm{C}}}\text{—}\mathrm{C} \sim\sim \longrightarrow \sim\sim \mathrm{C{=}O} + {}^{\cdot}\mathrm{C} \sim\sim \qquad (16.6)$$

The loss of the primary volatile products from the film will follow reaction (16.6), by polymer 'unzipping' where appropriate and by further primary attack on the backbone.

The fourth stage of the plasma oxidation process may seem to have little influence on overall stripping rates. However, the primary volatile products will probably vary with the polymer structure, so that further oxidation of the primary products may modify the plasma concentrations, etc., to different extents. Studies are being carried out in the author's laboratory aimed at identifying the primary volatile products and their reactions in plasmas using mass spectrometry with molecular-beam sampling.

The general ideas may now be applied to plasma oxidation of polymers of MMA and styrene. With poly(MMA), the backbone secondary C–H should be preferentially abstracted, leading to autoxidative cleavage through reactions (16.3) to (16.6) and summarized in reaction (16.7):

$$\begin{array}{c}
\begin{array}{ccccccc}
 & CH_3 & & CH_3 & & & \\
 & | & & | & & & \\
\sim\sim & C & -\dot{C}H- & C & -CH_2\sim\sim & & \\
 & | & & | & & & \\
 & CO & & CO & & & \\
 & | & & | & & & \\
 & OCH_3 & & OCH_3 & & &
\end{array}
\xrightarrow{O_2}
\begin{array}{cccccc}
 & CH_3 & & & CH_3 & \\
 & | & & & | & \\
\sim\sim & C & -C{=}O\ + & & \dot{C} & -CH_2\sim\sim \\
 & | & & & | & \\
 & CO & & & CO & \\
 & | & & & | & \\
 & OCH_3 & & & OCH_3 & \\
 & (I) & & & (II) &
\end{array}
\end{array} \qquad (16.7)$$

Cleavage product (I) is stable and requires further attack on the chain before volatile products are formed. However, radical (II) should unzip to monomer at sufficiently high temperatures.

With poly(styrene), the dominant attack on the polymer backbone will occur at the tertiary C–H bond leading to the following products:

$$\begin{array}{ccccc}
 & H & & & \\
 & | & & & \\
\sim\sim & C & -CH_2- & \dot{C} & -CH_2\sim\sim \\
 & | & & | & \\
 & Ph & & Ph &
\end{array}
\xrightarrow{O_2}
\begin{array}{cccccc}
 & H & & & & \\
 & | & & & & \\
\sim\sim & C & -\dot{C}H_2 + & O{=} & C & -CH_2\sim\sim \\
 & | & & & | & \\
 & Ph & & & Ph & \\
 & (III) & & & (IV) &
\end{array} \qquad (16.8)$$

In this case, (IV) is stable while radical (III) is unable to unzip readily, but will react with another molecule of oxygen, leading to another stable structure:

$$\begin{array}{ccc}
 & H & \\
 & | & \\
\sim\sim & C & -\dot{C}H_2\sim\sim \\
 & | & \\
 & Ph &
\end{array}
\xrightarrow{O_2}
\begin{array}{cccc}
 & H & & \\
 & | & & \\
\sim\sim & C & -C & {=}O + H^{\cdot} \\
 & | & | & \\
 & Ph & H &
\end{array} \qquad (16.9)$$

An understanding of the behaviour of polymer radicals in depropagation reactions (i.e., 'unzipping') was obtained in early studies of thermal decomposition of polymers (45–47). Relatively stable radicals unzip while unstable radicals (i.e., less substituted radicals) undergo alternative reactions more readily (47). McNeill and his colleagues have studied the thermal decompositions of poly(styrene) (48) and poly(MMA) (49) in detail.

The results illustrated in Figure 16.4 clearly show that copolymers containing more than one styrene unit to about six MMA units have almost the same resistance to plasma oxidation as poly(styrene). A reasonably plausible explanation might require the initial attack to be largely confined to the C–H on a styrene unit:

$$
\begin{array}{l}
\sim\sim \underset{\substack{|\\ CO\\ |\\ OCH_3}}{\overset{\substack{CH_3\\ |}}{C}}—CH_2—\underset{\substack{|\\ Ph}}{\dot{C}}—CH_2\sim\sim \xrightarrow{O_2} \sim\sim \underset{\substack{|\\ CO\\ |\\ OCH_3}}{\overset{\substack{CH_3\\ |}}{C}}—\dot{C}H_2 + O{=}\underset{\substack{|\\ Ph}}{C}—CH_2 \\
\qquad\qquad\qquad\qquad\qquad\qquad\qquad\qquad (V) \qquad\qquad\qquad (VI)
\end{array}
\qquad (16.10)
$$

Product (VI) will again act as a 'blocking' group, while radical (V) is the 'wrong' radical for an unzipping reaction (i.e., is too unstable) and will again block the chain end by further reactions with molecular oxygen.

Copolymers containing more than 15 mol % styrene would be expected to contain few long sequences of MMA units, as Monte Carlo simulations have confirmed (44). It is, therefore, not unreasonable that the overall stripping behaviour might be dominated by reactions at a junction of styrene and MMA units in such copolymers, as shown in reaction (16.10)

Why are aromatic polymers more resistant to plasma oxidation?

The preceding discussion makes extensive use of ideas relating to thermal and photo-oxidative (50) degradation of polymers. However, the conclusion reached cannot represent the full, or even the major, explanation for the relative stability of aromatic polymers. As Table 16.1 shows, both poly(α-methyl styrene) and poly(phenyl methacrylate) have stripping rates close to that for poly(styrene), and neither polymer has a tertiary C–H on the backbone. Poly(benzyl methacrylate) likewise has a very similar stripping rate (44) but has no tertiary C–H and the benzene ring is separated from the polymer backbone by three atoms.

The stability of aromatic polymers seems to depend on some property of the benzene ring which is substantially independent of its position in the polymer. We have recently suggested (44) that the explanation probably lies in the known ability of ground-state oxygen atoms to react with aromatic rings to form phenols in an addition reaction:

$$O(^3P) + \text{---C---}(C_6H_5) \longrightarrow \text{---C---}(C_6H_4OH) \qquad (16.11)$$

This reaction removes oxygen atoms without forming free radicals or other products which could assist cleavage of the polymer backbone.

Several gas-phase studies using the mercury-photosensitized decomposition of nitrous oxide as a source of oxygen atoms have shown that phenols are readily formed with aromatic compounds (51, 52). Sloane has examined the primary processes in the reaction of $O(^3P)$ with benzene, toluene and 1, 3, 5-trimethyl benzene in crossed molecular beams using mass-spectrometric detection (53). The only significant products detected in each case were the *adducts.*

Studies of the reactions of aromatic compounds in oxygen plasmas (54–56) have shown that phenols are formed along with other products. The interpretation of the results is complicated in these systems by extensive decomposition of the 'hot' primary addition products at low pressures.

The condensed-phase results of Mazur *et al.* (57, 58) are of more significance for the current argument. The earlier paper (57) showed that a microwave discharge in oxygen at 4 Torr could be effectively used as a preparative source of phenols when passed over neat liquids or solutions of arenes. More detailed later studies (58) examined the action of microwave discharges in CO_2/He (to minimize reactions involving O_2) on substituted benzenes in solution or adsorbed on silica gel. In almost all cases, the major products were substituted phenols or cyclohexadienones, characteristic of addition of oxygen atoms to the ring.

These results firmly establish the ability of aromatic rings to act as a 'sink' for oxygen atoms when incorporated into polymeric structures, and suggest strongly that the formation of phenols is likely to be a significant blocking process in the initiation stage of plasma oxidation. The overall effect will be to reduce the rate of initiation of backbone attack substantially, and will be largely independent of the position of the aromatic ring in the polymer structure. However, differences in reactivity arising from the nature of the radicals created on the polymer backbone will be superimposed on the effects of reduced initiation of chain cleavage reactions.

A variety of carbonyl, hydroperoxide, and other oxygen-containing groups have been identified in the photo-oxidation of poly(styrene) (50). Clark *et al.* have used ESCA in seeking to identify oxygen-containing groups on polymer surfaces exposed to oxygen plasmas; in each case, complex ESCA spectra must be deconvoluted (59, 60). Studies of plasma-oxidized poly(styrene) showed the presence of C—O, C=O, —(C=O)—O, and O—(C=O)—O groups. Some evidence was found for transformations of the aromatic ring, but it is not clear if phenolic structures contributed to the observed spectra.

Only poly(siloxanes) show much greater resistance to oxygen plasma than aromatic polymers because of the thin crust of silica formed in the plasma. Adesida *et al.* have recently demonstrated that the stripping rate of poly(methyl methacrylate) in an oxygen plasma may be reduced by a factor of more than 10 when the polymer surface is pre-exposed to beams of $^{28}Si^+$ ions

(61). A depth profile study of the plasma-treated polymer, using Auger spectroscopy, showed clearly a thin layer of SiO_2 formed from implanted silicon.

Polymers in semiconductor etchants

An essential property of useful resists is the ability to withstand etching processes carried out on the underlying substrates. All resist materials are attacked to some extent in reactive-gas plasma; to be useful as resists, the etch rate of the resist must be lower than that of the substrate by a factor of at least three. Most commonly used resists behave satisfactorily in the pure halocarbon plasmas used for substrate etching, but problems of resist degradation become more severe where oxygen is added to the etching plasma, exemplified by the CF_4/O_2 plasmas widely used in etching silicon. The section will deal mainly with studies of the reactions of resists and polymers in such plasmas.

The addition of oxygen to fluorocarbon plasmas generally serves two purposes, namely, to increase the concentration of fluorine atoms in the plasma and to suppress polymer formation on the wafer, both of which enhance the etch rate of the substrate (41, 62) (see Chapter 15). The great importance of such plasmas commercially has inspired many studies in the past few years. Smolinsky and Flamm used an alumina reactor with an effusive orifice to a mass spectrometer to study the chemical changes in CF_4 and CF_4/O_2 plasmas. The products from CF_4 alone were F, F_2, and C_2F_6 mainly. In the presence of O_2, C_2F_6 was replaced by CO, CO_2 and COF_2 and the concentrations of F and F_2 were greatly increased reaching a maximum at about 40–50 mol% O_2. Flamm (64) has reviewed the literature on CF_4/O_2 plasmas to 1979.

More recently, d'Agostino *et al.* (65) made a detailed study of this system (at a pressure of 1 Torr in an alumina reactor) by monitoring the emission from F, CF, CF_2, O, CO, and CO_2^+; the emission intensities were believed to relate directly to ground-state concentrations. Argon actinometry was used to compensate for changes in plasma conditions (66) and thereby obtain reliable estimates of relative particle densities. The emission intensity of the 750.4 nm Ar line decreased almost linearly with addition of O_2 (up to about 50% only reported) which implied a corresponding decrease in electron density. Other observations suggested that the energy distribution function did not change appreciably. Oxygen addition was also found to reduce CF and CF_2 emissions sharply while F, CO, and CO_2^+ all passed through a maximum at 15–20% O_2. An almost linear rise in O atom emission was observed over this range. Vibrational temperatures were estimated from band intensities to be around 4200 K, while rotational temperatures of ~ 500 K were found (this temperature should be close to the translational temperature of the gas).

A recent study of CF_4/O_2 plasmas (67) has shown that some caution must be used in making deductions about the concentrations of ground-state oxygen atoms from observations on the emission from excited states. Laser-induced fluorescence was used to measure ground-state concentrations with argon actinometry on two of the excited states (emitting at 844.6 nm and 777.4 nm). If the excited state is formed mainly by electron impact excitation of ground-state atoms, then the excited state serves as an indicator for the ground state. Formation of the excited state by dissociative excitation from molecular oxygen will invalidate this conclusion. The results showed that dissociative excitation seemed to be the dominant mechanism for the state giving 777.4 nm emission. However, the emission intensity ratio for O*(844.6)/Ar* followed the oxygen atom concentration, within certain limitations.

Ryan and Plumb have demonstrated that increase in fluorine atom concentrations in CF_4/O_2 plasmas arises from the fast reactions

$$:CF_2 + O: \rightarrow \dot{C}OF + F^{\cdot}$$
$$\dot{C}OF + O: \rightarrow CO_2 + F^{\cdot}$$

They studied the reaction between CF_2 and $O(^3P)$ directly in a fast-flow reactor sampled by a mass spectrometer (68), measuring the rate constants as well as observing the products.

Pederson examined the behaviour of several resists and polymers in the standard $CF_4 + 8\%\ O_2$ plasma etchant (69), and found that the superior etch

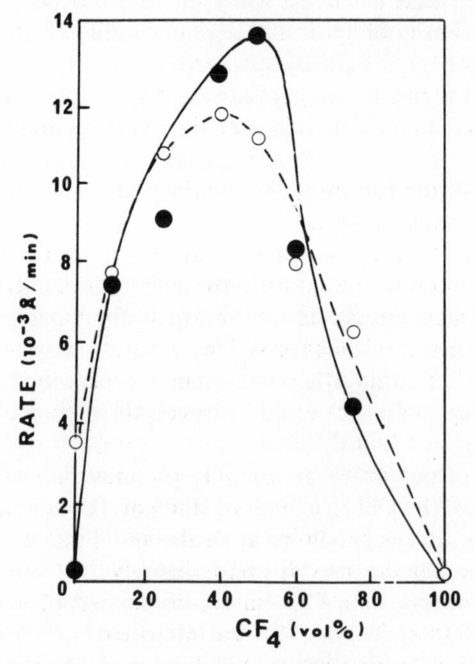

Figure 16.5 Absolute removal rates for etching of (○) poly(2, 3-dichloro-1-propyl acrylate) and (●) poly(styrene). Redrawn from Taylor and Wolf (39).

stability of materials containing aromatic polymers persisted in this plasma. However, Taylor and Wolf have shown that the marked difference in etch stability of poly(styrene) and dichloropropylacrylate is only observed at low concentrations of O_2 or CF_4 (39). At intermediate ranges, both polymers show greatly increased and comparable etch rates (see Figure 16.5).

Two studies of the degradation of poly(methyl methacrylate) in CF_4/O_2 plasma etchant have been carried out using gel permeation chromatography (gpc) to follow changes in molecular weight (70, 71) in the film. Harada exposed films of different thickness (0.5–3.5 μm) simultaneously to $CF_4 + 5\%\ O_2$ plasma for fixed periods of time (70). The whole of the remaining film was then dissolved away and the molecular weight distribution determined by gpc: the degradation profile in the film was obtained by subtracting the distributions found in successive layers. Harada concluded that significant degradation of the polymer occurred down to 3–4 μm after only short exposures to the plasma (5 minutes). This study also examined etch rates as a function of oxygen content. A pronounced maximum was observed in the rate at about 75% O_2, similar to that found by Taylor and Wolf (39) (see Figure 16.5).

The conclusion reached by Harada on the depth of degradation in the film has been criticized by Wu *et al.* (71). In this study, PMMA films of 1.7 μm thickness were exposed for 5 minutes to pure CF_4 and $CF_4 + 8\%\ O_2$ plasmas. The films were dissolved away in stages of about 0.5 μm and the molecular weight distribution thus determined at specific depths. Small reductions in molecular weight were observed throughout the films, but the degradation was high only in the top few tenths of a micrometre. The degradation profiles were almost identical in pure CF_4 and $CF_4 + 8\% O_2$, although the presence of oxygen greatly increased the etching rate. This observation suggests that the effects of oxygen are confined to the surface of the resists. A reasonably successful model of the process was developed using a statistical approach based on random chain scission.

Large amounts of fluorine-containing materials in the gas phase are clearly able to produce substantial increases in the etch rates of polymeric materials. Such pronounced increases in rate may be largely attributed to the increases in oxygen atom concentrations observed by several authors, and discussed earlier. However, the fluorine atoms present in fluorocarbon plasmas also play a part in the polymer etching by rapid hydrogen abstraction from the polymer and in other ways.

The etch rate of polyimide by CF_4/O_2 plasmas showed a pronounced maximum at about 20% CF_4 in a detailed study by Turban and Rapeaux (72). A pure SF_6 plasma etches polyimide at a substantial rate with added oxygen producing two weaker maxima at approximately 20% and 60% SF_6. The main reaction products were CO_2 in O_2-dominated plasmas and CF_4 in SF_6-dominated plasmas. The results were interpreted as showing a shift from oxygen-atom etching to fluorine-atom etching in the two regimes.

Egitto *et al.* (73) have recently suggested that the ratio [O]/[F] is most significant in the etching of polyimide and an oxirane-based copolymer by CF_4/O_2 in a diode reactor at around 350 mTorr. The etch rate of the polyimide passed through a maximum at about 25% CF_4 which coincided with the maximum in the emission intensity from excited O at 845 nm but did not follow precisely the emission curve. When [F] was reduced independently by placing silicon wafers close to the polyimide-coated wafer, the maximum shifted towards higher CF_4 concentrations. The dependence of the etch rate on [O]/[F] was complex, also passing through a maximum, but the importance of this ratio was suggested by the observation of the same maximum in the presence of Si and when SF_6 was used in place of CF_4 (in this case, the maximum occurs at about 5% SF_6).

The explanation probably lies in competition between F atoms and O_2 molecules for radical sites on the polymer backbone after abstraction by O or F. The same authors have also used x-ray photoelectron spectroscopy to show that the maximum rate for addition of CF_4 or SF_6 corresponds to the same surface composition (i.e., the same C:F ratio) (74).

Babu *et al.* (75) have suggested the following mechanism to describe the role of surface fluorination in polymer etching with CF_4/O_2 plasmas:

$$RH + F^{\cdot} \rightarrow R^{\cdot} + HF$$

$$RH + O, HO^{\cdot} \rightarrow R^{\cdot} + HO, H_2O$$

$$R^{\cdot} + F^{\cdot} \rightarrow RF$$

$$R^{\cdot} + O, O_2 \rightarrow RO^{\cdot}$$

$$RO^{\cdot} \rightarrow \text{products}$$

$$RO^{\cdot} + {}^{\cdot}CF_x \leftrightarrow ROCF_x$$

$$R^{\cdot} + {}^{\cdot}CF_x \rightarrow RCF_x$$

$$RF, RPCF_x, RCF_x, + O, O_2 \rightarrow \text{products (slow)}$$

Species such as RF, RCF_x, and $ROCF_x$, are assumed to react only very slowly with O or O_2, effectively protecting the surface, although the back reaction of $ROCF_x$ was thought to be important. A thorough test of the mechanism requires knowledge of the concentrations of species such as CF_x which are not readily available. However, the mechanism has recently been assessed using results from a large-scale kinetic model of CF_4 and O_2 plasmas (76, 77), where the concentrations of a large number of neutral and ionic species in the gas phase were calculated at the steady state. The model is related to the mechanism shown above by assuming that the etch rate is proportional to $[F][O + O_2]/[CF_3]$; this is clearly an approximation used of necessity. The scaled rate calculated by this expression as a function of added O_2 for polyimide etching is remarkably close in form to the experimental results, showing a maximum rate at 80–90% O_2. Although the model is by no means perfect, as the authors acknowledge, it represents an excellent attempt at a quantitative understanding of this complex process.

Reactive-ion etching of polymers

As stated earlier, all plasmas produce some bombardment of a substrate surface by more or less energetic ions; the use of a barrel reactor, especially with a perforated metallic etch tunnel, minimizes the bombardment. Effects from ion impacts may be emphasized by placing the substrate on the powered electrode, applying a dc bias to the electrode, operating at lower pressures, etc. The technique may then be referred to as *reactive-ion etching* (RIE), but the distinction between plasma etching and reactive-ion etching is often a fine one which Coburn (78) has addressed (see Chapter 15 for an extensive discussion). Under normal plasma etching conditions, ion bombardment may play an invaluable part in the etching process by assisting the attack of neutral species on the substrate. This interplay of ions and neutrals is largely responsible for the highly directional etching achievable by plasmas in some substrates, leading to the highly desirable characteristics of negligible undercutting and the resultant ability to etch sub-micron features (see Chapter 5).

The extreme example of ion bombardment is encountered in the use of ion beams from suitable sources at low pressures (to eliminate gas-phase collisions). Highly energetic ions are able to expel material from a substrate by purely physical effects, i.e., by momentum transfer from the ion to substrate atoms, a process known as *sputtering*. With less energetic ions, the chemical reactivity of the ion may also be important. Ion bombardment has played a relatively minor role in the systems discussed so far which relate to polymer etching in 'normal' plasmas at relatively high pressures. This final section briefly reviews some studies using ion beams or RIE in oxygen where substrates are exposed simultaneously to energetic O_2^+ ions and oxygen atoms.

Reactive-ion etching may be used to produce deep features with vertical walls and high aspect ratios in thick polymer films (79, 80). This property of RIE is of great value in producing sub-micron patterns in resists which may then be transferred to underlying substrates by further dry etching processes. A possible disadvantage of RIE is that most polymers have similar etch rates under such conditions so that selectivity cannot be easily achieved. Gokan *et al.* made a comparative study of the etch rates of a number of polymers exposed to Ar^+ and O_2^+ ion beams, as well as typical plasma etching conditions (81, 82). The etch rates in 300 eV Ar^+ and O_2^+ beams correlated well with the 'effective' carbon content in the polymer, i.e., with $N/(N_C - N_O)$, where N is the total number of atoms in the monomer, and N_C and N_O are the numbers of carbon and oxygen atoms respectively. The etch rate in the O_2^+ beam was about 15 times larger than in the Ar^+ beam. There is no such simple correlation between etch rate and effective carbon content in normal oxygen plasma etching. The results clearly suggest that different etching mechanisms operate under ion beam and plasma conditions.

A more recent detailed study of photoresist etching by RIE has been carried out by Steinbruchel *et al.* (83) using a range of diagnostic techniques including optical emission actinometry, mass spectrometry, and Langmuir probe measurements. They focus attention on the etch yield (Y_c), the net number of carbon atoms removed per impinging ion (i.e., after 'correcting' for oxygen atoms present in the film). There was no good correlation between Y_c and [O], yet mass spectrometry showed that O_2 was clearly consumed in the process. The results suggested that polymer etching under RIE conditions is ion enhanced, by one or more of four possible mechanisms:

(i) Physical sputtering to remove products held on the surface
(ii) Surface-damage-promoted etching
(iii) Chemical sputtering where the ions provide energy to drive the surface reactions
(iv) Direct reactive-ion etching by dissociation of the ions on the surface followed by reaction.

The authors concluded that mechanisms (i) and (ii) made the major contributions to the etching process.

The ability to etch fine lines in thick polymer layers by RIE is of great importance in multilevel resist techniques where a relatively thick polymer layer is used to planarize complex topography on a substrate wafer (see Chapter 9). The thick polymer film is typically covered with a thin metallic layer in which a pattern is etched using a thin third layer of suitable photoresist; RIE is then used to etch the pattern through the metallic mask into the underlying polymer (84, 85). Kakuchi *et al.* recently showed that amorphous carbon films may be used as a resist mask in a tri-level system which was able to produce patterns with features as small as 40 nm (86).

However, multilevel techniques are complicated and have other disadvantages, so that simpler but effective alternatives have been sought. One such approach takes advantage of the high etch resistance in oxygen RIE of polymers containing silicon by the formation of a very thin layer of SiO_2 on the polymer surface (26, 27, 87, 88, 89). Such resists have been used in bi-level systems. A very promising variation on this approach has recently been described which requires only a single photoresist layer (90, 91); the process is called DESIRE (diffusion-enhanced silylating resist) by its originators. A single planarizing layer of the resist is exposed to UV through a mask and the resist then treated with a silylating reagent (HMDS, hexamethyldisilazane) in the gas phase. The HMDS is selectively adsorbed on the UV-exposed surface and protects those regions of the resist film on subsequent development by oxygen RIE. The process is already capable of patterning with sub-micron features, and much further effort is promised in developing the process for the next generation of device processing.

References

1. S. M. Irving, *Proc. Kodak Photoresist Seminar*, Vol. 2 (1968) 26.
2. R. L. Bersin, *Solid State Technol.* (1970) 39.

3. J. F. Battey, *IEEE Trans. Electron Devices* **ED-24** (1977) 140.
4. J. F. Battey, *J. Electrochem. Soc.* **124** (1977) 147.
5. J. F. Battey, *J. Electrochem. Soc.* **124** (1977) 437.
6. R. F. Reichelderfer, J. M. Welty and J. F. Battey, *J. Electrochem. Soc.* **124** (1977) 1926.
7. J. M. Cook, J. J. Hannon and B. W. Benson, *Sixth Int. Symp. Plasma Chem.*, 616 (1983, Montreal); J. M. Cook and B. W. Benson, *J. Electrochem. Soc.* **130** (1983) 2459.
8. E. O. Degenkolb, C. J. Mogab, M. R. Goldrick, and J. E. Griffiths, *Appl. Spec.* **30** (1976) 520.
9. E. O. Degenkolb and J. E. Griffiths, *Appl. Spec.* **31** (1977) 40.
10. J. E. Griffiths and E. O. Degenkolb, *Appl. Spec.* **31** (1977) 134.
11. R. W. B. Pearse and A. G. Gaydon, *The Identification of Molecular Spectra*, 4th edn., Chapman and Hall, London (1976) 177.
12. R. G. Poulsen, G. M. Smith and W. D. Westwood, US Patent 4528438 (9th July 1985).
13. J. P. Roland, P. J. Marcoux, G. W. Ray and G. H. Rankin, *J. Vac. Sci. Technol.* **A3** (1985) 631.
14. J. Tamano, S. Hattori, S. Morita and K. Yoneda, *Plasma Chem. Plasma Processing* **1** (1981) 261.
15. H. G. Hughes and J. V. Keller, Ger. Offen, 2726813 (29th Dec., 1977).
16. W. R. Goodner. T. E. Wood, H. G. Hughes, J. N. Smith and J. V. Keller, *Proc. Kodak Microelec. Seminar* (1979) 51–9.
17. E. A. Chandross, C. A. Pryde, W. J. Tomlinson and H. P. Weber, *Appl. Phys. Lett.* **24** (1974) 72.
18. G. N. Taylor, T. M. Wolf and M. R. Goldrick, *J. Electrochem. Soc.* **128** (1981) 361.
19. G. N. Taylor and T. M. Wolf, *J. Electrochem Soc.* **127** (1981) 2665.
20. M. Tsuda, S. Oikawa, W. Kanai, A. Yokota, I. Hijikata, A. Uehara and H. Nakane, *J. Vac. Sci. Technol.* **19** (1981) 259.
21. M. Tsuda, S. Oikawa, W. Kanai, K. Hashimoto, A. Yokota, K. Nuino, I. Hijikata, A. Uehara and H. Nakane, *J. Vac. Sci. Technol.* **19** (1981) 1351.
22. M. Tsuda and S. Oikawa, *Jap. J. Appl. Phys.* **21** Suppl. 21–1 (1982) 135.
23. M. Tsuda, S. Oikawa, A. Yokota, M. Yabuta, W. Kanai, K. Kashiwagi, I. Hijikata and H. Nakane, *Polym. Eng. Sci.* **23** (1983) 993.
24. M. Tsuda, *J. Imag. Technol.* **11** (1985) 158.
25. G. N. Taylor, T. M. Wolf and J. M. Moran, *J. Vac. Sci. Technol.* **19** (1981) 872.
26. M. Suzuki, K. Saigo, H. Gokan and Y. Ohnishi, *J. Electrochem. Soc.* **130** (1983) 1962.
27. E. Reichmanis and G. Smolinsky, *J. Electrochem. Soc.* **132** (1985) 1178.
28. A. T. Bell and K. Kwong, *A. I. Chem. Eng. J.* **18** (1972) 990.
29. B. Eliasson, M. Hirth and U. Kogelschatz, *16th Int. Conf. Phenomena in Ionised Gases* (Dusseldorf, 1983) **4**, 558.
30. S. Pfau, A. Rutscher, H. Sabadil and H. E. Wagner, *Seventh Int. Symp. Plasma Chem.* (Eindhoven, 1985) 1392.
31. B. E. Cherrington, *Plasma Chem. Plasma Processing* **2** (1982) 113.
32. M. Venugopalan, *Plasma Chem. Plasma Processing* **3** (1983) 275.
33. J. L. Vossen, *J. Electrochem. Soc.* **126** (1979) 319.
34. S. M. Ojha, *Vacuum* **27** (1977) 65.
35. B. Ranby and J. F. Rabek (eds.) *Singlet Oxygen: Reactions with Organic Compounds and Polymers*, Wiley, New York (1978).
36. J. R. MacCallum and C. T. Rankin, *Makromol. Chem.* **175** (1974) 2477.
37. R. H. Hansen, J. V. Pascale, T. de Benedictis and P. M. Rentzepis, *J. Polym. Sci.* **A3** (1965) 2205.
38. E. L. Lawton, *J. Polym. Sci.* **A10** (1972) 1857.
39. G. N. Taylor and T. M. Wolf, *Polym. Eng. Sci.* **20** (1980) 1087.
40. D. L. Flamm, *Plasma Chem. Plasma Processing* **1** (1981) 37.
41. D. L. Flamm and V. M. Donnelly, *Plasma Chem. Plasma Processing* **1** (1981) 317.
42. J. A. Kerr and S. J. Moss (eds.) *Handbook of Bimolecular and Termolecular Gas Reactions*, Vol. **1**, CRC Press (1982).
43. S. J. Moss, A. M. Jolly and B. J. Tighe, *Sixth Int. Symp. Plasma Chem.* (Montreal, 1983) 621.
44. S. J. Moss, A. M. Jolly and B. J. Tighe, *Plasma Chem. Plasma Processing*, **6** (1986) 401.
45. N. Grassie and H. W. Melville, *Proc. Roy. Soc.* **A 190** (1949) 1.
46. P. J. Blatz and A. V. Tobolsky, *J. Phys. Chem.* **49** (1945) 77.
47. R. Simha, *Trans. New York Acad. Sci.* **14** (1952) 151.
48. I. C. McNeill and M. A. J. Mohammed, *Eur. Polym. J.* **8** (1972) 975.
49. A. Jamieson and I. C. McNeill, *Eur. Polym. J.* **10** (1973) 217.

50. J. Lucki and B. Ranby, *Polymer Degradation and Stability* **1** (1979) 1.
51. E. Grovenstein and A. J. Mosher, *J. Am. Chem. Soc.* **92** (1970) 3810.
52. J. S. Gaffney, R. Atkinson and J. N. Pitts, *J. Am. Chem. Soc.* **98** (1976) 1828.
53. T. M. Sloane, *J. Chem. Phys.* **67** (1977) 2267.
54. T. Yajima, A. Tsuchiya and M. Tezuka, *Nippon Kagaku Kaishi* (1984) 1677.
55. M. Tezuka, T. Yajima and A. Tsuchiya, *Chem. Letts.* (1982) 1437.
56. M. Tezuka, T. Yajima and A. Tsuchiya, *Sixth Int. Symp. Plasma Chem* (Montreal, 1983) 746.
57. E. Zadok, B. Sialom and Y. Mazur, *Angew. Chem. Int. Ed. Engl.* **19** (1980) 1004.
58. E. Zadok, S. Rubinrant, F. Frulow and Y. Mazur, *J. Am. Chem. Soc.* **107** (1985) 2489.
59. D. T. Clark and R. Wilson, *J. Polym. Sci. Polym. Chem. Ed.* **21** (1983) 837.
60. D. T. Clark and A. Dilks, *J. Polym. Sci. Polym. Chem. Ed.* **17** (1979) 957.
61. I. Adesida, J. D. Chinn, L. Rathbun and E. D. Walf, *J. Vac. Sci. Technol.* **21** (1982) 666.
62. C. J. Mogab, A. C. Adams and D. L. Flamm, *J. Appl. Phys.* **49** (1978) 3798.
63. G. Smolinsky and D. L. Flamm, *J. Appl. Phys.* **50** (1979) 4982.
64. D. L. Flamm, *Solid State Technol.* **22** (1979) 109.
65. R. d' Agostino, F. Cramarossa, S. de Benedictis and G. Ferrarro, *J. Appl. Phys.* **52** (1981) 1259.
66. J. W. Coburn and M. Chen, *J. Appl. Phys.* **51** (1980) 3134.
67. R. E. Walkup, K. L. Saenger and G. S. Selwyn, *J. Chem. Phys.* **84** (1986) 2668.
68. K. R. Ryan and I. C. Plumb, *Plasma Chem. Plasma Processing* **4** (1984) 271.
69. L. A. Pederson, *J. Electrochem Soc.* **129** (1982) 205.
70. K. Harada, *J. Appl. Polym. Sci.* **26** (1981) 1961.
71. B. J. Wu, D. W. Hess, D. S. Soong and A. T. Bell, *J. Appl. Phys.* **54** (1983) 1725.
72. G. Turban and M. Rapeaux, *J. Electrochem. Soc.* **130** (1983) 2231.
73. (a) F. Egitto, V. Vukanovic, R. Horwath and F. Emmi, *Seventh Int. Symp. Plasma Chem.* (Eindhoven, 1985) 983. (b) F. Egitto, V. Vukanovic, F. Emmi and R. Horwath, *J. Vac. Sci. Technol.* **B3** (1985) 893.
74. F. Emmi, F. Egitto, R. Horwath and V. Vukanovic, *Proc. Electrochem. Soc.* **85–1** (Plasma Process) (1983) 193.
75. S. V. Babu L. A. Tiemann and R. E. Partch, *Seventh Int. Symp. Plasma Chem.* (Eindhoven, 1985) 1025.
76. V. Srinavasan, M. S. Sivasubramanian and S. V. Babu, *Seventh Int. Symp. Plasma Chem.* (Eindhoven, 1985) 1405.
77. J. F. Rembetski, S. V. Babu, N. H. Lu and J. G. Hoffarth, *Proc. Electrochem. Soc.* **85–1** (Plasma Process) (1985) 184.
78. J. W. Coburn, *Plasma Etching and Reactive-ion Etching*, American Vacuum Society (Monograph, 1982).
79. I. S. Goldstein and F. Falk, *J. Vac. Sci. Technol.* **19** (1981) 743.
80. P. D. DeGraff and D. C. Flanders, *J. Vac. Sci. Technol.* **16** (1979) 1906.
81. H. Gokan and S. Esho, *J. Electrochem. Soc.* **131** (1984) 1105.
82. H. Gokan, S. Esho and Y. Ohnishi, *J. Electrochem. Soc.* **130** (1983) 143.
83. Ch. Steinbruchel, B. J. Curtis, H. W. Lehmann and R. Widmer, *IEEE Trans. Plasma Sci.* **PS-14** (1986) 137.
84. H. Gokan, M. Itoh and S. Esho, *J. Vac. Sci. Technol.* **B2(1)** (1984) 34.
85. P. Unger, A. Krings, N. Gellrich and S. Beneking, *Seventh Int. Symp. Plasma Chem.* (Eindhoven, 1985) 1084.
86. M. Kakuchi, M. Hikita and T. Tamamura, *Appl. Phys. Lett.* **48** (1986) 835.
87. T. Venkatesan, G. N. Taylor, A. Wagner, B. Wilkens and D. Barr, *J. Vac. Sci. Technol.* **19** (1981) 1379.
88. F. Watanable and Y. Ohnishi, *J. Vac. Sci. Technol.* **B4(1)** (1986) 422.
89. K. Saigo, F. Watanable and Y. Ohnishi, *J. Vac. Sci. Technol.* **B4(3)** (1986) 692.
90. B. Roland and F. Coopmans, *18th Int. Conf. on Solid State Devices and Materials* (Tokyo, August 1986) 33.
91. B. Roland, R. Lombaerts and F. Coopmans, *Dry Process Symposium* (Tokyo, November 1986).

Index